TE D

The Alkaloids

Volume 12

A Specialist Periodical Report

The Alkaloids
Volume 12

A Review of the Literature Published
between July 1980 and June 1981

Senior Reporter
M. F. Grundon *School of Physical Sciences, New University of
Ulster, Coleraine, Northern Ireland*

Reporters
K. W. Bentley *University of Technology, Loughborough*
A. S. Chawla *University College, Cardiff*
R. Dharanipragada *University of West Virginia, U.S.A.*
G. Fodor *University of West Virginia, U.S.A.*
H. Guinaudeau *University of Limoges, France*
D. M. Harrison *New University of Ulster*
R. B. Herbert *University of Leeds*
A. H. Jackson *University College, Cardiff*
J. A. Lamberton *C.S.I.R.O., Melbourne, Australia*
J. R. Lewis *University of Aberdeen*
S. W. Page *University of Georgia, U.S.A.*
S. W. Pelletier *University of Georgia, U.S.A.*
A. R. Pinder *Clemson University, South Carolina, U.S.A.*
D. J. Robins *Glasgow University*
J. E. Saxton *University of Leeds*
M. Shamma *Pennsylvania State University, U.S.A.*

The Royal Society of Chemistry
Burlington House, London W1V 0BN

ISBN 0-85186-357-4
ISSN 0305-9707

Printed in Great Britain by
Spottiswoode Ballantyne Ltd.,
Colchester and London

Foreword

The whole of the alkaloid literature is reviewed for the year with the exception of *Lycopodium* Alkaloids, which will be given a two-year coverage in Volume 13. As in previous volumes of the series we aim to include all references to the isolation, structure, chemistry, synthesis, and biosynthesis of alkaloids and to provide an adequate account of biological properties. The annual comprehensive treatment is unique in the alkaloids literature and has received much support. Rising costs, however, have put the future of the series in doubt and we are experimenting with cheaper methods of production; alternative means of producing chemical structures were used in Volume 10 (Chapters 6, 7 and 10) and in this Volume (Chapters 1, 6, 7, 10 and 14), I believe without significant reduction in quality, and Volume 13 will be published from camera-ready copy. These changes place an added burden on our authors and I would like to express my gratitude for their continuing commitment to the principles of this series and for their willingness to take part in new production techniques.

September 1982   M. F. GRUNDON

Contents

1
Biosynthesis

BY R. B. HERBERT

The established practice of including references to earlier Reports in this series for background information is continued. Two comprehensive reviews are also cited.[1,2] An authoritative account of the biosynthesis of fungal metabolites has been published,[3] as has an introductory text which includes a survey of the biosynthesis of alkaloids and nitrogenous microbial metabolites.[4]

1 Pyrrolidine and Piperidine Alkaloids

Simple Pyrrolidine Alkaloids.—It is well established that ornithine (1) is a key precursor in the biosynthesis of pyrrolidine alkaloids. Notably, the amino-acid (1) is utilized for the biosynthesis of nicotine (5) *via* the symmetrical intermediate putrescine (3), whereas the biosynthesis of tropane alkaloids, *e.g.* scopolamine (6), avoids any symmetrical intermediate[1,2] (*cf.* Vol. 11, p. 1).

The first intermediate beyond ornithine in the biosynthesis of tropane alkaloids has been deduced to be δ-N-methylornithine (2). Recently, (2) was identified as a natural constituent for the first time in a plant, namely *Atropa belladonna*, which produces tropane bases.[5] The (2) was labelled by radioactive ornithine (1), but, unfortunately, the alkaloids were not, so correlation between the formation of (2) and the biosynthesis of alkaloids has not yet been achieved.

The biosynthetic pathway to both nicotine (5) and the tropane alkaloids includes N-methylputrescine (4) as a probable intermediate. New results[6] obtained for nicotine (5) and scopolamine (6) with [1-^{13}C,^{14}C;*methylamino*-^{15}N]-N-methyl putrescine [(4); labels as shown] nicely confirm this. The specific incorporation of both stable isotopes was closely similar to that of the ^{14}C label in both alkaloids, indicating intact incorporation of the precursor. The labelling patterns deduced are illustrated (● = ^{13}C, * = ^{15}N), and they are in accord with earlier results that were

[1] R. B. Herbert, in 'Comprehensive Organic Chemistry', ed. D. H. R. Barton and W. D. Ollis, Pergamon, Oxford, 1978, Vol. 5, p. 1045.
[2] R. B. Herbert, in 'Rodd's Chemistry of Carbon Compounds', 2nd edn., ed. S. Coffey, 1980, Vol. IV, Part L, p. 291.
[3] 'The Biosynthesis of Mycotoxins', ed. P. S. Steyn, Academic Press, New York, 1980.
[4] R. B. Herbert, 'The Biosynthesis of Secondary Metabolites', Chapman and Hall, London, 1981.
[5] S. H. Hedges and R. B. Herbert, *Phytochemistry*, 1981, **20**, 2064.
[6] E. Leete and J. A. McDonell, *J. Am. Chem. Soc.*, 1981, **103**, 658.

HOOC   NH$_2$ NHR

Ornithine (1) R = H   Putrescine (3)   (4)

(2) R = Me

(5)

(6)

obtained with ^{14}C-labelled precursors.[1,2] [2-^{14}C]Ornithine labels C-1' of the tropane nucleus [as (6)].[7] The assignment of C-5' in (6), rather than C-1', as the site that is labelled by the *N*-methyl[^{13}C]putrescine is in accord with this; the n.m.r. assignment was not of itself completely unambiguous.

The labelling of the alkaloids was apparent from a small doublet in their ^{13}C n.m.r. spectra, flanking a central natural-abundance ^{13}C singlet for the relevant carbon atom. This technique of using the two contiguous stable isotopes, *i.e.* ^{13}C and ^{15}N, which give rise to ^{13}C n.m.r. doublets that are separate from natural-abundance singlets and which are detectable at much higher dilution than would be the case for a single ^{13}C label, is ingenious and notable. It has found previous application in experiments with two adjacent ^{13}C labels[8] (*cf.* Vol. 11, pp. 1 and 19).

The tiglic acid moieties that are found in some tropane alkaloids, *e.g.* (8), derive from isoleucine *via* 2-methylbutanoic acid[9,10] (*cf.* Vol. 5, p. 12; Vol. 6, p. 11; Vol. 9, p. 3). Since 2-hydroxy- and 3-hydroxy-derivatives of the latter are not involved in biosynthesis,[10,11] the conversion of 2-methylbutanoic acid (possibly *via* its CoA ester) into tiglic acid must be a dehydrogenation reaction. Incorporation of (2RS,3S,4S)-[2-^{14}C;4-^{3}H$_1$]isoleucine (7) into the tigloyl moieties of (8) in *Datura innoxia* with almost complete retention of tritium, and of (2S,4R)-[4-^{3}H$_1$]isoleucine into the tigloyl moiety of meteloidine (9) in *D. meteloides* with almost complete loss of tritium, shows that the dehydrogenation is achieved by an antiperiplanar elimination within (2S)-2-methylbutanoic acid, involving the loss of the 4-*pro-R*-isoleucine proton [see (10)].[12]

[7] E. Leete, *Tetrahedron Lett.*, 1964, 1619.

[8] E. Leete and M.-L. Yu, *Phytochemistry*, 1980, **19**, 1093; E. Leete, *J. Nat. Prod. (Lloydia)*, 1980, **43**, 130.

[9] E. Leete, *Phytochemistry*, 1973, **12**, 2203.

[10] K. Basey and J. G. Woolley, *Phytochemistry*, 1973, **12**, 2197.

[11] B. A. McGraw and J. G. Woolley, *Phytochemistry*, 1977, **16**, 1711.

[12] R. K. Hill, S.-W. Rhee, E. Leete, and B. A. McGaw, *J. Am. Chem. Soc.*, 1980, **102**, 7344.

(7)

(8) $R^1 = H$, $R^2 = -O-\overset{O}{\underset{}{C}}$ ...Me

(9) $R^1 = R^2 = OH$

(10)

(11)

An apparently similar, antiperiplanar elimination with loss of the isoleucine 4-*pro-S* proton, to generate a double-bond of opposite geometry in alkaloidal necic acid fragments, has been observed[13] (*cf.* Vol. 11, p. 2).

The structure of cocaine (11) suggests a biogenesis from ornithine and one that is closely similar to that of the tropane alkaloids, *e.g.* scopolamine (6), which was discussed above. Until recently, however, the only significant incorporation of labelled precursors to be obtained was with [3-^{14}C]phenylalanine: it specifically labelled the carboxy-group of the benzoyl moiety of cocaine.[2,14] Eventually, a change of feeding technique to one where a solution of the precursor was painted on the leaves of *Erythroxylon coca* resulted in a significant incorporation of DL-[5-^{14}C]ornithine [as (1)] into cocaine (11).[15] Further, the radioactivity that was present in the cocaine (11) was shown to be located specifically at one, or both, bridgehead carbon atoms, indicating a normal direct biosynthesis from ornithine.

Radioactive cuscohygrine (12) was also isolated in this experiment. Results obtained using other plants had shown that the biosynthesis of tropane alkaloids and of cuscohygrine (12) was related: both were formed *via* hygrine (13)[1,2] (*cf.* Vol. 1, p. 9). The formation of tropane alkaloids from ornithine without intervention of a symmetrical intermediate (see above) implies that cuscohygrine is formed similarly. The cuscohygrine that formed from [5-^{14}C]ornithine in *E. coca* was specifically labelled, but, surprisingly, one quarter of the radioactivity was deduced to be located at each of C-2, C-2', C-5 and C-5', indicating formation from ornithine *via* a symmetrical intermediate, presumably putrescine (3). Biosynthesis of cuscohygrine in *E. coca* appears to be different in detail then from that in other plants that

[13] R. Cahill, D. H. G. Crout, M. B. Mitchell, and U. S. Müller, *J. Chem. Soc., Chem. Commun.*, 1980, 419.

[14] D. Gross and H. R. Schütte, *Arch. Pharm. (Weinheim, Ger.)*, 1963, **296**, 1.

[15] E. Leete, *J. Chem. Soc., Chem. Commun.*, 1980, 1170.

(12)   (13)

have so far been examined. Obviously, this needs further examination, and it will be most interesting to see, in addition, whether cocaine is formed through a symmetrical intermediate too, resulting in labelling of both bridgehead carbon atoms.

Phenanthroindolizidine Alkaloids.—The phenanthroindolizidine alkaloids, *e.g.* tylophorine (14) and tylophorinine (15), are assembled in *Tylophora asthmatica* from fragments derived from ornithine, phenylalanine, and tyrosine (*cf.* Vol. 8, p. 6; Vol. 9, p. 5). The last amino-acid is the source of ring B plus C-9 and C-10. It has now been shown that dopa is a better precursor than tyrosine for this fragment. Label from [2-^{14}C]dopa was specifically incorporated into C-10 of (14) and (15).[16]

The diphenylhexahydroindolizine (16) has been identified as a key intermediate in the biosynthesis of tylophorine (14), tylophorinine (15), and tylophorinidine in *T. asthmatica*.[17] In support, material bearing a ^{14}C label at C-5 was found to be a specific precursor for (14) and (15); it was clearly the best precursor of eight diphenylhexahydroindolizines tested.[16] From the combined results[16-18] it is clear that earlier diphenylhexahydroindolizines have a hydroxylation and methylation pattern on ring A corresponding to that in (16). The result with [2-^{14}C]dopa indicates that ring B is dioxygenated when the first indolizine is formed. However, (19) is an intact alkaloid precursor, and is nearly as well incorporated as (16).[17]

Tetra-oxygenated precursors corresponding to (16), but with varying extents of methylation, and differing in methylation pattern, have been examined as precursors for (14) and (15).[16] Unfortunately, unlike (16), they were all less well incorporated than dopa. However, the results do allow tentative identification of (17) and (18) as possible intermediates, lying earlier than (16) on the biosynthetic pathway to (14) and (15). Interestingly, (20), which cannot give tylophorinine (14) by phenol oxidative coupling[17] without alteration of its methylation pattern to that of (16), was incorporated into both (14) and (15) at a similar level to that of (17) and (18).[16]

Pyrrolizidine Alkaloids.—Retronecine (23), the most common base portion found in the pyrrolizidine alkaloids, has been shown to derive from two molecules of ornithine (1) *via* putrescine (3) by the use of ^{14}C-labelled precursors[19] (*cf.* Vol. 10, p. 13). Unequivocal evidence on the manner of incorporation of putrescine comes from experiments in *Senecio isatideus* with ^{13}C-labelled samples of putrescine.[20]

---

[16] D. S. Bhakuni and V. K. Mangla, *Tetrahedron*, 1981, **37**, 401.
[17] R. B. Herbert and F. B. Jackson, *J. Chem. Soc., Chem. Commun.*, 1977, 955.
[18] R. B. Herbert, F. B. Jackson, and I. T. Nicolson, *J. Chem. Soc., Chem. Commun.*, 1976, 865.
[19] D. J. Robins and J. R. Sweeney, *J. Chem. Soc., Chem. Commun.*, 1979, 120.
[20] H. A. Khan and D. J. Robins, *J. Chem. Soc., Chem. Commun.*, 1981, 146.

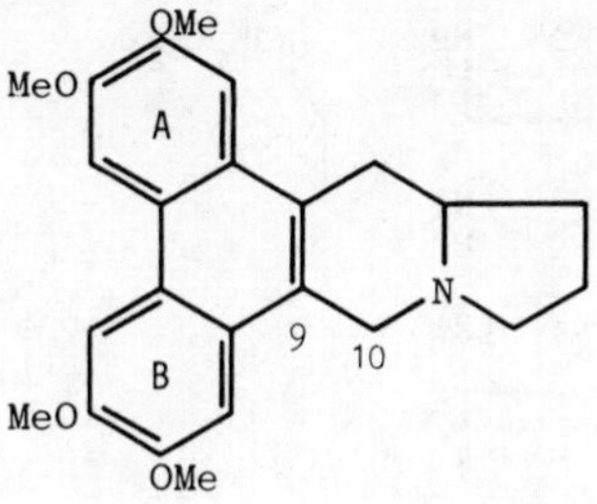

(14)

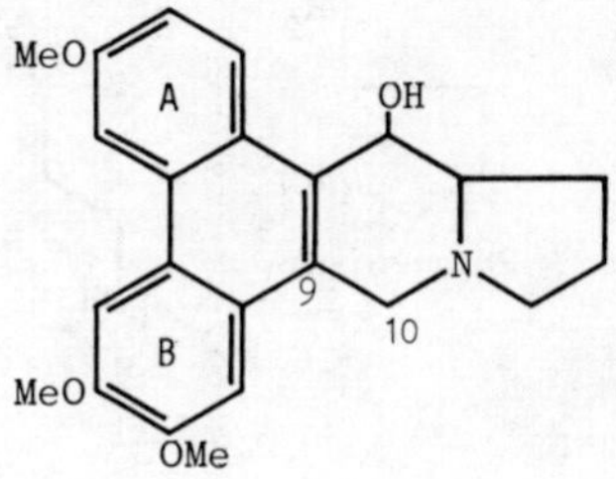

(15)

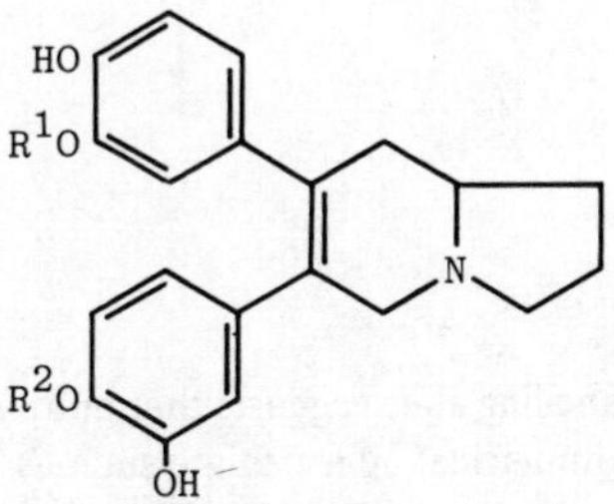

(16) $R^1 = R^2 = Me$

(17) $R^1 = R^2 = H$

(18) $R^1 = Me, R^2 = H$

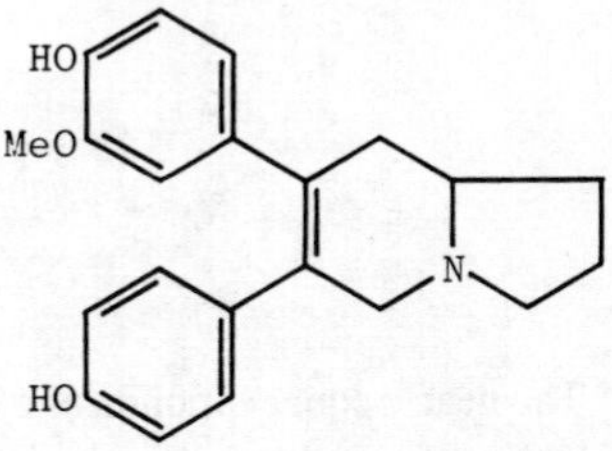

(19)

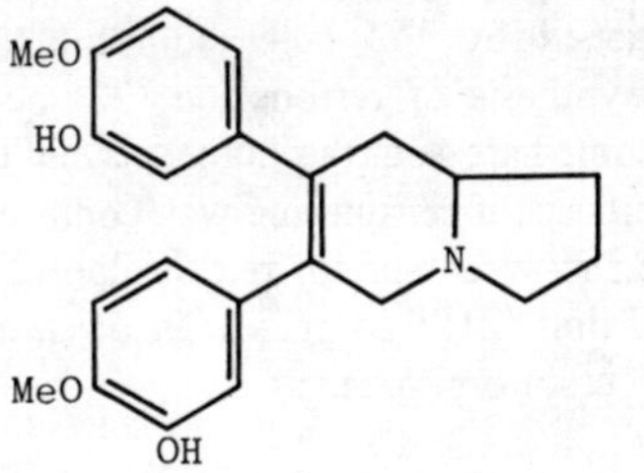

(20)

First, [1,4-$^{13}C_2$]putrescine [as (3)] was found to label C-5, C-9, C-3, and C-8 of (23) and to label each atom to a nearly equal extent. Secondly, [2,3-$^{13}C_2$]putrescine [as (3)] gave retronecine (23), the proton-decoupled ^{13}C n.m.r. spectrum of which displayed doublets (superimposed on natural-abundance singlets) for signals corresponding to C-1 and C-2 ($J = 71$ Hz) and for those corresponding to C-6 and C-7 ($J = 34$ Hz); the enrichment levels for all the doublets were similar. The derivation of retronecine (23) from the intact carbon skeletons of two putrescine units is, therefore, quite clear, and the pattern for their utilization in biosynthesis is shown in Scheme 1.

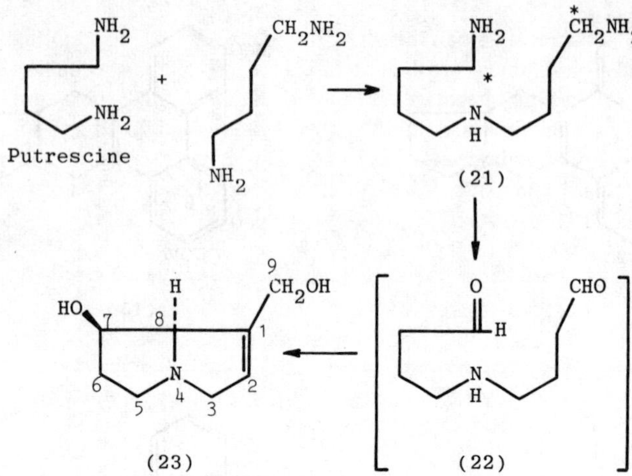

Scheme 1

The nearly equal enrichment of all four ^{13}C-labelling sites suggests that putrescine is biotransformed into retronecine (23) *via* a symmetrical intermediate such as (22) and/or homospermidine (21). This possibility has been explored in experiments with [1-*amino*-^{15}N;1-^{13}C]putrescine in *S. vulgaris*[21] and *S. isatideus*.[22] In both sets of experiments, the ^{13}C n.m.r. signals for C-3 and C-5 of retronecine (23) showed doublets of approximately equal intensity, arising from ^{13}C–^{15}N coupling superimposed on a singlet for each carbon, made up of natural-abundance ^{13}C enriched by ^{13}C label adjacent to ^{14}N. This is entirely consistent with the biosynthesis of retronecine (23) occurring by way of at least one symmetrical intermediate such as homospermidine (21). The intermediacy of (21) in the bio-synthesis of retronecine was confirmed by showing that (21; * = ^{14}C label) was a specific precursor for retronecine (23) (labelling sites deduced to be C-9 and C-8) and that (21) was present in *S. isatideus* plants, being labelled by the retronecine precursor ornithine.[22]

Lysine.—Lysine is a common precursor of piperidine alkaloids. Of the two enantiomers of this amino-acid, the L-isomer is the more direct precursor, in plants, for piperidine alkaloids, *e.g.* anabasine, whereas D-lysine is more directly implicated in the biosynthesis of pipecolic acid (24)[1,2,23] (*cf.* Vol. 7, p. 7). It has now been shown that a pathway exists in the plant *Nicotiana glauca*,[24] and also in the micro-organism *Neurospora crassa*,[25] which transforms D-lysine into L-lysine by way of L-pipecolic acid (24).

[21] G. Grue-Sørensen and I. D. Spenser, *J. Am. Chem. Soc.*, 1981, **103**, 3208.
[22] H. A. Khan and D. J. Robins, *J. Chem. Soc., Chem. Commun.*, 1981, 554.
[23] E. Leistner, R. N. Gupta, and I. D. Spenser, *J. Am. Chem. Soc.*, 1973, **95**, 4040.
[24] N. Fangmeier and E. Leistner, *J. Chem. Soc., Perkin Trans. 1*, 1981, 1769.
[25] N. Fangmeier and E. Leistner, *J. Biol. Chem.*, 1980, **255**, 10 205.

Quinolizidine Alkaloids.—Important new information (*cf.* Vol. 11, p. 4) has been obtained on the biosynthesis of quinolizidine alkaloids such as lupanine (27) in experiments with enzyme preparations from *Lupinus polyphyllus* cell suspension cultures[26] and with chloroplasts.[27] These alkaloids are formed from three molecules of lysine by way of cadaverine (25),[1,2] and the enzymic evidence[26,27] is that conversion of cadaverine into these alkaloids occurs without release of intermediates until 17-oxosparteine (26) is generated; the enzyme is a transaminase and not a diamine oxidase.

The enzyme, *i.e.* lysine decarboxylase, that is required for the conversion of lysine into cadaverine, and thus the first step of alkaloid biosynthesis, has been isolated from chloroplasts of *L. polyphyllus*.[28] Like the majority of amino-acid decarboxylases, this enzyme is dependent on pyridoxal 5'-phosphate. Its activity was found not to be affected by the presence or absence of quinolizidine alkaloids. Control of the enzyme by simple product feedback inhibition therefore seems unlikely. The operational parameters of this enzyme resemble those of the 17-oxosparteine synthase. Co-operation between the two enzymes would explain why cadaverine is almost undetectable *in vivo*.

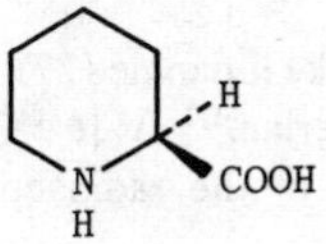

(24) 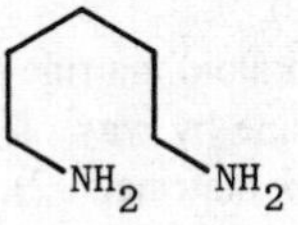   (25)

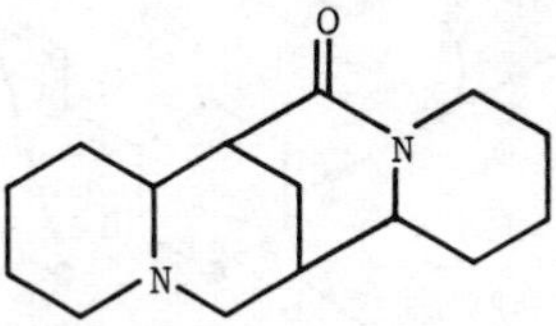

(26) 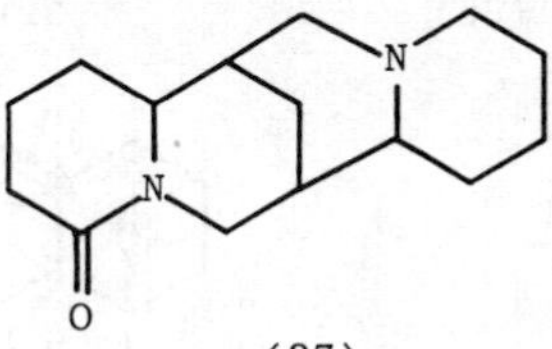   (27)

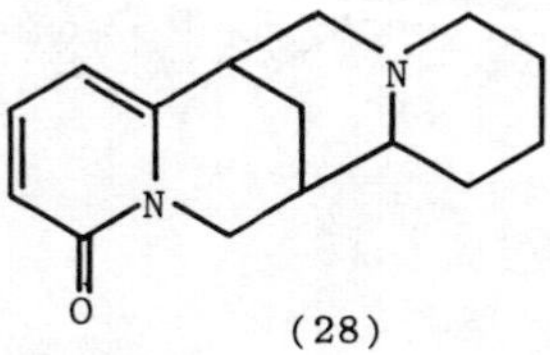

(28) 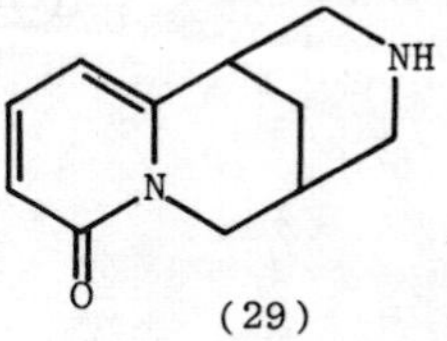   (29)

[26] M. Wink, T. Hartmann, and H.-M. Schiebel, *Z. Naturforsch., Teil. C*, 1979, **34**, 704; M. Wink and T. Hartmann, *FEBS Lett.*, 1979, **101**, 343.
[27] M. Wink, T. Hartmann, and L. Witte, *Z. Naturforsch., Teil. C*, 1980, **35**, 93.
[28] T. Hartmann, G. Schoofs, and T. M. Wink, *FEBS Lett.*, 1980, **115**, 35.

Essentially only lupanine (27) is accumulated in cell suspension cultures of *Lupinus polyphyllus*,[29,30] *Sarothamnus scoparius*,[29] and *Baptista australis*,[29,31] whereas the intact plants accumulate other alkaloids. It is reasonable to assume that the cultures will accumulate alkaloids early rather than late in a biosynthetic pathway. Thus lupanine (27) is identified as a likely intermediate in the biosynthesis of the other alkaloids of these plants. In the case of *B. australis*, these alkaloids are of the pyridone type, *e.g.* anagyrine (28) and cytisine (29).[31] Earlier results with $^{14}CO_2$ had indicated that lupanine (27) is an intermediate in the biosynthesis of pyridone alkaloids, *e.g.* (28) and (29), in *Thermopsis rhombifolia* and *T. caroliniana*[32] (*cf.* Vol. 3, p. 30), and this is supported by these new results. In addition, ^{14}C-labelled lupanine was found to be incorporated, *inter alia*, into (28) and (29).[31]

Biosynthesis of piperidine alkaloids from lysine/cadaverine commonly occurs *via* Δ^1-piperideine (31). Three molecules are utilized for the construction of lupanine (27), and an attractive biosynthetic route involving the all-*trans*-isomer of isotripiperideine has been hypothesized[33] (*cf.* Vol. 8, p. 3).

The enzymic evidence alluded to above (*cf.* Vol. 11, p. 4) indicates that Δ^1-piperideine (31) may not be a normal intermediate in lupanine biosynthesis, but may be utilized *via* (30).

The quinolizidine alkaloid matrine (32), like lupanine (27), is biosynthesized from three molecules of lysine by way of cadaverine.[1,2] Δ^1-[6-^{14}C]Piperideine [as (31)] was incorporated into matrine (32); 10% of the radioactivity was found to be

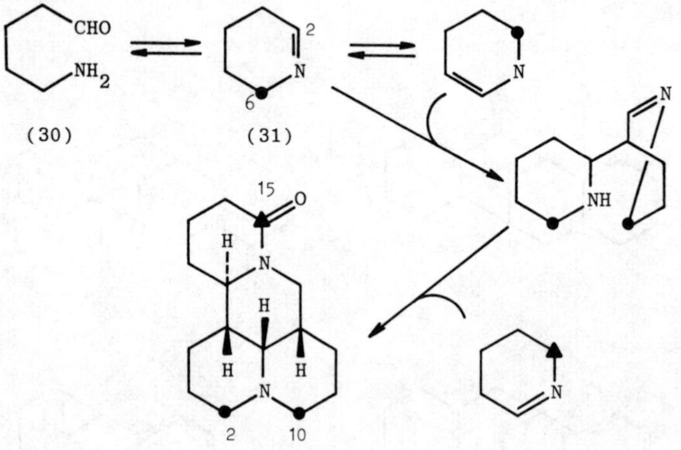

Scheme 2

[29] M. Wink and T. Hartmann, *Planta Med.*, 1980, **40**, 149.
[30] M. Wink, L. Witte, H.-M. Schiebel, and T. Hartmann, *Planta Med.*, 1980, **38**, 238.
[31] M. Wink, T. Hartmann, L. Witte, and H. M. Schiebel, *J. Nat. Prod.* (*Lloydia*), 1981, **44**, 14.
[32] Y. D. Cho and R. O. Martin, *Can. J. Biochem.*, 1971, **49**, 971.
[33] W. M. Golebiewski and I. D. Spenser, *J. Am. Chem. Soc.*, 1976, **98**, 6726.

located at C-15 and the remainder was distributed over C-2 and C-10.[34] The incorporation of Δ^1-[6-^{14}C]piperideine [as (31)] into matrine has been re-examined[35] in two plant species, *Sophora tetraptera* and *S. microphylla*, with similar but more detailed results: It was deduced that C-2, C-10, and C-15 were labelled, showing that three units of Δ^1-piperideine (31) are used specifically for the elaboration of matrine (32). In the matrine obtained from each of the two *Sophora* species, C-2 and C-10 showed the same molar specific activity whereas C-15 showed a different, lower, molar specific activity. This means that two units of Δ^1-piperideine (the two with the same molar specific activity: ●) combine first, this being followed by combination with the third (▲) after dilution with unlabelled material in the plant (see Scheme 2).

One third of the label from Δ^1-[6-^{14}C]piperideine that was found in the matrine (32) was located elsewhere than at C-2, C-10, and C-15. A plausible explanation for the partial randomization that was observed is that reversible transamination converted the Δ^1-[6-^{14}C]piperideine [as (31)], *via* its ring-opened form (30), into cadaverine (25). Subsequent re-conversion into (31) gave material that was labelled at C-6 and also at C-2.[35]

The model scheme developed for the biosynthesis of lupanine from Δ^1-piperideine and isotripiperideine[33] has been adapted for the biosynthesis of matrine (32).[35] At the moment, the two hypothetical pathways[26,33,35] for the biosynthesis of quinolizidine alkaloids are manifestly different (*cf.* Vol. 11, p. 4 and Vol. 8, p. 3): one uses Δ^1-piperideine (31) as an intermediate; the other does not. Where the points of fundamental agreement between the two models lie, and which model is a more accurate picture of what is really happening, are questions that remain to be answered.

In accord with a general body of evidence on the biosynthesis of alkaloids as against that of pipecolic acid (see above), L-lysine has been shown to be the preferred precursor for lupanine (27) and D-lysine the preferred precursor for L-pipecolic acid (24) in *Lupinus angustifolia*.[36] A high retention of tritium, present at C-4 and C-5 in the lysine, on formation of (27) is to be noted.

2 Isoquinoline Alkaloids

Papaver Alkaloids.—Biosynthesis of morphine (36) occurs, in *Papaver somniferum*, through reticuline (33) by way of thebaine (35). The sequence from (35) to (36) involves, *inter alia*, two O-demethylations, with that at the methoxy-group at C-6 occurring first.[1,2] Confirmation that the other methoxy-group is not demethylated first in this *Papaver* species obtains from the failure to detect oripavine (37), which is found in other *Papaver* species, as a natural constituent of *P. somniferum*. The experiment involved attempted isolation of radioactive (37), using inactive alkaloid as carrier, following a feeding experiment with radioactive reticuline (33).[37]

[34] S. Shibata and U. Sankawa, *Chem. Ind.* (*London*), 1963, 1161.
[35] F. J. Leeper, G. Grue-Sørensen, and I. D. Spenser, *Can. J. Chem.*, 1981, **59**, 106.
[36] E. Leistner and I. D. Spenser, *Pol. J. Chem.*, 1979, **53**, 49.
[37] E. Brochmann-Hanssen and Y. Okamoto, *J. Nat. Prod.* (*Lloydia*), 1980, **43**, 731.

Oripavine 3-ethyl ether (38), an unnatural analogue of thebaine (35), was found to be metabolized to morphine 3-ethyl ether and to morphine (36). The efficiency of the conversion was comparable to that of natural biosynthesis. (For an examination of other unnatural compounds as substrates for the enzymes of the biosynthesis of morphine alkaloids, see Vol. 4, p. 15.)

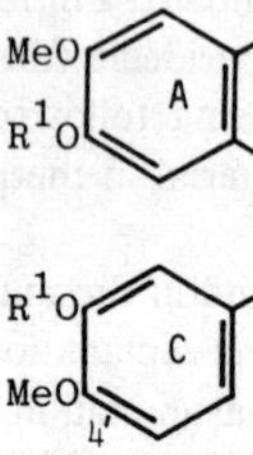

(33) R^1 = H, R^2 = Me   (35)   (36)
(34) R^1 = Me, R^2 = H

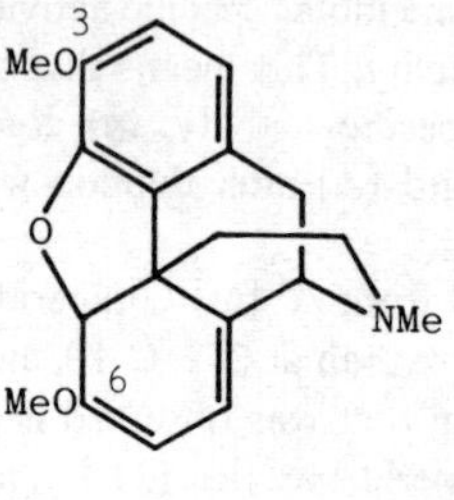

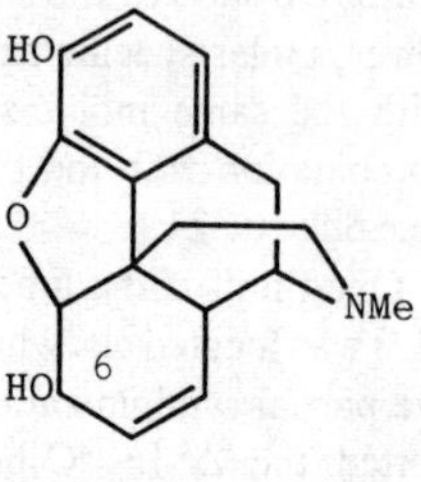

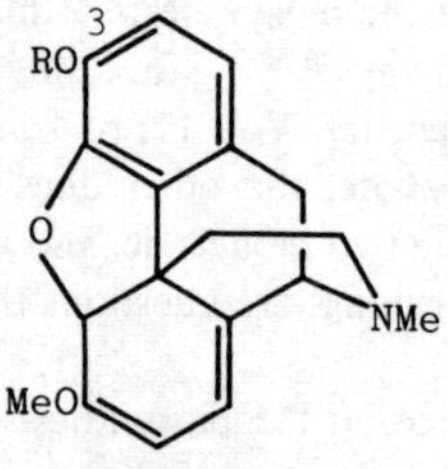

(37) R = H
(38) R = Et

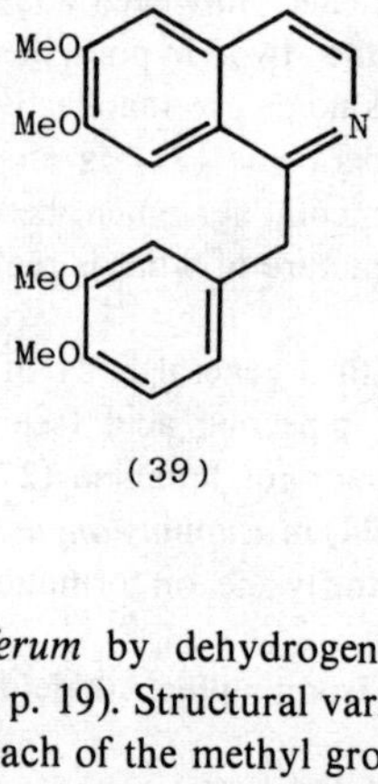

(39)

Papaverine (39) is formed in *P. somniferum* by dehydrogenation of tetra-hydropapaverine (34) (*cf.* Vol. 7, p. 6; Vol. 8, p. 19). Structural variations on (34), with a single ethyl group replacing (in turn) each of the methyl groups, have been examined as substrates for the dehydrogenation sequence. A conversion efficiency comparable to that of (34) into (39) was observed if the *O*-ethyl group was on ring A, but a considerably less efficient conversion was observed if the *O*-ethyl group was on ring C.[38]

Further work on the biosynthesis of alkaloids by fractions of the latex of seed capsules of *P. somniferum* has been published.[39]

Aporphine and Berberine Alkaloids.—The aporphine alkaloid magnoflorine (42) has been shown to be formed specifically from reticuline (33) in *Aquilegia* species[40]

[38] E. Brochmann-Hanssen, C.-Y. Chen, and E. E. Linn, *J. Nat. Prod.* (*Lloydia*), 1980, **43**, 736.
[39] J. W. Fairbairn and M. J. Steele, *Phytochemistry*, 1980, **20**, 1031.
[40] E. Brochmann-Hanssen, C.-H. Chen, H.-C. Chiang, and K. McMurtrey, *J. Chem. Soc., Chem. Commun.*, 1972, 1269.

(*cf.* Vol. 4, p. 17). It has been found also to derive from reticuline (33), rather than from nororientaline or norprotosinomenine, in *Cocculus laurifolius*.[41] Norlaudanosoline (40) and nor-reticuline (41), which are precursors for reticuline, were also significantly incorporated.

Laurifoline (43), isolated along with the magnoflorine (42), was also found to be formed specifically from the same precursors. Both alkaloids were shown to be formed from reticuline (33) without loss of the 4'-methoxyl and *N*-methyl groups. Furthermore, stereochemical integrity is maintained in the bioconversion of reticuline into laurifoline (43) and magnoflorine (42) because (+)-(*S*)-reticuline [same configuration as (42)] was a much better precursor than its enantiomer for both alkaloids. The presence of reticuline (unknown configuration) in these plants was confirmed.[41]

(40) R = H
(41) R = Me

(42) R^1 = OH, R^2 = H
(43) R^1 = H, R^2 = OH

It is known that berberine (46) and the protoberberine alkaloid (*S*)-stylopine (44) are elaborated from an intact molecule of (*S*)-reticuline [as (33)], the *N*-methyl group providing C-8 in each alkaloid[1,2,42] (*cf.* Vol. 7, p. 12). Exactly similar findings were obtained for tetrahydropalmatine (45) and palmatine (47) in *C. laurifolius*.[43] Of several benzylisoquinolines tested, only reticuline (33), nor-reticuline (41), and norlaudanosoline (40) were significantly incorporated. Tetrahydropalmatine (45) was found to be converted into palmatine (47) and the conversion was irreversible.

Bisbenzylisoquinoline Alkaloids.—The biosynthesis of a number of bisbenzylisoquinoline alkaloids has been investigated (*cf.* Vol. 11, p. 11; Vol. 10, p. 16; Vol. 9, p. 11). Appropriate to the oxygenation patterns of the alkaloids examined, they have all been found to derive through coclaurine (48), and also, in some cases, *N*-methylcoclaurine (49); *e.g.*, tiliacorinine (50) in *Tiliacora racemosa*[44] (*cf.* Vol. 9, p. 11). The biosynthesis of nortiliacorinine A (51), produced by the same plant, has been investigated and found to be from two molecules of coclaurine (48).[45]

[41] D. S. Bhakuni, S. Jain, and R. S. Singh, *Tetrahedron*, 1980, **36**, 2525.
[42] A. R. Battersby, R. J. Francis, M. Hirst, E. A. Ruveda, and J. Staunton, *J. Chem. Soc., Perkin Trans. 1*, 1975, 1140; A. R. Battersby, J. Staunton, H. R. Wiltshire, R. J. Francis, and R. Southgate, *ibid.*, p. 1147; and refs. cited therein.
[43] D. S. Bhakuni, S. Jain, and S. Gupta, *Tetrahedron*, 1980, **36**, 2491.
[44] D. S. Bhakuni, A. N. Singh, S. Jain, and R. S. Kapil, *J. Chem. Soc., Chem. Commun.*, 1978, 226.
[45] D. S. Bhakuni, A. N. Singh, and S. Jain, *Tetrahedron*, 1981, **37**, 2651.

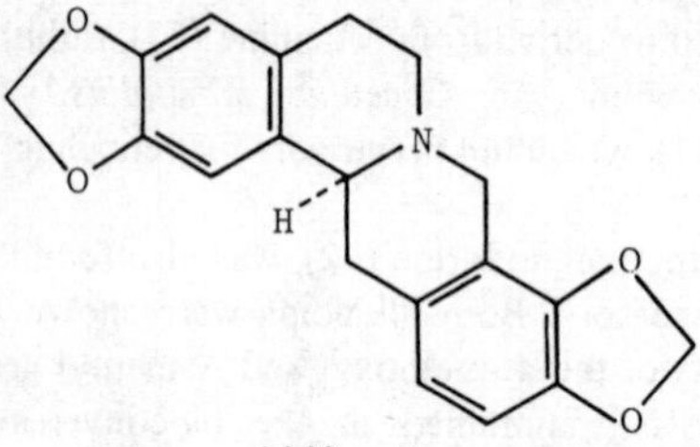

(44)

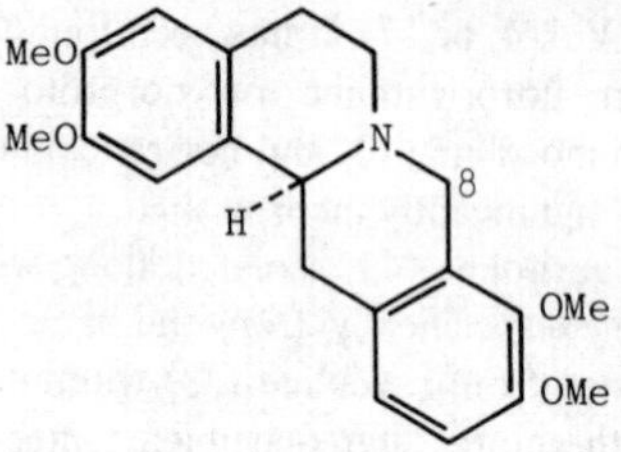

(45)

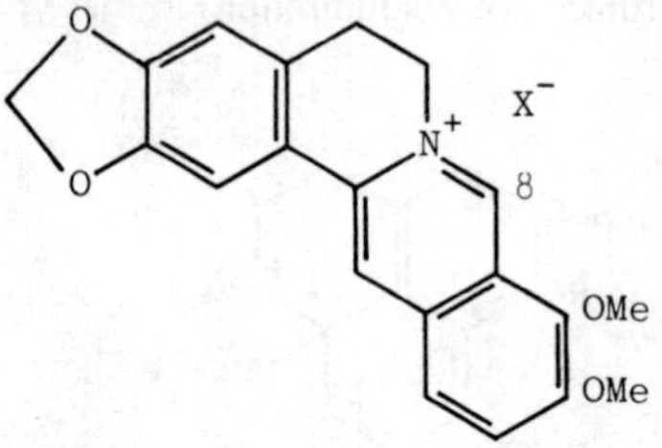

(46)

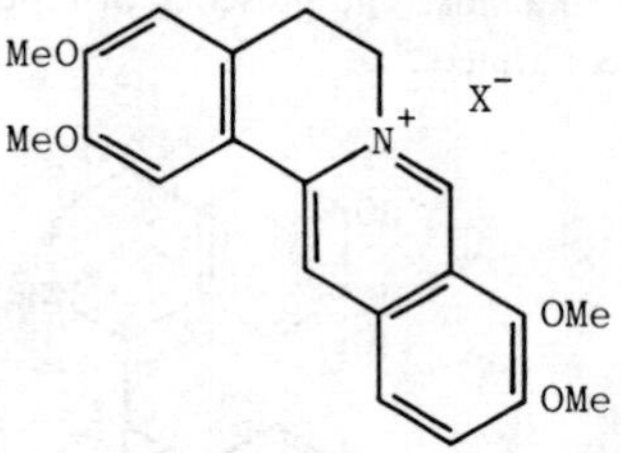

(47)

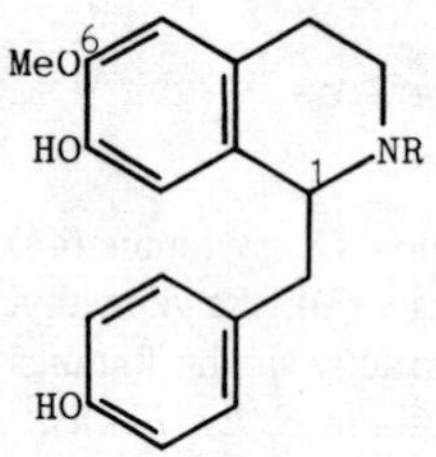

(48) R = H

(49) R = Me

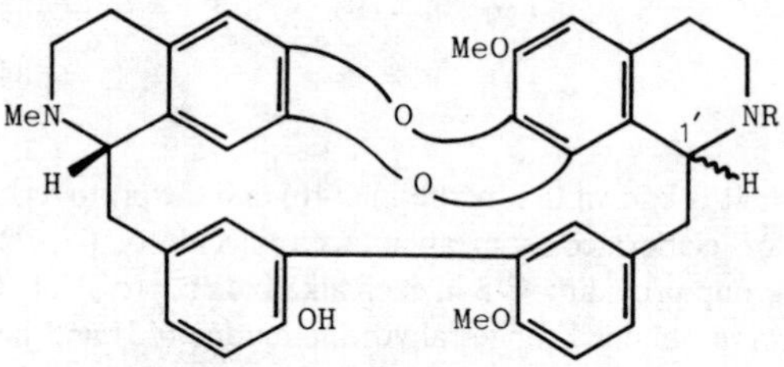

(50) R = Me; 1' *R*

(51) R = H; 1'*S*

N-Methylcoclaurine (49) only provides the left-hand half, and the necessary loss of its 6-methoxy-group was observed in an appropriate labelling experiment. Furthermore, (*S*)-coclaurine and (*S*)-*N*-methylcoclaurine, rather than their anti-podes, were utilized for biosynthesis; the proton at C-1 of (49) was shown not to be lost on incorporation into (51). The natural occurrence of *N*-methylcoclaurine (49) in *T. racemosa* has been demonstrated,[44] and coclaurine is also a constituent of this plant.[45]

It may be noted that results with the various precursors neatly allowed the structure of tiliacorinine A (51) to be clarified and the stereochemistry at the two asymmetric centres to be defined.[45] It is interesting that the right-hand half of (51) does not form from *N*-methylcoclaurine (by demethylation), unlike the similar *N*-methylated fragment in (50), which has the opposite stereochemistry at C-1'.

Tetrandrine (52) has been shown[46] to be formed in *Cocculus laurifolius* from two molecules of (+)-(S)-N-methylcoclaurine [as (49)], a natural constituent of this plant (*cf.* the biosynthesis of the isomeric isotetrandrine in Vol. 11, p. 11).

The oxygenation and methylation pattern in the aporphine–benzylisoquinoline alkaloid thalicarpine (53) corresponds to that of reticuline (33), and not coclaurine (48). Appropriately, (R,S)-[1-^{14}C]reticuline [as (33)] was found to be an efficient precursor for both halves of thalicarpine (53).[47] Presumably, formation of the aporphine skeleton found in the right-hand half of (53) occurs before junction with reticuline (33) to give the thalicarpine skeleton, since there would be no free hydroxy-group available for phenol oxidation in a bisbenzylisoquinoline precursor. Interestingly, the level of radioactivity in the right-hand half was very marginally higher than in the other.

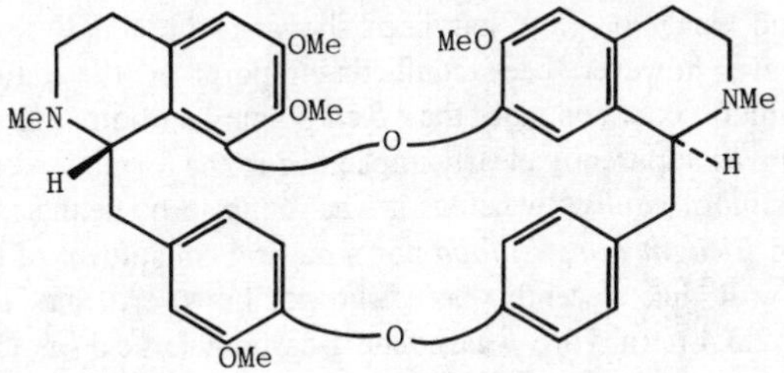

(52)

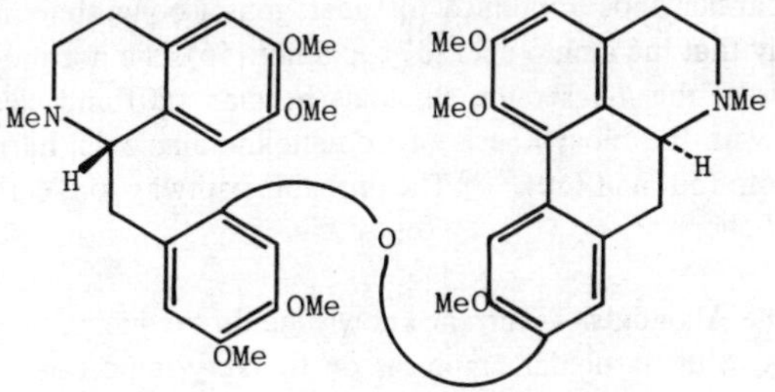

(53)

Several bisbenzylisoquinolines that are found in *Berberis* species of northern Pakistan can be derived, formally at least, from the naturally occurring spiro-dienone alkaloid pakistanamine (54).[48] It is notable that the relative abundance of alkaloids derived by formal dienone–phenol rearrangement in (54) by mechanism (*a*) is greater than that of more hindered alkaloids formed by mechanism (*b*).

[46] D. S. Bhakuni, S. Jain, and A. N. Singh, *Phytochemistry*, 1980, **19**, 2347.
[47] N. L. Marekov and A. K. Sidjimov, *Tetrahedron Lett.*, 1981, **22**, 2311.
[48] S. F. Hussain, M. T. Siddiqui, and M. Shamma, *Tetrahedron Lett.*, 1980, **21**, 4573.

(54)

3 Alkaloids Derived from Tryptophan

β-Carboline Alkaloids.—The biosynthesis of simple β-carboline alkaloids, *e.g.* harman (60) and eleagnine (59), has been shown to be simply from tryptamine (55).[49] There have, however, been conflicting reports on the nature of the first intermediate which contains all the β-carboline carbon atoms. *N*-Acetyl-tryptamine (56) was apparently clearly implicated in the formation of harman (60), *via* (58), in *Passiflora edulis*,[49] whereas it was found to be neither a precursor for eleagnine (59) in *Eleagnus angustifolia* nor a natural constituent of this plant[50] (*cf.* Vol. 3, p. 14). It has recently been shown, however, that [8-^{3}H; *methyl-*^{14}C]-1-methyl-1,2,3,4-tetrahydro-β-carboline-1-carboxylic acid [as (57)] is both an efficient and an intact precursor for harman (60) in *P. edulis*; the amino-acid (57) was also shown to be a natural constituent of this plant, being formed from the harman precursor tryptophan.[51] Similar results were obtained for eleagnine (59).[52] These results parallel those obtained for analogous isoquinoline alkaloids,[1,2] and indicate strongly that the amino-acid (57), and not (56), is a normal intermediate in the biosynthesis of the β-carboline alkaloids harman (60) and eleagnine (59). In further accord with the biosynthesis of isoquinoline alkaloids, harmalan (58) is a precursor for both (59) and (60).[49,50] The probable pathway to β-carboline alkaloids is illustrated in Scheme 3.

Terpenoid Indole Alkaloids.—Current knowledge on the biosynthesis of terpenoid indole alkaloids, with particular emphasis on the very important results obtained with enzyme preparations from tissue cultures of *Catharanthus roseus*, has been authoritatively reviewed.[53] Further work on cell lines of *C. roseus* that are able to produce *Aspidosperma*-type alkaloids has been published[54] (*cf.* Vol. 11, p. 19).

[49] M. Slaytor and I. J. MacFarlane, *Phytochemistry*, 1968, **7**, 605.
[50] I. J. MacFarlane and M. Slaytor, *Phytochemistry*, 1972, **11**, 229.
[51] R. B. Herbert and J. Mann, *J. Chem. Soc., Chem. Commun.*, 1980, 841.
[52] R. B. Herbert and J. Mann, unpublished work.
[53] A. I. Scott, S.-L. Lee, M. G. Culver, W. Wan, T. Hirata, F. Guéritte, H. Nordlöv, C. A. Dorschel, H. Mizukami, and N. E. Mackenzie, *Heterocycles*, 1981, **15**, 1257; M. Zenk, *J. Nat. Prod. (Lloydia)*, 1980, **43**, 438.
[54] W. G. W. Kurz, K. B. Chatson, F. Constabel, J. P. Kutney, L. S. L. Choi, P. Kolodziejczyk, S. K. Sleigh, K. L. Stuart, and B. R. Worth, *Phytochemistry*, 1980, **19**, 2583; J. P. Kutney, L. S. L. Choi, P. Kolodziejczyk, S. K. Sleigh, K. L. Stuart, B. R. Worth, W. G. W. Kurz, K. B. Chatson, and F. Constabel, *ibid.*, p. 2589.

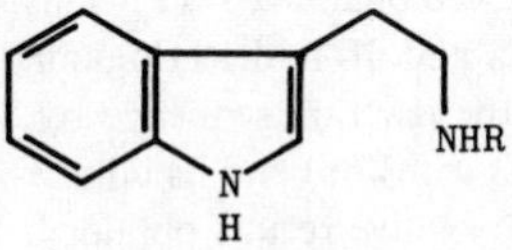

(55) R = H

(56) R = COMe

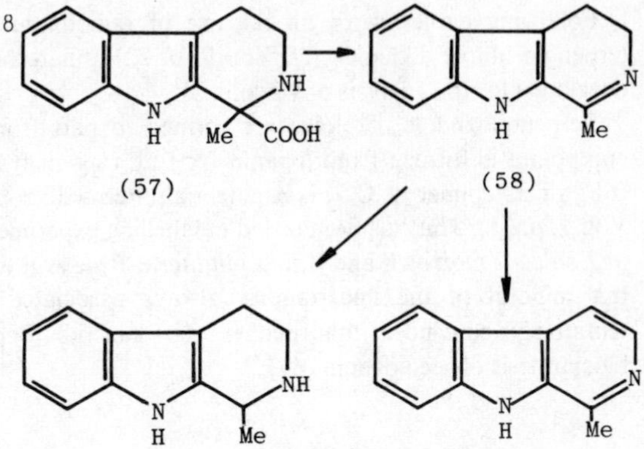

(57)

(58)

(59)

(60)

Scheme 3

Further information is available on the biosynthesis of vinblastine-type alkaloids (*cf.* Vol. 11, p. 19; Vol. 10, p. 19). Anhydrovinblastine (61) was found to be metabolized to leurosine (63) and catharine (64) in cultures of a *C. roseus* cell-line that do not normally produce these alkaloids.[55] Ring-opening in the conversion of the skeleton of anhydrovinblastine (61) into that of catharine (64) has been suggested to occur by Baeyer–Villiger-type oxidation of an imine [as (62)]. The alternative 21'-imine could give catharinine.[56]

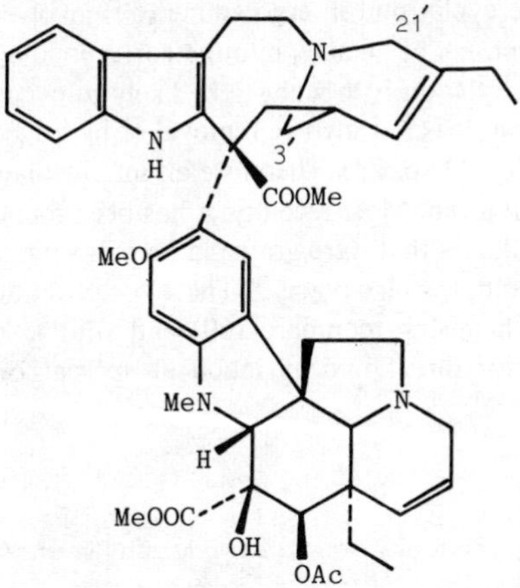

(61)

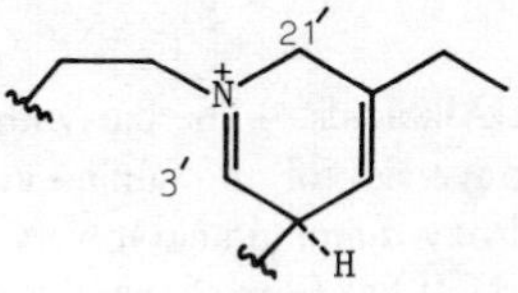

(62)

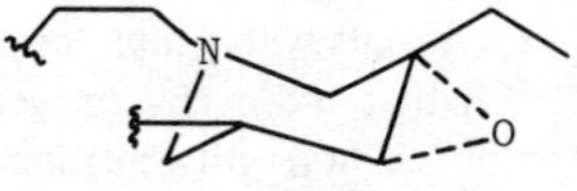

(63)

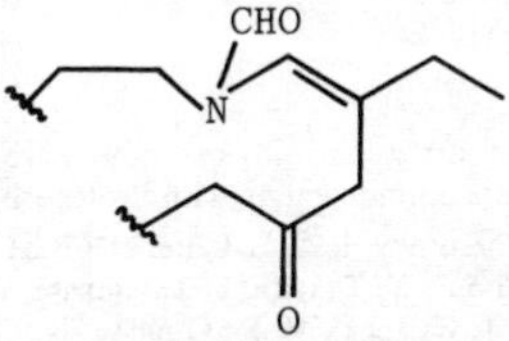

(64)

[55] J. P. Kutney, B. Aweryn, L. S. L. Choi, P. Kolodziejczyk, W. G. W. Kurz, K. B. Chatson, and F. Constabel, *Heterocycles*, 1981, **16**, 1169.
[56] N. Murugesan and M. Shamma, *Heterocycles*, 1981, **16**, 257.

Following earlier work on the use of radioimmunoassay for the analysis of terpenoid indole alkaloids (*cf.* Vol. 9, p. 22), this technique has been successfully developed for the analysis of vindoline.[57]

Terpenoid indole alkaloids are formed in part from secologanin (67).[1,2] This compound is formed from loganin (65) by ring-cleavage, and 10-hydroxyloganin (66), or its epimer at C-7, is a potential intermediate in the reaction sequence (*cf.* Vol. 1, p. 31). This has been tested in labelling experiments with (66) and its epimer in *Lonicera morrowii* and *Adina pilulifera*. Based on the negative results obtained, but mindful of the uncertainties always associated with such results, it was tentatively concluded that neither (66) nor its 7-epimer is implicated in the biosynthesis of secologanin (67).[58]

(65) R = H

(66) R = OH

(67)

Ergot Alkaloids.—The biosynthesis of the cyclol unit in ergotamine (69) involves the conversion of an α-amino-acid fragment, *i.e.* of alanine, into the corresponding α-hydroxy-α-amino-acid moiety; the hydroxylation is thought to be likely to occur on (68). It has been shown that the reaction does not involve removal of hydrogen from the methyl group of alanine[59] (*cf.* Vol. 11, p. 22). Therefore an intermediate with a 2,3-dehydroalanine moiety is not involved. Most recently, it has been found in experiments with *Claviceps purpurea* cultures that were grown in an atmosphere of $^{18}O_2$ that the cyclol oxygen arises from aerial oxygen.[60] The hydroxylation reaction occurs with retention of stereochemistry [compare (69) and (68)]*, in agreement with that generally observed for direct hydroxylation at sp³ carbon atoms.[61]

* The structures (68) and (69) were inadvertently drawn with incorrect stereochemistry in ref. 60: personal communication with Professor H. G. Floss.

[57] J. P. Kutney, L. S. L. Choi, and B. R. Worth, *Phytochemistry*, 1980, **19**, 2083.

[58] K. Inoue, Y. Takeda, T. Tanahashi, and H. Inouye, *Chem. Pharm. Bull.*, 1981, **29**, 981.

[59] C. M. Belzecki, F. R. Quigley, H. G. Floss, N. Crespi-Perellino, and A. Guicciardi, *J. Org. Chem.*, 1980, **45**, 2215.

[60] F. R. Quigley and H. G. Floss, *J. Org. Chem.*, 1981, **46**, 464.

[61] R. Bentley, 'Molecular Asymmetry in Biology', Academic Press, New York, 1969, Vol. 1: 1970, Vol. 2; A. R. Battersby, J. Staunton, H. R. Wiltshire, B. J. Bircher, and C. Fuganti, *J. Chem. Soc., Perkin Trans. 1*, 1975, 1162; and refs. cited therein.

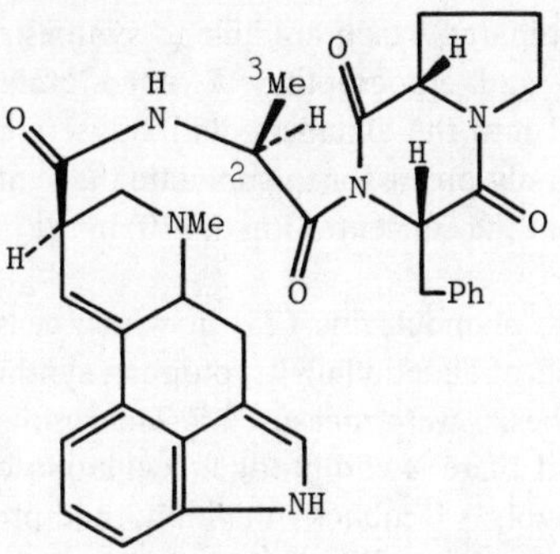

(68)(Alanine numbering) 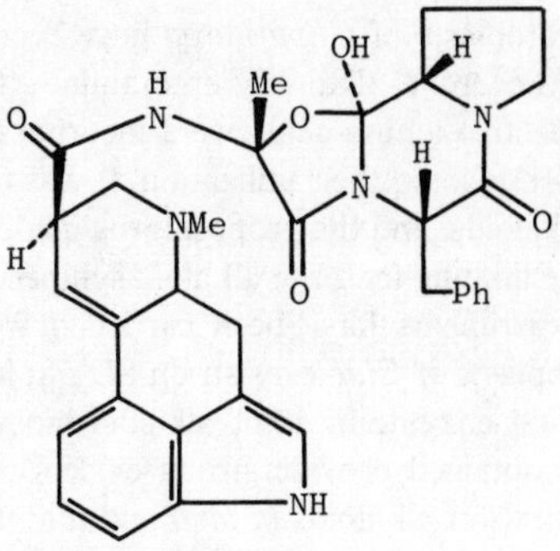　　　　　　(69)

The lysergic acid moiety [as (70)] in ergotamine (69) is formed from elymoclavine (73), which is formed in turn by hydroxylation of (72), the source of the oxygen being molecular oxygen.[62] The aldehyde (74) is a plausible intermediate beyond (73). Its conversion into ergotamine (69) can, in principle, be (*a*) by way of lysergic acid (70) and subsequent activation to (71) for the formation of an amide bond, leading to ergotamine, or (*b*) by direct formation of an activated intermediate on the aldehyde (74). Barring exchange reactions on (74), alternative (*a*) would result in loss of half of the oxygen that was originally present in the aldehyde, whereas none should be lost with alternative (*b*). Since, in the above experiment, the level of ^{18}O label in the lysergyl carboxamide group of ergotamine was found to be similar to that in the cyclol oxygen, alternative (*b*) is the more likely course of biosynthesis.[60]

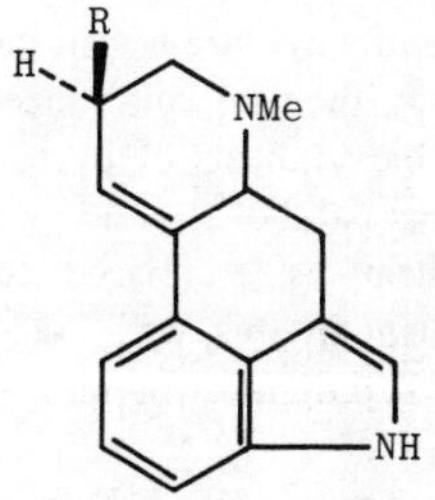

(70) R = COOH

(71) R = COX

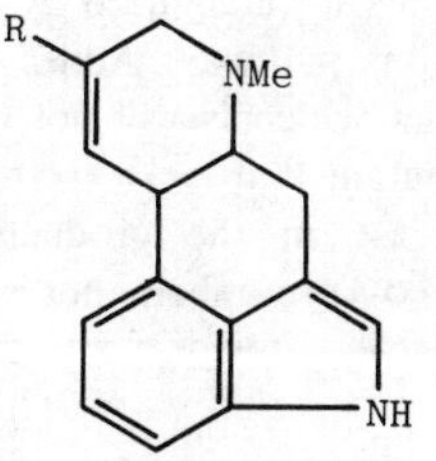

(72) R = Me

(73) R = CH$_2$OH

(74) R = CHO

[62] H. G. Floss, H. Günther, D. Gröger, and D. Erge, *J. Pharm. Sci.*, 1967, **56**, 1675.

Protoplasts of *C. purpurea* have been prepared which are able to synthesize the peptidic ergot alkaloids ergotamine (69) and ergocryptine *de novo*.[63] Various radioactive amino-acids were incorporated into the alkaloids, and D-lysergic acid (70) stimulated their utilization. It was the only precursor to stimulate the synthesis of alkaloids, and the proposal was made that the concentration of (70) in the cells is a rate-limiting factor in alkaloid synthesis.

Experiments have been carried out with elymoclavine (73) in whole cells and protoplasts of *Claviceps* strain SD 58; levels of dimethylallyltryptophan synthetase, the first enzyme in ergot alkaloid biosynthesis, were measured.[64] The results that were obtained provide strong evidence that there is end-product regulation of the synthesis of alkaloids *in vivo* and that it involves feedback inhibition; end-product repression of the synthesis of enzymes appears to be of lesser importance.

Pyrrolnitrin.—The *Pseudomonas* metabolite pyrrolnitrin (77) has been shown to derive from tryptophan.[2] Some of the early information on utilization of tryptophan[65] has now been published in full, together with new results.[66] It is clear that the indole nitrogen of tryptophan is the source of the nitro-group in (77), that the amino nitrogen provides the pyrrole nitrogen, that C-3′ of the amino-acid gives rise to C-3 of (77), and that the hydrogen from C-2 of the indole ring is retained. The last result excludes oxindole intermediates.

In cultures of *P. aureofaciens*, L-tryptophan was incorporated into (77) with high retention of labels that were present on the nitrogen in the side-chain and on the proton at C-2′ whereas D-tryptophan showed the reverse.[66] This indicates that the L-isomer is the more immediate precursor of pyrrolnitrin, with incorporation of D-tryptophan occurring *via* its enantiomer. It was found that L-tryptophan was more rapidly taken up by the organism than was the D-isomer, but ^{14}C-labelled D-tryptophan was more efficiently incorporated into pyrrolnitrin (77) than was the L-isomer. The latter was true irrespective of when the precursors were added during the growth of the culture, and so the lower relative incorporation of L-tryptophan is unlikely to be the consequence of competitive utilization for protein synthesis.

The catabolism of D- and of L-tryptophan in *P. aureofaciens* is different.[67] Only the latter isomer is catabolized by the kynurenine pathway, and it also induces the enzymes of this pathway. Added L-tryptophan may then be catabolized by this pathway. That which is will not be available for biosynthesis of pyrrolnitrin. In support, a mutant that lacks the first enzyme of this catabolic pathway showed a 30-fold increase in the production of pyrrolnitrin as compared to normal organisms.[67] D-Tryptophan, not suffering catabolism in this way, will be more readily available, through slow conversion into L-tryptophan, for biosynthesis of

[63] U. Keller, R. Zocher, and H. Kleinkauf, *J. Gen. Microbiol.*, 1980, **118**, 485.
[64] L.-J. Cheng, J. E. Robbers, and H. G. Floss, *J. Nat. Prod.* (*Lloydia*), 1980, **43**, 329.
[65] H. G. Floss, P. E. Manni, R. L. Hamill, and J. A. Mabe, *Biochem. Biophys. Res. Commun.*, 1971, **45**, 781; L. L. Martin, C. J. Chang, H. G. Floss, J. A. Mabe, E. W. Hagaman, and E. Wenkert, *J. Am. Chem. Soc.*, 1972, **94**, 8942.
[66] C. J. Chang, H. G. Floss, D. J. Hook, J. A. Mabe, P. E. Manni, L. L. Martin, K. Schröder, and T. L. Shieh, *J. Antibiot.*, 1981, **34**, 555.
[67] O. Salcher and F. Lingens, *J. Gen. Microbiol.*, 1980, **121**, 465.

pyrrolnitrin. This, in simple terms, will account for the different relative incorporations of D- and L-tryptophan, provided also that there is strict compartmentation of endogenous and exogenous pools of tryptophan.[66]

The proton at C-2' of L-tryptophan is retained at C-2 of (77).[66] Therefore the loss of the carboxy-group from C-2' of tryptophan must occur prior to, or simultaneous with, the formation of a double-bond at this carbon atom. This excludes a compound like (78), which has been isolated from *Pseudomonas* cultures,[68] from being an intermediate on this pathway.

The two pyrroles (76) and (79) have been tested as precursors of pyrrolnitrin.[66] The amino-compound (76) was efficiently incorporated, but the nitro-compound (79) was poorly utilized, indicating a favoured, if not obligatory, sequence involving halogenation before oxidation of the amino-group. The isomeric alcohols (80), analogues of (81) (previously postulated to be an intermediate in pyrrolnitrin biosynthesis[2]), were found not to act as precursors. An alternative sequence to pyrrolnitrin (77), based tenuously on this finding and on enzymic and chemical analogy, has been proposed (Scheme 4).[66] Alternatives involving early chlorination are possible in view of the broad substrate specificity that is seen in this pathway.[69] 7-Chlorotryptophan [as (75)] can act as a precursor of pyrrolnitrin,[69] and the isolation of its catabolic products (*cf.* Vol. 10, p. 23) from *Pseudomonas aureofaciens* indicates its existence in the organism; 7-chlorotryptophan is catabolized at a lower rate than tryptophan on the kynurenine pathway.[67, 70]

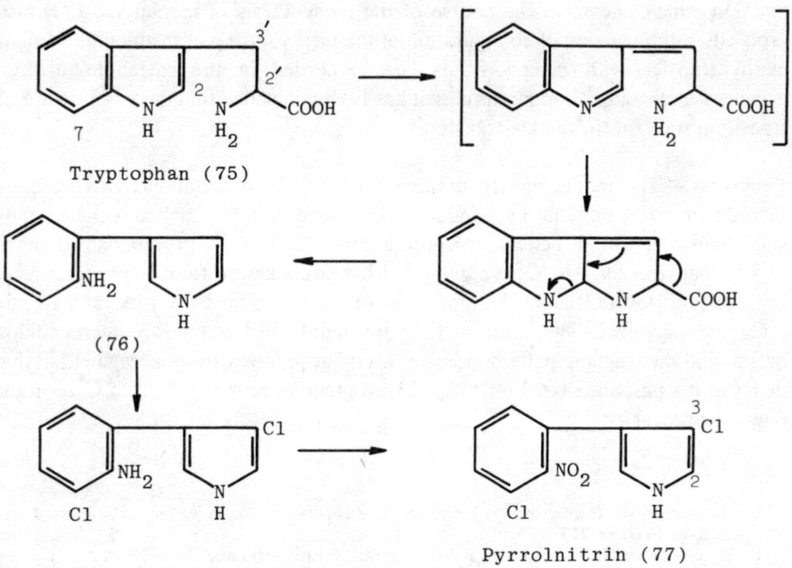

Scheme 4

[68] O. Salcher, F. Lingens and P. Fischer, *Tetrahedron Lett.*, 1978, 3097.
[69] R. L. Hamill, R. P. Elander, J. A. Mabe, and M. Gorman, *Appl. Microbiol.*, 1970, **19**, 721.
[70] K.-H. van Pée, O. Salcher, and F. Lingens, *Liebigs Ann. Chem.*, 1981, 233.

(78)

(79)

(80) R = OH

(81) R = Cl

(82)

Indolmycin.—It has been shown that the C-methyl group in indolmycin (82) originates from the methyl group of methionine with inversion of configuration (*cf.* Vol. 8, p. 23). Previously published in preliminary form,[71] the results are now available in a full paper.[72] In addition, it has been shown that the N-methylation reaction which occurs in the course of the biosynthesis of indolmycin (82) also proceeds with inversion of configuration of the methyl group of methionine. Similar methyl-transfer with inversion has been recorded in the catechol-O-methyl-transferase reaction,[73] and in this case it has been concluded that there is a tight $S_N 2$ transition state for the methyl-transfer.[74]

Penitrems.—The indole moiety of penitrem A (83), a metabolite produced by *Penicillium crustosum*, has been found to be formed from tryptophan with loss of its side-chain {eight-fold better incorporation of $(2S)$-[3'-14C]tryptophan than of $(2RS)$-[*benzene-ring*-U-14C]tryptophan}.[75] It was concluded from incorporations of [13C]acetate [see (83)] that the remainder of (83) is formed from mevalonate, six units being involved. The fragment C-18 through C-38 is manifestly diterpenoid in origin, and the labelling pattern indicates a similar pathway for this fragment to that deduced for paspaline (*cf.* Vol. 11, p. 23); a methyl group is lost from C-23 in the diterpene precursor.

[71] L. Mascaro, Jr., R. Hörhammer, S. Eisenstein, L. K. Sellers, K. Mascaro, and H. G. Floss, *J. Am. Chem. Soc.*, 1977, **99**, 273.

[72] R. W. Woodward, L. Mascaro, Jr., R. Hörhammer, S. Eisenstein, and H. G. Floss, *J. Am. Chem. Soc.*, 1980, **102**, 6314.

[73] R. W. Woodward, M.-D. Tsai, and J. K. Coward, unpublished work quoted in ref. 72; see also H. G. Floss and M.-D. Tsai, *Adv. Enzymol.*, 1979, **50**, 243.

[74] M. F. Hegazi, R. T. Borchardt, and R. L. Schowen, *J. Am. Chem. Soc.*, 1979, **101**, 4359.

[75] A. E. de Jesus, P. S. Steyn, F. R. van Heerden, R. Vleggaar, P. L. Wessels, and W. E. Hull, *J. Chem. Soc., Chem. Commun.*, 1981, 289.

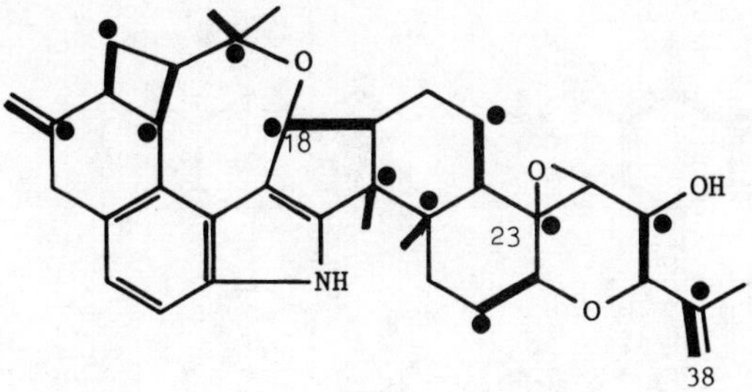

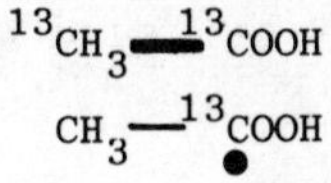

(83)

Anthramycin.—Anthramycin (85) is formed from tryptophan and tyrosine.[76] Utilization of the latter amino-acid is with cleavage of the aromatic ring. New results confirm earlier ones[76] that C-1 of (85) derives from C-3 of tyrosine and without loss of the protons at C-3.[77] It has also been shown that N-10 of anthramycin (85) has its origin in the indole nitrogen of tryptophan.[78] The experiment which gave this result involved incorporation of L-[*indole*-15N]-tryptophan and of DL-[1-13C]tyrosine into (85) and analysis essentially by 13C n.m.r. and mass spectrometry [C-1 of tyrosine (84) is the source of C-11 in (85)].

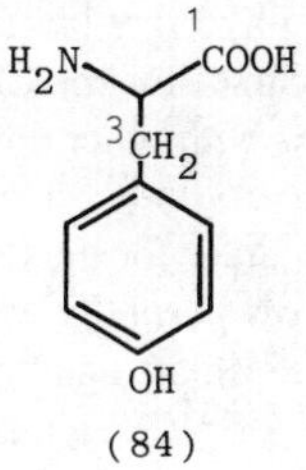

(84)

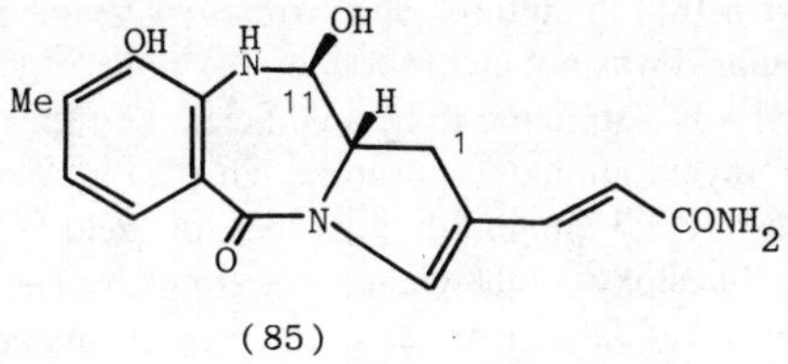

(85)

4 Miscellaneous

Ansamycins, Mitomycins, and Pactamycin.—At last, the nature of the long-sought C_7N intermediate [thickened bonds in (86), (87), and (88)] in the biosynthesis of the ansamycins [*e.g.* actamycin (86) and rifamycin B (87)], of the mitomycins [*e.g.* porfiromycin (88)], and of related metabolites is known! It is 3-amino-5-hydroxybenzoic acid (91).

[76] L. H. Hurley, *Acc. Chem. Res.*, 1980, **13**, 263.
[77] R. K. Malhotra, J. M. Ostrander, L. H. Hurley, A. G. McInnes, D. G. Smith, J. A. Walter, and J. L. C. Wright, *J. Nat. Prod.* (*Lloydia*), 1981, **34**, 38.
[78] J. M. Ostrander, L. H. Hurley, A. G. McInnes, D. G. Smith, J. A. Walter, and J. L. C. Wright, *J. Antibiot.*, 1980, **33**, 1167.

(86)

It had previously been deduced that the C_7N unit of the ansamycins was formed from an unknown compound which diverts from the shikimate pathway between 3-deoxy-D-*arabino*-heptulosonic acid 7-phosphate (DAHP) (89) and shikimic acid (*cf.* Vol. 9, p. 34; for similar conclusions on the biosynthesis of mitomycin, see Vol. 11, p. 28). Analysis of the structures of the known ansamycins and maytansinoids suggested that this elusive C_7N compound could be 3-amino-5-hydroxybenzoic acid (91), which is found, strikingly, in unmodified form as part of ferrimycin A, being linked by carboxy- and amino-groups to other residues in this Streptomycete metabolite.[79] In an important experiment, 3-amino-5-hydroxy-[*carboxy*-^{14}C]-benzoic acid was found to be an efficient and specific precursor for the C_7N unit of actamycin (86) in cultures of a *Streptomyces* species.[80] As with other ansamycins, shikimic acid was not incorporated. 3-Amino-5-hydroxy-[*carboxy*-^{13}C]benzoic acid [as (91)] was also found to be an efficient and specific precursor for the C_7N unit in the mitomycin antibiotic porfiromycin (88) in *Streptomyces verticillatus*.[81] In this experiment the 3-amino-5-hydroxybenzoic acid (91) was re-isolated and purified as methyl 3-acetoxy-5-(acetylamino)benzoate.[82] The excess of ^{13}C over its natural abundance had fallen from 85 atom % in the material that was fed to 52 atom % in the re-isolated material. This must have arisen by dilution with endogenous 3-amino-5-hydroxy-benzoic acid (91), and this key compound is therefore proved to be a natural constituent of *S. verticillatus*. The evidence that it is a normal intermediate in the biosynthesis of porfiromycin is, at the very least, very strong.

Independent evidence has been obtained that 3-amino-5-hydroxybenzoic acid (91) is also involved in the biosynthesis of rifamycins.[83,84] A compound, P8/1-OG (92),[84] has been isolated as a metabolite of several u.v. mutants (*P*⁻ mutants) of a strain of *Nocardia mediterranei* which normally produced rifamycins, *e.g.* rifamycin B (87). Another mutant series accumulated (92) and shikimic acid. These

[79] H. Bickel, P. Mertens, V. Prelog, J. Seibl, and A. Walser, *Tetrahedron*, 1966, Suppl. 8, p. 171.
[80] J. J. Kirby, I. A. McDonald, and R. W. Rickards, *J. Chem. Soc., Chem. Commun.*, 1980. 768.
[81] M. G. Anderson, J. J. Kirby, R. W. Rickards, and J. M. Rothschild, *J. Chem. Soc., Chem. Commun.*, 1980, 1277.
[82] J. J. Kirby and R. W. Rickards, *J. Antibiot.*, 1981, **34**, 605.
[83] O. Ghisalba and J. Nüesch, *J. Antibiot.*, 1981, **34**, 64.
[84] O. Ghisalba, H. Fuhrer, W. J. Richter, and S. Moss, *J. Antibiot.*, 1981, **34**, 58.

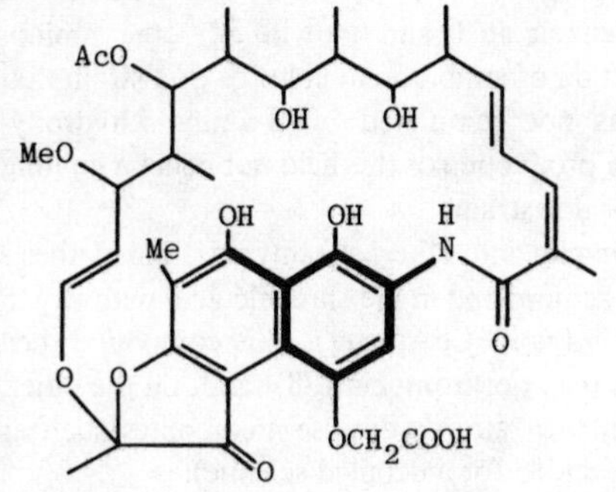

(87)

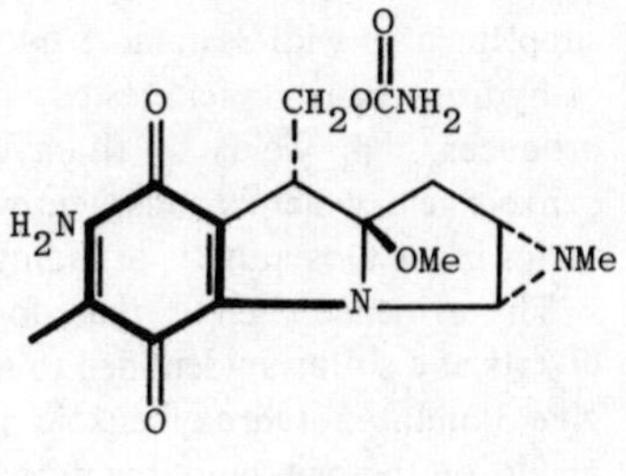

(88)

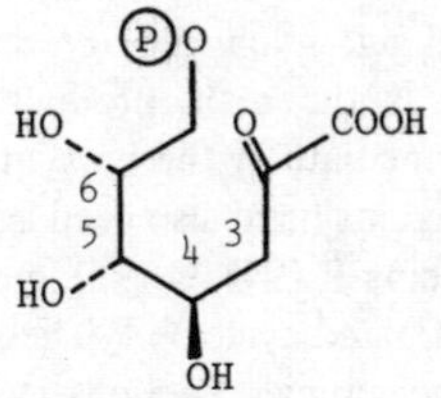

(89)

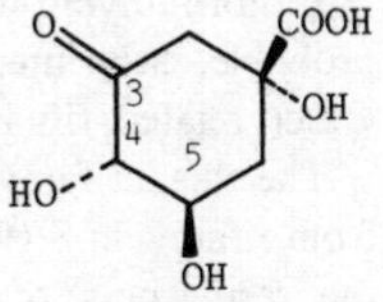

(90)

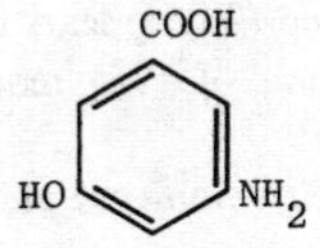

(91)

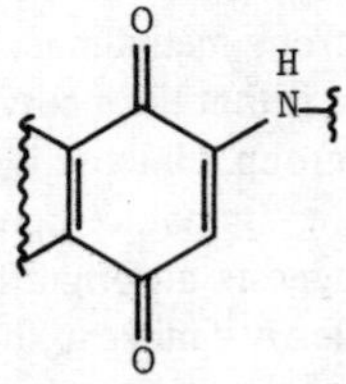

(92)

(93)

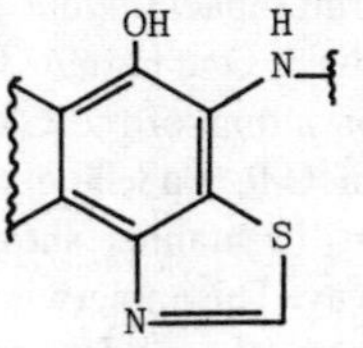

(94)

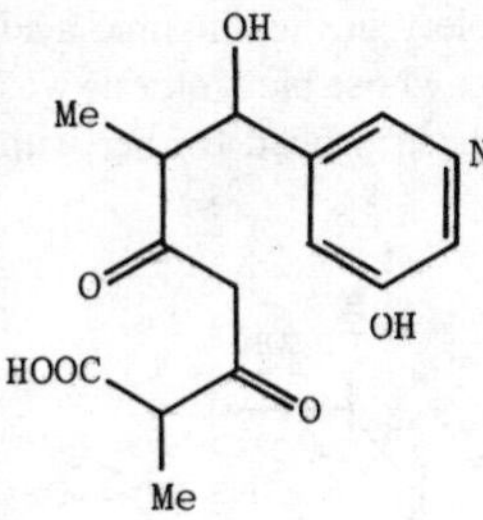

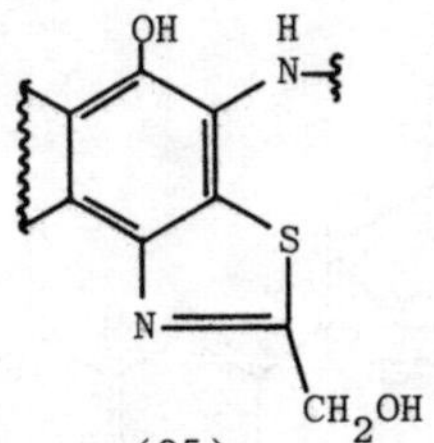

(95)

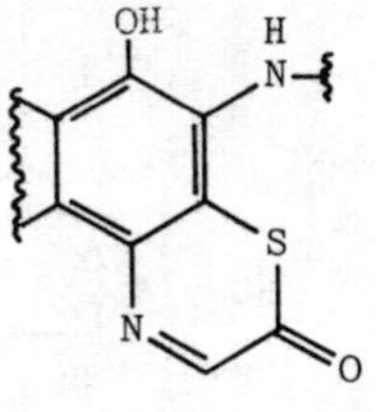

(96)

results show that P8/1-OG (92) arises from blocks that are beyond the point of diversion from the shikimic acid pathway but before the formation of rifamycins. It seems clear that P8/1-OG is an early intermediate in rifamycin biosynthesis which derives from 3-amino-5-hydroxybenzoic acid (91) (probably *via* its CoA ester). A mixed culture of *N. mediterranei* and two mutants, *i.e.* A8 (*transketolase⁻* mutant, having drastically reduced rifamycin production) and P14 (*P⁻* mutant; no production of rifamycins, but accumulation of P8/1-OG), showed enhanced production of rifamycin B which was not due to P8/1-OG.[83] A strong stimulation of the production of rifamycin B was found when cultures of strain A8 were

supplemented with 3-amino-5-hydroxybenzoic acid (and not with any other amino-
or hydroxy-benzoic acid tested); production of antibiotic in cultures of a strain that
produces high yields of rifamycin was not stimulated by 3-amino-5-hydroxy-
benzoic acid, which is consistent with the production of this acid not being a limiting
factor in the biosynthesis of rifamycin for this strain.

The evidence then is that, for rifamycin and other ansamycins, biosynthesis
diverts at a so-far unidentified (but early) compound in the shikimic acid pathway to
give 3-amino-5-hydroxybenzoic acid (91) (as its CoA ester). This compound then
yields, on the one hand, the mitomycins [e.g. porfiromycin (88)] and, on the other,
the CoA ester of P8/1-OG (92), which then affords diverse metabolites such as
rifamycin B (87) and actamycin (86) (cf. ref. 83 for a detailed scheme).

Protorifamycin I is the main product of cultures of *N. mediterranei* and a
probable early intermediate in the biosynthesis of rifamycins (cf. Vol. 10, p. 29).
Other, related, rifamycins have also been isolated from cultures of this organism.[85]

The thiazorifamycins P (94), Q (95), and Verde (96) have been shown to derive
from rifamycin S (93) and cysteine.[86] The formation of (94) and (96) appears to be
the result only of chemical reactions, whereas the formation of (95) requires
enzymic assistance.

Using ^{13}C-labelled precursors, it has been shown[87] that *Streptomyces pactum*
elaborates pactamycin (97) from methionine, acetate, and glucose as shown in
Scheme 5. One notable feature is that the methyl group of methionine provides both
carbon atoms of the *C*-ethyl group. Glucose is the source of the cyclopentane ring
and of C-9. Labelling of the *m*-amino-acetophenone moiety by D-[6-^{13}C]glucose,
and in the manner shown, suggests an origin for this moiety in the shikimic acid
pathway. This moiety is manifestly similar to the C_7N units whose biosynthesis was
discussed above. The origin of the pactamycin unit has been probed further with

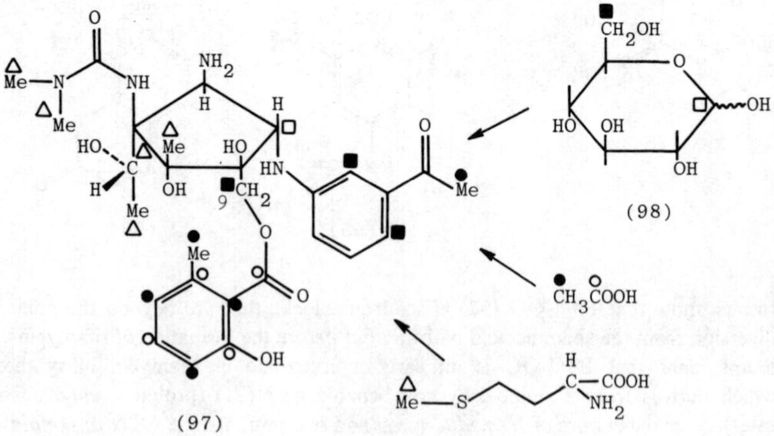

Scheme 5

[85] O. Ghisalba, P. Traxler, H. Fuhrer, and W. J. Richter, *J. Antibiot.*, 1980, **33**, 847.
[86] R. Cricchio, P. Antonini, and G. Sartori, *J. Antibiot.*, 1980, **33**, 842.
[87] D. D. Weller and K. L. Rinehart, Jr., *J. Am. Chem. Soc.*, 1978, **100**, 6757.

[U-^{13}C]glucose in an experiment in which it was administered to cultures of *S. pactum* with a large amount of unlabelled glucose.[88] Analysis of the n.m.r. couplings for a derivative of pactamycin (97) revealed intact glucose fragments still present and indicated clearly that nitrogen in the *m*-aminoacetophenone moiety is attached to C-3 of a hypothetical dehydroquinate (90) precursor [or C-6 of DAHP (89)]. This is at variance with conclusions that were recently reached on the C$_7$N unit of mitomycin (amination of C-4 of DAHP)[89] (*cf.* Vol. 11, p. 28).

[*ring*-U-^{14}C]-*m*-Aminobenzoic acid has been found to be a very efficient precursor of pactamycin, and this amino-acid is likely therefore to be a normal intermediate in biosynthesis. It has a close structural kinship to the ansamycin and mitomycin precursor 3-amino-5-hydroxybenzoic acid (91). It will be most interesting to discover their biosynthetic relationship and to resolve the apparently different formation from intermediates in the shikimate pathway. Conversion of *m*-aminobenzoic acid into *m*-aminoacetophenone presumably occurs by way of (99).

β-Lactam Antibiotics.—Important recent results on the biosynthesis of penicillins have been obtained, using cell-free preparations of *Cephalosporium acremonium*. The results show that the penicillins, *e.g.* penicillin V (108), are formed by way of (L-*α*-amino-*δ*-adipyl)-L-cysteinyl-D-valine (100), with isopenicillin N (104) as the first of these metabolites to be formed; the tripeptide (100) was incorporated

(99)

(100) ● = ▲ = ^{12}C, ✱ = ^{1}H

(101) ● = ^{13}C, ▲ = ^{12}C, ✱ = ^{1}H

(102) ● = ▲ = ^{13}C, ✱ = ^{1}H

(103) ● = ▲ = ^{12}C, ✱ = ^{2}H

[88] K. L. Rinehart, Jr., M. Potgieter, and D. L. Delaware, *J. Am. Chem. Soc.*, 1981, **103**, 2099.
[89] U. Hornemann, J. H. Eggert, and D. P. Honor, *J. Chem. Soc., Chem. Commun.*, 1980, 11.

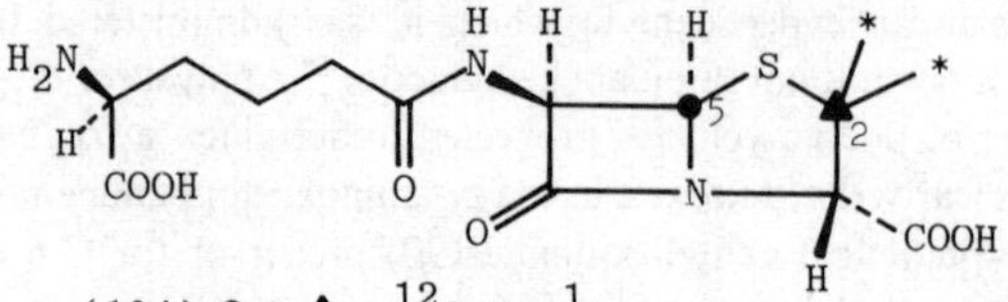

(104) ● = ▲ = ^{12}C, ✱ = ^{1}H

(105) ● = ^{13}C, ▲ = ^{12}C, ✱ = ^{1}H

(106) ● = ▲ = ^{13}C, ✱ = ^{1}H

(107) ● = ▲ = ^{12}C, ✱ = ^{2}H

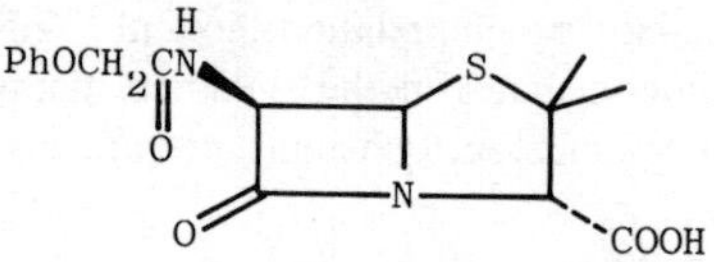

(108)

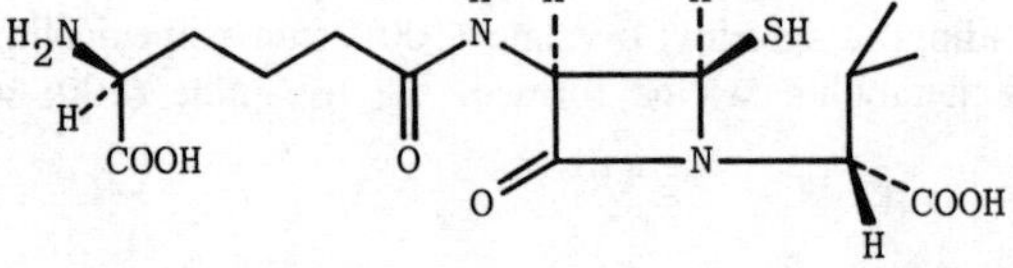

(109)

intact[90] (see also Vol. 8, p. 35). When the tripeptide that was labelled with ^{13}C as shown in (101) was incubated with cell-free extracts of *C. acremonium* in the probe of an n.m.r. spectrometer, the appearance of a single new enriched ^{13}C signal due to C-5 in the labelled isopenicillin N (105) was observed; a high conversion into (105) was obtained, and the titre of antibiotic against time was proportional to the growth of the new ^{13}C signal.[91] Similar results were obtained with the tripeptide that was labelled as shown in (102): new enriched signals for C-2 and C-5 in (106) were observed. It is apparent from these results that the conversion of (100) into (104) is direct, since no intermediates could be detected. Additional information on the conversion of (100) into (104) is that the heptadeuterio-tripeptide (103) gave isopenicillin N with retention of all the deuterons, and at the expected positions, as shown in (107).[92]

[90] J. O'Sullivan, R. C. Bleaney, J. A. Huddleston, and E. P. Abraham, *Biochem. J.*, 1979, **184**, 421; T. Konomi, S. Herchen, J. E. Baldwin, M. Yoshida, N. A. Hunt, and A. L. Demain, *ibid.*, p. 427.
[91] J. E. Baldwin, B. L. Johnson, J. J. Usher, E. P. Abraham, J. A. Huddleston, and R. L. White, *J. Chem. Soc., Chem. Commun.*, 1980, 1271.
[92] J. E. Baldwin, M. Jung, J. J. Usher, E. P. Abraham, J. A. Huddleston, and R. L. White, *J. Chem. Soc., Chem. Commun.*, 1981, 246.

The conversion, in a cell-free preparation of *Penicillium chrysogenum*, of the tripeptide (100) into what was identified as isopenicillin N (104) has been reported.[93] The penicillin was formed with retention of a tritium label sited at C-2 of the valine moiety (*cf.* ref. 94). Also detected as a product of the metabolism of (100) in this system was apparently (109); appropriately, tritium that was originally present at C-2 and C-3 of the valine moiety in (100) was retained, and so was half of a tritium label from C-3 of the cysteine fragment. However, the structure assigned to this metabolite has not yet been rigorously proved, and it remains to be seen whether (109) is a normal free intermediate in the biosynthesis of penicillins, but evidence has been provided that the first ring-closure in (100) is to give the β-lactam ring.

L-Valine is one of three amino-acid precursors for the penicillins. It has recently been shown that L-valine is incorporated into penicillin V (108) with loss of the ^{18}O label that is sited on one of the oxygen atoms of the carboxy-group (only a singly labelled penicillin species was formed, even though there was substantial double labelling in the precursor).[95] It is reasonable to conclude that an acyl derivative of valine is formed at some stage in the biosynthesis of penicillins [either before or after the formation of (100)] which may be associated with the conversion of L-valine into the D-isomer.

It is known that valine is incorporated into the penicillins with overall retention of configuration at C-3. It has been shown that L-$[1-^{14}C,3-^3H]$valine is incorporated into the disulphide dimer of (100), in cultures of a mutant of *C. acremonium*, with complete retention of tritium at C-3 of the valine moiety.[96] This indicates that valine is built into (100) with retention of configuration and that therefore no double inversion occurs during the biotransformation of valine into the penicillins.

It has been shown that nocardicin A (113) is formed from L-serine (112), from L-homoserine, and from two units of L-tyrosine with loss of C-1 (*cf.* Vol. 9, p. 33). New results confirm these findings and extend them.[97] The manner of utilization of tyrosine (L-phenylalanine was not a precursor) suggested that (*p*-hydroxy-phenyl)glycine [as (110)] would be a more immediate precursor than tyrosine. This was confirmed: (110) and its D-isomer were incorporated without loss of a ^{14}C label that was sited at C-1, and DL-$[1-^{13}C]$-(*p*-hydroxyphenyl)glycine [as (110)] gave nocardicin A (113) that was labelled at C-10 and at C-1′.

Tritium at C-2 of either (110) or its D-isomer was lost on formation of (113), and the L-isomer (110) was the preferred precursor of nocardicin. These results parallel those for the utilization of valine in the biosynthesis of penicillins, and it has been suggested that the configurational inversion of L-(*p*-hydroxyphenyl)glycine (110) which necessarily occurs in the course of the biosynthesis of nocardicin may also parallel the inversion of L-valine which occurs in the biosynthesis of penicillins (*cf.* above).

[93] B. Meesschaert, P. Adriaens, and H. Eyssen, *J. Antibiot.*, 1980, **33**, 722.
[94] P. A. Fawcett, J. J. Usher, J. A. Huddleston, R. C. Bleaney, J. J. Nisbet, and E. P. Abraham, *Biochem. J.*, 1976, **157**, 651.
[95] J. S. Delderfield, E. Mtetwa, R. Thomas, and T. E. Tyobeka, *J. Chem. Soc., Chem. Commun.*, 1981, 650.
[96] J. E. Baldwin and T. S. Wan, *Tetrahedron*, 1981, **37**, 1589.
[97] C. A. Townsend and A. M. Brown, *J. Am. Chem. Soc.*, 1981, **103**, 2873.

HOOC—SMe

NH₂ (111) → NH_2

L -(*p*-Hydroxyphenyl)-
glycine (110)

(112)

(113)

L-Methionine (111) was a significantly better precursor than homoserine for nocardicin A (113), indicating that it is sulphur rather than oxygen which is displaced in the formation of an ether leading to (113).

Neither L-alanine nor L-cysteine was incorporated into (113), but both L-serine (112) and glycine were. Good evidence was obtained that utilization of glycine is by way of L-serine. L-Serine was utilized without loss of the tritium that was sited at C-3, so the construction of the β-lactam ring occurs without change in the oxidation state at this carbon atom,[97] in contrast to the parallel situation in the biosynthesis of penicillins.[2] For nocardicin A (113), direct nucleophilic displacement of a (presumably activated) hydroxy-group of a seryl unit by amide nitrogen apparently occurs.

Aranotins and Sirodesmin.—*cyclo*-(L-Phenylalanyl-L-seryl) is an efficient and intact precursor for gliotoxin (114) (*cf.* Vol. 10, p. 27). Bisdethiobis(methylthio)-acetylaranotin (115) is related to gliotoxin: in the biosynthesis of this metabolite, it is believed that ring-opening of an arene oxide [as (116)] occurs, followed by further epoxidation and cyclization, as in the biosynthesis of gliotoxin (114).[2]

It has been shown conclusively that *cyclo*-(L-phenylalanyl-L-phenylalanyl) is an essentially intact precursor for (115) in *Aspergillus terreus*, and is present in the cultures of this organism being labelled by phenylalanine, a precursor for (115).[98] The cyclic dipeptide was incorporated into (115) with high efficiency and with low

[98] M. I. Pita Boente, G. W. Kirby, and D. J. Robins, *J. Chem. Soc., Chem. Commun.*, 1981, 619.

(114)

(115)

(116)

(117)

dilution. Neither *cyclo*-(L-phenylalanyl-D-phenylalanyl) nor *cyclo*-(D-phenylalanyl-D-phenylalanyl) was at all efficiently incorporated.

Of the set of experiments carried out, one using *cyclo*(-L-[^{15}N]Phe-L-[1-^{13}C]Phe-) (117) is the most notable and informative. Incorporation of the precursor into (115) was examined by ^{13}C n.m.r. spectroscopy. The spectrum of (115) showed enrichment of the carbonyl signal of the dioxopiperazine moiety, and of this one only. Further, this signal was split by the adjacent ^{15}N label, establishing that there is (essentially) intact incorporation of the precursor. In the feeding experiment, the labelled precursor was administered together with unlabelled dipeptide, to ensure that the labelled and the unlabelled phenylalanine that could have been formed would have been adequately mixed before recombination.

L-[U-^{14}C]Tyrosine and L-[U-^{14}C]serine have been found to be incorporated into sirodesmin PL (120) in *Phoma lingam*.[99] Efficient incorporations of *cyclo*(-L-[U-^{14}C]Tyr-L-Ser-) [as (118)] and of similarly labelled phomamide [as (119)] were also observed. Phomamide itself was labelled by radioactive serine and tyrosine in cultures of *P. lingam*. Results obtained with [1-^{13}C]- and [1,2-^{13}C$_2$]-acetate established that the remaining skeletal carbon atoms were mevalonoid in origin. The combined results lead to the pathway shown in Scheme 6.

Phenazines.—Evidence had been obtained with some micro-organisms, but not with others (*Pseudomonas* and related organisms), that phenazine-1,6-dicarboxylic acid (121) is the first phenazine to be formed, *i.e.* the one from which all others derive (*cf.* Vol. 10, p. 28). The negative results that were obtained with (121) were

[99] J.-P. Ferezou, A. Quesneau-Thierry, C. Servy, E. Zissmann, and M. Barbier, *J. Chem. Soc., Perkin Trans. 1*, 1980, 1739.

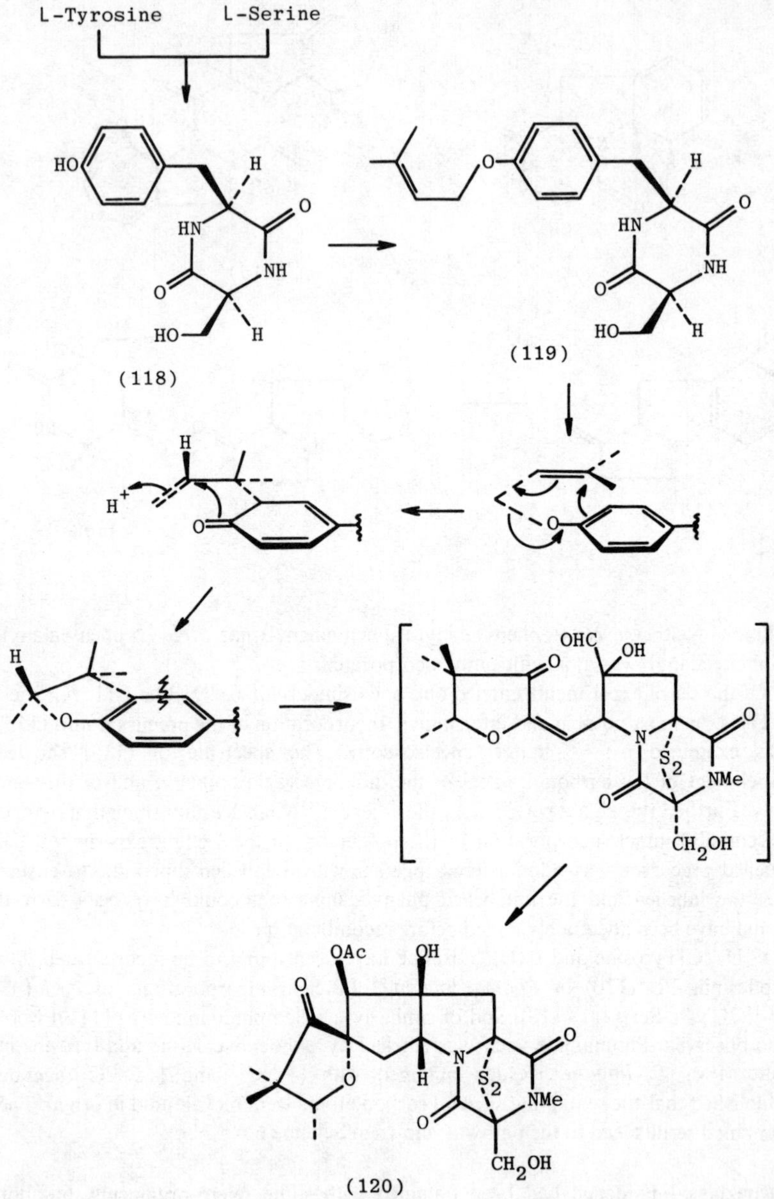

(118)

(119)

(120)

Scheme 6

attributed to poor cell permeability in *Pseudomonas* species and in related organisms. Bacterial cell walls, including those of *Pseudomonas* species, have been made permeable to peptidoglycan and DNA precursors, which are not normally assimilated, by treatment of the cells with toluene or diethyl ether.[100] The technique of washing cells with ether has been successfully applied to this problem in the biosynthesis of phenazines, and it should find application elsewhere in problems involving the non-permeability of membranes to any precursors that are administered.[101]

Using ether-treated cells of *P. aureofaciens*, [*dicarbonyl*-$^{14}C_2$]phenazine-1,6-dicarboxylic acid [as (121)] was found to be an efficient and specific precursor for phenazine-1-carboxylic acid (123), and also for 2-hydroxyphenazine-1-carboxylic acid (124). The rate of growth of the organism appeared to be important, because an incorporation was also recorded of the labelled (121) into (123), albeit at a lower level, with cultures that had been grown rapidly. The position of phenazine-1,6-dicarboxylic acid (121) as a universal intermediate in the biosynthesis of phenazines now seems secure. The previously reported failure of dimethyl phenazine-1,6-dicarboxylate (122) to act as a precursor of phenazines (*cf.* Vol. 10, p. 28; Vol. 9, p. 29) has been confirmed with ether-treated cells of *P. aureofaciens*. Efficient hydrolysis of (125) to (123) did, however, occur.[101]

COOR N

COOR

(121) R = H

(122) R = Me

$COOR^2$ N

R^1

(123) $R^1 = R^2 =$ H

(124) $R^1 =$ OH, $R^2 =$ H

(125) $R^1 =$ H, $R^2 =$ Me

Cytochalasins.—Previous results have shown that the cytochalasins, *e.g.* cytochalasin B (126), are partially polyketide in origin (*cf.* Vol. 7, p. 29; Vol. 6, p. 44). It has now been shown that [2-$^2H_3$,2-^{13}C]acetic acid is incorporated into cytochalasin B (126) in *Phoma exigua* and into cytochalasin D in *Zygosporium masonii*. Labelling of the expected sites by ^{13}C was observed for both metabolites, but, in keeping with results on other acetate metabolites, most of the deuterium was lost, being retained only in the polyketide chain at C-11, which is part of the starter acetate unit.[102]

Cycloheximide.—It is known that the carbon skeleton of cycloheximide (127) is derived from malonate, with the two methyl groups (C-15 and C-16) arising from

[100] W. P. Schrader and D. P. Fan, *J. Biol. Chem.*, 1974, **249**, 4815; D. Mirelman, Y. Yashouv-Gan, and U. Schwarz, *Biochemistry*, 1976, **15**, 1781; D. Mirelman and V. Nuchamowitz, *Eur. J. Biochem.*, 1979, **94**, 541; R. E. Moses and C. C. Richardson, *Proc. Natl. Acad. Sci. USA*, 1970, **67**, 674; H.-P. Vosberg and H. Hoffmann-Berling, *J. Mol. Biol.*, 1971, **58**, 739.
[101] P. R. Buckland, R. B. Herbert, and F. G. Holliman, *Tetrahedron Lett.*, 1981, **22**, 595.
[102] R. Wyss, Ch. Tamm, and J. C. Vederas, *Helv. Chim. Acta*, 1980, **63**, 1538.

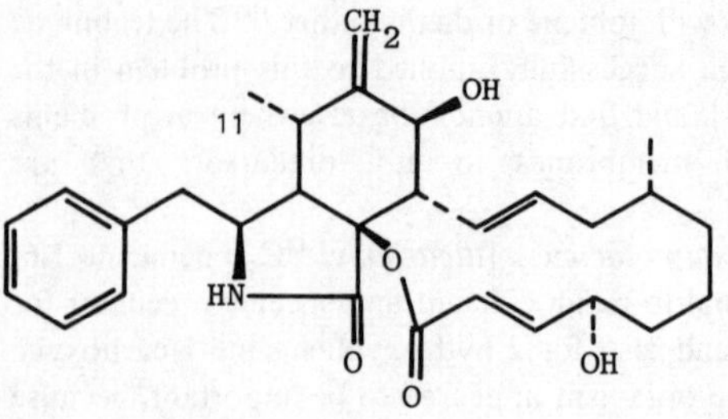

(126) 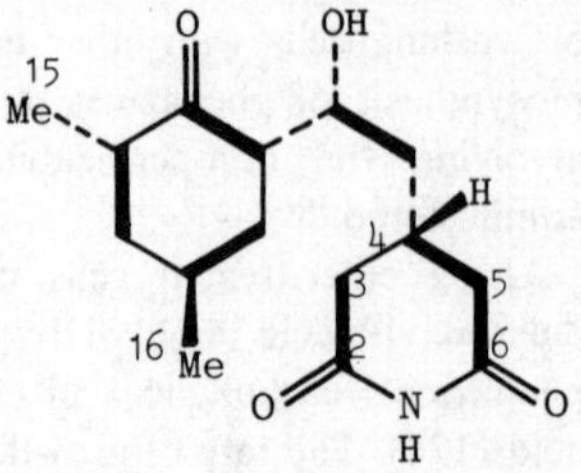   (127)

methionine.[103] Interestingly, one of the malonate units, the starter unit, is uniquely utilized as a C_3 moiety. Evidence was obtained that the glutarimide ring was assembled in a partially stereospecific manner, with the C_3 malonate unit appearing predominantly at C-4, C-5, and C-6 rather than at C-4, C-3, and C-2. The biosynthesis of cycloheximide (127) has been further probed in *Streptomyces naraensis* with $[1,2,3-^{13}C_3]$malonate.[104] The pattern of labelling that was deduced

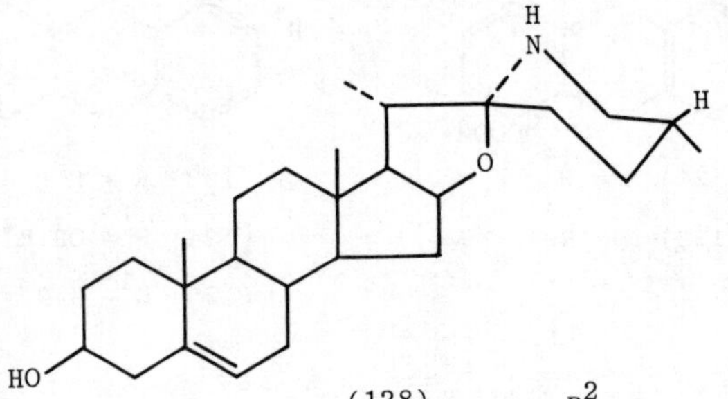

(128)

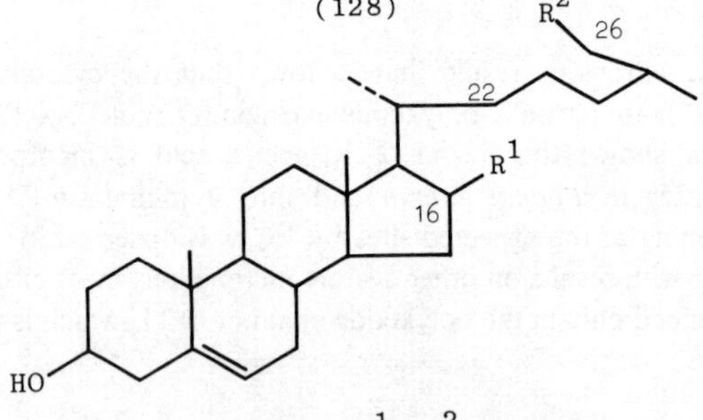

(129) $R^1 = R^2 = H$

(130) $R^1 = H, R^2 = NH_2$

(131) $R^1 = OH, R^2 = NH_2$

[103] Z. Vaněk, M. Půža, J. Cudlín, M. Vondráček, and R. W. Rickards, *Folia Microbiol.*, 1969, **14**, 388; F. Johnson, in 'Fortschritte der Chemie Organischer Naturstoffe', ed. W. Herz, H. Grisebach, and G. W. Kirby, Springer-Verlag, Wien, 1971, Vol. 29, p. 140.
[104] H. Shimada, H. Noguchi, Y. Itaka, and U. Sankawa, *Heterocycles*, 1981, **15**, 1141.

from the ^{13}C spectra that were obtained is illustrated in (127). Notably, it was apparent that the labelling of the glutarimide nucleus is completely stereospecific, C-4, C-5, and C-6 deriving exclusively from the C_3 malonate unit. The coupled signal for C-6 was smaller than that for C-2, consistent with exchange of the carboxyl carbon of malonyl-CoA with unlabelled carbon dioxide through the interconversion of malonyl-CoA and acetyl-CoA.

Steroidal Alkaloids.—In the biosynthesis of alkaloids such as solasodine (128), from cholesterol (129), it appears that the cholesterol side-chain is first functionalized at C-26 with the introduction of a hydroxy-group (*cf.* Vol. 8, p. 28; Vol. 7, p. 32). The 26-amino-compound, (25*R*)-26-aminocholesterol (130), has been found to act as a significant precursor for solasodine (128) in *Solanum laciniatum*, whereas (25*R*)-26-aminocholest-5-ene-3β,16β-diol (131) was poorly utilized.[105] This indicates that replacement of the hydroxy-group at C-26 by an amino-group may occur before further oxygenation elsewhere in the steroid nucleus (particularly at C-16). It may also be concluded from this and other evidence (*cf.* Vol. 9, p. 27) that oxidation at C-22 precedes hydroxylation at C-16.

Betalains.—The effect of various treatments on the production of betalains in seedlings of *Amaranthus candatus* has been studied.[106] The regulation of production of betanin and betaxanthin appears to be somewhat different to that of amaranthin.

Capsaicin.—It is known that phenylalanine acts as the source of the vanillylamine moiety of capsaicin (132) in *Capsicum annuum* and that valine (136) is used for the construction of the acyl part.[2] It has been noted[107] that whilst radioactive valine (136) was incorporated into capsaicin (132) and its dihydro-derivative (133), L-[U-^{14}C]leucine [as (137)] labelled the capsaicin analogues (134) and (135) in *Capsicum* fruits; both amino-acid precursors were very considerably better incorporated into capsaicinoid metabolites by using spheroplasts prepared from the placentas of *Capsicum* fruits. α-Ketoisovaleric acid (138) and α-ketoisocaproic acid

(132) R =

(133) R =

(134) R =

(135) R =

[105] R. Tschesche and H. R. Brennecke, *Phytochemistry*, 1980, **19**, 1449.
[106] J. Bianco-Colomas, *Planta*, 1980, **149**, 176.
[107] T. Suzuki, T. Kawada, and K. Iwai, *Plant Cell Physiol.*, 1981, **22**, 23.

(139), which are potential metabolites of (136) and (137) respectively, were shown to be present in the spheroplasts, and each was shown to be formed from the corresponding amino-acid in a cell-free preparation. Metabolism of L-[U-^{14}C]valine in spheroplasts gave radioactive isobutyric acid and, by chain-elongation, (140); metabolism of L-[U-^{14}C]leucine gave radioactive isovaleric acid and (141). The results obtained are consistent with the pathway to capsaicinoids shown in Scheme 7.

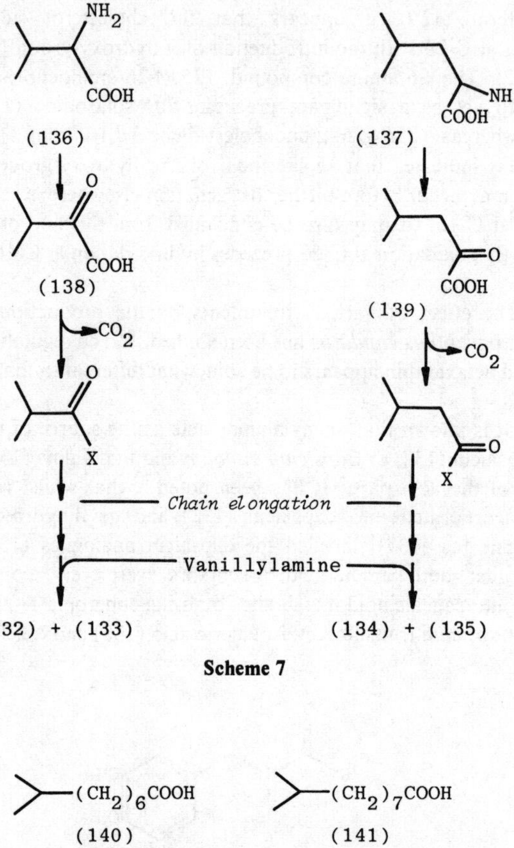

Scheme 7

2

Pyrrolidine, Piperidine, and Pyridine Alkaloids

BY A. R. PINDER

Two new monographs on alkaloid chemistry have appeared; both contain sections on pyrrolidine, piperidine, and pyridine alkaloids.[1,2]

1 Pyrrolidine Alkaloids

Species of marine sponge of the genus *Laxosuberites* contain several pyrrole-2-aldehydes with a long side-chain at position 5. Their structures (1), (2), and (3) have been settled by spectral study and by chemical degradation. One, (3), proved to be a cyanhydrin of unusual stability.[3]

(1) $n = 14$, 15, 16, or 18   (2)

(3)

Certain 2,5-dialkyl-pyrrolidines, which are trail pheromones of the pharaoh ant (*Monomorium pharaonis* L.), have been synthesized by application of the Hofmann–Löffler reaction to 5-aminotridecane and 7-aminopentadecane. The same type of compound results from Seebach alkylation of *N*-nitrosopyrrolidines.[4]

[1] G. A. Cordell, 'Introduction to Alkaloids', Wiley–Interscience, Somerset, New Jersey, 1981.
[2] M. Hesse, 'Alkaloid Chemistry', Wiley–Interscience, Somerset, New Jersey, 1981.
[3] D. B. Stierle and D. J. Faulkner, *J. Org. Chem.*, 1980, **45**, 4980.
[4] E. Schmitz, H. Sonnenschein, and C. Gründemann, *J. Prakt. Chem.*, 1980, **322**, 261.

Jatropham (surely better called jatrophine) is an alkaloid occurring in *Jatropha macrorhiza* (Euphorbiaceae) which has inhibitory activity towards leukaemia. Its racemic form (4) has been synthesized from 3-methyl-2-furoic acid, as outlined in Scheme 1.[5] This synthesis vindicates an earlier suggestion that the structure be revised to 5-hydroxy-3-methyl-3-pyrrolin-2-one (4), as opposed to the 4-methyl isomer.[6]

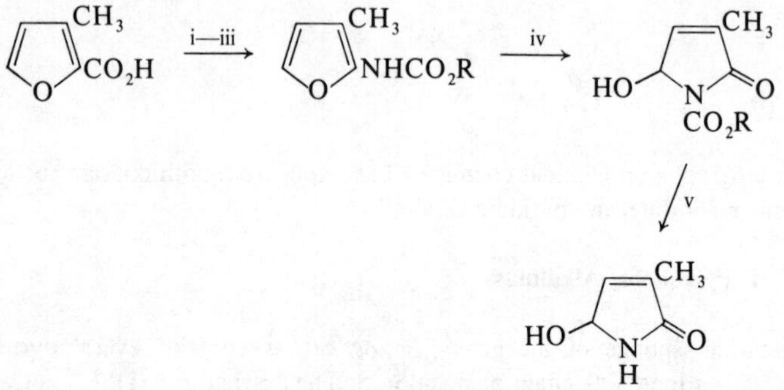

Reagents: i, $ClCO_2Et$; ii, NaN_3; iii, ROH (R = benzyl, *p*-methoxybenzyl, or t-butyl); iv, autoxidation; v, CF_3CO_2H, at 0 °C.

Scheme 1

Odorine and roxburghilin are identical. The stereochemistry of this alkaloid has been established as (5) by synthesis of the (−)-base from L-proline, and by stereochemical correlations with compounds of known absolute configuration.[7]

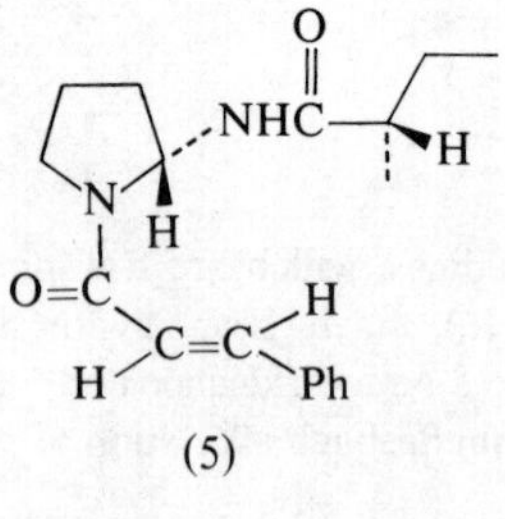

[5] K. Yakushijin, R. Suzuki, R. Hattori, and H. Furukawa, *Heterocycles*, 1981, **16**, 1157.
[6] K. Yakushijin, M. Kozuka, Y. Ito, R. Suzuki, and H. Furukawa, *Heterocycles*, 1980, **14**, 1073.
[7] P. J. Babidge, R. A. Massy-Westropp, S. G. Pyne, D. Shiengthong, A. Ungphakorn, and G. Veerachat, *Aust. J. Chem.*, 1980, **33**, 1841.

A review of recent chemistry of Indian *Piper* spp. has appeared: it includes some pyrrolidine alkaloids.[8] The synthesis of tricholeine (6), an alkaloid of *P. trichostachyon*, has been reported; the pathway is summarized in Scheme 2.[9]

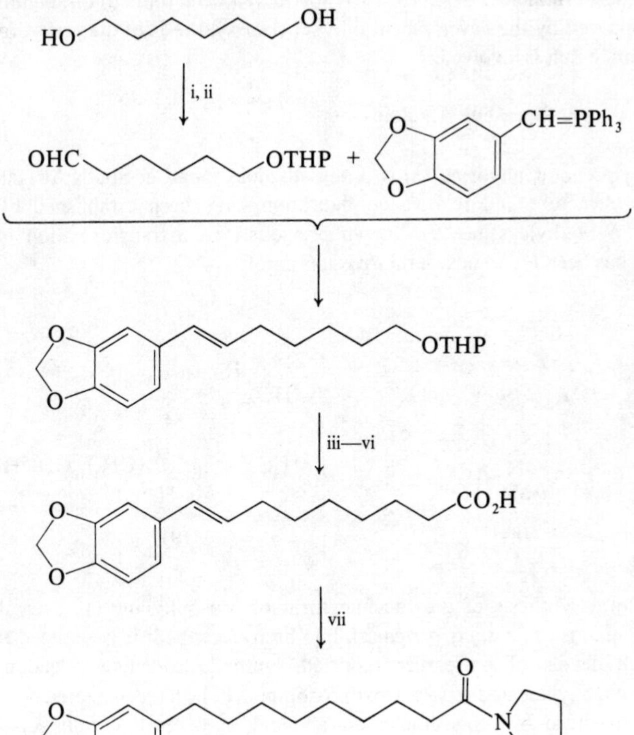

Reagents: i, DHP, H⁺; ii, pyridinium chlorochromate, NaOAc; iii, *p*-TSA, MeOH; iv, PBr₃, py; v, NaCH(CO₂Et)₂; vi, NaOH, MeOH; vii, POCl₃, then pyrrolidine, Et₃N

Scheme 2

(−)-Dihydrocuscohygrine (7) has been isolated from Peruvian coca leaves.[10]

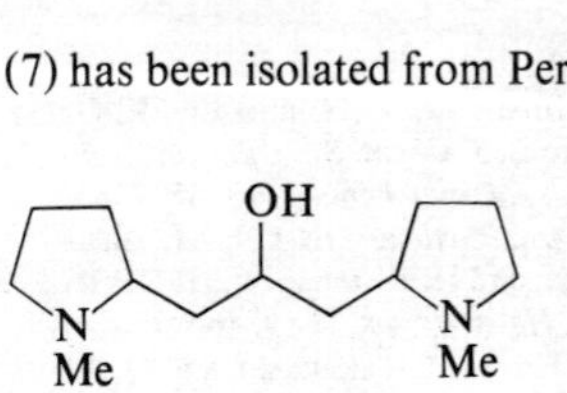

(7)

[8] B. S. Joshi, *Heterocycles*, 1981, **15**, 1309.
[9] O. P. Vig, R. C. Aggarwal, C. Shekher, and S. D. Sharma, *Indian J. Chem., Sect. B*, 1979, **17**, 560.
[10] C. E. Turner, M. A. Elsohly, L. Hanus, and H. N. Elsohly, *Phytochemistry*, 1981, **20**, 1403.

***Sceletium* Alkaloids.**—A new synthesis of (+)-mesembrine has been reported; it uses a chiral γ-lactone as a synthon.[11] An alternative synthesis of ($\pm$)-sceletium alkaloid A_4 has also been described.[12]

***Dendrobium* Alkaloids.** —An earlier reported synthetic route to (+)-dendrobine has been improved by the development of a superior synthesis of the perhydroindenone intermediate that is involved.[13]

2 Piperidine Alkaloids

N-Methylpseudoconhydrine (8) is a new alkaloid found in South African *Conium* spp.; its structure and relative configuration have been established by spectral study.[14] *N*-Methylcassine (9), known previously as a transformation product of cassine, has been found in several *Prosopis* spp.[15]

(8)   (9)

The total synthesis of the racemic form of prosophylline (10), an alkaloid of African mimosa (*Prosopis africana*), has been described; it is outlined in Scheme 3.[16] Full details of an earlier reported synthetic endeavour, leading to (−)-deoxoprosophylline and to (−)-deoxoprosopinine (11), have appeared.[17]

The structure of (−)-sedinine, an alkaloid of several *Sedum* spp., must be modified to (12), with the double-bond at C(3)–C(4) rather than at C(4)–C(5), as a consequence of *X*-ray diffraction analyses of the base and of its hydrochloride; (12) also represents its absolute configuration.[18] Sedacryptine is a new minor base from *S. acre*; a single-crystal *X*-ray analysis of the free base points to structure and relative stereochemistry (13).[19] The alkaloid could be identical with 'hydroxy-sedinone', isolated earlier.[20]

[11] S. Takano, C. Murakata, Y. Imamura, N. Tamura, and K. Ogasawara, *Heterocycles*, 1981, **16**, 1291.
[12] C. P. Forbes and G. L. Wenteler, *J. Chem. Soc., Perkin Trans. 1*, 1981, 29.
[13] W. R. Roush and H. R. Gillis, *J. Org. Chem.*, 1980, **45**, 4283.
[14] M. F. Roberts and R. T. Brown, *Phytochemistry*, 1981, **20**, 447.
[15] I. B. Gianinetto, J. L. Cabrera, and H. R. Juliani, *J. Nat. Prod.*, 1980, **43**, 632.
[16] M. Natsume and M. Ogawa, *Heterocycles*, 1981, **16**, 973.
[17] Y. Saitoh, Y. Moriyama, H. Hirota, T. Takahashi, and Q. Khuong-Huu, *Bull. Chem. Soc. Jpn.*, 1981, **54**, 283.
[18] C. Hootelé, B. Colau, F. Halin, J. P. Declercq, G. Germain, and M. Van Meerssche, *Tetrahedron Lett.*, 1980, **21**, 5063.
[19] C. Hootelé, B. Colau, F. Halin, J. P. Declercq, G. Germain, and M. Van Meerssche, *Tetrahedron Lett.*, 1980, **21**, 5061.
[20] J. H. Kooy, *Planta Med.*, 1976, **30**, 295.

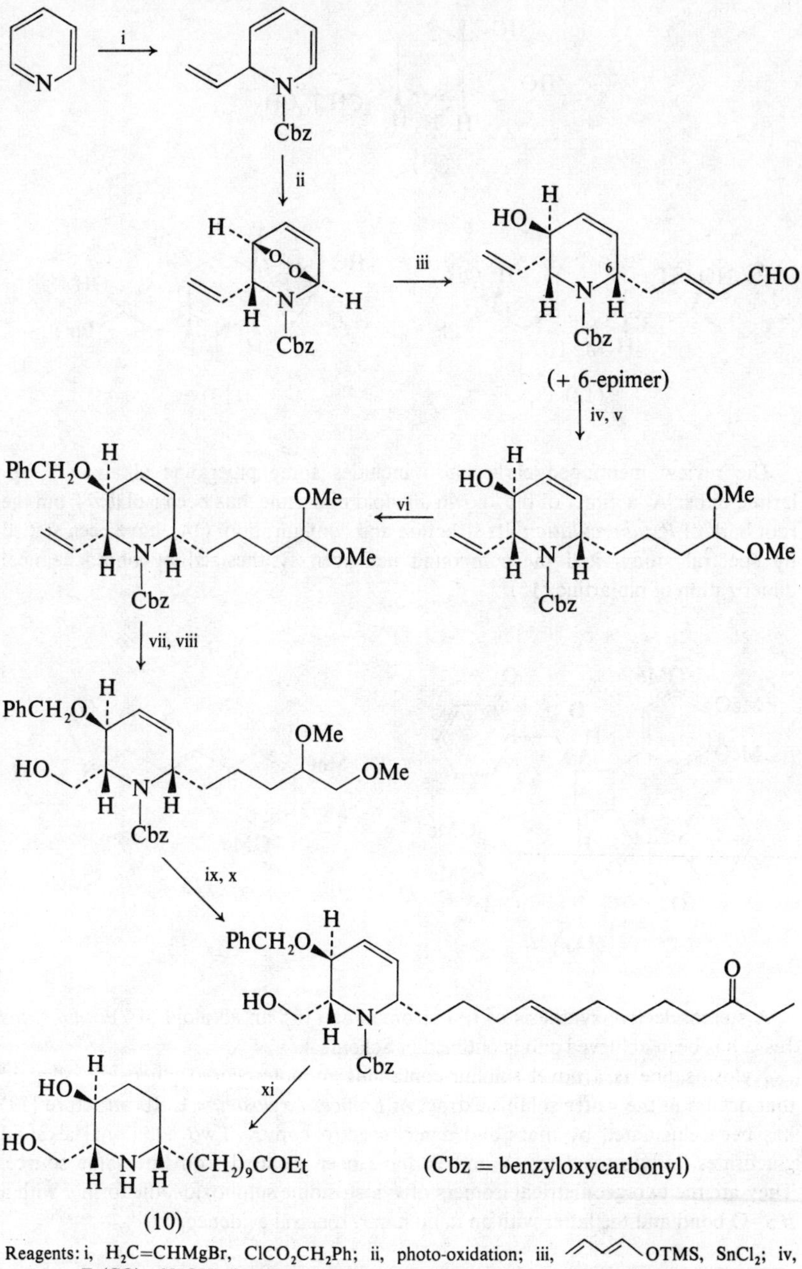

(10)

(Cbz = benzyloxycarbonyl)

Reagents: i, $H_2C{=}CHMgBr$, $ClCO_2CH_2Ph$; ii, photo-oxidation; iii, [diene]OTMS, $SnCl_2$; iv, $Fe(CO)_5$, NaOH, then I_2; v, MeOH, *p*-TSA; vi, $PhCH_2Cl$, NaH; vii, $KMnO_4$, $Bu^n{}_4NBr$, CH_2Cl_2, H_2O, then $NaIO_4$; viii, $NaBH_4$; ix, aq. HCl; x, $Ph_3P{=}CH(CH_2)_4C(OLi)_2Et$; xi, $2H_2$, Pd/C, H^+

Scheme 3

(11)

(12)   (13)

The review mentioned earlier also includes some piperidine alkaloids.[8] Piplartine-dimer A, a dimer of the known alkaloid piplartine, has been isolated from the root bark of *P. tuberculatum*. Its structure and configuration, (14), have been settled by spectral study, and the compound has been synthesized by photochemical dimerization of piplartine (15).[21]

(14)   (15)

A stereoselective synthesis of (±)-tecomanine (16), an alkaloid of *Tecoma stans* Juss., has been achieved and is outlined in Scheme 4.[22]

Xylostosidine is a novel sulphur-containing monoterpenoid glucosidic alkaloid that occurs in the water-soluble extract of *Lonicera xylosteum* L. Its structure (17) has been elucidated by mass and n.m.r. spectroscopy.[23] Two additional alkaloid glucosides, loxylostosidines A and B, have been isolated from the same source. They are the two geometrical isomers of xylostosidine sulphoxide, the former with a β S—O bond and the latter with an α, on n.m.r. spectral evidence.[24]

[21] R. Braz Filho, M. P. DeSouza and M. E. O. Mattos, *Phytochemistry*, 1981, **20**, 345.
[22] T. Imanishi, N. Yagi, and M. Hanaoka, *Tetrahedron Lett.*, 1981, **22**, 667.
[23] R. K. Chaudhuri, O. Sticher, and T. Winkler, *Helv. Chim. Acta*, 1980, **63**, 1045.
[24] R. K. Chaudhuri, O. Sticher, and T. Winkler, *Tetrahedron Lett.*, 1981, **22**, 559.

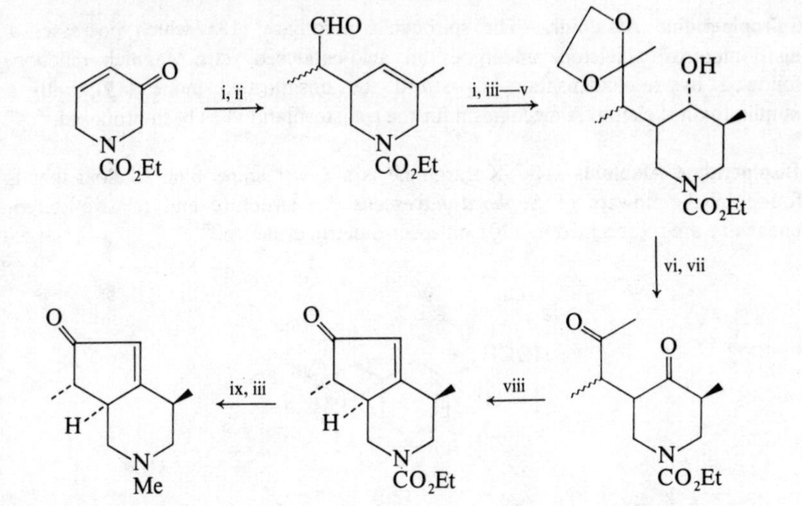

(16)

Reagents: i, MeMgI; ii, MeCH=CHOEt, Hg(OAc)$_2$, at 200 °C; iii, pyridine chlorochromate; iv, (HOCH$_2$)$_2$, H$^+$; v, B$_2$H$_6$, then H$_2$O$_2$, $^-$OH; vi, H$_3$O$^+$; vii, Jones oxidation; viii, K$_2$CO$_3$, EtOH, at 45—50 °C; ix, LiAlH$_4$

Scheme 4

Decahydroquinoline Alkaloids.—A review on the total synthesis of lycopodium alkaloids has appeared; it includes syntheses of luciduline.[25] Additional total syntheses of (±)-gephyrotoxin[26] and (±)-perhydrogephyrotoxin[27] have been described, and (±)-depentylperhydrogephyrotoxin has been synthesized *via* an interesting cyclization of a vinylogous *N*-acyliminium ion.[28]

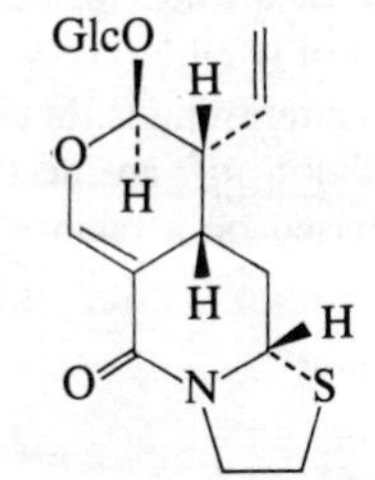

(17) Glc = glucosyl 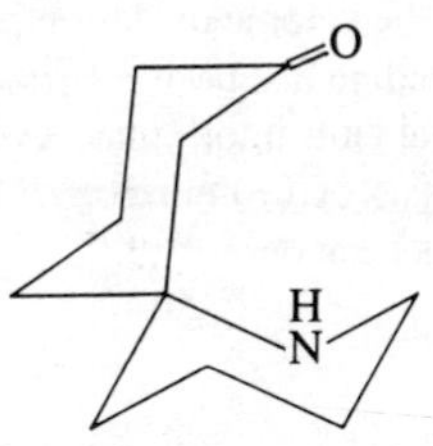   (18) 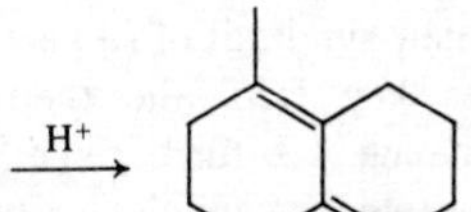   (19)

[25] Y. Inubushi and T. Harayama, *Heterocycles*, 1981, **15**, 611.
[26] D. J. Hart, *J. Org. Chem.*, 1981, **46**, 3576.
[27] L. E. Overman and R. L. Freerks, *J. Org. Chem.*, 1981, **46**, 2835.
[28] D. J. Hart, *J. Org. Chem.*, 1981, **46**, 367.

Spiropiperidine Alkaloids.—The spirocyclic keto-base (18), which possesses a histrionicotoxin skeleton, undergoes an acid-catalysed retro-Mannich reaction, followed by re-condensation, to afford the unsaturated imine (19), with a pumiliotoxin skeleton. A mechanism for the transformation has been proposed.[29]

Bispiperidine Alkaloids.—(+)-Kuraramine is a new bispiperidine alkaloid that is found in the flowers of *Sophora flavescens*. Its structure and relative stereochemistry are represented by (20), on spectrometric evidence.[30]

(20)

3 Pyridine Alkaloids

The synthesis of the glucosidic alkaloid buchananine has been reported (Scheme 5). The penultimate product (21), though crystalline, was shown by n.m.r. spectroscopy to be a mixture of α- and β-anomers, confirmed chemically by the last step, which led to a separable mixture of two tetra-acetates. This same mixture of acetates resulted when 1,2,3,4-tetra-O-acetyl-D-glucopyranose was esterified with nicotinyl chloride. It is concluded that buchananine is itself a mixture of α- and β-anomers (21), as a consequence of its ability to catalyse its own mutarotation.[31]

Melochinine (22) has been correlated with cassine (23) by transformation into the pyridine ketone (24), obtainable by dehydrogenation of cassine. It follows that the ring substituents in the two alkaloids have the same arrangement (Scheme 6).[32] Other constituents of the leaves of *Melochia pyramidata* L. include melochinine D-glucoside and melochinone [the ketonic oxidation product of (22)].[33]

A new synthesis of (±)-actinidine has been reported; it is interesting in that it has, as its key step, the thermal intramolecular cycloaddition of an acetylenic pyrimidine.[34] A further synthesis of (±)-muscopyridine, based on a regioselective cyclopentenone annulation, has been described.[35]

[29] J. J. Venit and P. Magnus, *Tetrahedron Lett.*, 1980, **21**, 4815.
[30] I. Murakoshi, E. Kidoguchi, J. Haginiwa, S. Ohmiya, K. Higashiyama, and H. Otomasu, *Phytochemistry*, 1981, **20**, 1407.
[31] R. Somanathan, H. D. Tabba, and K. M. Smith, *J. Org. Chem.*, 1980, **45**, 4999.
[32] E. Medina and G. Spiteller, *Chem. Ber.*, 1981, **114**, 814.
[33] E. Medina and G. Spiteller, *Liebigs Ann. Chem.*, 1981, 538.
[34] L. B. Davies, S. G. Greenberg, and P. G. Sammes, *J. Chem. Soc., Perkin Trans. 1*, 1981, 1909.
[35] H. Saimoto, T. Hiyama, and H. Nozaki, *Tetrahedron Lett.*, 1980, **21**, 3897.

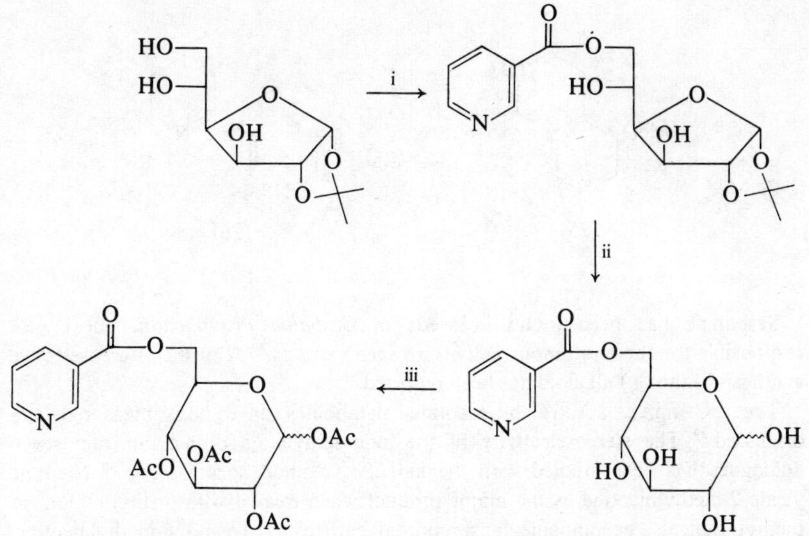

α- and β-anomers
(separated by h.p.l.c.)

(21) α- and β-anomers

Reagents: nicotinyl chloride, pyridine; ii, 5M-HCl (aq.); iii, Ac₂O, pyridine

Scheme 5

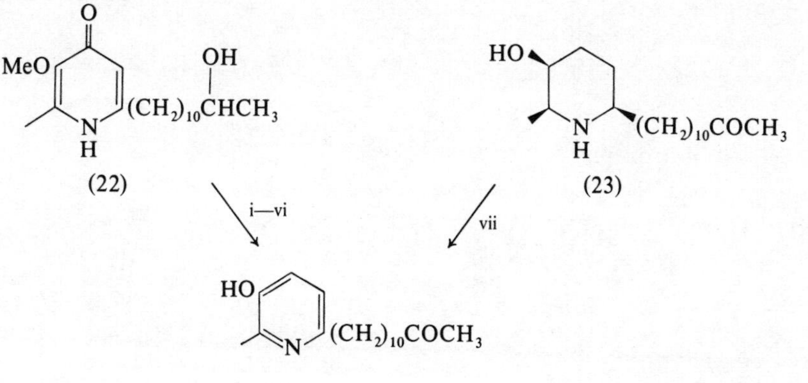

(22)

(23)

(24)

Reagents: i, CrCO₃, pyridine; ii, POCl₃; iii, H₂/Pt, AcOH; iv, (CH₂OH)₂, H⁺; v, NaOMe, MeOH, at 170 °C; vi, H₃O⁺; vii, 3H₂, Pd

Scheme 6

Acanthicifoline, a new alkaloid of *Acanthus ilicifolius*, has been formulated as (25) on spectral evidence.[36]

[36] K. P. Tiwari, P. K. Minocha, and M. Masood, *Pol. J. Chem.*, 1980, **54**, 857.

(25)   (26)

Sesbanine has been found in seeds of *Sesbania drummondii*, but is not responsible for the antitumour activity of seed extracts.[37] A further stereoselective synthesis of the ($\pm$)-alkaloid has been reported.[38]

The ^{13}C n.m.r. spectra of nicotine metabolites have been measured and discussed.[39] The stereoselectivity of the iodomethylation of nicotine and seven analogues has been studied with the aid of ^{13}C n.m.r. spectroscopy.[40] Nicotine yields 2-methylnicotine as the major product when treated with methyl-lithium or methyl radicals, accompanied, as reported earlier, by 4- and 6-methylnicotine. Nicotine *N*-oxide and methylmagnesium bromide afford 2- and 6-methyl-nicotines.[41] Anabaseine (26), a well-known minor alkaloid of tobacco, has been identified as a poison-gland product in ants of the genus *Aphaenogaster*, which use it as an attractant.[42]

[37] R. G. Powell and C. R. Smith, Jr., *J. Nat. Prod.*, 1981, **44**, 86.
[38] M. J. Wanner, G.-J. Koomen, and U. K. Pandit, *Heterocycles*, 1981, **15**, 377.
[39] T. Nishida, Å. Pilotti, and C. R. Enzell, *Org. Magn. Reson.*, 1980, **13**, 434.
[40] J. I. Seeman, H. V. Secor, C. G. Chavdarian, E. B. Sanders, R. L. Bassfield, and J. D. Whidby, *J. Org. Chem.*, 1981, **46**, 3040.
[41] H. V. Secor, C. G. Chavdarian, and J. I. Seeman, *Tetrahedron Lett.*, 1981, **22**, 3151.
[42] J. W. Wheeler, O. Olubajo, C. B. Storm, and R. M. Duffield, *Science*, 1981, **211**, 1051.

3

Tropane Alkaloids

BY G. FODOR AND R. DHARANIPRAGADA

1 Structure of New Alkaloids

The genus *Knightia* of the family Proteaceae comprises three species, two of which grow in New Caledonia and the third in New Zealand. Several new tropane alkaloids have been isolated[1,2] from *K. deplanchei*. Quite recently, twelve new tropane esters were obtained from *K. strobilina* Labill.; the structures of all of them became known[3,9] this year.

Compound A, 'acetylknightinol', is laevorotatory (M^+ 331; empirical formula $C_{19}H_{25}NO_4$). It loses an AcO· group under electron impact. The i.r. spectrum shows two ester carbonyl groups, at 1735 and 1725 cm^{-1}. The ^{1}H n.m.r. spectrum, combined with double irradiation, led to the conclusion that the alkaloid is representative of the small class of hydroxybenzyl-tropanes[4] that were formerly isolated from *K. deplanchei*, and that structure (1) reflects all of the properties that have been observed.

The major product, compound D, named 'strobiline', is strongly dextrorotatory (M^+ 191; $C_{11}H_{13}NO_2$). The fragmentation pattern resembles that of bellendine,[5] from other Australian members of the Proteaceae (*i.e. Bellendena montana* and *Darlingia darlingiana*). The i.r. spectrum shows a carbonyl group at 1650 cm^{-1}; the group is conjugated (as inferred from the u.v. spectrum). The ^{1}H and ^{13}C n.m.r. spectra are in excellent accord with the structure of the fused γ-pyrono-trop-2,3-ene (3). The authors mistakenly omit the nitrogen atom in the tropane ring (No. 8) in their numbering of the system.

A dihydro-derivative (4) of strobiline (M^+ 193) was also obtained. In the mass spectrometer, the pyrrolidine ring is opened to give fragments of m/z 165 and 164 (base peak), followed by a retro-Diels–Alder reaction to give fragments of m/z 137 and 136, respectively. The location of the two additional hydrogen atoms was deduced from the relative u.v. extinctions of (3) and (4), which is consistent with the dihydro-γ-pyrone skeleton in the latter compound. This was confirmed by ^{1}H n.m.r. spectra that showed no vinylic protons.

[1] See these Reports, Vol. 10, p. 42.
[2] C. Kan-Fan and M. Lounasmaa, *Acta Chem. Scand.*, 1973, **27**, 1039.
[3] M. Lounasmaa, J. Pusset, and T. Sévenet, *Phytochemistry*, 1980, **19**, 949.
[4] M. Lounasmaa, *Planta Med.*, 1975, **27**, 83.
[5] W. D. S. Motherwell, N. W. Isaacs, O. Kennard, I. R. C. Bick, J. B. Bremner, and J. Gillard, *J. Chem. Soc., Chem. Commun.*, 1971, 133.

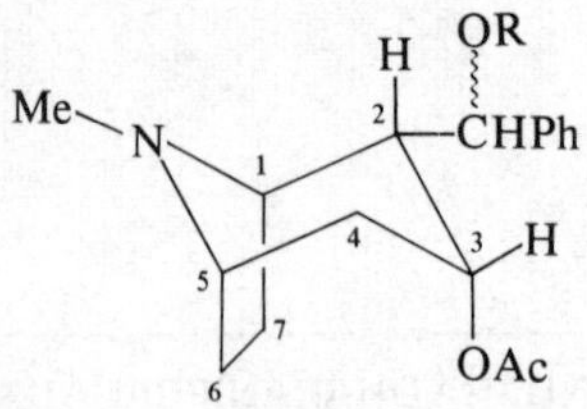

(1) R = Ac
(2) R = H

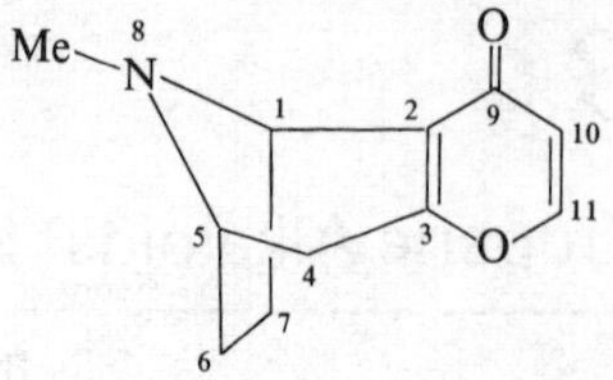

(3)

Knightoline, the fourth alkaloid, has structure (6), based on its mass spectrum (M^+ 289; $C_{17}H_{23}NO_3$). The fragmentation is typical[8] for tropane-$3\alpha,6\beta$-diols, *i.e.* loss of vinyl alcohol from the pyrrolidine moiety to give the ion of m/z 245. Loss of the acetate radical, on the other hand, from M^+ gives an ion of m/z 230, which is consistent with an acylated and a free hydroxyl group, logically at C-3 and C-6, respectively. The i.r. and ^{1}H n.m.r. spectra proved the presence of both a carbonyl in an ester group and the benzylic (aromatic) protons.

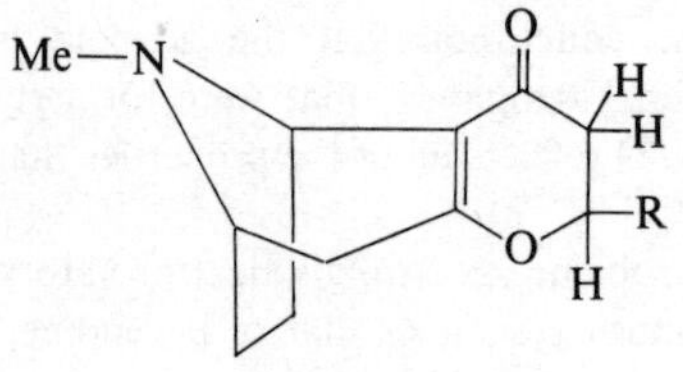

(4) R = H
(5) R = Ph

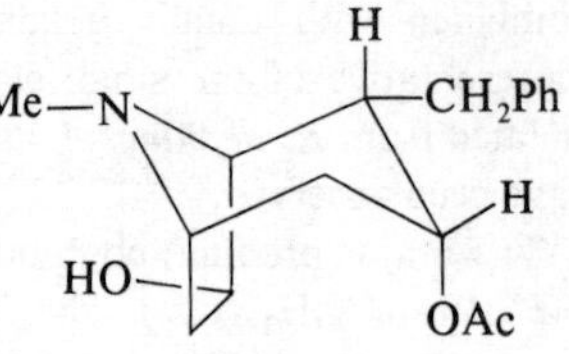

(6)

Product G proved to be the benzoyl 6-ester (7) of optically active tropane-$3\alpha,6\beta$-diol. Three other esters of tropane-$3\alpha,6\beta$-diol had earlier been isolated from *Bellendena montana* by another research group.[6,7]

Product H, 'knightinol', is a 2'-monodeacetylated form of the diester (1) ('acetylknightinol'), *i.e.* a 2-(2-hydroxybenzyl)-3α-acetoxytropane (2). This assignment is based on the mass spectrum (M^+ 289; $C_{17}H_{23}NO_3$). The i.r. spectrum confirmed the presence of a free hydroxyl group (at 3460 cm^{-1}) and an ester carbonyl group (at 1715 cm^{-1}). The ^{1}H n.m.r. spectrum is in full agreement with this assignment.

[6] R. L. Clarke, in 'The Alkaloids', ed. R. F. Manske, Academic Press, New York, 1970, Vol. XVI, Ch. 2, pp. 84–180.
[7] I. R. C. Bick, personnal communication to M. Lounasmaa.
[8] E. C. Blossey, H. Budzikiewicz, M. Ohashi, G. Fodor, and C. Djerassi, *Tetrahedron*, 1964, **20**, 585.

The seventh new alkaloid of *K. strobilina* is the cinnamoyl 3α-ester of $3\alpha,6\beta$-dihydroxytropane (8), the assignment being based mostly on mass-spectral evidence (M^+ 287; $C_{17}H_{21}NO_3$); the radical ions of m/z 156 (PhCH=CH−COO$^+$) and m/z 140 (PhCH=CH−CO$^+$) are crucial, in addition to the typical[8] 6β-tropanol fragments.

(7) R^1 = PhCO, R^2 = H
(8) R^1 = H, R^2 = COCH=CHPh

The same group of researchers succeeded[9] in elucidating the structures of five more alkaloids, mostly benzyl-pseudotropanols and pyrono-tropanes. Compound (5) is a 10,11-dihydro-11-phenyl-2,3-γ-pyrono-trop-2,3-ene (M^+ 269; $C_{17}H_{29}NO_2$), and it was given the name 'strobamine'. The i.r. spectrum is significant, the absorption of an $\alpha\beta$-unsaturated carbonyl group being at 1660 cm^{-1}, and the ^{1}H n.m.r. spectrum shows the presence of aromatic protons. The phenyl group has been arbitrarily located on C-11. The assignment was based on double resonance and on the biogenetic consideration that a cinnamoyl residue, rather than a benzoylethenyl group, may be the precursor [see compound (9)]. The ninth alkaloid, formulated as (9), is the enolic form of 2-cinnamoyltropan-3-one. It is isomeric with (5) (M^+ 269; $C_{17}H_{19}NO_2$). The fragmentation pattern under electron impact is similar to that of its cyclic isomer. The authors intended to indicate the analogy between a chalcone and a flavanone by the name 'chalcostrobamine' for (9). Details of the ^{1}H n.m.r. spectrum were discussed earlier.[4] Alkaloid J, the tenth of this series, gave spectral data (i.r., u.v., and ^{1}H n.m.r.) that are in harmony with a 7-hydroxy-strobamine (10), which is the first representative of a pyranotropane with a hydroxyl function in the pyrrolidine ring (M^+ 285; $C_{17}H_{19}NO_3$); it was named 'strobolamine'.

(9)

[9] M. Lounasmaa, J. Pusset, and T. Sévenet, *Phytochemistry*, 1980, **19**, 953.

Product K, named 'knightalbinol' is a 2α-hydroxybenzyl-tropane-$3\beta,7\beta$-diol (11) that is acetylated on C-7 (or C-6). This is based on the assumption that the acetoxyvinyl radical is easily split off during mass-spectral fragmentation and gives an ion of m/z 219. The assignment of a 3β-hydroxyl is based on the large coupling constant between H-2 and H-3, and from double irradiation.

The last alkaloid, compound L, named 'knightolamine' (12), is closely related to (11) except for the lack of a carbonyl group in an acetyl ester and the presence of a benzoyloxy-group (1740 cm^{-1}); the parent ion (M^+ 367; $C_{22}H_{25}NO_4$) is cleaved to an ion of m/z 246 and a benzoyloxyl radical. The fragment of m/z 94 usually appears in the mass spectrum of those tropanes[8] that have a hydroxy-group in the pyrrolidine moiety.

(10)

(11) $R^1 = Ac, R^2 = H$
(12) $R^1 = H, R^2 = COPh$

The preliminary configurational assignments in most of the twelve alkaloids were made on biogenetic grounds (by analogy with bellendine). All alkaloids of *K. strobilina*, unlike those isolated from *K. deplanchei*, are optically active.

Atropoylnortropan-3α-ol (aponoratropine) and atropoylscopine (aposcopolamine) were detected[10] in the leaves of *Anthotroche myoporoides* and *Anthocercis genistoides*, respectively. 3α-Isobutyryloxytropan-6β-ol and 3α-butyryloxytropane were characterized from aerial parts of *Anthocercis albicans*.

2 Synthesis and Reactions

A route to 6β-aryl-tropan-6-ols has been devised[11] *via* 6β-hydroxytropan-3-one and 6β-hydroxytropane.[12] A new series of potential analgesics have been made.

The action of aryl Grignard reagents upon tropan-6-one affords 6-aryl-tropan-6-ols (13). The assignment of the β-orientation of the aryl groups was based on analogy and on the ^{1}H n.m.r. spectra of 6-phenyltropan-6-ol and its methiodide. One of the quaternary $\overset{+}{N}$—CH$_3$ signals in the latter occurs at $\delta\,2.67$, owing to anisotropic shielding by the phenyl ring in the β-position. The corresponding N-acyl-6β-aryl-nortropan-6α-ols have also been synthesized as potential analgesics.

[10] W. C. Evans and K. P. A. Ramsey, *Phytochemistry*, 1981, **20**, 497.
[11] G. H. Dewar, S. Leung, and R. T. Parfitt, *J. Pharm. Pharmacol.*, 1980, **32**, (Suppl.), 71P.
[12] J. B. Jones and A. R. Pinder, *Chem. Ind.* (*London*), 1958, 1000; *cf.* G. Fodor, in 'The Alkaloids', ed. R. F. Manske, Academic Press, New York, 1960, Vol. VI, p. 170.

An interesting rearrangement took place[13] when $1\alpha,4\beta$-di-iodo-2α-acetoxy-tropan-3-one (14; R = I) and 1α-iodo-2α-acetoxytropan-3-one (14; R = H) [products of the oxidation of tropan-3α-ol by lead(IV) acetate in the presence of iodine] were subject to reduction to give (15).

(13)   (14)   (15)

Substituted (*e.g.* β-methyl, *p*-t-butyl, *p*-chloro, and *p*-amino)benzoate esters of nortropine were synthesized[14] by *N*-demethylation of the corresponding tropanyl esters with 2,2,2-trichloroethyl chloroformate.

The Katritzky synthesis[15] of trop-3-en-2-ones was applied[16] to acrylonitrile and the *N*-methyl and *N*-phenyl 3-hydroxypyridinium betaines, leading to the 6- and 7-*exo*- and -*endo*-cyano-derivatives (16).

(16)

A total synthesis of diastereoisomers of anisodamine, an alkaloid from *Anisodus tanguticus*, has been claimed[17] by Chinese authors. ($\pm$)-Acetyltropoylation of ($\pm$)-β-acetoxy-3α-hydroxytropane, followed by partial deacetylation, led to a diastereoisomeric mixture of 6β-hydroxy-atropines. (−)-(*S*)-6β-Hydroxyhyoscyamine, a natural tropane, had earlier been obtained by total synthesis.[18] In view of this and of the non-identity of the i.r. spectra of 'synthetic' mixture and 'natural anisodamine', it remains uncertain whether the latter is identical with natural (−)-(*S*)-6β-hydroxyhyoscyamine.[18]

[13] Sh. Sarel and E. Dykman, *Heterocycles*, 1981, **15**, 719.
[14] J. L. Wallace, M. R. Kidd, S. E. Cauthen, and J. D. Woodyard, *J. Pharm., Sci.*, 1980, **69**, 1357.
[15] A. R. Katritzky and Y. Takeuchi, *J. Chem. Soc. C*, 1971, 874.
[16] J. A. Lepoivre, R. A. Dommisse, E. L. Esmans, J. J. Van Luppen, E. M. Merckx, and F. C. Alderweireldt, *Bull. Soc. Chim. Belg.*, 1981, **90**, 49 (*Chem. Abstr.*, 1981, **94**, 192 083).
[17] X. Jingxi, Z. Jin, J. Xiaoxian, and L. Chunxue, *Acta Pharm. Sinica*, 1980, **15**, 409 (*Chem. Abstr.*, 1981, **94**, 121 757).
[18] G. Fodor, I. Koczor, and G. Janzso, *Arch. Pharm.* (*Weinheim, Ger.*), 1962, **295**, 91.

The 80 MHz ^{1}H n.m.r spectrum of *N*-(cylcopropylmethyl)scopolaminium bromide in D_2O has been analysed and the conclusions have been compared with those from *X*-ray crystal-structure studies.[19] They were consistent with the equatorial position of the cyclopropylmethyl group, as shown in (17). All protons in symmetrical positions (1 and 5, 6 and 7, 2 and 4) of the tropane ring were found to be magnetically non-equivalent, as in scopolamine hydrobromide and in atropine. This was attributed to the different shielding effects that arise from the aromatic substituent which is attached to the asymmetric centre of the tropic acid moiety. The carboxyl group of the tropate residue was found to be opposite to H-3.

The preferred conformations of a pharmacologically interesting series of *N*-substituted nortropane spirohydantoins (18) were determined[20] from ^{1}H and ^{13}C n.m.r. and *X*-ray data, to permit a structure–activity relationship to be investigated. The piperidine ring, as expected,[21] has a flattened chair conformation. Surprisingly, however, the nitrogen atom in the piperidine ring shows a pyramidal geometry, with the methyl group in an axial position. The dihedral angle C(6,7)–C(1,5)–N(8) has a smaller value than in pseudotropine.

(17)   (18)

1,3-Cycloaddition of diazomethane to methyl atropate, apoatropine, and aposcopolamine gave[22] the expected 3-phenyl-3-alkoxycarbonyl-1-pyrazolines.

New 3β-acylamino-nortropanes (19) with a variety of A—CO (aroyl, pyrimidine-carboxylic, *etc.*) and R (benzyl, heteroarylmethyl, *etc.*) groups have been synthesized as potential pharmaceutical agents.[23]

(19)

[19] A. Gallazzi, P. C. Vanoni, M. Mondoni, M. Gaetani, and R. Stradi, *Farmaco, Ed. Sci.*, 1980, **35**, 913 (*Chem. Abstr.*, 1981, **94**, 84 341).
[20] G. G. Trigo, M. Martinez, and E. Galvez, *J. Pharm. Sci.*, 1981, **70**, 87.
[21] R. J. Bishop, G. Fodor, A. R. Katritzky, F. Soti, L. E. Sutton, and F. J. Swinbourne, *J. Chem. Soc. C*, 1966, 74.
[22] K. Kagei, Y. Ogata, T. Kunii, K. Hayashi, S. Toyoshima, and S. Matsuura, *Yakugaku Zasshi*, 1980, **101**, 193 (*Chem. Abstr.*, 1981, **95**, 25 357).
[23] P. Dostert, T. Imbert, and B. Bucher, Ger. Offen. 3 001 328, 1979 (*Chem. Abstr.*, 1981, **94**, 65 926).

3 Pharmacology

Cocaine was, and still is, in the focus of investigations.[24-27, 29-39] Amongst others, the mechanism of acute cocaine intoxication,[24] behavioural sensitization,[25] the effects of electric shock on responding, when maintained by cocaine,[26] its effects in sleep-deprived humans,[27] and the discriminative stimulus properties of cocaine, norcocaine, and N-allylnorcocaine[32] have been studied. The effect of benzotropine mesylate on the prolactin response to haloperidol was observed.[28]

The pharmacology of scopolamine also received due attention.[39-49] Amongst other effects, the effect of scopolamine hydrobromide on the electrical activity of rat cerebellum was analysed.[39] An increase in the electrical resistance of the skin of white rats was observed following injection with scopolamine.[40] Scopolamine is only marginally effective in eliminating behavioural depression,[41] and it was found to impair the performance of animals in a spatial maze.[48] A comparative study on N-butylscopolaminium bromide as an alternative to atropine has been undertaken.[49]

[24] J. D. Catravas and I. W. Waters, *J. Pharmacol. Exp. Ther.*, 1981, **271**, 350.

[25] R. M. Post, A. Lockfeld, K. M. Squillace, and N. R. Contel, *Life Sci.*, 1981, **28**, 755 (*Chem. Abstr.*, 1981, **94**, 132 266).

[26] J. Bergman and C. E. Johanson, *Pharmacol. Biochem. Behav.*, 1981, **14**, 423.

[27] M. W. Fischman and C. R. Schuster, *Psychopharmacology (Berlin)*, 1980, **72**, 1 (*Chem. Abstr.*, 1981, **95**, 35 575).

[28] U. Halbreich, E. J. Sacher, G. M. Gregory, and K. S. Halpern, *Psychopharmacology (Berlin)*, 1980, **72**, 61 (*Chem. Abstr.*, 1981, **94**, 58 432).

[29] H. Brewitt, E. Bonatz, and H. Honegger, *Ophthalmologica*, 1980, **180**, 198 (*Chem. Abstr.*, 1981, **94**, 58 270).

[30] D. Nigro and M. A. Enero, *Gen. Pharmacol.*, 1981, **12**, 255.

[31] R. W. Freeman and R. D. Harbison, *J. Pharmacol. Exp. Ther.*, 1981, **218**, 566.

[32] J. A. Bedford, G. L. Nail, R. B. Borne, and M. C. Wilson, *Pharmacol. Biochem. Behav.*, 1981, **14**, 81.

[33] M. E. A. Reith, H. Sershen, and A. Lajtha, *Neurochem. Res.*, 1980, **5**, 1291.

[34] I. P. Stolerman and G. D. D'Mello, *Psychopharmacology (Berlin)*, 1981, **73**, 295 (*Chem. Abstr.*, 1981, **95**, 574).

[35] G. Puma and R. Ramos-Aliaga, *Res. Commun. Subst. Abuse*, 1981, **2**, 47 (*Chem. Abstr.*, 1981, **95**, 35 180).

[36] R. W. Steger, A. Y. Silverman, A. Johns, and R. K. Asch, *Fertil. Steril.*, 1981, **35**, 567 (*Chem. Abstr.*, 1981, **95**, 54 796).

[37] G. Barnett, R. Hawks, and R. Resnick, *J. Ethnopharmacol.*, 1981, **3**, 353 (*Chem. Abstr.*, 1981, **95**, 217).

[38] J. Schrold and O. A. Nedergaard, *Acta Pharmacol. Toxicol.*, 1981, **48**, 233 (*Chem. Abstr.*, 1981, **94**, 185 446).

[39] G. Gogolak and R. H. Jindra, *Wien, Klin. Wochenschr.*, 1980, **92**, 737.

[40] R. J. Horschburgh, *Pharmacol. Biochem. Behav.*, 1981, **14**, 1.

[41] R. J. Katz and S. Hersh, *Neurosci. Biobehav. Rev.*, 1981, **5**, 265.

[42] M. Tonini, G. Frigo, S. Lecchini, L. D'Angelo, and A. Crema, *Eur. J. Pharmacol.*, 1981, **71**, 375.

[43] E. M. Joyce and G. F. Koob, *Psychopharmacology (Berlin)*, 1981, **73**, 311 (*Chem. Abstr.*, 1981, **95**, 35 575).

[44] X. Niu, W.-J. Wang, and Y. Chin, *Proc.–US–China Pharmacol. Symp.*, 1979, (publ. 1980), p. 235 (*Chem. Abstr.*, 1981, **94**, 95 983).

[45] X. Niu, W.-J. Wang, and Y.-C. Jin, *Chung-kuo Yao Li Hsueh Pao*, 1981, **2**, 4 (*Chem. Abstr.*, 1981, **94**, 167 776).

[46] K. P. Satinder, *Pharmacol. Biochem. Behav.*, 1981, **14**, 121.

[47] R. J. Horsburgh and R. N. Hughes, *Psychopharmacology (Berlin)*, 1981, **73**, 388 (*Chem. Abstr.*, 1981, **95**, 54 801).

[48] R. Stevens, *Physiol. Behav.*, 1981, **27**, 385.

[49] A. M. Soeterboek, H. Wesseling, and M. Van Thiel, *Pharm. Int.*, 1980, **1**, 165.

Studies with atropine were quite extensive.[50–67] Atropine effects REM (rapid-eye-movement) sleep and nocturnal penil tumescence.[51] Peripheral vascular effects of atropine and methylatropine in the autoperfused hind limbs of rats[57] and the hypotensive response to atropine in hypertensive rats[58] were observed. The antiviral effect of atropine has been investigated.[59] Experimental evidence was found for the cytoprotective effect of atropine on the rat gastric mucosa.[61] Atropine was found to lower blood-pressure in normotensive rats through the blockage of α-adrenergic receptors.[63] Inhibition of the release of acetylcholine by activation through atropine has been reviewed.[64]

A study of the bronchodilator influence of *N*-isopropylatropinium bromide (ipratropine) has been completed.[65–67]

Tropane derivatives were found not only to decrease the effect of opiate analgesics but also to prevent the development of the analgesic action of opiates.[68]

4 Analytical Aspects

Gas–liquid chromatography, using a mass spectrometer as a detector, has been used for the quantitative detection of cocaine hydrochloride in powders.[69] The procedure has a relative deviation of 2.3% and is applicable to samples that contain a wide variety of diluents and adulterants.

Methods for detecting cocaine and its metabolites in biological material have been reviewed.[70] N.m.r. and i.r. spectroscopies, electron-impact mass spectroscopy,

[50] I. Szelenyi and H. Engler, *Agents Actions*, 1980, **10**, 411 (*Chem. Abstr.*, 1981, **94**, 58 244).
[51] B. Dervent, I. Karacan, and C. Ware, *Sleep: Proc. Eur. Congr. Sleep Res., 4th*, 1978 (publ. 1980), p. 451 (*Chem. Abstr.*, 1981, **94**, 25 184).
[52] B. R. Cotton and G. Smith, *Br. J. Anaesth.*, 1981, **53**, 875.
[53] H. A. Schuil, J. R. Brunsting, H. Van der Molen, and W. G. Zijlstra, *Eur. J. Pharmacol.*, 1981, **69**, 229.
[54] B. J. Richardson and R. W. Welch, *Gastroenterology*, 1981, **81**, 85.
[55] V. Santucci, A. Glatt, H. Demieville, and H. R. Olpe, *Eur. J. Pharmacol.*, 1981, **73**, 113.
[56] S. Toetterman, S. Santavirta, and H. Antila, *Res. Exp. Med.*, 1980, **178**, 37.
[57] B. A. Merrick and T. L. Holcslaw, *J. Pharmacol. Exp. Ther.*, 1981, **218**, 771.
[58] S. Abraham, E. H. Cantor, and S. Spector, *J. Pharmacol. Exp. Ther.*, 1981, **218**, 662.
[59] Z. Yamazaki and I. Tagaya, *J. Gen. Virol.*, 1980, **50**, 429 (*Chem. Abstr.*, 1981, **94**, 114 323).
[60] K. A. Conrad, *Circulation*, 1981, **63**, 371 (*Chem. Abstr.*, 1981, **94**, 96 274).
[61] G. Mozsik, L. Lovasz, G. Kutor, L. Nagy, and F. Tarnok, *Acta Med. Acad. Sci. Hung.*, 1980, **37**, 401.
[62] I. Onnen and G. Olive, *J. Pharmacol. Methods*, 1981, **5**, 249.
[63] S. Abraham, E. H. Cantor, and S. Spector, *Life Sci.*, 1981, **28**, 315.
[64] P. C. Molenaar and R. L. Polak, *Prog. Pharmacol.*, 1980, **3**, 39 (*Chem. Abstr.*, 1981, **94**, 131 722).
[65] G. W. Sybrecht, T. Fluegee, and H. Fabel, *Prog. Respir. Res.*, 1980, **14**, (*Asthma*), p. 155.
[66] L. Diamond and M. O'Donnell, *J. Pharmacol. Exp. Ther.*, 1981, **216**, 1.
[67] S. Bongrani, P. Schiantarelli, M. Papotti, and G. C. Foleo, *Arzneim.-Forsch.*, 1981, **31**, 970 (*Chem. Abstr.*, 1981, **95**, 54 608).
[68] V. V. Zakusov, *Trends Pharmacol.*, 1980, **1**, 452 (*Chem. Abstr.*, 1981, **94**, 95 917).
[69] C. C. Charles, *J. Assoc. Off. Anal. Chem.*, 1981, **64**, 884 (*Chem. Abstr.*, 1981, **95**, 109 567).
[70] J. E. Lindgren, *J. Ethnopharmacol.*, 1981, **3**, 337 (*Chem. Abstr.*, 1981, **95**, 39).

high-performance liquid chromatography, t.l.c. microcrystalline tests, melting points, and gas-chromatographic techniques have been used for the differentiation of diastereoisomers of cocaine.[71]

A quantitative gas-chromatographic–mass-spectrometric assay has been developed for the determination of the anti-Parkinsonian drug benztropine in human urine and plasma.[72]

[71] A. C. Allen, D. A. Cooper, W. O. Kiser, and R. C. Cottrell, *J. Forens. Sci.*, 1981, **26**, 21 (*Chem. Abstr.*, 1981, **95**, 109 565).
[72] S. P. Jindal, Th. Lutz, C. Hallstrom, and P. Vestergaard, *Clin. Chim. Acta*, 1981, **112**, 267.

4

Pyrrolizidine Alkaloids

BY David J. ROBINS

Plant species known to contain pyrrolizidine alkaloids have been listed by Smith and Culvenor.[1] These sources are classified according to the toxicity of the alkaloids that are present. It is estimated that the potential number of these alkaloid-containing species is as high as 6000, or 3% of the world's flowering plants.[2]

1 Syntheses of Necine Bases

More synthetic routes to the 1-(hydroxymethyl)pyrrolizidines have been reported. Two groups[3,4] have published the same route to (±)-isoretronecanol (3) (Scheme 1). The key step is the nucleophilic attack of succinimide anion on the cyclopropylphosphonium salt (1), followed by intramolecular Wittig reaction to generate the unsaturated pyrrolizidinone ester (2). Catalytic reduction and reduction with a hydride yielded (±)-isoretronecanol (3). Flitsch and Wernsmann achieved a higher overall yield (62%) by carrying out the first step in boiling xylene.[3]

Another route to (±)-isoretronecanol (3) and (±)-trachelanthamidine (6) is outlined in Scheme 2.[5] This Reporter can hardly agree with the authors' extravagant claim that this is 'the most direct and operationally convenient synthesis [of these bases] yet reported'. Unfortunately, the intramolecular alkylation step is not stereospecific, and a mixture of diastereoisomers (4) and (5), in a ratio of 1:4, was formed. This mixture was separated chromatographically, in order to effect the synthesis of the bases (3) and (6).

Robins and Sakdarat have published full details of their route to optically active (+)- and (−)-forms of isoretronecanol (3), trachelanthamidine (6), and supinidine (8).[6] In addition, the synthesis of two new optically active bases (7) and (9) (isolated as its diacetate) is described.

[1] L. W. Smith and C. C. J. Culvenor, *J. Nat. Prod.* (*Lloydia*), 1981, **44**, 129.
[2] C. C. J. Culvenor, in 'Toxicology in the Tropics', ed. R. L. Smith and E. A. Bababunmi, Taylor and Francis, London, 1980.
[3] W. Flitsch and P. Wernsmann, *Tetrahedron Lett.*, 1981, **22**, 719.
[4] J. M. Muchowski and P. H. Nelson, *Tetrahedron Lett.*, 1980, **21**, 4585.
[5] G. A. Kraus and K. Neuenschwander, *Tetrahedron Lett.*, 1980, **21**, 3841.
[6] D. J. Robins and S. Sakdarat, *J. Chem. Soc., Perkin Trans. 1*, 1981, 909.

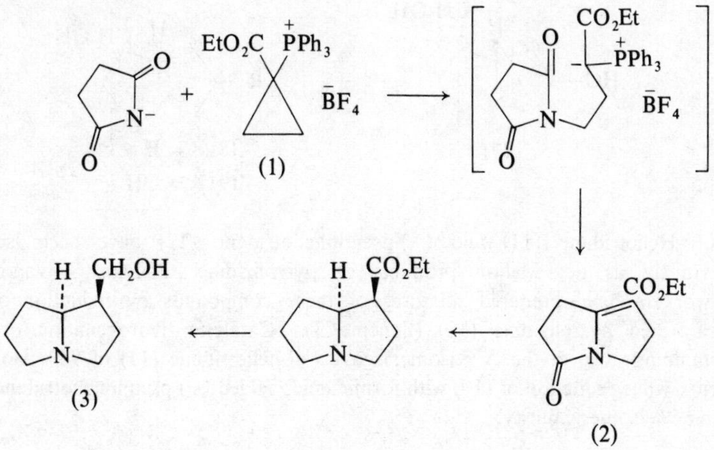

Scheme 1

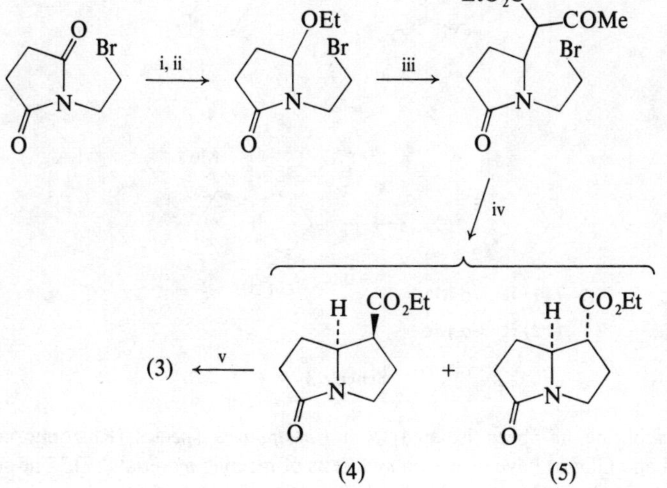

Reagents: i, NaBH₄; ii, EtOH, HCl; iii, CH₃COCH₂CO₂Et; iv, NaOEt, EtOH; v, LiAlH₄

Scheme 2

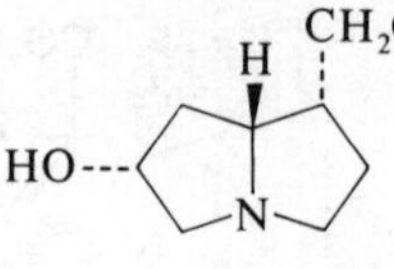

(7)

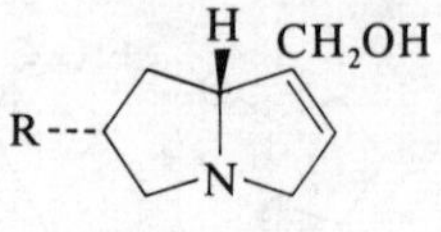

(8) R = H
(9) R = OH

($\pm$)-Heliotridane (11) and ($\pm$)-pseudoheliotridane (12) have been isolated previously as degradation products of pyrrolizidine alkaloids. Miyano and co-workers have prepared mixtures of these compounds by reduction of the unsaturated pyrrolizidine (10) (Scheme 3).[7] Catalytic hydrogenation of (10) (containing 10% of the $\Delta^{7,8}$-isomer) gave ($\pm$)-heliotridane (11) of 93% isomeric purity, while reduction of (10) with formic acid yielded ($\pm$)-pseudoheliotridane (12) with 66% isomeric purity.

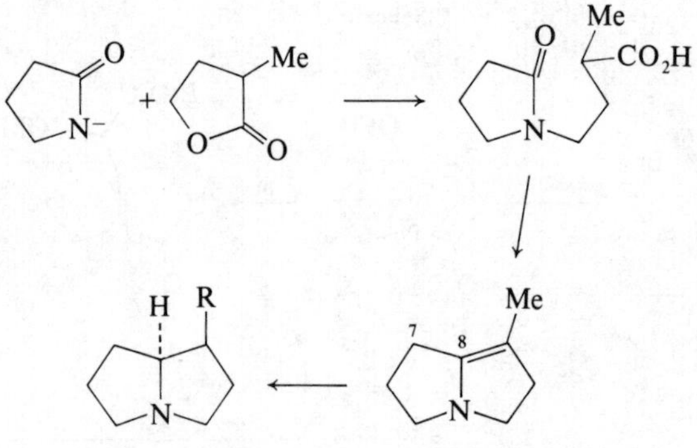

(11) R = β-Me
(12) R = α-Me

(10)

Scheme 3

Cassipourine has been isolated from *Cassipourea* species (Rhizophoraceae).[8] Wróbel and Gliński have devised a synthesis of racemic material (18).[9] The starting material was the β-epoxide (13), which was converted into the α-epithio-compound (14) in three steps (see Scheme 4). Thiobenzyl alcohol reacted regio- and stereo-specifically with this material (14) to give the thiol (15), which oxidized in air to give dimeric products. The racemate (16) and the *meso*-form were separated by column chromatography. Reduction of the racemate (16) with sodium in liquid ammonia yielded the dithiol (17), which afforded ($\pm$)-cassipourine (18) on aerial oxidation.

[7] S. Miyano, S. Fujii, O. Yamashita, N. Toraishi, K. Sumoto, F. Satoh, and T. Masuda, *J. Org. Chem.*, 1981, **46**, 1737.
[8] W. G. Wright and F. L. Warren, *J. Chem. Soc. C*, 1967, 283.
[9] J. T. Wróbel and J. A. Gliński, *Can. J. Chem.*, 1981, **59**, 1101.

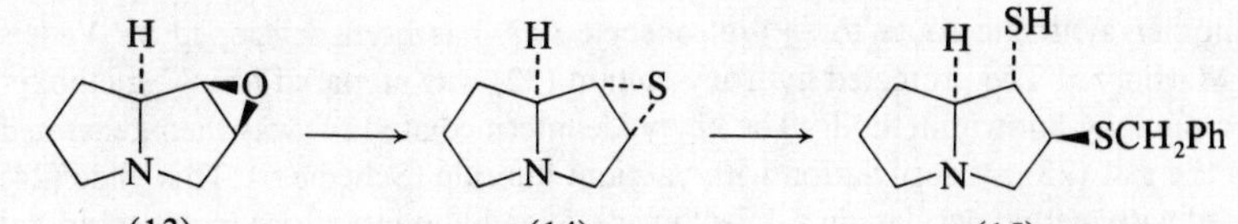

(13)   (14)   (15)

Scheme 4

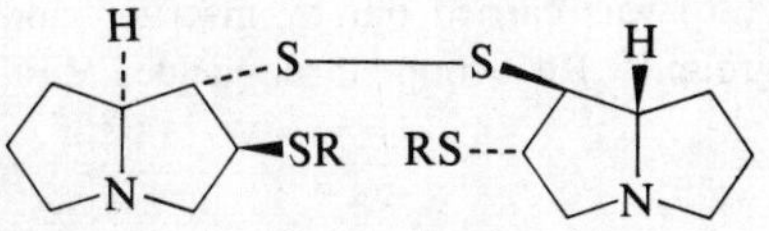

(16) R = CH₂Ph
(17) R = H

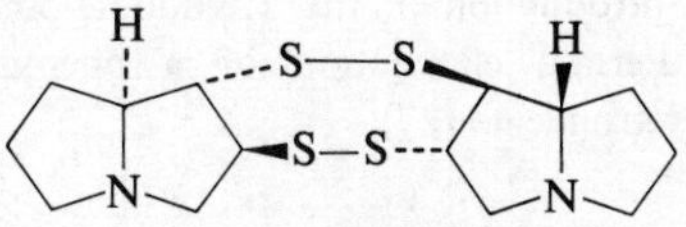

(18)

Bohlmannn and co-workers have synthesized the necine (21),[10] which they had previously isolated from *Senecio* species (*cf.* Vol. 8, p. 54). They used the Bestmann reaction of the pyrrole derivative (19) with triphenylphosphoranylideneketen to give the pyrrolizinone (20) (Scheme 5). Hydrogenation of one double-bond and hydrolysis of the protecting groups in (20) afforded the necine (21) in 40% overall yield.

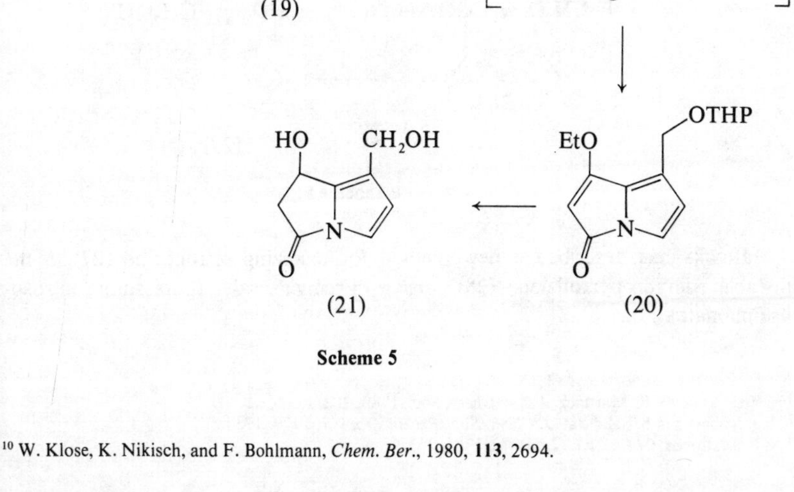

(19)

(21)   (20)

Scheme 5

[10] W. Klose, K. Nikisch, and F. Bohlmann, *Chem. Ber.*, 1980, **113**, 2694.

Another synthetic route to (±)-retronecine (27) has been developed by Vedejs and Martinez.[11] The protected hydroxy-lactam (22) was prepared from 2-methoxy-1-pyrroline by known methods. The key ylide intermediate (24) was then generated from the salt (23) by desilylation with caesium fluoride (Scheme 6). This ylide (24) reacted with methyl acrylate in a 1,3-dipolar cycloaddition to afford the unsaturated pyrrolizidine (25) in 57% yield from the lactam (22). Catalytic hydrogenation of the ester (25) gave an unstable *endo*-product, which epimerized to the *exo*-form (26). Introduction of the 1,2-double-bond into (26) was carried out by insertion and thermal elimination of a phenylseleno-group.[12] Reduction then yielded (±)-retronecine (27).

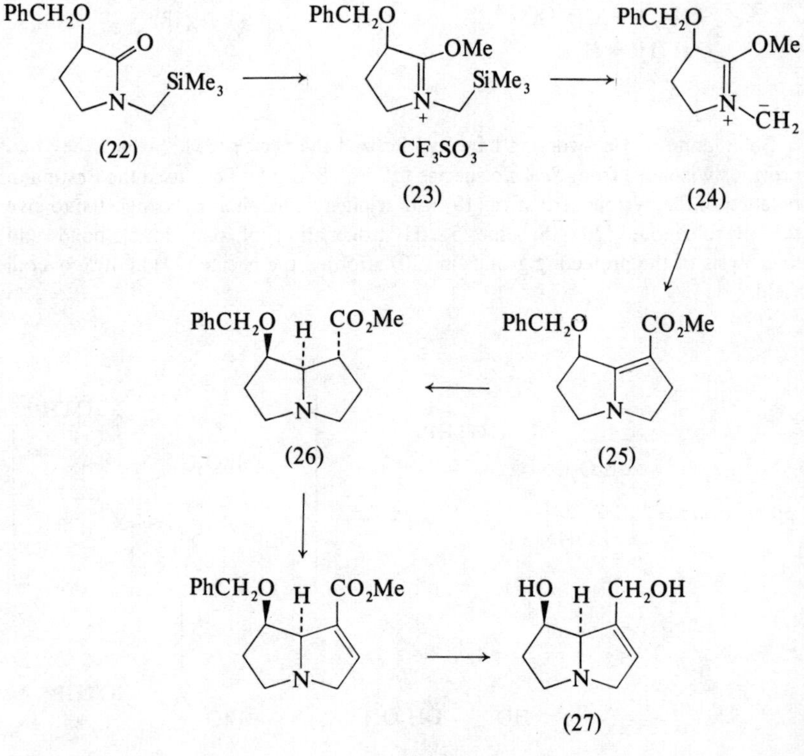

Scheme 6

Mattocks has described a new method for oxidizing retronecine (27) to the unstable dihydropyrrolizine (28), using Fremy's salt (potassium nitroso-disulphonate).[13]

[11] E. Vedejs and G. R. Martinez, *J. Am. Chem. Soc.*, 1980, **102**, 7994.
[12] D. J. Robins and S. Sakdarat, *J. Chem. Soc., Perkin Trans. 1*, 1979, 1734.
[13] A. R. Mattocks, *Chem. Ind. (London)*, 1981, 251.

HO   CH₂OH

(28)

2 Synthesis of Necic Acids

The synthesis of senecivernic acid (29)[14] and nemorensic acid in the open-chain form (30)[15] has been claimed. These acids were presumably formed as part of mixtures of diastereoisomeric racemates, since no steric control was exercised in either route, nor were any of the mixtures separated. No comparisons were made with natural material.

$$H_2C=C-CH-CH-C-OH$$

(29)

(30)

3 Synthesis of Macrocyclic Diesters

A synthesis of an eleven-membered macrocyclic pyrrolizidine diester, namely crobarbatine acetate, together with a diastereoisomer [(34) and (36)], has been described by Huang and Meinwald.[16] The lactone (31) was opened as its t-butyl thiolester, and the hydroxy-group was protected as the acetate (Scheme 7). The remaining free carboxylic acid group was activated as the imidazolide (32). Treatment of this material (32) with (+)-retronecine (27) in the presence of a catalytic amount of sodium hydride yielded the monoester (33). The intramolecular lactonization step was effected with copper(I) trifluoromethanesulphonate–benzene complex to give two diastereoisomeric products (34) and (36), which were separated by column chromatography. (The configuration of the acid portion of each of these compounds was established by acid hydrolysis of each diastereoisomer, and observation of the Cotton effect in each lactonic acid.) It could not be demonstrated which one of the two products [(34) or (36)] is the acetate of natural crobarbatine [(35) or (37)]. Only one of the esters (36) could be hydrolysed to the corresponding base (37), in low yield. The authors could not prove that this base (37) corresponded to natural crobarbatine.

[14] U. Pastewka, H. Wiedenfeld, and E. Röder, *Arch. Pharm.* (*Weinheim, Ger.*), 1980, **313**, 785.
[15] E. Röder, H. Wiedenfeld, and M. Frisse, *Arch. Pharm.* (*Weinheim, Ger.*), 1980, **313**, 803.
[16] J. Huang and J. Meinwald, *J. Am. Chem. Soc.*, 1981, **103**, 861.

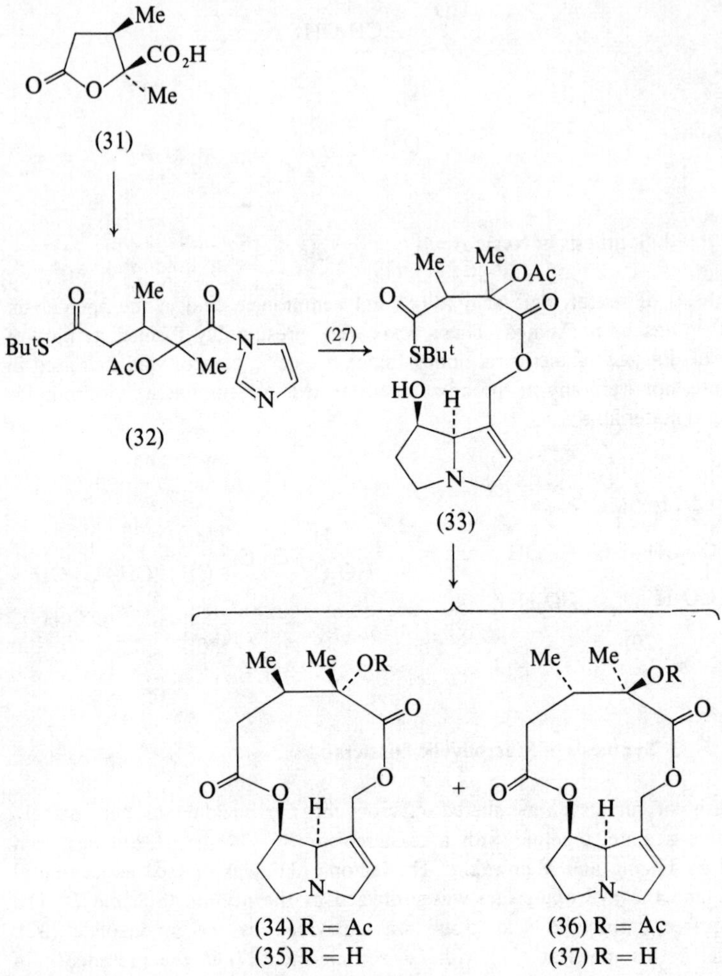

Scheme 7

A different type of macrocyclic diester has been prepared by Drewes and Pitchford.[17] They coupled the dipotassium salt of retronecic acid with 2,6-bis(bromomethyl)pyridine to give a 9.7% yield of 'pyridine-retronecate' (38). The authors claim that this product contains many of the characteristics responsible for the toxicity of pyrrolizidine alkaloids. However, the results of their biological tests with (38) on mice were inconclusive.

[17] S. E. Drewes and A. T. Pitchford, *J. Chem. Soc., Perkin Trans. 1*, 1981, 408.

(38)

4 Alkaloids of the Boraginaceae

A new type of pyrrolizidine alkaloid has been isolated from the leaves of *Ehretia aspera* Willd. by Suri *et al.*[18] The structure (39) for ehretinine was established by spectroscopic evidence and by hydrolysis to give retronecanol and *p*-methoxy-benzoic acid. Confirmation of the structure (39) was obtained by synthesis of ehretinine from retronecanol and *p*-methoxybenzoyl chloride. This is the first reported occurrence of a retronecanol ester, although it should be pointed out that retronecanol esters can be produced by hydrogenolysis of retronecine diesters. The authors do not appear to have considered the possibility that ehretinine could have been formed by hydrogenolysis of the common type of ester (40) on treatment with the zinc and sulphuric acid that are used in their extraction procedure.

(39)

(40)

A chemotaxonomic study of the pyrrolizidine alkaloids present in the subfamilies Heliotropioideae and Boraginoideae has been carried out, using t.l.c.[19] No alkaloids were identified. Lycopsamine (41) and 9-angelylretronecine have been isolated from *Messerschmidia sibirica*.[20] 9-Angelylretronecine has only been found previously in *Bhesa archboldiana* (Celastraceae).[21] Heliotrine *N*-oxide is present in *Heliotropium ramosissimum*.[22] A full paper describing the alkaloids identified in *H. curassavicum*

[18] O. P. Suri, R. S. Jamwal, K. A. Suri, and C. K. Atal, *Phytochemistry*, 1980, **19**, 1273.
[19] H. J. Huizing and T. M. Malingré, *Plant Syst. Evol.*, 1981, **137**, 127.
[20] M. Hikichi, Y. Asada, and T. Furuya, *Planta Med.*, 1980, (Suppl.), p. 1.
[21] C. C. J. Culvenor, S. R. Johns, J. A. Lamberton, and L. W. Smith, *Aust. J. Chem.*, 1970, **23**, 1279.
[22] M. A. Khan and A. S. Khan, *Planta Med.*, 1980. **40**, 383.

has appeared.[23] The rotation of homoviridifloric acid has been corrected to (−) (*cf.* Vol. 9, p. 58). The alkaloids that are present in *Symphytum* species (comfrey) have been estimated by t.l.c. and h.p.l.c. methods.[24] The amounts of lasiocarpine that are present in various types of extract from comfrey have been determined.[25] The variation in alkaloid content in different-sized leaves of comfrey has also been studied.[26] Typical cups of comfrey tea contain at least 8.5 mg of pyrrolizidine alkaloids.[27]

5 Alkaloids of the Compositae

Eupatorieae Tribe.—Nine *Eupatorium* species that are found in the U.S.A. have been screened for pyrrolizidine alkaloids by Herz *et al.*[28] Intermedine (42) (crystalline for the first time) and echinatine were isolated from *Conoclidium coelestinum* (L.) DC. (syn. *Eupatorium coelestinum* L.). Intermedine (42) and lycopsamine (41) were present in *E. compositifolium* Walt. These diastereoisomeric diols were separated as their borate complexes.[29] Rinderine and 7-angelyl-heliotridine (43) were isolated from *E. altissimum* L.[28]

(41) R^1 = OH, R^2 = H
(42) R^1 = H, R^2 = OH

(43)

Senecioneae Tribe.—A new dihydropyrrolizinone alkaloid has been isolated from *Jacmaia incana* (Sw.) B. Nord. by Bohlmann and co-workers.[30] The structure (44) for this alkaloid was deduced from ^{1}H n.m.r. and mass-spectral data. The 7β-configuration was assumed after comparison of the rotation of (44) with those of other related alkaloids. Previously isolated dihydropyrrolizinone alkaloids have all contained macrocyclic rings, as in senaetnine (45). Senaetnine was present in *Odontocline hollickii* (Britt. ex Greenm.) B. Nord.[30]

[23] P. S. Subramanian, S. Mohanraj, P. A. Cockrum, C. C. J. Culvenor, J. A. Edgar, J. L. Frahn, and L. W. Smith, *Aust. J. Chem.*, 1980, **33**, 1357.
[24] H. Wagner, U. Neidhardt, and G. Tittel, *Planta Med.*, 1981, **41**, 232.
[25] W. Debska, A. Owczarska, and R. Madalinska, *Herba Pol.*, 1980, **26**, 47 (*Biol. Abstr.*, 1980, **71**, 68 583).
[26] A. R. Mattocks, *Lancet*, 1980, **2**, (No. 8204), 1136.
[27] J. N. Roitman, *Lancet*, 1981, **1**, (No. 8226), 944.
[28] W. Herz, P. Kulanthaivel, P. S. Subramanian, C. C. J. Culvenor, and J. A. Edgar, *Experientia*, 1981, **37**, 683.
[29] J. L. Frahn, C. C. J. Culvenor, and J. A. Mills, *J. Chromatogr.*, 1980, **195**, 379.
[30] F. Bohlmann, R. K. Gupta, and J. Jakupovic, *Phytochemistry*, 1981, **20**, 831.

Me$_2$C=CHCOO   CH$_2$OCO—OCOCH=CMe$_2$

(44)

(45)

Pterophorine and inaequidenine, two other dihydropyrrolizinone alkaloids, were previously isolated from *Senecio inaequidens* DC. (*cf.* Vol. 8, pp. 54–56). Senecionine and retrorsine have now also been shown to be present in this species.[31,32] 7-Angelylheliotridine (43) has been isolated from *S. ovirensis* ssp. *gaudinii*.[32,33] The cause of poisoning in dairy cattle in Switzerland was found to be *S. alpinus*.[34] Nine known pyrrolizidine alkaloids were isolated from this species, and the major constituent was seneciphylline. Otosenine, floridanine, and doronine are present in *S. othonnae*.[35] The extraction of seneciphylline and platyphylline from *S. platyphylloides* has been studied by a variety of methods.[36]

Matarique (*Cacalia decomposita*) is commonly used as a herbal medicine in America. Roots of this material contain at least seven compounds which give a positive test with Ehrlich's reagent.[37] The presence of toxic pyrrolizidine alkaloids should be confirmed in order to establish this species as a possible health hazard. Another herbal remedy, widely used in China, is *Gynura segetum* (Lour.) Merr. A pyrrolizidine alkaloid, C$_{18}$H$_{25}$NO$_5$, which is also an anti-malarial compound, is present in this species.[38] Senecionine, platyphylline, and neopetasitenine were isolated from the roots of *Ligularia japonica*.[39]

[31] E. Röder, H. Wiedenfeld, and P. Stengl, *Planta Med.*, 1981, **41**, 412.

[32] H. Wiedenfield, U. Pastewka, P. Stengl, and E. Röder, *Planta Med.*, 1981, **41**, 124.

[33] E. Röder, J. Wiedenfeld, and P. Stengl, *Planta Med.*, 1980, (Suppl.), p. 182.

[34] J. Lüthy, U. Zweifel, B. Karlhuber, and C. Schlatter, *J. Agric. Food Chem.*, 1981, **29**, 302; J. Lüthy, U. Zweifel, and C. Schlatter, *Mitt. Geb. Lebensmittelunters. Hyg.*, 1981, **72**, 55 (*Chem. Abstr.*, 1981, **95**, 92 054).

[35] D. S. Khalilov, M. V. Telezhenetskaya, and S. Yu. Yunusov, *Khim. Prir. Soedin.*, 1980, 262 (*Chem. Abstr.*, 1980, **93**, 110 570).

[36] S. I. Kocherga, V. G. Vygon, N. G. Larionov, and G. A. Mikheeva, *Khim.-Farm. Zh.*, 1980, **14**, 65 (*Chem. Abstr.*, 1980, **93**, 120 309); N. G. Larionov, S. L. Akhnazarova, S. I. Kocherga, V. G. Vygon, and O. V. Rusinovskaya, *ibid.*, 1981, **15**, 66 (*Chem. Abstr.*, 1981, **94**, 180 561); E. A. Vdoviko, O. R. Pryakhin, and S. A. Pokhmelkina, *Khim. Prir. Soedin.*, 1979, 674 (*Chem. Abstr.*, 1981, **95**, 30 458); O. V. Pogorelova, *Khim.-Farm. Zh.*, 1980, **14**, 52 (*Chem. Abstr.*, 1981, **94**, 44 009).

[37] G. Sullivan, *Vet. Human Toxicol.*, 1981, **23**, 6.

[38] S.-R. Tang, Y.-F. Wu, and C.-S. Fang, *Chung Ts'ao Yao*, 1980, **11**, 193 (*Chem. Abstr.*, 1981, **94**, 36 187).

[39] Y. Asada, T. Furuya, and N. Murakami, *Planta Med.*, 1981, **42**, 202.

6 Alkaloids of the Gramineae

The bases (46) and (47), derived from the loline group of alkaloids, have been used as stimulants of plant growth.[40]

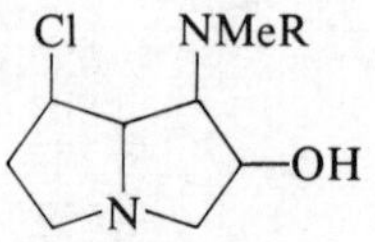

(46) R = H
(47) R = Ac

7 Alkaloids of the Leguminosae

The active principle of the Chinese anti-tumour herbal medicine *Crotalaria sessiflora* L. has been shown to be monocrotaline (48).[41] Monocrotaline is also present in Chinese *C. retusa*[42] and in the seeds of *C. nitens*.[43] Spectrophotometric estimation of crispatine and fulvine in *C. madurensis* R. Wight was achieved by complexing these alkaloids with bromocresol purple.[44]

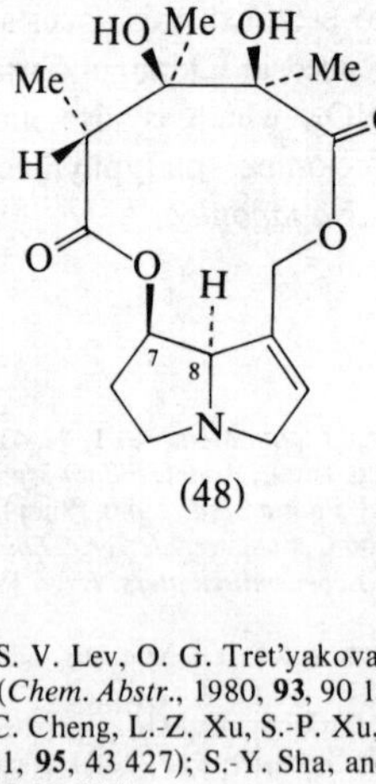

(48)

[40] S. Yu. Yunusov, A. A. Umarov, S. V. Lev, O. G. Tret'yakova, S. T. Akramov, V. M. Malikov, and E. Kh. Batirov, USSR P. 728 813 (*Chem. Abstr.*, 1980, **93**, 90 194).

[41] L. Huang, K.-M. Wu, Z. Xue, J.-C. Cheng, L.-Z. Xu, S.-P. Xu, and Y.-G. Xi, *Yao Hsueh Hsueh Pao*, 1980, **15**, 278 (*Chem. Abstr.*, 1981, **95**, 43 427); S.-Y. Sha, and C.-Y. Tseng, *Yao Hsueh T'ung Pao*, 1980, **15**, 4 (*Chem. Abstr.*, 1981, **94**, 90 416).

[42] G.-Q. Han, S.-M. Xu, F.-Q. Liu, and X. Wang, *Pei-ching I Hsueh Yuan Hsueh Pao*, 1981, **13**, 15 (*Chem. Abstr.*, 1981, **94**, 162 632).

[43] P. Hoet, M. Astigueta, and A. M. Frisque, *Rev. Latinoam. Quim.*, 1981, **12**, 34 (*Chem. Abstr.*, 1981, **94**, 188 670).

[44] A. A. M. Habib and N. A. El-Sebakhy, *Egypt. J. Pharm. Sci.*, 1978, **19**, 71 (*Chem. Abstr.*, 1981, **95**, 30 464).

8 Alkaloids from Animals

The presence of pyrrolizidine alkaloids in arctiid moths that had been reared on *Senecio* and *Crotalaria* species has been established by Rothschild *et al.*[45] These alkaloids are stored in the moths, and serve as a deterrent to vertebrate predators and as precursors for insect sex pheromones. A pyrrolizidine alkaloid metabolite from the Cinnabar moth (*Tyria jacobaea* L.), named callimorphine, has been shown to have the structure (49) on the basis of mass-spectral and degradative evidence.[46] The structure (49) was confirmed by synthesis of callimorphine and a diastereo-isomer by treatment of 9-chlororetronecine with the sodium salt of (±)-2-acetoxy-2-methylbutanoic acid.

HO   H   CH₂OCOC(OAc)MeEt

(49)

The venomous constituent of the cryptic thief ant, *Solenopsis xenovenenum*, has been identified as the 3-heptyl-5-methylpyrrolizidine (50) from its mass spectrum and the fact that a related pyrrolidine (51) has been isolated from another species of ant.[47] This is the first reported occurrence of a 3,5-dialkyl-pyrrolizidine, and its structure was confirmed by synthesis. Reductive amination of the known triketone (52) with sodium cyanoborohydride and ammonium acetate gave a mixture of four isomers of 3-heptyl-5-methylpyrrolizidine, which were separated by preparative g.l.c. The stereochemistry of the ring-junction of each isomer was established from its i.r. and n.m.r. spectra.

(50)   (51)

$Me(CH_2)_6CO(CH_2)_2CO(CH_2)_2COMe$

(52)

[45] M. Rothschild, R. T. Aplin, P. A. Cockrum, J. A. Edgar, P. Fairweather, and R. Lees, *Biol. J. Linn. Soc.*, 1979, **12**, 305.
[46] J. A. Edgar, C. C. J. Culvenor, P. A. Cockrum, L. W. Smith, and M. Rothschild, *Tetrahedron Lett.*, 1980, **21**, 1383.
[47] T. H. Jones, M. S. Blum, H. M. Fales, and C. R. Thompson, *J. Org. Chem.*, 1980, **45**, 4778.

9 General Studies

Separation of mono- and di-ester pyrrolizidine alkaloids has been achieved by ion-pair adsorption t.l.c., using chloride (or iodide) as the counter-ion.[48] Chloranil has been used to oxidize pyrrolizidine alkaloids on t.l.c. The pyrrole derivatives that were formed were then detected with Ehrlich's reagent[49] or sulphuric acid.[50] Mixtures of pyrrolizidine alkaloids have been separated by h.p.l.c. on a reversed-phase styrene–divinylbenzene resin.[51] In a sensitive method for the detection of pyrrolizidine alkaloids, the protonated alkaloids were complexed with aqueous methyl orange. The dye was then released from the complex and estimated spectrophotometrically.[52]

Indicine *N*-oxide (53) is an anti-cancer drug that is now undergoing clinical trials. Methods for its estimation include oxidation to the pyrrole derivative and treatment with Ehrlich's reagent,[53] g.l.c. analysis of its trimethylsilyl derivatives,[54] and differential pulse radiography.[55] The aqueous degradation of indicine *N*-oxide (53) to retronecine *N*-oxide and (−)-trachelanthic acid has been studied.[56] Indicine *N*-oxide is reduced to indicine by Fe^{III}, by reduced cytochrome *c*, by ascorbic acid, and by denatured haemoglobin.[57]

Further ^{13}C n.m.r. spectral data on pyrrolizidine alkaloids have been reported,[58] including a re-assignment for the signals due to C-7 and C-8 in monocrotaline (48) and retronecine (27).[59] This re-assignment is supported by ^{13}C n.m.r. spectra that were obtained after biosynthetic enrichment of retronecine with ^{13}C.[60]

(53)

(54)

[48] H. J. Huizing and T. M. Malingré, *J. Chromatogr.*, 1981, **205**, 218.
[49] R. J. Molyneux and J. N. Roitman, *J. Chromatogr.*, 1980, **195**, 412.
[50] H. J. Huizing, F. De Boer, and T. M. Malingré, *J. Chromatogr.*, 1980, **195**, 407.
[51] H. S. Ramsdell and D. R. Buhler, *J. Chromatogr.*, 1981, **210**, 154.
[52] H. Birecka, J. L. Catalfamo, and R. N. Eisen, *Phytochemistry*, 1981, **20**, 343.
[53] J. B. D'Silva and R. E. Notari, *J. Pharm. Sci.*, 1980, **69**, 471.
[54] J. V. Evans, S. K. Daley, G. A. McClusky, and C. J. Nielsen, *Biomed. Mass Spectrom.*, 1980, **7**, 65.
[55] M. McComish, I. Bodek, and A. R. Branfman, *J. Pharm. Sci.*, 1980, **69**, 727.
[56] J. B. D'Silva and R. E. Notari, *Int. J. Pharm.*, 1981, **8**, 45.
[57] G. Powis and C. L. De Graw, *Res. Commun. Chem. Pathol. Pharmacol.*, 1980, **30**, 143.
[58] S. E. Drewes, I. Antonowitz, P. T. Kaye, and P. C. Coleman, *J. Chem. Soc., Perkin Trans. 1*, 1981, 287.
[59] E. J. Barreiro, A. De Lima Pereira, L. Nelson, L. F. Gomes, and A. J. R. Da Silva, *J. Chem. Res. (S)*, 1980, 330.
[60] H. A. Khan and D. J. Robins, *J. Chem. Soc., Chem. Commun.*, 1981, 146, 554; G. Grue-Sørensen and I. D. Spenser, *J. Am. Chem. Soc.*, 1981, **103**, 3208.

The X-ray structure of bulgarsenine (54) as its (+)-D-hydrogen tartrate has been determined.[61] The presence of the uncommon thirteen-membered macrocyclic ring is confirmed. In all pyrrolizidine alkaloids with twelve-, thirteen-, or fourteen-membered rings that have been studied, the ester carbonyl groups are antiparallel and the C-9 ester carbonyl bond is directed above the plane of the macro-ring. The X-ray structures of retrorsine,[62] monocrotaline (48),[63] and its N-oxide[64] have been determined again.

10 Pharmacological and Biological Studies

Several reviews of the pharmacology of pyrrolizidine alkaloids have appeared.[65,66] A number of plant species which contain carcinogenic pyrrolizidine alkaloids are used as human food and for herbal remedies in Japan. These species include *Petasites japonicus*, *Tussilago farfara* (coltsfoot), and *Symphytum officinale* (comfrey).[67] Possible health hazards incurred in the ingestion of comfrey have been discussed.[26,27] The carcinogenicity of a number of pyrrolizidine alkaloids and their analogues has been demonstrated, using the mammalian-cell-transformation test *in vitro*,[68] the hepatocyte primary culture/DNA-repair test,[69] and the transplacental micronucleus test.[70] Clivorine is carcinogenic.[71]

Pyrrolizidine alkaloids have been detected in the honey produced from stands of *Echium plantagineum* L. The major alkaloidal constituent was echimidine.[72] Consumption of the common fireweed (*Senecio lautus*) was responsible for mortalities and poor growth of cattle in New South Wales, Australia.[73] Dairy goats that had been fed large quantities of *S. jacobaea* (tansy ragwort) secreted pyrrolizidine alkaloids in their milk.[74] The toxicity of tansy ragwort in rats has been studied.[75] The levels of amino-acids in the plasma of horses were determined after feeding them with hay containing *S. vulgaris*.[76]

[61] H. Stoeckli-Evans, *Acta Crystallogr., Sect. B*, 1980, **36**, 3150.
[62] P. C. Coleman, E. D. Coucourakis, and J. A. Pretorius, *S. Afr. J. Chem.*, 1980, **33**, 116.
[63] S.-D. Wang, *Sci. Sin. (Engl. Ed.)*, 1981, **24**, 497 (*Chem. Abstr.*, 1981, **95**, 16 382).
[64] S.-T. Wang and N.-H. Hu, *K'o Hsueh T'ung Pao*, 1980, **25**, 1071 (*Chem. Abstr.*, 1981, **94**, 157 121).
[65] J. O. Dickinson, *Food Drug Adm., Bur. Vet. Med.*, (*Tech. Rep.*), FDA/BVM-80/132, PB80-221773, p. 45.
[66] R. J. Huxtable, *Trends Pharmacol. Sci.*, 1980, **1**, 299; *Gen. Pharmacol.*, 1979, **10**, 159.
[67] I. Hirono, H. Mori, M. Haga, M. Fujii, K. Yamada, Y. Hirata, H. Takanashi, E. Uchida, S. Hosaka, I. Ueno, T. Matsushima, K. Umezawa, and A. Shirai, *Proc. Int. Symp. Princess Tatematsu Cancer Res. Fund*, 1979, 79 (*Chem. Abstr.*, 1980, **93**, 180 283).
[68] J. Styles, J. Ashby, and A. R. Mattocks, *Carcinogenesis*, 1980, **1**, 161.
[69] G. M. Williams, H. Mori, I. Hirono, and M. Nagao, *Mutat. Res.*, 1980, **79**, 1.
[70] C. J. Stoyel and A. M. Clark, *Mutat. Res.*, 1980, **74**, 393.
[71] K. Kuhara, H. Takanashi, I. Hirono, T. Furuya, and Y. Asada, *Cancer Lett.*, 1980, **10**, 117.
[72] C. C. J. Culvenor, J. A. Edgar, and L. W. Smith, *J. Agric. Food Chem.*, 1981, **29**, 958.
[73] K. H. Walker and P. D. Kirkland, *Aust. Vet. J.*, 1981, **57**, 1.
[74] J. O. Dickinson, *Proc. West. Pharmacol. Soc.*, 1980, **23**, 377.
[75] C. L. Miranda, P. R. Cheeke, J. A. Schmitz, and D. R. Buhler, *Toxicol. Appl. Pharmacol.*, 1980, **56**, 432.
[76] B. A. Gulick, K. M. L. Irwin, C. W. Qualls, D. H. Gribble, and Q. R. Rogers, *Am. J. Vet. Res.*, 1980, **41**, 1894.

Many pyrrolizidine alkaloids are metabolized to toxic pyrrole metabolites in the liver by mixed-function oxidases. The structural and chemical features necessary for the formation of these metabolites have been discussed.[77] The most important features, in addition to the 3-hydroxymethyl-3-pyrroline system, are steric hindrance to hydrolysis of the ester, lipophilic character (favouring attack by the hepatic microsomal enzymes), and the presence of a conformation that allows preferential oxidation of the pyrroline ring rather than *N*-oxidation. The alkylating activities of a series of these pyrrole derivatives have been examined.[78]

The effects of pyrrolizidine alkaloids on the mixed-function oxidase enzyme system in rat liver have been studied.[79,80] Dehydroheliotridine and heliotrine (at higher dose rates) have similar effects on pregnant rats and their embryos.[81] The development of pulmonary hypertension and obstructive lesions in rats after administration of monocrotaline (48) has been studied.[82] Butylated hydroxyanisole protects young mice against the acute toxicity of monocrotaline.[83] Reduced levels of pyrrole metabolites were observed.

Tritium-labelled *N*-(4-phenylphenacyl)heliotridanium bromide (55) has been prepared. The absorption, tissue distribution, and excretion of this radioactive antispasmodic drug in rats were then studied.[84]

(55)

[77] A. R. Mattocks, *Chem.-Biol. Interact.*, 1981, **35**, 301.
[78] J. J. Karchesy and M. L. Deinzer, *Heterocycles*, 1981, **16**, 631.
[79] D. F. Eastman and H. J. Segall, *Toxicol. Lett.*, 1981, **8**, 217; 1980, **5**, 369.
[80] C. L. Miranda, P. R. Cheeke, and D. R. Buhler, *Biochem. Pharmacol.*, 1980, **29**, 2645; *Res. Commun. Chem. Pathol. Pharmacol.*, 1980, **29**, 573.
[81] J. E. Peterson and M. V. Jago, *J. Pathol.*, 1980, **131**, 339.
[82] F. Ghodsi and J. A. Will, *Am. J. Physiol.*, 1981, **240**, H 149.
[83] C. L. Miranda, R. L. Reed, P. R. Cheeke, and D. R. Buhler, *Toxicol. Appl. Pharmacol.*, 1981, **59**, 424.
[84] U. Zutshi, P. G. Rao, A. Soni, and C. K. Atal, *Arzneim.-Forsch.*, 1981, **31**, 44.

5

Indolizidine Alkaloids

BY J. A. LAMBERTON

1 *Castanospermum* Alkaloids

Castanospermine, a new alkaloid isolated from toxic seeds of the Australian legume *Castanospermum australe*, has been shown to be $8a\beta$-indolizidine-$1\alpha,6\beta,7\alpha,8\beta$-tetraol (1). The structure and relative configuration were determined by X-ray crystallography.[1]

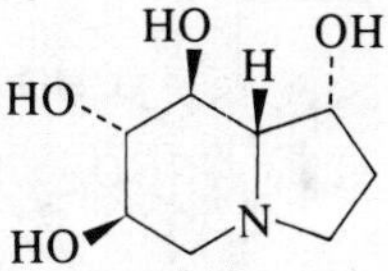

(1)

2 *Prosopis* Alkaloids

Juliprosopine from *Prosopis juliflora* A. DC. is a new type of indolizidine alkaloid, and on the basis of an extensive spectroscopic and chemical study it has been assigned the structure (2), in which relative configurations are shown for the substituents on the piperidine ring. Juliprosopine may be derived biosynthetically from combination of a dihydropyrrole unit with two *Prosopis* piperidine alkaloids that have C_{12} side-chains.[2]

3 *Elaeocarpus* Alkaloids

The ester (3) has been shown to be a useful starting point for the synthesis of *Elaeocarpus* alkaloids. Ester (3) is converted by several steps into the protected keto-aldehyde (4); this, by reaction with n-propylmagnesium bromide followed by a Jones oxidation of the product and then deprotection, gives the diketone (5). As

[1] L. D. Hohenschutz, E. A. Bell, P. J. Jewess, D. P. Leworthy, R. J. Pryce, E. Arnold, and P. J. Clardy, *Phytochemistry*, 1981, **20**, 811.
[2] R. Ott-Longoni, N. Viswanathan, and M. Hesse, *Helv. Chim. Acta*, 1980, **63**, 2119.

(2)

diketone (5) has already been converted into (±)-elaeokanine C (6), this method constitutes a formal synthesis of (6). Alternatively, compound (7), in which the aldehyde group is selectively protected, instead of the ketone, can be converted into the unsaturated aldehyde (8), and thence into (±)-elaeokanine A (9) and (±)-elaeokanine B (10).[3]

(3)   (4)   (5)

(6)   (7)   (8) R = CHO
                                            (9) R = COPr^n
                                            (10) R = CH(OH)Pr^n

The protected keto-aldehyde (4) has also been converted, in several steps, into the dihydro-γ-pyrones (11) and (12). A Birch reduction of the pyrone (11) with lithium in liquid ammonia afforded (±)-elaeokanine E (13) as the only product. A similar reduction of the epimeric pyrone (12) gave a product that is not identical with elaeokanine D, and it is considered to be (±)-epielaeokanine D (14).[4]

[3] T. Watanabe, Y. Nakashita, S. Katayama, and M. Yamauchi, *Heterocycles*, 1980, **14**, 1433.
[4] T. Watanabe, Y. Nakashita, S. Katayama, and M. Yamauchi, *Heterocycles*, 1981, **16**, 39.

(11) R¹ = H, R² = Me
(12) R¹ = Me, R² = H

(13)

(14)

In an alternative approach to this group of alkaloids, (±)-elaeokanine A (9) has been made by an intramolecular imino-Diels–Alder reaction. Pyrolysis of (15) in toluene solution yielded a mixture of the diastereoisomers (16), which, by a sequence of reactions, were converted into (±)-elaeokanine A (9).[5] The possible derivation of the *Elaeocarpus* alkaloids from a common biosynthetic intermediate 3-(1-pyrrolinium)propionaldehyde (17), and the use of (17) in a synthetic approach to these alkaloids, have been discussed.[6]

(15)

(16)

(17)

4 *Dendrobates* Alkaloids

The dendrobatid toxin 251D (18) has been synthesized by a new approach which utilizes an iminium–vinylsilane cyclization to produce the (Z)-6-alkylidene-indolizidine ring-system in a stereospecific manner. The key intermediate (19) was prepared from L-proline and converted into toxin 251D (18) by reaction with paraformaldehyde and (+)-camphor-10-sulphonic acid in refluxing ethanol. The method is potentially a general one for forming unsaturated azacyclic rings, and it provides a convenient route to the pumiliotoxin A alkaloids.[7]

A highly stereoselective synthesis of (±)-gephyrotoxin (20) has been carried out. An interesting feature is the reversal of the stereochemical course of the hydrogenation of the vinylogous amide (21) by the use of an alumina support. Hydrogenation of (21) over palladium on charcoal gives the amino-alcohol (22) as

[5] H. F. Schmitthenner and S. M. Weinreb, *J. Org. Chem.*, 1980, **45**, 3372.
[6] G. W. Gribble and R. M. Soll, *J. Org. Chem.*, 1981, **46**, 2433.
[7] L. E. Overman, and K. L. Bell, *J. Am. Chem. Soc.*, 1981, **103**, 1851.

(18)   (19)

the only product, whereas hydrogenation over platinum on alumina gives a 12 : 1 mixture of (23) and (22).[8] A total synthesis of (±)-perhydrogephyrotoxin has also been achieved.[9]

(20)   (21)

(22)   (23)

5 Phenanthroindolizidines

The phenanthroindolizidine alkaloid antofine has been synthesized by a new method, based on the direct metallation of the amide of a phenanthrenecarboxylic acid.[10] Tylophorine[11] and septicine[12] have been synthesized by a modification of an earlier biosynthetic route to the phenanthroindolizidines.

[8] R. Fujimoto, Y. Kishi, and J. F. Blount, *J. Am. Chem. Soc.*, 1980, **102**, 7154.
[9] L. E. Overman and R. L. Freerks, *J. Org. Chem.*, 1981, **46**, 2833.
[10] M. Iwao, M. Watanabe, S. O. de Silva, and V. Snieckus, *Tetrahedron Lett.*, 1981, **22**, 2349.
[11] V. K. Mangla and D. S. Bhakuni, *Tetrahedron*, 1980, **36**, 2489.
[12] V. K. Mangla and D. S. Bhakuni, *Indian J. Chem., Sect. B*, 1980, **19**, 748.

6
Quinolizidine Alkaloids

BY M. F. GRUNDON

Although the detection and isolation of bicyclic, tricyclic, and tetracyclic quinolizidine alkaloids and stereochemical studies, increasingly aided by X-ray analysis, proceeds apace, the main emphasis this year is on synthesis of the *Nuphar*, azaphenalene, and phenanthroquinolizidine alkaloids.

1 The Lupinine–Cytisine–Sparteine–Matrine–*Ormosia* Group

Occurrence.—Kinghorn and co-workers[1] determined the distribution of quinolizidine alkaloids by g.l.c.–m.s. in the following *Lupinus* species, indigenous to North and South America: *L. aduncus*†, *L. agardhianus*†, *L. arboreus*, *L. arizonicus*†, *L. chamissonis*, *L. concinnus*†, *L. densiflorus*†, *L. greenei*†, *L. hartwegii*, *L. hilarianus*, *L. hirsutissimus*†, *L. latifolius*, *L. longifolius*, *L. multiflorus*, *L. nanus*, *L. oscar-haughtii*†, *L. polycarpus*†, *L. polyphyllus*, *L. rivularis*, *L. sparsiflorus*†, and *L. truncatus*†. Nineteen quinolizidine alkaloids were identified, including aphyllidine and *N*-methylcytisine, which have not previously been found in the genus.

Other isolation studies are summarized in Table 1;[2–7] five new alkaloids have been obtained this year. Cell suspension cultures of *Baptisia australis*, *Lupinus polyphyllus*, and *Sarothamnus scoparius* produce lower yields of alkaloids than the differentiated plants, with lupanine as the main component[8] (*cf.* Vol. 11, p. 63). Examination of the leaf alkaloids of *B. australis* by g.l.c. and by g.l.c.–m.s. resulted in the identification of eleven constituents, including two new alkaloids.[2] The structures and the distribution of some quinolizidine alkaloids that may be used as systematic markers in the Leguminosae have now been supplemented by more recent data.[9]

† Species in which alkaloids were reported for the first time.
[1] A. D. Kinghorn, M. A. Selim, and S. J. Smolenski, *Phytochemistry*, 1980, **19**, 1705.
[2] M. Wink, T. Hartmann, L. Witte, and H. M. Schiebel, *J. Nat. Prod.*, 1981, **44**, 14.
[3] S. Ohmiya, O. Otomasu, J. Haginiwa, and I. Murakoshi, *Chem. Pharm. Bull.*, 1980, **28**, 546.
[4] W. J. Keller, *Phytochemistry*, 1980, **19**, 2233.
[5] S. Kuchkarov, Yu, K. Kushmuradov, Kh. A. Aslanov, and A. S. Sadykov, *Tezisy Dokl.-Sov.-Indiiski Simp. Khim. Prir. Soedin*, 5th, 1978, 44 (*Chem. Abstr.*, 1980, **93**, 164 311).
[6] G. Kavalali, *J. Nat. Prod.*, 1981, **44**, 236.
[7] I. Murakoshi, E. Kidoguchi, M. Nakamura, J. Haginiwa, S. Ohmiya, K. Higashiyama, and M. Otomasu, *Phytochemistry*, 1981, **20**, 1725.
[8] M. Wink and T. Hartmann, *Planta Med.*, 1980, **40**, 149.
[9] A. Salatino and O. R. Gottlieb, *Biochem. Syst. Ecol.*, 1980, **8**, 133.

Table 1 *Isolation of alkaloids of the lupinine–cytisine–sparteine–matrine group*

Species	Alkaloid (Structure)	Ref.
Baptisia australis	*13-Acetoxyanagyrine stereoisomer *Isotinctorine Tinctorine (6)	2
Euchresta japonica (cf. Vol. 10, p. 67)	(−)-Baptifoline (−)-N-Formylcytisine (−)-N-Methylcytisine (+)-Sophoranol (+)-Sophoranol N-oxide (−)-Sophoridine (−)-Sophoridine N-oxide (11)	3
Lupinus holosericeus	Anagyrine α-Isolupanine (+)-Lamprolobine (1) Lupanine	4
Sophora alopecuroides (cf. Vol. 11, p. 64; Vol. 9, p. 70)	N-Methylaloperine	5
S. jaubertii	Matrine	6
S. tomentosa (cf. Vol. 5, p. 94)	(±)-Ammodendrine *(−)-Epilamprolobine (2) *(+)-Epilamprolobine N-oxide (3) (−)-N-formylcytisine (+)-Sophocarpine N-oxide *Compound (4)	7

* New alkaloids.

Structural and Stereochemical Studies.—The unusual alkaloid lamprolobine (1), first isolated from *Lamprolobium fruticosum* (*cf.* Vol. 1, p. 89), has been shown to be the major alkaloid of *Lupinus holosericeus*[4] and the stereoisomer, (−)-epilamprolobine (2), was obtained from *Sophora tomentosa*.[7] The gross structure of epilamprolobine was indicated by comparison with synthetic (±)-lamprolobine; the absolute configuration of the new alkaloid was (5R,6S), since its optical rotation was of opposite sign to that of a synthetic epilamprolobine formed from (−)-(5R,6R)-lupinine (5; R = OH). *S. tomentosa* contains two other new alkaloids, *i.e.* epilamprolobine N-oxide (3), which was converted into epilamprolobine by reduction, and the quinolizidine N-oxide (4); the latter was readily formed from alkaloid (3) by reaction with methanol and is regarded as an artefact.[7]

An *X*-ray study of the perchlorate salt showed that protonation of iodolupinine (5; R = I) did not result in conformational inversion.[10]

One of the new alkaloids of *Baptisia australis* has a mass spectrum differing from that of tinctorine (6) only in the intensity of fragmentation peaks; it is believed to be a stereoisomer of tinctorine and has been named isotinctorine. Another alkaloid, $C_{17}H_{22}N_2O_3$, from the same source is apparently a 13-acetoxy-anagyrine, since alkaline hydrolysis gives a compound with a mass spectrum identical with that of 13-hydroxyanagyrine (baptifoline) (7).[2] The structures of the new *Baptisia* alkaloids clearly need further study.

[10] A. E. Koziol and Z. Kosturkiewicz, *Acta. Crystallogr.*, *Sect. B*, 1980, **36**, 2483.

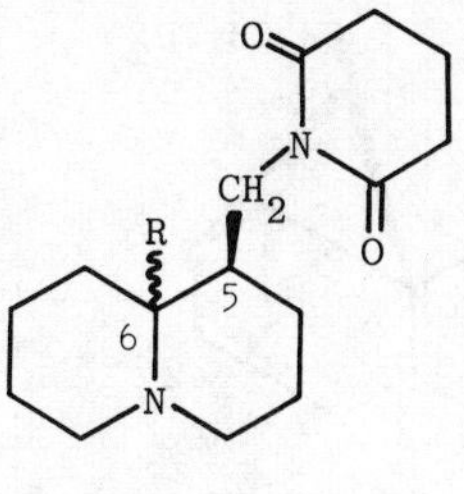

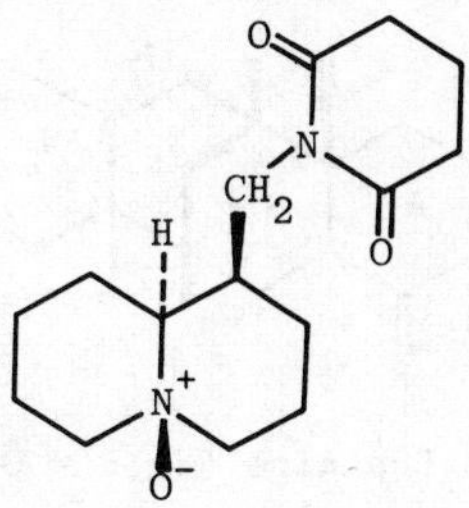

Lamprolobine (1) R = β-H
(−)-Epilamprolobine (2) R = α-H

(+)-Epilamprolobine *N*-oxide (3)

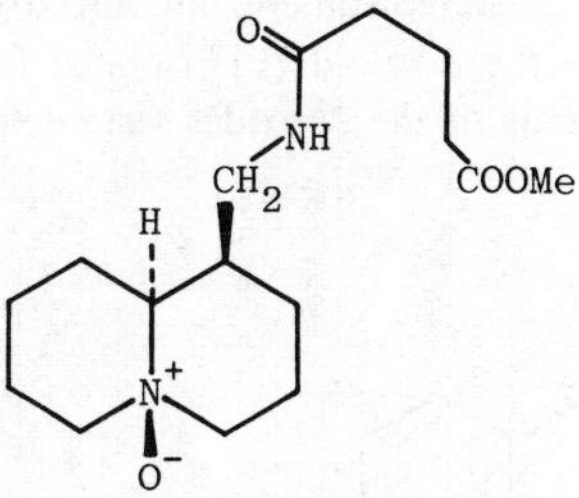

(4)

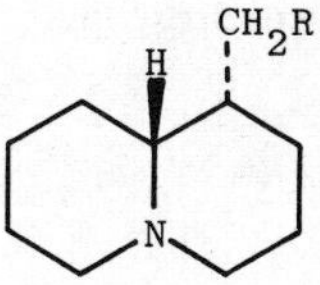

(5)

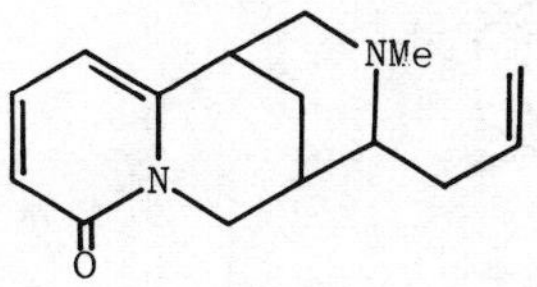

(6)

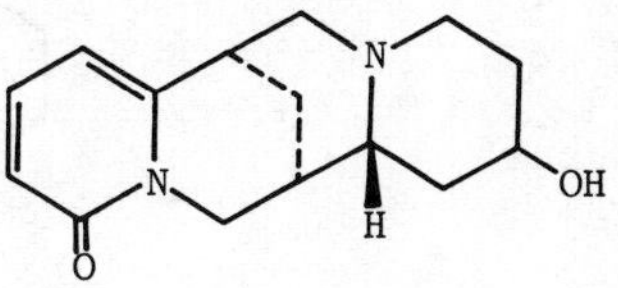

(7)

An *X*-ray analysis of salts of lupanine, (8)·HCl·2H$_2$O and (8)·HClO$_4$·H$_2$O, showed that ring C was in a boat conformation (*cf.* Vol. 10, p. 68); infrared studies of the salts and of the anhydrous hydrochloride indicate that this conformation is retained in solution, but that ring C is changed to a chair confirmation in anhydrous lupanine perchlorate. Conformations of mono- and di-protonated salts of 17β-methyl-lupanine, α-isolupanine, and 17β-methyl-α-isolupanine were also discussed on the basis of i.r. data.[11]

15-Oxosparteine perchlorate hemihydrate, *cf.* (9), has been shown by *X*-ray crystallography to have rings A, B, C, and D in chair, chair, boat, and half-chair conformations, respectively;[12] the corresponding rings of 17-oxosparteine, *cf.* (9), are in chair, chair, sofa, and chair conformations.[13]

[11] A. Perkowska, G. Pieczonka, and M. Wiewiorowski, *Bull. Acad. Pol. Sci., Ser. Sci. Chim.*, 1979, **27**, 637.
[12] A. Hoser, A. Katrusiak, Z. Kaluski, and A. Perkowska, *Acta Crystallogr.*, Sect. B, 1981, **37**, 281.
[13] A. Katrusiak, A. Hoser, E. Grzesiak, and Z. Kaluski, *Acta Crystallogr.*, Sect. B, 1980, **36**, 2442.

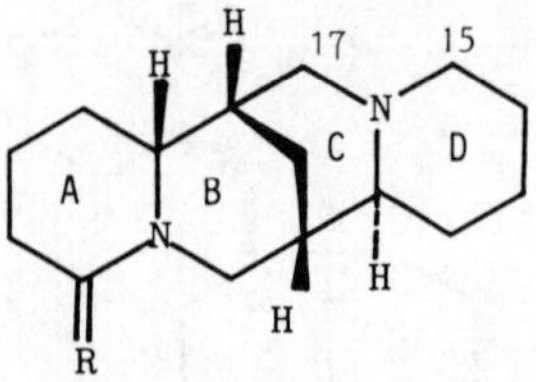

Lupanine (8) R = O

Sparteine (9) R = H

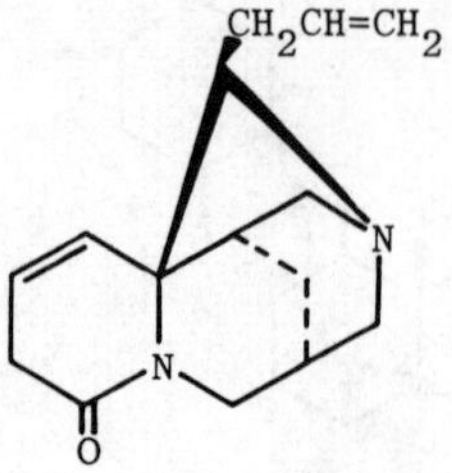

(10)

Oxidation of (−)-sophoridine gave a diastereoisomeric mixture of *N*-oxides, the minor component being identical with the alkaloid (11) isolated from *Euchresta japonica* (see Table 1); the configurations of the *N*-oxides were determined by ¹H and ¹³C n.m.r. spectroscopy.[3]

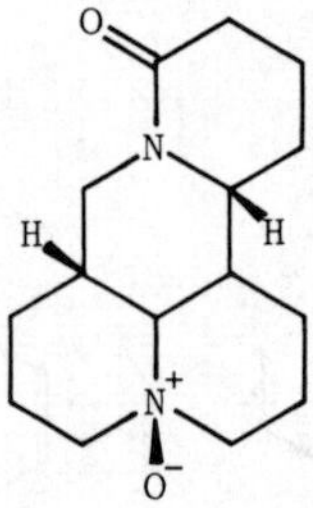

(11)

The absolute configuration of the unusual tetracyclic alkaloid tsukushinamine-A (10) (*cf.* Vol. 10, p. 69) has been established by *X*-ray crystallography.[14]

A new investigation[15] of the *Ormosia* alkaloids of *Podopetalum ormondii* (*cf.* Vol. 4, p. 114; Vol. 8, p. 70; and Vol. 10, p. 70) led to the isolation of two quasi-racemates, consisting of co-crystals of (−)-podopetaline and (−)-ormosanine; separation was effected through the formaldehyde adducts, *i.e.* (−)-homopodopetaline and (−)-homo-ormosanine [(−)-jasmine]. The absolute configuration of (−)-ormosanine (12) was determined from the known configuration of (−)-podopetaline (13). The co-occurrence of two related alkaloids of 'opposite' configuration was discussed; possibilities are that (±)-ormosanine is formed first and (±)-ormosanine is selectively oxidized to (−)-podopetaline or that (±)-podopetaline is selectively reduced to one isomer of ormosanine.

[14] J. Bordner, S. Ohmiya, H. Otomasu, and J. Haginiwa, *Chem. Pharm. Bull.*, 1980, **28**, 1965.

[15] R. Misra, W. Wong-Ng, P-K. Cheng, S. McLean, and S. C. Nyburg, *J. Chem. Soc., Chem. Commun.*, 1980, 659; S. McLean, R. Misra, V. Kumar, and J. A. Lamberton, *Can. J. Chem.*, 1981, **59**, 34.

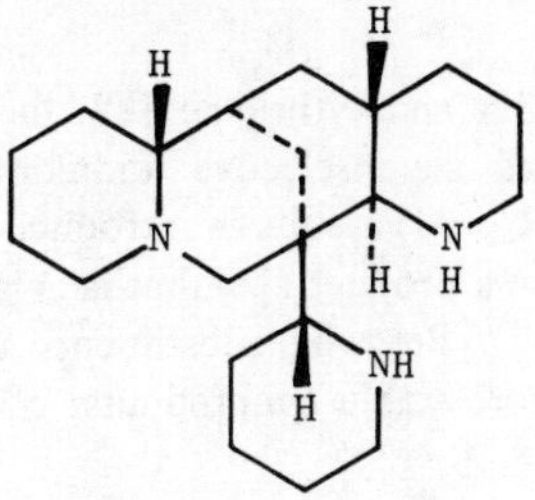

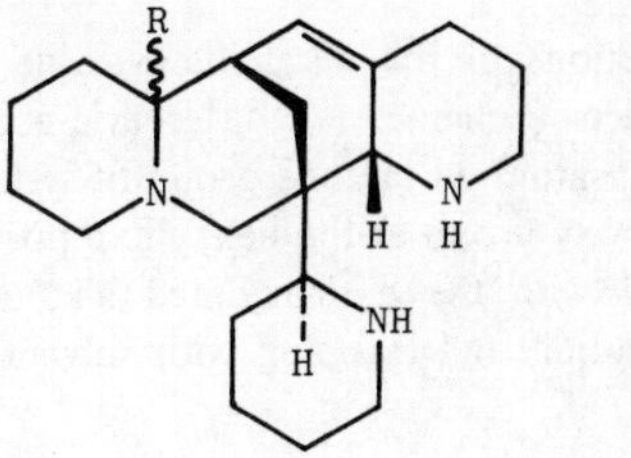

(−)-Ormosanine (12)

(−)-Podopetaline (13) R = α-H

6-*epi*-Podopetaline (14) R = β-H

A hydrated hydrobromide salt obtained from *P. ormondii*, apparently as a racemate, yielded optically active crystals that were shown by *X*-ray analysis to be (6-*epi*-podopetaline)$_2$·6HBr·$\frac{7}{2}$H$_2$O. This work confirmed the absolute configuration of the alkaloid (14).[16]

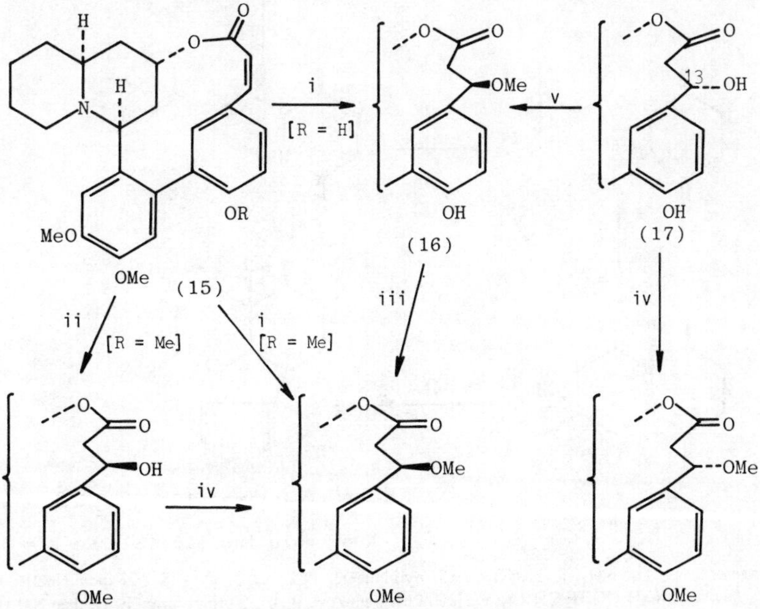

Reagents: i, Hg(OAc)$_2$, MeOH, then NaBH$_4$; ii, Hg(OAc)$_2$, aq. THF, then NaBH$_4$; iii, Me$_2$SO$_4$, NaOH; iv, MeI, NaH; v, BF$_3$, CH(OMe)$_3$

Scheme 1

[16] M. F. Mackay and B. J. Poppleton, *Cryst. Struct. Commun.*, 1980, **9**, 805.

2 Alkaloids of the Lythraceae

Reactions of the alkaloids lythrine (15; R = H) and lythridine (17) have been studied (Scheme 1).[17] Mercuric-acetate-induced stereoselective addition to the $\alpha\beta$-unsaturated lactone group of lythrine (15; R = H) results in introduction of a methoxy-group at the benzylic β-position to give a product (16) that is epimeric at C-13 with the hydroxylated alkaloids, *cf.* (17). Benzylic substituents undergo ionization and trapping with solvent to give more stable *epi*-products; *cf.* (17) → (16).

3 Sesquiterpenoid Alkaloids of *Nuphar* Species

A review of reactions of sulphur-containing *Nuphar* alkaloids has been published.[18]

An earlier synthesis of the piperidine alkaloid anhydronupharamine by Beckmann rearrangement of a cyclopentanone derivative has now been extended to

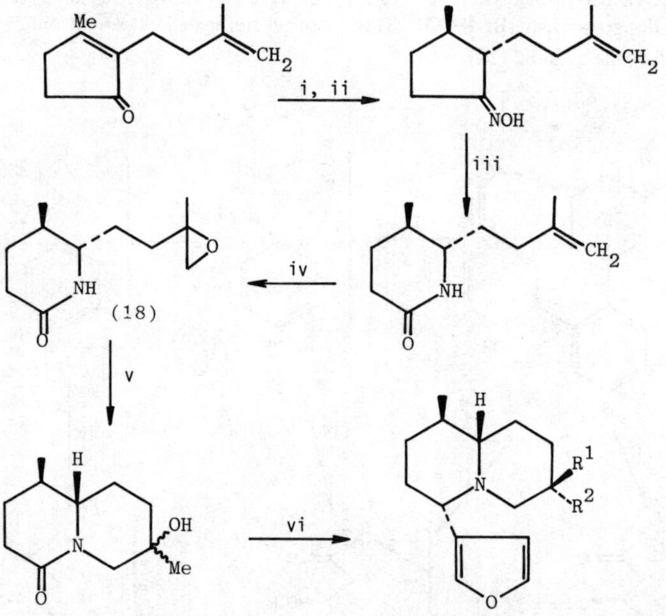

Nupharolutine (19) $R^1 = OH, R^2 = Me$

7-*epi*-Nupharolutine (20) $R^1 = Me, R^2 = OH$

Reagents: i, Li, liq. NH_3; ii, $H_2NOH \cdot HCl$, pyridine; iii, PCl_5, Et_2O, at −78 °C, then NaOH; iv, m-$ClC_6H_4CO_3H$, CH_2Cl_2; v, NaH, PhH, reflux; vi, Bu^{n}Li, 3-bromofuran, Et_2O, then $NaBH_4$, EtOH

Scheme 2

[17] I. Lantos, C. Razgaitis, B. Loev, and B. Douglas, *Can. J. Chem.*, 1980, **58**, 1851.
[18] J. T. Wrobel, H. Bielawska, A. Iwanow, and J. Ruszkowska, *Nat. Sulphur Compd.* [*Proc. Int. Meet.*] *3rd*, 1979, (publ. 1980), 353.

the synthesis of *Nuphar* quinolizidine alkaloids (Scheme 2).[19] The second ring was formed by base-catalysed cyclization of a piperidinone epoxide (18), and resulted in the synthesis of (±)-nupharolutine (19) and (±)-7-*epi*-nupharolutine (20) in a ratio of 1 : 9.

The reaction of endoperoxides of derivatives of 1,2-dihydropyridine with vinyl ethers, effected by stannous chloride (*cf.* Vol. 10, p. 35), has been used in another synthesis of (±)-nupharolutine (19) (Scheme 3).[20] The crystalline quinolizidine (21) was separated from its isomer that is epimeric at C-4.

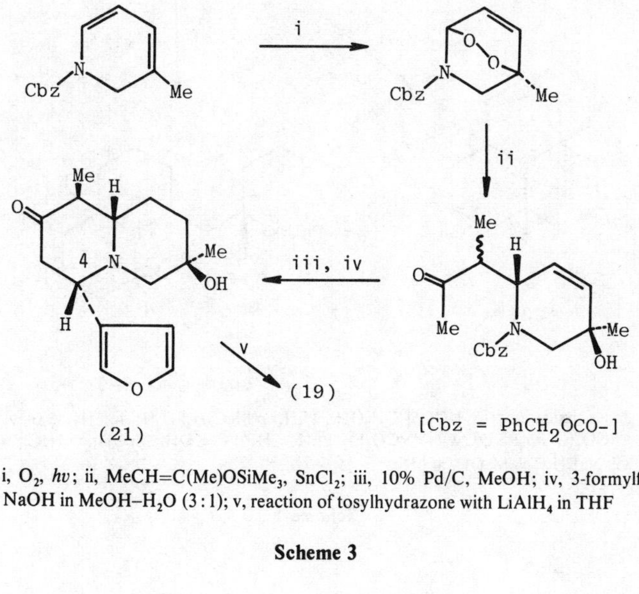

Reagents: i, O_2, *hν*; ii, MeCH=C(Me)OSiMe$_3$, SnCl$_2$; iii, 10% Pd/C, MeOH; iv, 3-formylfuran, 1% NaOH in MeOH–H$_2$O (3 : 1); v, reaction of tosylhydrazone with LiAlH$_4$ in THF

Scheme 3

A full account is now available of the synthesis of the alkaloids (±)-7-*epi*-deoxynupharidine and (±)-1-*epi*-7-*epi*-deoxynupharidine (*cf.* Vol. 8, p. 71); it has been extended to the synthesis of (±)-deoxynupharidine and 1-*epi*-deoxynupharidine.[21]

4 Porantherilidine and Porantheridine

Modified syntheses of porantherilidine (24) and porantheridine (23) (*cf.* Vol. 11, p. 68) from the common intermediate (22) have been described (Scheme 4).[22]

[19] R. T. LaLonde, N. Muhammad, C. F. Wong, and E. R. Sturiale, *J. Org. Chem.*, 1980, **45**, 3664.
[20] M. Natsume and M. Ogawa, *Heterocycles*, 1981, **15**, 237.
[21] S. Yasuda, M. Hanaoka, and Y. Arata, *Chem. Pharm. Bull.*, 1980, **28**, 831.
[22] E. Gössinger, *Monatsh. Chem.*, 1980, **111**, 783.

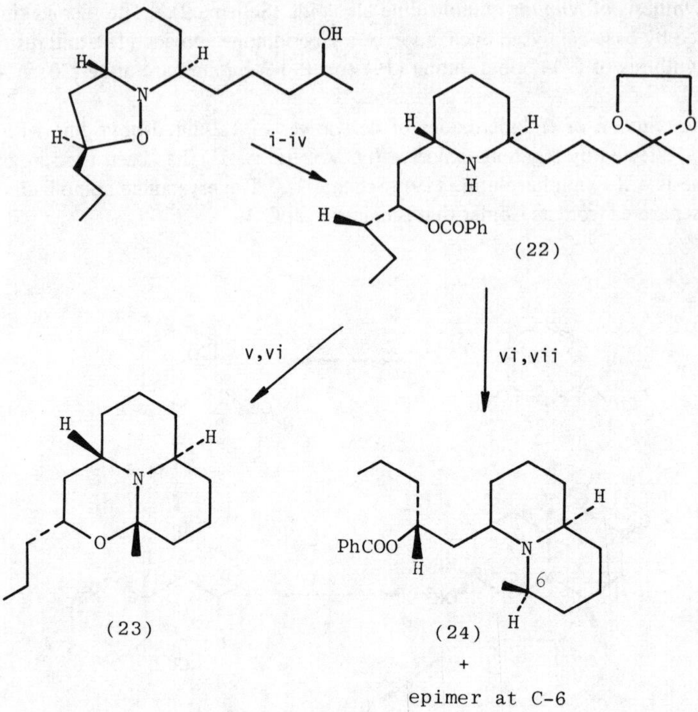

Reagents: Jones oxidation; ii, $HOCH_2CH_2OH$, PhH, p-$MeC_6H_4SO_3H$; iii, H_2, Raney nickel; iv, $PhCO_2H$, Ph_3P, $EtO_2CN=NCO_2Et$, PhH, THF; v. KOH, MeOH; vi, HCl, THF, H_2O; vii, $NaBH_3CN$, MeOH, at pH 6

Scheme 4

5 9b-Azaphenalene Alkaloids

Intense interest in the synthesis of the defensive substances of ladybird beetles continues (*cf.* Vol. 11, p. 68), and Mueller and Thompson have recently reported the extension of their perhydroboraphenalene methodology to an alternative synthesis of ($\pm$)-hippodamine (29) and its *N*-oxide (convergine) and to the first syntheses of ($\pm$)-hippocasine (26) and of ($\pm$)-hippocasine *N*-oxide (28),[23] and of propyleine (31)[24] (Scheme 5). The ketone (25) was the common intermediate; it gave hippodamine *via* the thioketal, and was converted into hippocasine by means of a Bamford–Stevens reaction on the tosylhydrazone. The mesylate (27) reacted with K_2CO_3 in DMSO (*E*1 conditions), apparently through the immonium cation (30), to give a mixture of olefins which was shown by spectroscopy to consist of (31)

[23] R. H. Mueller and M. E. Thompson, *Tetrahedron Lett.*, 1980, **21**, 1093.
[24] R. H. Mueller and M. E. Thompson, *Tetrahedron Lett.*, 1980, **21**, 1097.

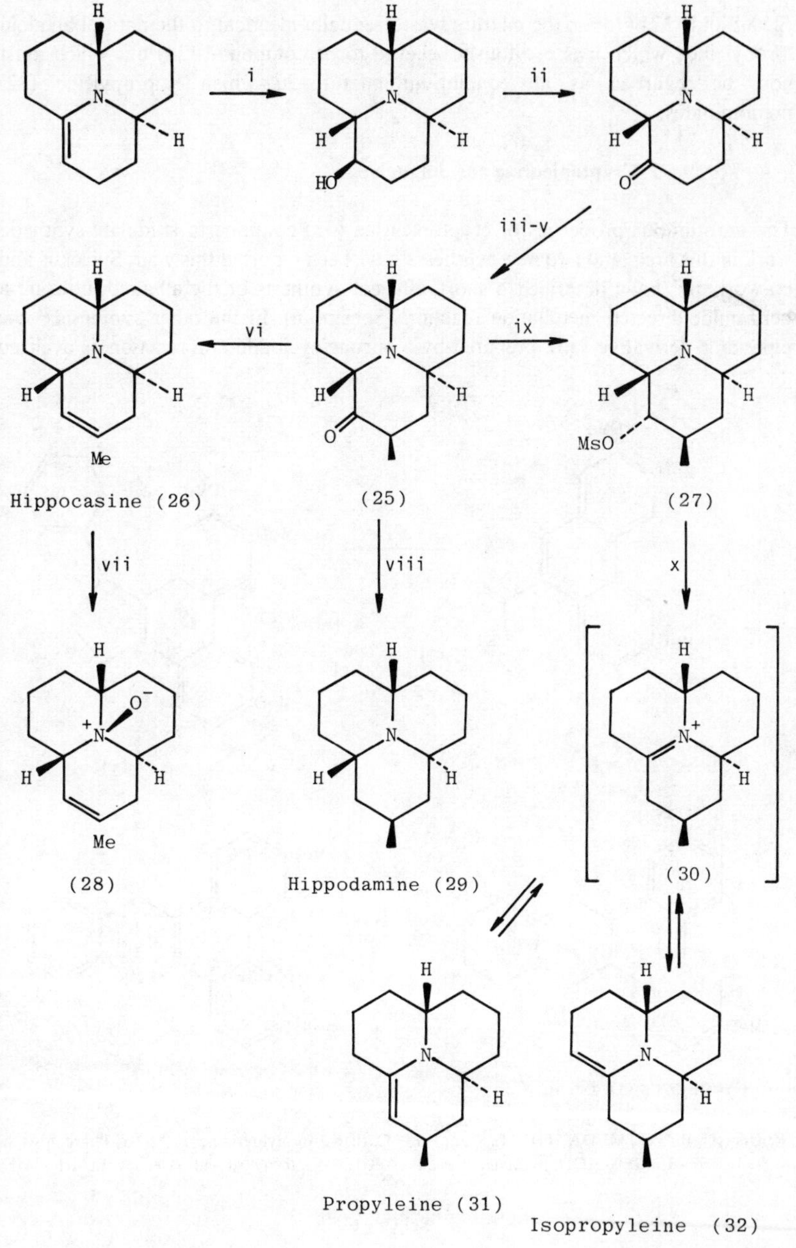
i
ii
iii-v
vi
ix
vii
viii
x
Hippocasine (26)
(25)
(27)
(28)
Hippodamine (29)
(30)
Propyleine (31)
Isopropyleine (32)

Reagents: i, $BH_3 \cdot DMS$, then H_2O_2, NaOH; ii, CrO_3, AcOH, H_2SO_4; iii, $MeOCH(NMe_2)_2$; iv, $Li \cdot 4NH_3$, Bu^tOH; v, H_2, Pd; vi, H_2NNH_2, then TsCl, then Bu^tNHLi; vii, H_2O_2, MeOH; viii, $(HSCH_2)_2$, BF_3, then Li, EDA; ix, $LiAlH_4$, then MsCl, Et_3N; x, K_2CO_3, DMSO, at 115 °C

Scheme 5

(25%) and (32) (75%); the mixture was essentially identical to the natural alkaloid 'propyleine', which was previously believed to be compound (31) but which must now be regarded as an equilibrium mixture in which isopropyleine (32) predominates.

6 Cryptopleurine and Julandine

The anti-tumour properties of cryptopleurine (33) continue to stimulate synthetic work in this area, and two new syntheses have been reported this year. Snieckus and co-workers[25] have described a short, efficient synthesis of the alkaloid, utilizing a benzamide-directed metallation reaction (Scheme 6). In the other synthesis,[26] the piperidine derivative (35), prepared by a nitrone cycloaddition reaction, is cyclized

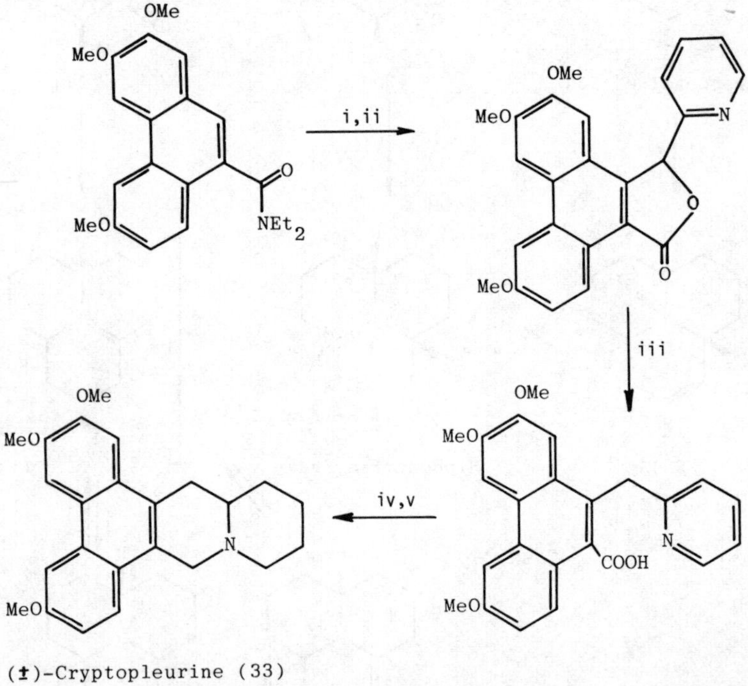

(±)-Cryptopleurine (33)

Reagents: i, Bu^{s}Li, TMEDA, THF, Et$_2$O, at −78 °C, then 2-formylpyridine; ii, TsOH, PhMe, heat; iii, Zn–Cu, 10% KOH, pyridine; iv, H$_2$, Pt, AcOH, HCl, then heat with xylene; v, LiAlH$_4$, THF

Scheme 6

[25] M. Iwao, M. Watanabe, S. O. de Silva, and V. Snieckus, *Tetrahedron Lett.*, 1981, **22**, 2349.
[26] H. Iida and C. Kibayashi, *Tetrahedron Lett.*, 1981, **22**, 1913.

to the quinolizidinone (34), which serves as an intermediate for the synthesis of (±)-julandine (36) and for (±)-cryptopleurine (Scheme 7).

Silicon(IV) chloride is an effective Lewis acid in the cyclization and subsequent dehydration steps of Herbert's synthesis of julandine and of cryptopleurine (*cf.* Vol. 10, p. 72).[27] An early cryptopleurine synthesis has been improved.[28]

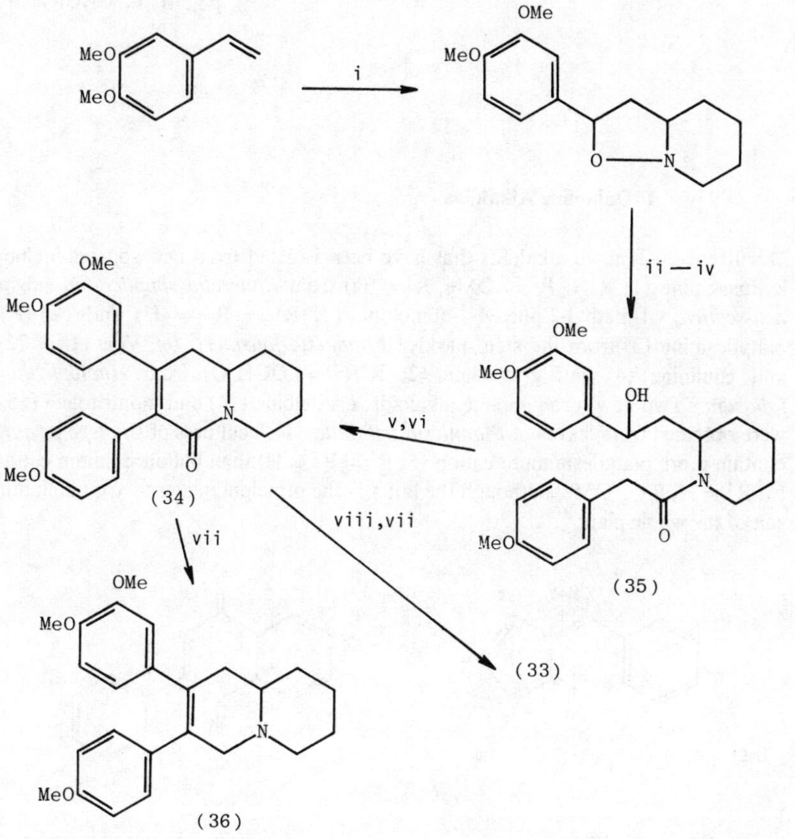

Reagents: i, 2,3,4,5-tetrahydropyridine 1-oxide, boiling PhMe; ii, Zn, aq. AcOH; iii, *p*-MeOC₆H₄COCH₂Cl, K₂CO₃, CH₂Cl₂, at 0 °C; iv, K₂CO₃, aq. MeOH, reflux; v, Collins oxidation, CH₂Cl₂, at r.t., for 2 h; vi, NaOEt, EtOH, reflux; vii, LiAlH₄, THF–Et₂O, reflux for 1 h; viii, *hv*, I₂, dioxan

Scheme 7

[27] J. E. Cragg, S. H. Hedges, and R. B. Herbert. *Tetrahedron Lett.*, 1981, **22**, 2127.
[28] G. G. Trigo, E. Gálvez, and M. M. Söllhuber. *J. Heterocycl. Chem.*, 1980, **17**, 69.

7

Quinoline, Quinazoline, and Acridone Alkaloids

BY M. F. GRUNDON

1 Quinoline Alkaloids

Occurrence.—Known alkaloids that have been isolated from new sources include kokusaginine (1; $R^1 = R^2 = OMe$, $R^3 = H$) from *Bauerella simplicifolia* subsp. *neo-scotica*,[1] 1-methyl-2-phenyl-4-quinolone (2; $R^1 = R^2 = H$) and (+)-(*R*)-platydesmine (3) from the stem bark of *Flindersia fournieri*[2] (*cf*. Vol. 11, p. 72), and edulinine (4) and graveoline (2; $R^1R^2 = OCH_2O$) from *Haplophyllum foliosum*.[3] Two new furoquinoline alkaloids, *i.e.* delbine (14) and montrifoline (15), were obtained from leaves of *Monnieria trifoliata*.[4] Cell cultures of *Choisya ternata* contain more platydesminium cation (5; $R^1 = R^2 = H$) than balfourodinium cation (5; $R^1 = H$, $R^2 = OMe$), although the latter is the principal quaternary quinolinium salt of the whole plant.[5]

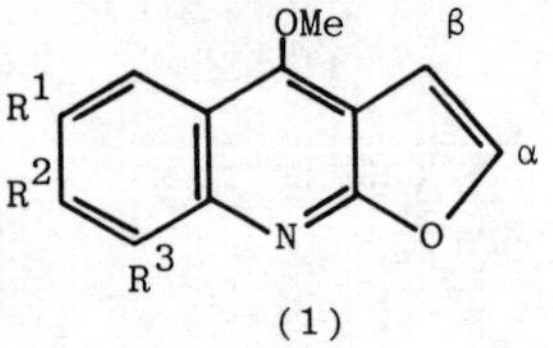

(1)

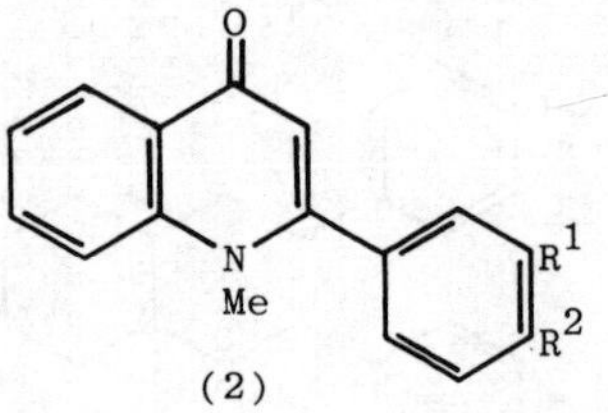

(2)

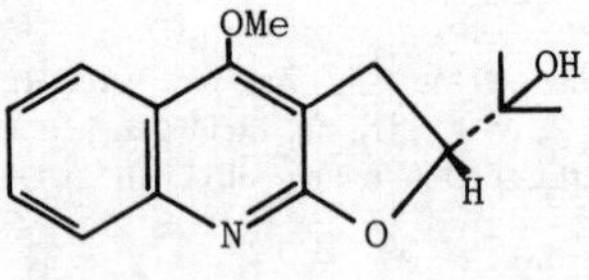

(3)

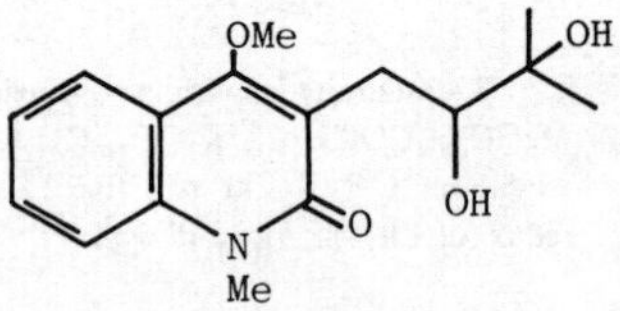

(4)

[1] F. Tillequin, G. Baudouin, M. Koch, and T. Sévenet, *J. Nat. Prod.*, 1980, **43**, 498.
[2] F. Tillequin, M. Koch, and T. Sévenet, *Planta Med.*, 1980, **39**, 383.
[3] V. I. Akmedzhanova, I. A. Bessonova, and S. Yu. Yunusov, *Khim. Prir. Soedin.*, 1980, 803 (*Chem. Abstr.*, 1981, **94**, 136 157).
[4] J. Battacharyya and L. M. Serur, *Heterocycles*, 1981, **16**, 371.
[5] M. Sejourne, C. Viel, J. Bruneton, M. Rideau, and J. C. Chenieux, *Phytochemistry*, 1981, **20**, 353.

Carbon-13 N.M.R. Spectroscopy.—A study of the ^{13}C n.m.r. spectra of twenty-five hemiterpenoid quinoline alkaloids and related prenylquinolines, including *C*-, *O*-, and *N*-prenyl-quinoline and -quinolone derivatives, hydroxyisopropyldihydro-furoquinolinones, hydroxydimethyldihydropyranoquinolinones, and furoquino-lines, has been carried out;[6] only isolated examples were reported previously.[7,8]

The ^{13}C chemical shifts of prenyl side-chains of quinoline alkaloids correspond, in general, to those of related aromatic hemiterpenes, but the following observations may be of value in resolving structural problems associated with hemiterpenoid quinolines. (*a*) In 2-quinolones, the presence of a hydroxyl group at C-4 produces a signal at 156.8–157.5 p.p.m. whereas substitution by a methoxyl group shifts the resonance of C-4 to a lower field (160.4–162.1 p.p.m.). (*b*) When the homocyclic ring of an *N*-methylquinolone derivative is unsubstituted at C-8, *e.g.* (6; R = H), (8;

(5)   (6)

R^1 = H), and (12; R^1 = R^2 = H), then the signal for the *N*-methyl group appears at *ca* 30 p.p.m., but, when a methoxyl group is present at C-8, *e.g.* (8; R^1 = OMe) and (12; R^1 = OMe, R^2 = H), then the signal for the *N*-methyl group occurs at lower field (*ca* 36 p.p.m.). (*c*) Comparison of the ^{13}C n.m.r. spectra of hydroxy-isopropyldihydrofuro-4-quinolinone alkaloids with 3-hydroxy-2,2-dimethyl-dihydropyrano-4-quinolinone analogues reveals a large difference in chemical shifts between corresponding carbons, *e.g.* + 23.8 p.p.m. between C-11 of the compounds (10; R^1 = R^2 = H) and (6; R = H) and −12.0 p.p.m. between C-12 in the same pair; this appears to be one of the most unequivocal spectroscopic methods of characterizing such furo- and pyrano-isomers. (*d*) In the tricyclic series of alkaloids, an important difference between the angular compounds (2-quinolones), *e.g.* (12) and (13), and linear compounds, *e.g.* (6) and (10), is that the carbon of the carbonyl group (C-2) in the former group has a chemical shift that is higher by *ca* 10 p.p.m. compared to that in the 4-quinolones; this criterion supplements existing spectroscopic methods. (*e*) The ^{13}C n.m.r. spectra of skimmianine (1; R^1 = H, R^2 = R^3 = OMe)[8] and other furoquinolines[6] show that the high-intensity signals for furanoid α-carbon (142.6–143.6 p.p.m.) and for furanoid β-carbons (104.6–105.3 p.p.m.) are characteristic of furoquinoline alkaloids and occur at higher fields than in compounds containing furan rings that are attached directly to benzene rings.

[6] N. M. D. Brown, M. F. Grundon, D. M. Harrison, and S. A. Surgenor, *Tetrahedron*, 1980, **36**, 3579.
[7] F. R. Stermitz and I. A. Sharifi, *Phytochemistry*, 1977, **16**, 2003.
[8] A. Ahond, F. Picot, P. Potier, C. Poupat, and T. Sévenet, *Phytochemistry*, 1978, **17**, 166.

Hemiterpenoid Tricyclic Alkaloids.—A full account has been published of the synthesis of ptelefolone (9) from the *N*-methyl-4-quinolone (8; $R^1 = R^2 = OMe$, $R^3 = H$)[9] (*cf.* Vol. 9, p. 84). ($\pm$)-*O*-Methylribaline (10; $R^1 = H$, $R^2 = OMe$) was synthesized for the first time; since the latter compound has been correlated with ribaline (10; $R^1 = H$, $R^2 = OH$), ribalinidine (6; R = OH), and ribalinium cation (5; $R^1 = OH$, $R^2 = H$), the preparation confirms the structures of these three alkaloids of *Balfourodendron riedelianum*. The same paper includes a discussion of the ring-closure of epoxides of 3-prenylquinolones. Under non-basic conditions, 4-methoxy-3-prenyl-2-quinolones give a mixture of dihydrofuro- and dihydro-pyrano-quinolines on treatment with peracids, apparently by 5-*exo*- or 6-*endo*-ring-closure of an intermediate epoxide (Scheme 1). *N*-methyl-2-quinolones (8) behave differently with peracids, and give *N*-methyldihydrofuroquinolinones only; the reluctance to form pyrano-derivatives may be due to non-bonded interactions between the *N*-methyl group and the terminal $=CMe_2$ group in the transition state. Studies of rearrangements of furoquinolinones (10) (*cf.* Vol. 10, p. 79) show that base-catalysed ring-closure of epoxides (11) leads to furo- and pyrano-quinolinones with angular annelation (Scheme 2); in this case there is clearly no stereochemical interaction with the *N*-methyl group, inhibiting the formation of pyrano-compounds.

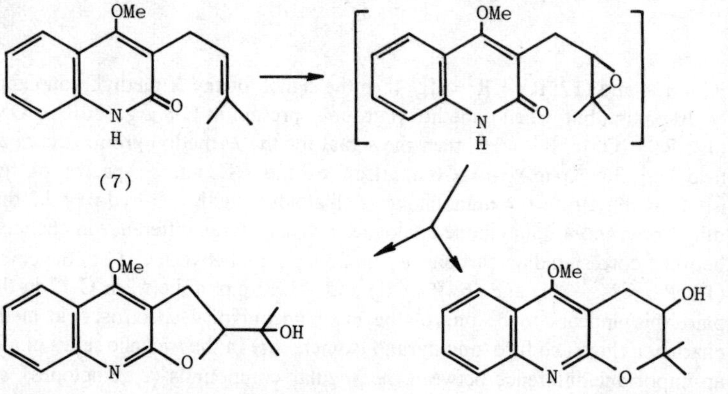

Scheme 1

Furoquinoline Alkaloids.—Delbine and montrifoline, the new alkaloids of *Monnieria trifolia*, were shown by spectroscopic studies to be 4-methoxy-furoquinolines. Delbine is a hydroxy-methoxy-dictamnine, and, as it gives kokusaginine (1; $R^1 = R^2 = OMe$, $R^3 = H$) on reaction with diazomethane, the substituents are in positions 6 and 7; since it is not identical with helipavifoline (1; $R^1 = OMe$, $R^2 = OH$, $R^3 = H$) (*cf.* Vol. 7, p. 83), delbine apparently is 6-hydroxy-7-methoxydictamnine (14). Montrifoline was assigned structure (15) on

[9] J. L. Gaston and M. F. Grundon, *J. Chem. Soc., Perkin Trans. 1*, 1980, 2294.

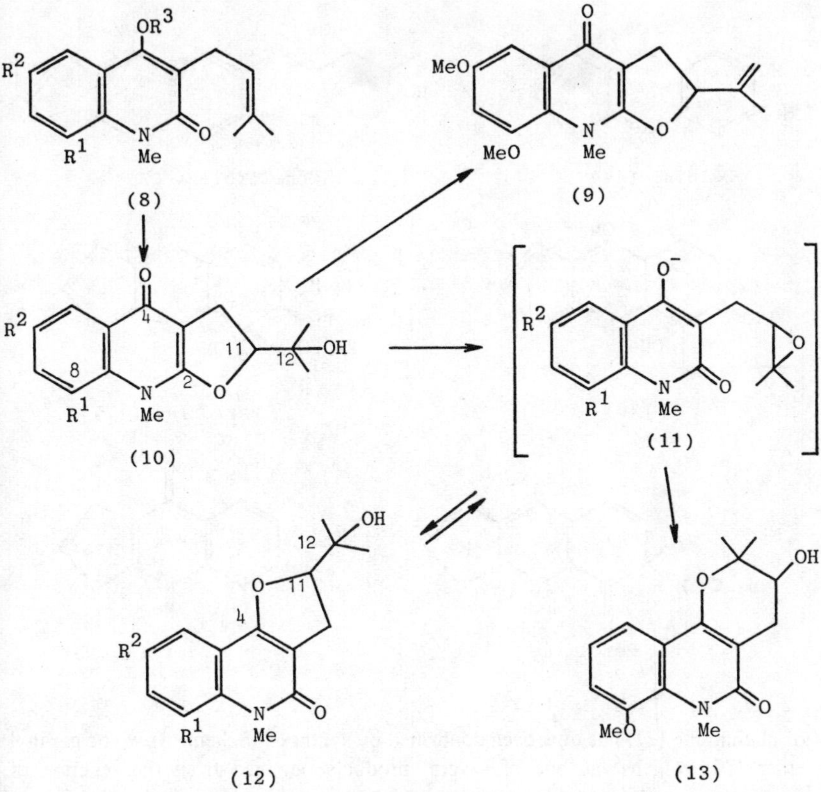

Scheme 2

the basis of its spectroscopic properties, of its non-identity with evolatine [1; $R^1 =$ OMe, $R^2 = OCH_2CH(OH)C(OH)Me_2$, $R^3 = H$], and of its conversion (by alkaline fusion) into delbine (14).[4]

X-Ray analysis of anhydroperforine (16) appears to establish the stereochemistry of the *Haplophyllum* alkaloid perforine (17).[10] A related alkaloid, perfamine (18), reacts with methyl iodide at 250 °C to give an isodictamnine derivative (19).[11]

Monoterpenoid Quinoline Alkaloids.—The first examples of monoterpenoid quinoline alkaloids were isolated from *Haplophyllum bucharicum* some time ago, but the structures proposed on the basis of chemical and spectroscopic studies are not yet secure (*cf.* Vol. 1, p. 97; Vol. 2, p. 88; and Vol. 4, p. 119). The constitution

[10] Z. Karimov, I. A. Bessonova, M. R. Yagudaev, and S. Yu. Yunusov, *Khim. Prir. Soedin.*, 1979, 805 (*Chem. Abstr.*, 1980, **93**, 150 428).

[11] D. M. Razakova, I. A. Bessonova, and S. Yu. Yunusov, *Khim. Prir. Soedin.*, 1979, 738 (*Chem. Abstr.*, 1981, **94**, 175 337).

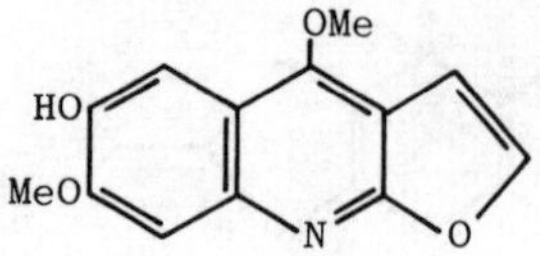

Delbine (14)

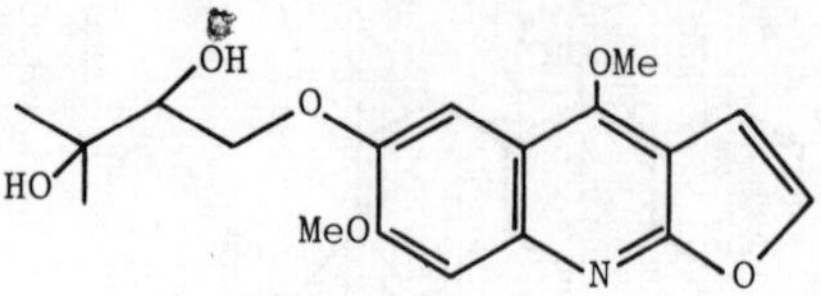

Montrifoline (15)

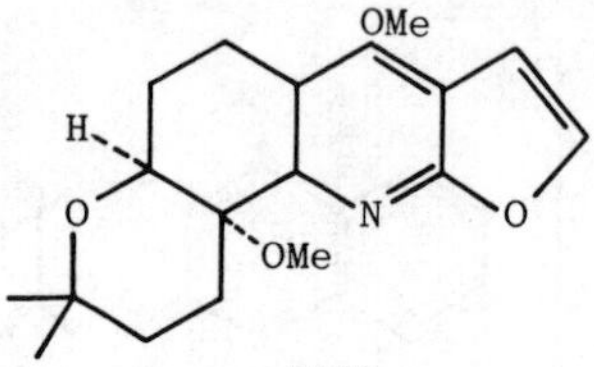

(16)

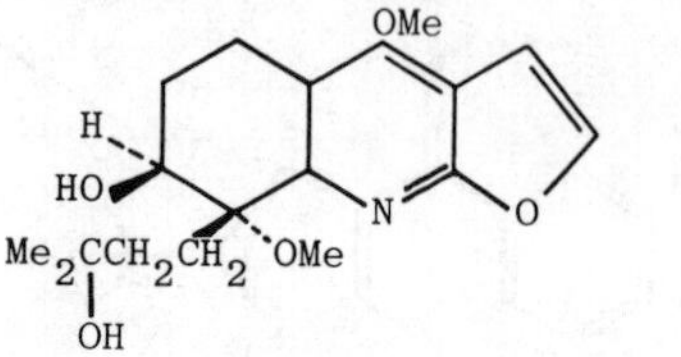

Perforine (17)

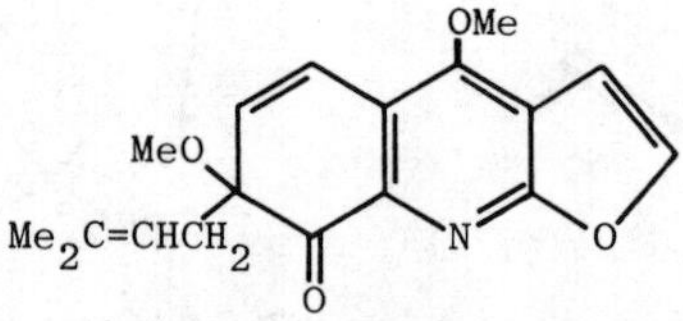

Perfamine (18)

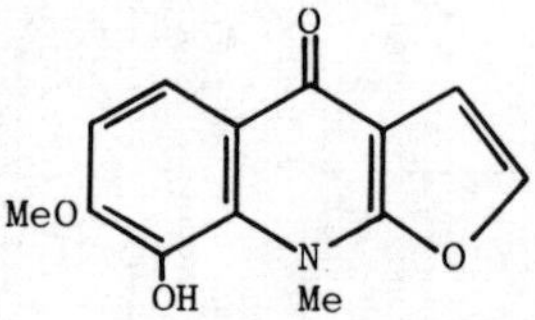

(19)

of bucharaine (21) has now been confirmed by synthesis (Scheme 3).[12] The geranyl ether (20), which was one of several products obtained from the reaction of 4-hydroxy-2-quinolone with geranyl chloride, was converted into bucharaine (21) in 60% yield by successive mono-epoxidation, reaction with formic acid, and hydrolysis of the formate ester; bucharaine was obtained less efficiently by regiospecific hydroxylation of the geranyl ether (20) with osmium tetroxide in pyridine–ether. Although it contains a chiral centre, bucharaine was apparently isolated from *H. bucharicum* as a racemate.

2 Quinazoline Alkaloids

Carbon-13 N.M.R. Spectroscopy.—The [13]C n.m.r. spectra of ten 4-quinazolinones have been reported, and are useful for determining their predominant tautomeric forms and for resolving controversial structural problems.[13] Thus, the resonance of C-4 in compound (22), containing a double-bond between C-2 and N-3, appears at *ca* 7 p.p.m. to lower field (167.7–168.9 p.p.m.) than in compound (23), in which there is a double-bond between N-1 and C-2. There is also a significant difference between resonances of C-8a in the two groups of compounds, those in compounds

[12] M. F. Grundon, V. N. Ramachandran, and M. E. Donnelly, *J. Chem. Soc., Perkin Trans 1*, 1981, 633.
[13] J. Bhattacharyya and S. C. Pakrashi, *Heterocycles*, 1980, **14**, 1469.

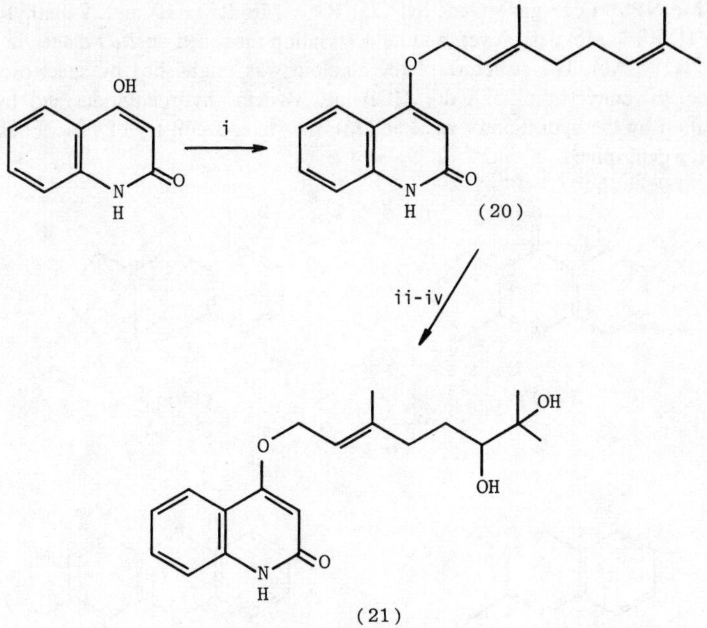

Reagents: i, $Me_2C=CH(CH_2)_2C(Me)=CHCH_2Cl$, K_2CO_3, Me_2CO, reflux; ii, $m\text{-}ClC_6H_4CO_3H$, $CHCl_3$; iii, HCO_2H, at 20 °C; iv, 5% aq. KOH, MeOH

Scheme 3

(23) appearing at a lower field by 7—9 p.p.m. On this basis, glycosminine exists predominantly in the tautomeric form (23; $R^1 = H$, $R^2 = CH_2Ph$) (*cf.* Vol. 10, p. 80).

Occurrence and Structural Studies.—Roots of *Adhotida vasica* were shown to contain deoxyvasicinone (24; R = H),[14] vasicinone (24; R = OH),[15] and a new, optically active alkaloid, vasicol (25; $R^1 = R^2 = H$),[15] in addition to the quinaz-olines that had previously been isolated from this species. The reaction of vasicol

---

[14] M. P. Jain, S. K. Koul, K. L. Dhar, and C. K. Atal, *Phytochemistry*, 1980, **19**, 1880.
[15] K. L. Dhar, M. P. Jain, S. K. Koul, and C. K. Atal, *Phytochemistry*, 1981, **20**, 319.

with Me_3NPh^+ OH^- gave *O*-methyl (25; R^1 = Me, R^2 = H) and *N*-methyl (25; R^1 = H, R^2 = Me) derivatives, and its acetylation furnished an *N,O*-diacetate (25; $R^1 = R^2$ = Ac). The structure of the alkaloid was established by spectroscopic studies, by conversion (with dry HCl) into vasicine hydrochloride, and by its formation by the hydration of vasicine (26); the stereochemistry of vasicol has not yet been determined.

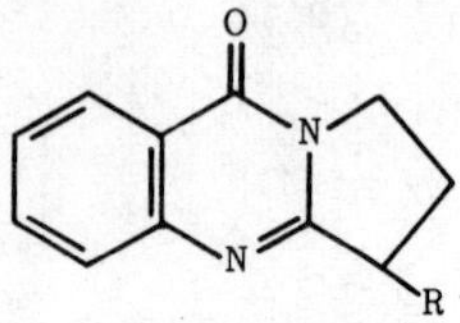

(24)

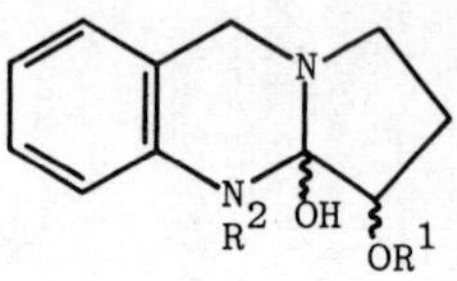

(25)

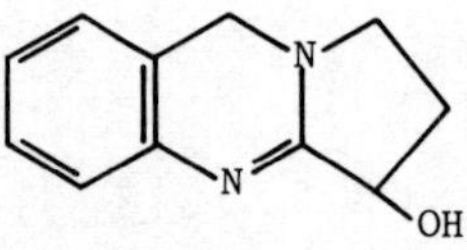

(26)

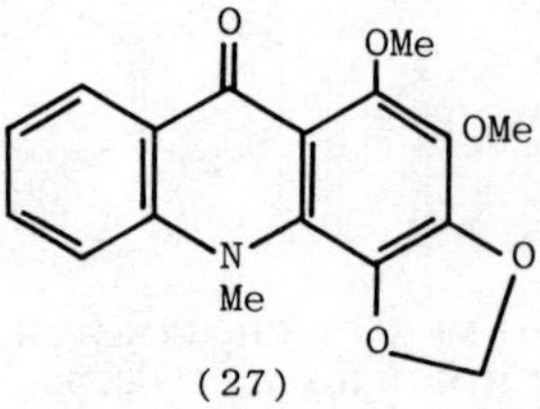

(27)

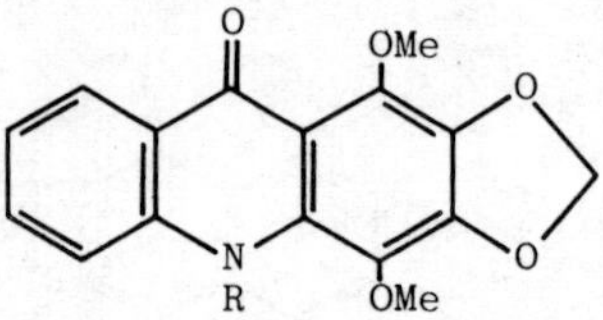

(28)

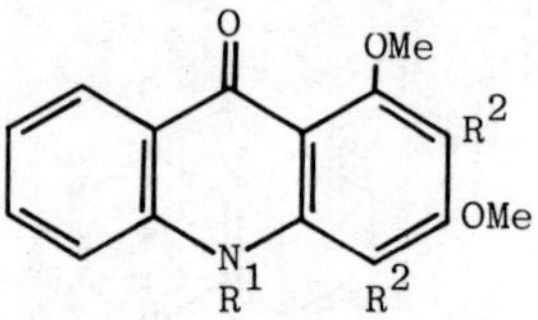

(29)

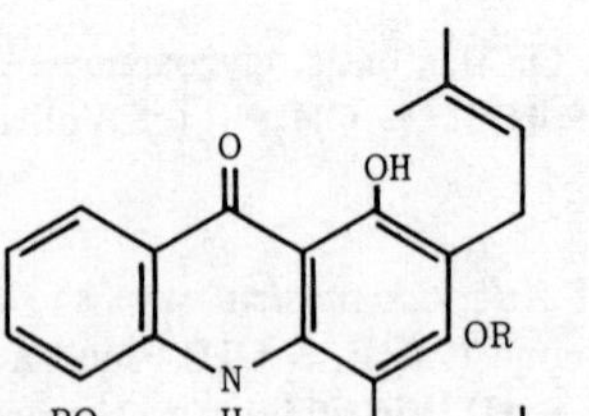

(30)

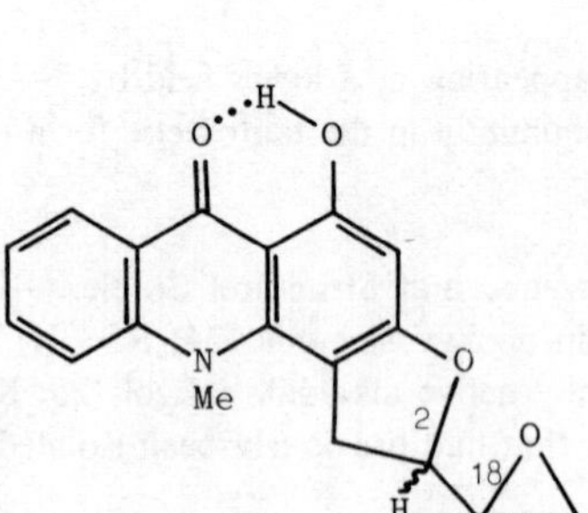

Rutacridone epoxide (31)

3 Acridone Alkaloids

Occurence and New Alkaloids.—Six acridone alkaloids have been isolated from the leaves of *Bauerella simplicifolia* subsp. *neo-scotica*.[1] Four of the alkaloids, *i.e.* melicopine (27), melicopidine (28; R = Me), 1,3-dimethoxy-*N*-methylacridone (29; R^1 = Me, R^2 = H), and xanthovodine (28; R = H), have been obtained from other sources, but the alkaloids (29; R^1 = H, R^2 = OMe) and (29; R^1 = R^2 = H) were known previously only as synthetic compounds. The new alkaloids were identified by spectroscopy, by chemical correlation, and by synthesis.

Another new diprenylacridone, atalaphylline 3,5-dimethyl ether (30; R = Me), has been obtained from *Atlantia monophylla* (*cf.* Vol. 2, p. 95; Vol. 6, p. 108; Vol. 7, p. 91; Vol. 8, p. 85). The structure of the alkaloid was determined by spectroscopy and by its formation by the reaction of atalaphylline (30; R = H) with diazomethane.[16]

An antimicrobial alkaloid that was isolated from roots of *Ruta graveolens* and from callus tissue cultures proved to be rutacridone epoxide (31).[17] The structure was determined by ^{1}H and ^{13}C n.m.r. and by mass spectroscopy, although the configurations at C-2 and C-18 are not known. The epoxide, rather than rutacridone (37), is a major root alkaloid; clearly, the plant is chemically different from that studied previously, but the reason for the variation is unknown.

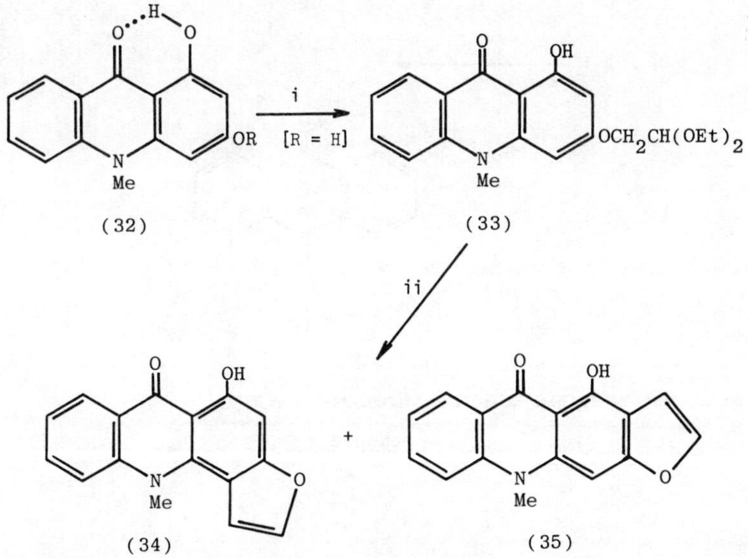

Reagents: i, BrCH$_2$CH(OEt)$_2$, NaH, DMF, at 120 °C; ii, 0.5N-H$_2$SO$_4$, dioxan, reflux

Scheme 4

[16] G. H. Kulkarni and B. K. Sabata, *Phytochemistry*, 1981, **20**, 867.
[17] A. Nahrstedt, V. Eilert, B. Wolters, and V. Wray, *Z. Naturforsch., Teil. C*, 1981, **36**, 200.

Synthesis.—Reisch, Mester, and co-workers have made important contributions this year by synthesizing the alkaloids furacridone (34) and ($\pm$)-rutacridone (37) for the first time. Regioselective etherification of 1,3-dihydroxy-N-methylacridone (32; R = H) gave the acetal (33), which furnished furacridone (34) as the major product of acid-catalysed cyclization (Scheme 4). Claisen rearrangements of the 3-allyloxy-acridone (32; R = CH$_2$CH=CH$_2$) and the propargyl derivative (32; R = CH$_2$C≡CH) were also studied.[18]

A one-step synthesis of ($\pm$)-rutacridone (37) was achieved by the reaction of acridone (32; R = H) with isoprene dibromide; isorutacridone (36), the isomer with linear annelation, was a minor product of the reaction (Scheme 5).[19] A full account of the synthesis of acronycine from o-aminobenzophenones has now appeared (*cf.* Vol. 7, p. 90) and a modified route has been described (Scheme 6).[20] Cyclization of the benzophenone derivative (38; R = Me) gave acronycine (39) (38%) and isoacronycine (40) (38%); a similar ring-closure of compound (38; R = H) furnished des-N-methylacronycine.

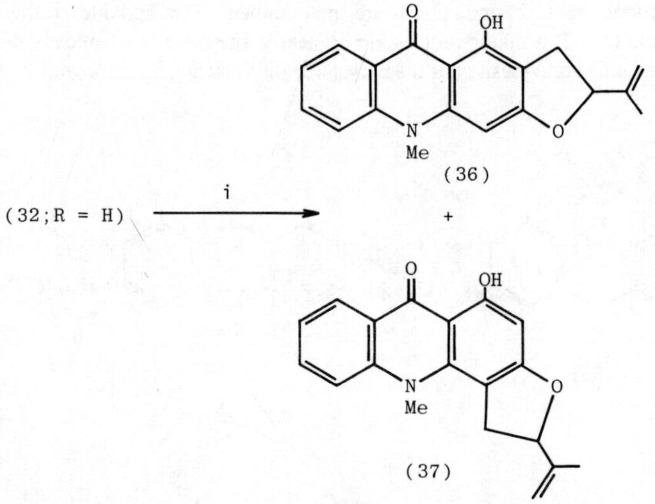

$(32; R = H)$ $\xrightarrow{\quad i \quad}$

Reagents: i, *trans*-BrCH$_2$C(Me)=CHCH$_2$Br, MeONa, MeOH, at 20 °C

Scheme 5

[18] J. Reisch, I. Mester, S. K. Kapoor, Z. Rózsa, and K. Szendrei, *Liebigs Ann. Chem.*, 1981, 85.
[19] I. Mester, J. Reisch, Z. Rózsa, and K. Szendrei, *Heterocycles*, 1981, **16**, 77.
[20] J. H. Adams, P. M. Brown, P. Gupta, M. S. Khan, and J. R. Lewis, *Tetrahedron*, 1981, **37**, 209.

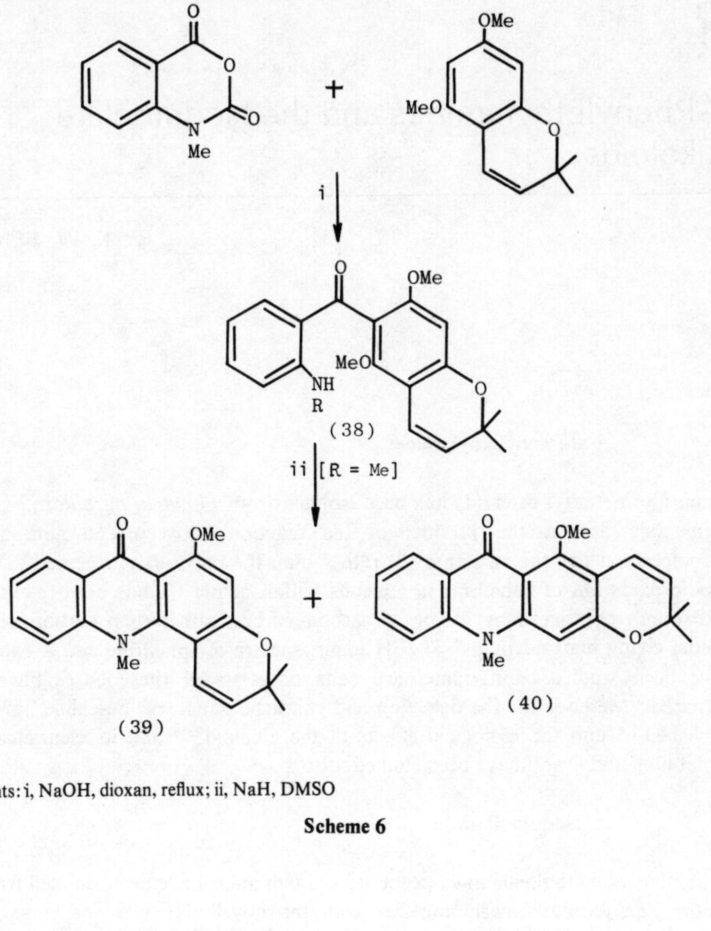

Reagents: i, NaOH, dioxan, reflux; ii, NaH, DMSO

Scheme 6

8

β-Phenylethylamines and the Isoquinoline Alkaloids

BY K. W. BENTLEY

1 β-Phenylethylamines

Fumariflorine ethyl ester (1) has been isolated from *Fumaria parviflora*.[1,2] It has been claimed that the product of the reaction between cotarnine and 6-nitropiperonal has the structure (2) rather than the previously accepted (3).[3] The anodic oxidation of ephedrine in aqueous buffer, at pH 10, has been re-examined and found to proceed by fission of carbon–carbon rather than carbon–nitrogen bonds, giving benzaldehyde.[4] The ^{1}H n.m.r. spectra of ephedrine and ψ-ephedrine have been studied[5] and quaternary salts of esters of these bases have been prepared.[6] Methods for the detection and characterization of mescaline have been published,[7,8] and the biological effects of the alkaloid[9,10] and its clearance from rabbit lung and liver[11] have been studied.

2 Isoquinolines

Heliamine, *N*-methylheliamine, pellotine, and tehuanine have been isolated from the cactus *Pachycereus weberi* together with the novel alkaloids nortehuanine (4; $R^1 = R^2 = OMe$, $R^3 = H$), lemaireocereine (4; $R^1 = R^2 = H$, $R^3 = OMe$), weberine (4; $R^1 = R^2 = R^3 = OMe$), and weberidine (4; $R^1 = R^2 = R^3 = H$), the structures of which, with the exception of weberine, were confirmed by synthesis.[12] Noroxyhydrastinine and 6,7-dimethoxy-2-methylisoquinolone have been isolated from

[1] S. F. Hussain and M. Shamma, *Tetrahedron Lett.*, 1980, **21**, 1693.
[2] S. F. Hussain, R. D. Minard, A. J. Freyer, and M. Shamma, *J. Nat. Prod.*, 1981, **44**, 169.
[3] H. Möhrle, W. Jäkel, and D. Wendisch, *Arch. Pharm.* (*Weinheim, Ger.*), 1980, **313**, 715.
[4] M. Masui, Y. Kamada, and S. Ozaki, *Chem. Pharm. Bull.*, 1980, **28**, 1619.
[5] A. Nakibova, A. A. Pashchenko, and M. Karimov, *Uzb. Khim. Zh.*, 1980, No. 2, p. 33.
[6] Yu. R. Khakimov, A. A. Abduvakhabov, A. A. Sadykov, and Kh. A. Aslanov, *Dokl. Akad. Nauk Uzb. SSR*, 1979, No. 1, p. 39.
[7] G. Guebitz and R. Wintersteiger, *J. Anal. Toxicol.*, 1980, **4**, 141.
[8] W. Gielsdorf, K. Schubert, and K. Allin, *Arch. Kriminol.*, 1980, **166**, 21.
[9] A. S. Moorthy and J. Mitra, *Nucleus* (*Calcutta*), 1979, **22**, 28.
[10] D. A. Gorelick and W. H. Bridger, *Biol. Psychiatry*, 1980, **15**, 619.
[11] K. S. Hilliker and R. A. Roth, *Biochem. Pharmacol.*, 1980, **29**, 253.
[12] R. Mata and J. L. McLaughlin, *Phytochemistry*, 1980, **19**, 673.

(1)

(2)

(3)

(4)

Thalictrum alpinum[13] and thalifoline from *Cryptocarya longifolia*.[14] Thalifoline has been synthesized from β-(3,4-dimethoxyphenyl)ethyl isocyanate by cyclization to the lactam with methyl fluorosulphonate, followed by regioselective cleavage of the 7-methoxy-group by methionine and methanesulphonic acid; related syntheses of corypalline and cherylline have been accomplished.[15] The ¹H n.m.r. spectra[5] and methods for the detection[16] of salsoline and salsolinol have been studied, as have the half-life of salsolinol in brain,[17, 18] its selective *O*-methylation in brain,[19, 20] and its effect on the consumption of alcohol.[21]

Macrostomine (8) has been synthesized from 4-hydroxy-6,7-dimethoxy-2-nitrosotetrahydroisoquinoline by a deprotonation reaction with 3,4-methylene-dioxybenzyl bromide and denitrosation to (5; R = H, X = H₂), *N*-benzoylation and then oxidation to (5; R = COPh, X = O), and subsequent reactions as shown in Scheme 1.[22]

¹³ W.-N. Wu, J. L. Beal, and R. W. Doskotch, *J. Nat. Prod.*, 1980, **43**, 372.
¹⁴ I. R. C. Bick, T. Sévenet, W. Sinchai, B. W. Skelton, and A. H. White, *Aust. J. Chem.*, 1981, **34**, 195.
¹⁵ H. Irie, A. Shiina, T. Fushimi, J. Katakawa, N. Fujii, and H. Yajima, *Chem. Lett.*, 1980, 875.
¹⁶ B. Sjöquist and E. Magnuson, *J. Chromatogr.*, 1980, **183**, 17.
¹⁷ C. L. Melchior, A. Mueller, and R. A. Dietrich, *Biochem. Pharmacol.*, 1980, **29**, 657.
¹⁸ C. L. Melchior and R. A. Dietrich, *Adv. Exp. Med. Biol.*, 1980, **126**, 121.
¹⁹ M. Bail, S. Miller, and G. Cohen, *Life Sci.*, 1980, **26**, 2051.
²⁰ T. C. Origitano and M. A. Collins, *Life Sci.*, 1980, **26**, 2061.
²¹ Z. W. Brown, Z. Amit, and B. Smith, *Adv. Exp. Med. Biol.*, 1980, **126**, 103.
²² W. Wykypiel and D. Seebach, *Tetrahedron Lett.*, 1980, **21**, 1927.

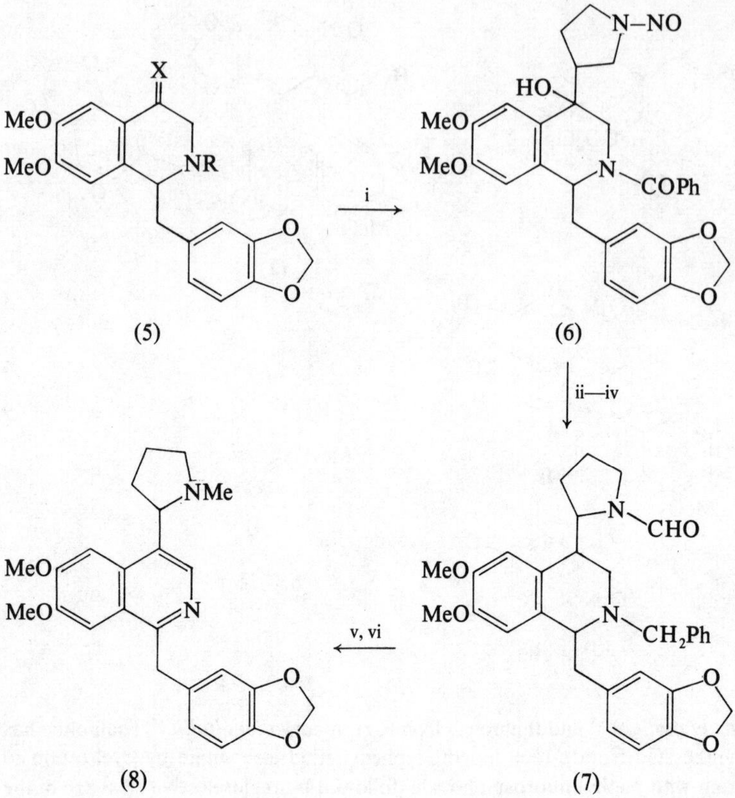

Reagents: i, [pyrrolidine-N-NO]-Li, THF, at −78 °C, acting on (5; R = COPh, X = O); ii, Raney nickel, H₂, MeOH

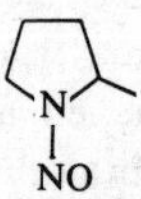

iii, LiAlH₄, THF; iv, MeCO₂CHO, THF, at r.t.; v, Pd/C, decalin, at 180 °C; vi, LiAlH₄.

Scheme 1

The α-naphthylisoquinoline alkaloids triphyophylline, isotriphyophylline, *N*-methyltriphyophylline, 8-*O*-methyltetradehydrotriphyophylline, and triphyopeltine [assigned the structure (9; R = H) (*cf.* Vol. 7, p. 107)] and its 5'-*O*-methyl ether (9; R = Me) have been isolated from *Dioncophyllum thollonii*.[23]

The tropoloisoquinoline grandirubrine[24] was described previously (*cf.* Vol. 11, p. 130).

[23] M. Lavault and J. Bruneton, *Planta Med.*, 1980 (Suppl.), p. 17.
[24] M. Menachery and M. P. Cava, *Heterocycles*, 1980, **14**, 943.

(9)

(10)

(11)

(12)

3 Benzylisoquinolines

Reticuline has been isolated from *Machilus duthei*[25] and, together with coclaurine and *N*-methylcoclaurine, from *Cryptocarya longifolia*.[14] The following new bases have been reported: norjuziphine (10; $R^1 = R^2 = H$) from *Fumaria vaillantii*,[26] longifolidine (10; $R^1 = R^2 = Me$) from *C. longifolia*,[14] higenamine (11) from *Annona squamosa*,[27] and longifolonine (12) from *C. longifolia*.[14] The structure of higenamine was confirmed by synthesis[27] and that of longifolonine by X-ray methods.[14] Glycomarine, a new alkaloid from *Papaver arenarium*, has been identified as the *β*-D-glucoside of sevanine.[28]

Prolonged heating of papaverine with methylamine at 110 °C affords 1,6-di(methylamino)-7-methoxy-2-(3,4-dimethoxyphenyl)naphthalene.[29] *N*-Methyl-papaverinium salts, on treatment with alkali, give the related pseudo-bases, whereas their 2′-hydroxymethylated analogues give the corresponding cyclic carbinolamine ethers, which are related to bases of the hypecorine group[30] (*cf.* Vol. 10, p. 105). Both *cis-* and *trans-N*-oxides of laudanosine have been prepared. Thermal decomposition of the *cis*-oxide affords a mixture of the product (13) from Cope degradation and the rearranged compounds (14), (15), and (16) whereas the *trans*-oxide gives almost wholly (13) under similar conditions.[31]

[25] S. F. Hussain, A. Amin, and M. Shamma, *J. Chem. Soc. Pak.*, 1980, **2**, 157.

[26] M. Alimova, I. A. Israilov, and S. Yu. Yunusov, *Khim, Prir. Soedin.*, 1979, 870.

[27] M. Leboeuf, A. Cavé, A. Touche, J. Provost, and P. Forgacs, *J. Nat. Prod.*, 1981, **44**, 53.

[28] I. A. Israilov, M. M. Manushakyan, V. A. Mnatsakayan, M. S. Yunusov, and S. Yu. Yunusov, *Khim. Prir. Soedin.*, 1980, 852.

[29] A. N. Kost, L. G. Yudin, R. Sagitullin, and V. I. Terenin, *Khim. Geterotsikl. Soedin.*, 1979, 1564.

[30] D. Walterova, V. Preininger, L. Doljes, F. Grambal, M. Kysely, I. Valka, and V. Simanek, *Collect. Czech. Chem. Commun.*, 1980, **45**, 956.

[31] J. B. Bremner and Le Van Thuc, *Aust. J. Chem.*, 1980, **33**, 379.

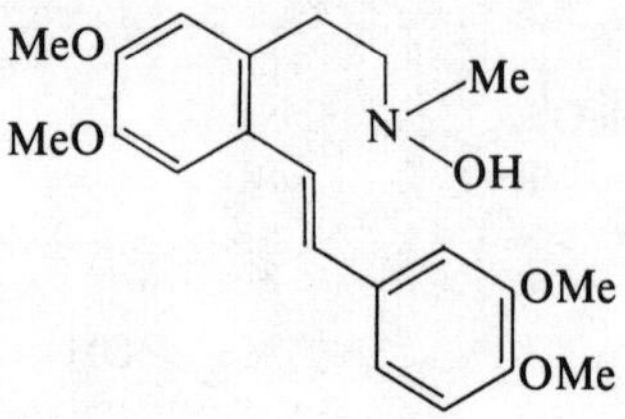

(13)

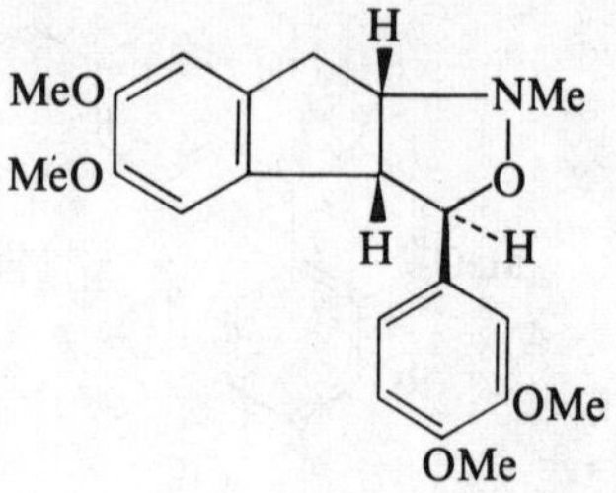

(14)

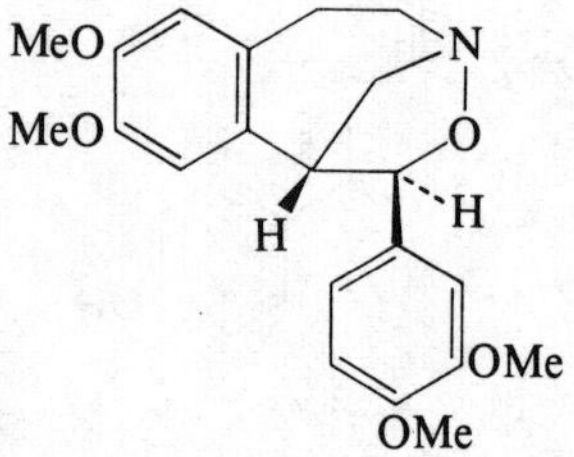

(15)

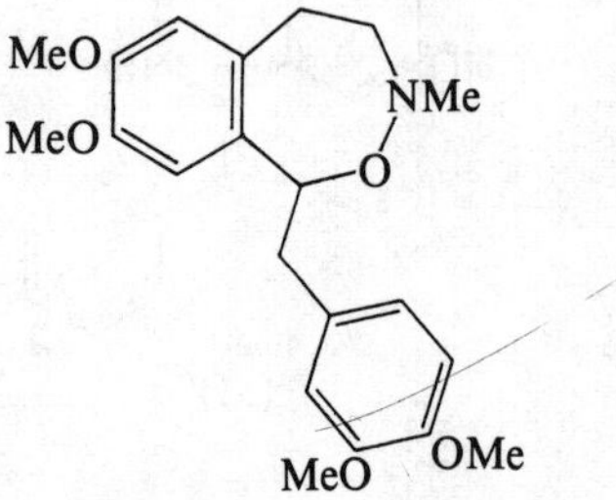

(16)

Chlorination and bromination of N-formyl- and N-ethoxycarbonyl-nor-reticuline have been shown to give 8,6'-dihalogenated compounds[32] and anodic oxidation of the bridged hexamethylene diether of reticuline has been found to give the proerythradienone (17) and the morphinandienone (18).[33]

Papaverine has been synthesized from the oxazole (19), obtained from the 2-lithio-derivative of 5-(3,4-dimethoxyphenyl)oxazole with veratric aldehyde, by catalytic reduction to the acyl-β-phenylethylamine followed by Bischler–

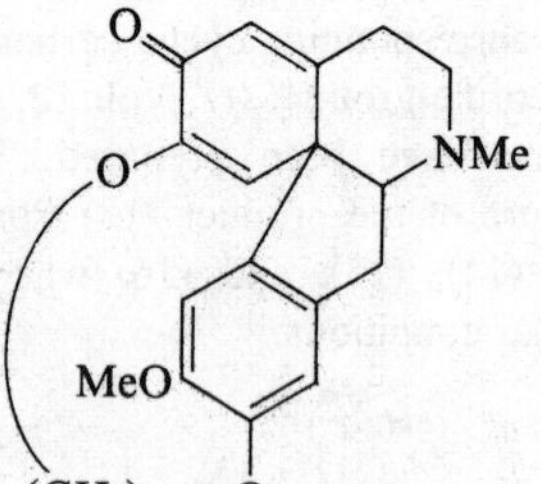

(17)

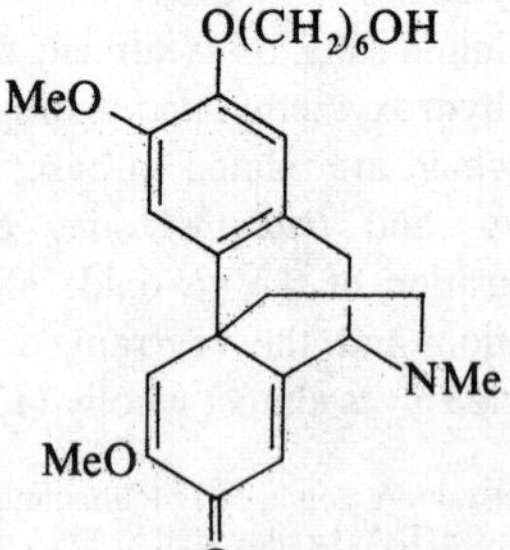

(18)

[32] C. Szantay, G. Blasko, M. Barczai-Beke, G. Dornyei, and L. Radics, *Heterocycles*, 1980, **14**, 1127.
[33] M. Murase and S. Tobinaga, *Heterocycles*, 1981, **15**, 1219.

Napieralksy ring-closure and dehydrogenation.[34] Norlaudanosine has been synthesized by the reduction of the product (20) of the reaction between 3,3′,4,4′-tetramethoxybenzoin and glycine in the presence of polyphosphoric acid.[35] Rugosinone (21) has been synthesized by the reaction of 2-benzyloxy-3,4-dimethoxybenzaldehyde with the Reissert compound that is obtained from 6,7-methylenedioxyisoquinoline followed by debenzylation[36] and also by the oxidation of 3,4-dihydrorugosinone, obtained by hydrolysis and aerial oxidation of (22), itself prepared from berberine chloride by oxidation with *m*-chloroperoxybenzoic acid;[37] a similar sequence of reactions, starting from coptisine chloride, resulted in the production of 1,2-dehydronorledecorine, which could be converted into norledecorine and ledecorine.[37]

(19)

(20)

(21)

(22)

Methods of purification[38] and estimation[39, 40] of papaverine and of estimation of ethaverine[41] have been described and the stereochemistry of the quaternization of laudanosine has been studied.[42] The effects of papaverine on cerebral circulation of

[34] A. P. Kozikowski and A. Ames, *J. Org. Chem.*, 1980, **45**, 2548.
[35] Y. Ichinohe, T. Sato, T. Takido, and H. Sakamaki, *Koen Yoshishu-Tennen Yuki Kagobutsu Toronkai 22nd*, 1979, 567.
[36] H.-Y. Cheng and R. W. Doskotch, *J. Nat. Prod.*, 1980, **43**, 151.
[37] M. Murugesan and M. Shamma, *Heterocycles*, 1980, **14**, 585.
[38] E. F. Zueva, S. Ya. Skachilova, Z. A. Ryzhova, V. G. Voronin, A. A. Bakhareva, and G. A. Dubskikh, USSR P. 767 100 (*Chem. Abstr.*, 1981, **94**, 84 360).
[39] S. R. Gautam, A. Nahum, J. Baechler, and D. W. A. Bourne, *J. Chromatogr.*, 1980, **182**, 482.
[40] Kh. K. Dzhalilov and S. M. Makhkamov, *Med. Zh. Uzb.*, 1980, No. 11, p. 79.
[41] R. R. Brodie, L. F. Chasseaud, L. M. Walmsley, H. H. Soegtrop, A. Darragh, and D. A. Kelly, *J. Chromatogr.*, 1980, **182**, 379.
[42] J. C. Lindon and A. G. Ferrige, *Tetrahedron*, 1980, **36**, 2157.

blood,[43-45] on the uptake and release of calcium in the taenia coli of the rabbit,[46,47] on the efflux of potassium in heart muscle[48] and of calcium in rat ileal muscle,[49] on the levels of prostaglandins E and F in brain and cerebrospinal fluid,[50] on induced breakdown of nucleotides,[51] on the consumption of glucose in placenta[52] and in brain,[53] on pancreatic exocrine secretion,[54] on Ehrlich ascites tumour,[55] on feline colonic muscle,[56] and on rat uterus[57] have been studied, as have the effects of ethaverine on cochlear microcirculation,[58] the effects of higenamine on the cardio-vascular system,[59,60] and the absorption and excretion[61] and the estimation of the metabolites of[62] drotaverine. A bis-quaternary salt of laudanosine with the dihalide $ClCH_2CH_2COO(CH_2)_5OCOCH_2CH_2Cl$ (atracurium) has been examined as a competitive neuromuscular blocking agent.[63,64]

In the 2-benzylisoquinoline series, sendaverine (23; R = Me) has been synthesized by the condensation of either 7-benzyloxy-6-methoxyisochromone[65,66] or ethyl 4-benzyloxy-2-bromoethyl-5-methoxyphenylacetate[67] with 4-methoxybenzylamine, reduction of the resulting lactams, and debenzylation. Repetition of the second of these syntheses, using 4-benzyloxybenzylamine, gave corgoine (23; R = H).[68]

[43] V. G. Duzhak, *Farmakol. Toksikol.* (*Kiev*), 1978, **13**, 56.
[44] S. A. Kryzhanovskii, *Farmakol. Toksikol.* (*Moscow*), 1980, **43**, 690.
[45] R. S. Conway and H. R. Weiss, *Eur. J. Pharmacol.*, 1980, **68**, 17.
[46] I. Takayanagi and T. Hisayama, *Jpn. J. Pharmacol.*, 1980, **30**, 641.
[47] I. Takayanagi, T. Hisayama, Y. Yoshida, and K. Koike, *J. Pharmacobio.-Dyn.*, 1980, **3**, 160.
[48] H. Nawrath, *Naunyn-Schmiedeberg's Arch. Pharmacol.*, 1980, **312**, 183.
[49] H. Huddart and K. H. M. Saad, *J. Exp. Biol.*, 1980, **86**, 99.
[50] A. G. Aivazyan and E. S. Gabrielyan, *Krovoobraschchenie*, 1979, **12**, 3.
[51] H. Sheppard, S. Sass, and W. H. Tsein, *Immunopharmacology*, 1980, **2**, 221.
[52] H. Zrubek, J. Oleszczuk, T. Panczyk, H. Sawulkicka-Oleszczuk, and W. Dabek, *Ginekol. Pol.*, 1980, **51**, 289.
[53] A. R. Dick, S. R. Nelson, and P. L. Turner, *Intracranial Pressure*, 1979, **4**, 261.
[54] K. Iwatsuki and S. Chiba, *Arch. Int. Pharmacodyn. Ther.*, 1980, **248**, 314.
[55] S. Shinozawa, Y. Araki, and T. Oda, *Physiol. Chem. Phys.*, 1980, **12**, 291.
[56] W. J. Snape, *Gastroenterology*, 1981, **80**, 498.
[57] I. Takayanagi, K. Koike, and T. Hisayama, *Eur. J. Pharmacol.*, 1981, **69**, 367.
[58] J. Prazma, W. P. Biggers, and N. D. Fischer, *Arch. Otolaryngol.*, 1981, **107**, 227.
[59] N.-H. Huang, Y.-P. Zhou, W.-H. Liu, and L.-L. Fan, *Chung-kuo Yao Li Hsueh Pao*, 1980, **1**, 34.
[60] H. Wagner, M. Reite, and W. Ferstl, *Planta Med.*, 1980, **40**, 77.
[61] G. Simon, Z. Vargay, M. Winter, and T. Szuets, *Eur. J. Drug Metab. Pharmacokinet.*, 1979, **4**, 213.
[62] Z. Vargay, G. Simon, M. Winter, and T. Szuets, *Eur. J. Drug. Metab. Pharmacokinet.*, 1980, **5**, 69.
[63] R. Hughes and D. J. Chapple, *Br. J. Anaesth.*, 1981, **53**, 31.
[64] G. G. Coker, G. H. Dewar, R. Hughes, T. M. Hunt, J. P. Payne, J. B. Stenlake, and R. Waigh, *Acta Anaesthesiol. Scand.*, 1981, **25**, 67.
[65] M. Masood, P. K. Minocha, and K. P. Tiwari, *Curr. Sci.*, 1980, **49**, 510.
[66] Sendai Heterocyclic Chemical Research Foundation, Jpn. Kokai Tokkyo Koho 80 111468 (*Chem. Abstr.*, 1981, **94**, 15 946).
[67] G. D. Pandey and K. P. Tiwari, *Indian J. Chem., Sect. B.*, 1980, **19**, 160.
[68] G. D. Pandey and K. P. Tiwari, *Pol. J. Chem.*, 1980, **54**, 763.
[69] R. Torres, F. Delle Monache, and G. B. Marini-Bettolo, *Gazz. Chim. Ital.*, 1979, **109**, 567.
[70] Z.-J. Tang, A.-N. Lao, X.-X. Zhang, G. C. Wang, and F. D. Zhong, *Yao Hsueh Hsueh Pao*, 1980, **15**, 506.
[71] F. Scheinmann, E. F. V. Scriven, and O. N. Ogbeide, *Phytochemistry*, 1980, **19**, 1837.
[72] W.-N. Wu, J. L. Beal, and R. W. Doskotch, *J. Nat. Prod.*, 1980, **43**, 372.
[73] W.-N. Wu, W.-T. Liao, Z. F. Mahmoud, J. L. Beal, and R. W. Doskotch, *J. Nat. Prod.*, 1980, **43**, 472.
[74] R. Ahmad, *Islamabad J. Sci.*, 1978, **5**, 38.

(23)

4 Bisbenzylisoquinolines

The following species have been shown to contain the alkaloids stated:

Berberis chilensis[69]	isothalicberine, *O*-methylisothalicberine, and 7-*O*-demethylisothalicberine
Cyclea barbata[70]	(−)-curine, homoaromoline, (+)-isochondodendrine, and (+)-tetrandrine
Synclisia scabrida[71]	cycleanine
Thalictrum alpinum[72]	thalpindione*, thalidasine, thalrugosamine, thalrugosidine, and *N*-demethylthalrugosidine*
Thalictrum minus (Race B)[73]	thalistine*, thalmirabine*, *O*-methylthalibrine*, thalphinine, and thalrugosine
Thalictrum rochebrunianum[74]	*O*-methylthalibrunamine* and 2'-northalibrunine*
Tiliacora funifera[75]	oblongine and funiferine dimethiodide*
Triclisia gilleti[76]	gilletine *N*-oxide*, isogilletine *N*-oxide*, obamegine, and stebisimine

* These alkaloids are reported for the first time.

Thalistine (24), thalmirabine (25; R = H), and *O*-methylthalibrine (25; R = Me) have antimicrobial properties.[73] Thalpindione has been assigned the structure (26; R = O), since *O*-methylation gives thalrugosinone, and demethylthalrugosidine the structure (26; R = H₂) after identification of the products of fission of the ethyl ether with sodium and liquid ammonia and following the *N*-methylation of the alkaloid to thalrugosidine.[72] The structures given for 2'-northalibrunine and *O*-methylthalibrunamine[74] doubtless need to be amended, following the re-assignment of structures of the bases of this group that was reported in Volume 11.

Cycleanine has been cleaved, with oxidation, by mercuric acetate in acetic acid to give the salt (27),[77] and the thermolytic demethylation of the alkaloid has been shown to proceed sequentially through norcycleanine and isochondodendrine to the tetra-*O*-demethyl compound.[78] The spatial conformation of the alkaloid has been studied by time-dependent n.m.r. spectroscopy.[79]

[75] A. N. Tackie, J. B. Reighard, M. M. El-Azizi, D. L. Slatkin, P. L. Schiff, and J. E. Knapp, *Phytochemistry*, 1980, **19**, 1882.
[76] P. Owusu, D. J. Slatkin, J. E. Knapp, and P. L. Schiff, *J. Nat. Prod.*, 1981, **44**, 61.
[77] O. P. Sheichenko and O. N. Tolkachev, *Khim. Prir. Soedin.*, 1979, 676.
[78] O. P. Sheichenko and O. N. Tolkachev, *Khim. Prir. Soedin.*, 1980, 263.
[79] V. I. Sheichenko, O. P. Sheichenko, and O. N. Tolkachev, *Khim. Prir. Soedin.*, 1980, 60.

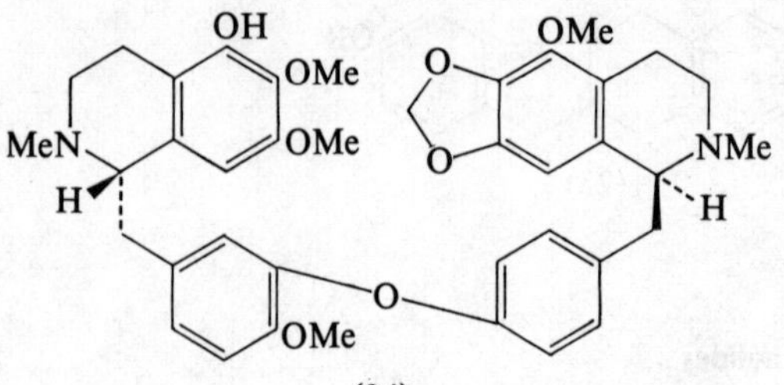

(24)

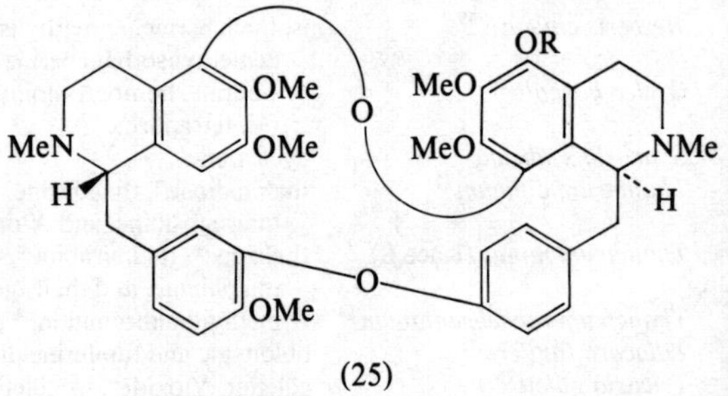

(25)

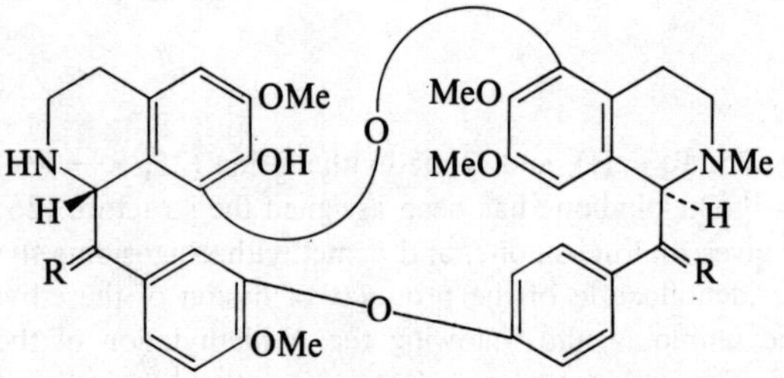

(26)

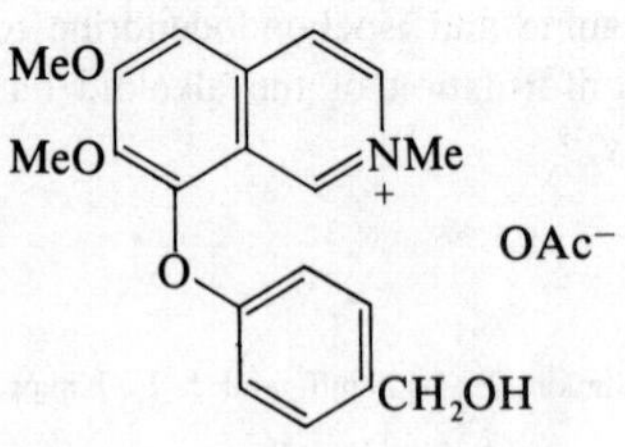

(27)

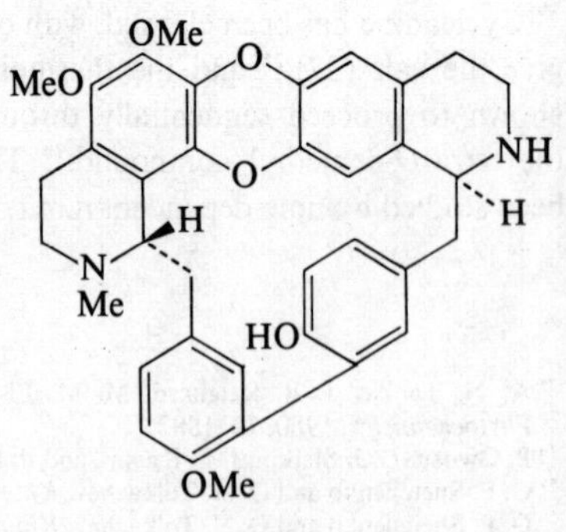

(28)

The ^{13}C n.m.r spectra[80] and the fluorescence and phosphorescence characteristics[81] of the berbamine–oxyacanthine group of alkaloids have been studied. A study of the specific rotations of 175 bisbenzylisoquinolines shows that the sign of rotation is characteristic of the stereochemical group, and this has led to the assignment of the configuration (28) to tiliamosine.[82]

Methods of synthesis of bisbenzylisoquinoline alkaloids have been reviewed[83] and the use of electrochemical methods for the reduction of the dihydro-isoquinolinium salts that result from Bischler–Napieralsky ring-closures in the synthesis of, for example, espinidine has been studied.[84]

The pharmacokinetics[85] of tubocurarine and its ability to cross the placenta[86] have been studied, as have the effects of the alkaloid on skeletal muscle,[87] the cardiovascular system,[88] receptors for acetylcholine,[89] levels of calcium and potassium,[90] heart muscle,[91] coronary circulation,[92] the olfactory cortex,[93] and the central nervous system.[94] The use of tubocurarine as a neuromuscular blocking agent[95] and in anaesthesia,[96] and the effect of temperature on its potency,[97] have also been studied. The effects of dimethyl-(−)-curine dimethochloride on neuro-muscular transmission,[98] of berbamine on blood pressure,[99] of cepharanthine on tumours,[100] on the formation of granulomas,[101] and on the peroxidation of lipids in membranes,[102] of dimethylcycleanine bromide on blood pressure[103] and cardiac blood-flow,[104] of tetrandrine on tumours[105] and on cardiac arrhythmias,[106] and of thalidasine on tumours[107] have also been studied.

[80] L. Koike, A. J. Marsaioli, E. A. Ruveda, F. de A. M. Reis, and I. R. C. Bick, *Tetrahedron Lett.*, 1979, 3765.

[81] E. P. Gibson and J. H. Turnbull, *J. Chem. Soc., Perkin Trans. 2*, 1980, 1696.

[82] B. K. Cassels and M. Shamma, *Heterocycles*, 1980, **14**, 211.

[83] O. N. Tolkachev, E. P. Nakova, and R. P. Evstigneeva, *Usp. Khim.*, 1980, **40**, 1617.

[84] T. Shono, Y. Usui, T. Miyamoto, and H. Hamaguchi, *Fukusokan Kagaku Toronkai Koen Yoshishu 12th*, 1979, 26.

[85] R. S. Matteo, K. Nishitateno, E. K. Pua, and S. Spector, *Anesthesiology*, 1980, **52**, 335.

[86] C. Melloni, G. Cantamessa, C. Macchiagodena, M. Bonora, and M. Zanello, *Minerva Anestesiol.*, 1980, **46**, 387.

[87] D. Colquhoun and R. E. Sheridan, *Br. J. Pharmacol.*, 1980, **68**, 143P.

[88] D. Tsoucaris-Kupper, L. Liblau, M. Legrand, and H. Schmidt, *Eur. J. Pharmacol.*, 1980, **65**, 301.

[89] J. J. Lambert, R. L. Volle, and E. G. Henderson, *Proc. Natl. Acad. Sci. USA*, 1980, **77**, 5003.

[90] B. E. Wand and D. R. Wand, *Br. J. Anaesth.*, 1980, **52**, 863.

[91] N. Iwatsuki, Y. Hashimoto, K. Amaha, S. Obara, and K. Iwatsuki, *Anesth. Analg. (Cleveland)*, 1980, **59**, 717.

[92] V. V. Buyanov, *Farmakol. Toksikol. (Moscow)*, 1980, **43**, 687.

[93] C. N. Scholfield, *Naunyn-Schmiedeberg's Arch. Pharmacol.*, 1980, **314**, 79.

[94] V. R. Dhumal and V. H. Bhavsar, *Arch. Int. Pharmacodyn. Ther.*, 1980, **248**, 148.

[95] R. S. Leeuwin, R. D. Veldsema-Currie, H. Van Wilgenburg, and M. Ottenhof, *Eur. J. Pharmacol.*, 1981, **69**, 165.

[96] T. J. Gal and S. K. Goldberg, *Anesthesiology*, 1981, **54**, 141.

[97] L. Farrell, M. J. Dempsey, B. E. Wand, and D. R. Wand, *Anesth. Analg. (Cleveland)*, 1981, **60**, 18.

[98] Q.-Z. Yang and L.-R. Lin, *Chung-kuo Yao Li Hsueh Pao*, 1981, **2**, 19.

[99] Z.-D. Zhou, C.-H. Han, and P. Wang, *Yao Hsueh Hsueh Pao*, 1980, **15**, 248.

[100] R. Fujiwara, M. Ono, N. Tanaka, H. Miura, T. Mannami, E. Konaga, and K. Orita, *Gan to Kagaku Ryoho*, 1980, **7**, 481.

[101] R. Fujiwara, M. Ono, and K. Orita, *Igaku no Ayumi*, 1980, **114**, 1056.

[102] N. Shiraishi, T. Arima, K. Aono, B. Inouye, Y. Morimoto, and K. Utsumi, *Physiol. Chem. Phys.*, 1980, **12**, 299.

[103] Z. Sun, X.-Y. Yang, Z.-L. Dai, G.-Z. Jin, & Z.-D. Zhang, *Chung-kuo Yao Li Hsueh Pao*, 1980, **1**, 23.

[104] W.-Z. Chen, Y.-L. Dong, and G. S. Ding, *Chung-kuo Yao Li Hsueh Pao*, 1980, **1**, 27.

[105] L.-L. H. Liao, *Proc. Natl. Sci. Counc. Repub. China*, 1980, **4**, 285.

[106] L. Cha, J.-Q. Qian, and F.-H. Liu, *Chung-kuo Yao Li Hsueh Pao*, 1981, **2**, 26.

[107] Z.-Q. Ma, S.-M. Hsing, and H.-C. Chen. *Chung Ts'ao Yao*, 1980, **11**, 217.

5 Pavines and Isopavines

Norargemonine and bisnorargemonine have been isolated from *Cryptocarya longifolia*[14] and amurensinine from *Papaver tauricola*.[108]

6 Berberines and Tetrahydroberberines

The following species have been shown to contain the alkaloids stated:

Anamirta cocculus[109]	berberine, columbamine, palmatine, and (−)-8-oxotetrahydropalmatine
Bocconia frutescens[110]	berberine, (−)-α-canadine, columbamine, coptisine, (−)-isocorypalmine, and (−)-scoulerine
Cryptocarya longifolia[14]	(−)-scoulerine
Fumaria judaica[111]	cheilanthifoline, coptisine, and stylopine
Fumaria schleicheri[112]	(±)-sinactine
Meconopsis cambrica[113]	(−)-mecambridine
Stephania glabra[114]	glabrine and glabrinine
Thalictrum alpinum[13]	berberine, columbamine, jatrorrhizine, oxyberberine, palmatine, and thalifendine
Thalictrum fauriei[115]	dehydrodiscretine and thalifaurine
Thalictrum minus (Race B)[73]	columbamine, jatrorrhizine, *N*-methylcorydaldine, and thalifendine
Zanthoxylum williamsii[116]	berberine

The chemistry of the alkaloids of this group has been reviewed.[117,118] Dehydrodiscretine (29; $R^1 = R^2 = Me$) and thalifaurine (29; $R^1R^2 = CH_2$) are novel alkaloids, and their structures have been determined by spectroscopic methods, reduction, and synthesis.[115] An alkaloid named fumajudaine has been shown to be identical with stylopine.[111]

(29)   (30)

[108] G. Sariyar and J. D. Phillipson, *Phytochemistry*, 1980, **19**, 2189.
[109] R. Verpoorte, J. Siwon, M. E. M. Tieken, and S. A. Baerheim, *J. Nat. Prod.*, 1981, **44**, 221.
[110] E. Taborska, F. Veznik, and J. Slavic, *Collect. Czech. Chem. Commun.*, 1980, **45**, 1301.
[111] A. H. A. Abou-Donia, S. El-Masry, M. R. I. Saleh, and J. D. Phillipson, *Planta Med.*, 1980, **40**, 295.
[112] Kh. Kiryakov, Z. Mardirosyan, and P. Panov, *Dokl. Bolg. Akad. Nauk*, 1980, **33**, 1377.
[113] S. R. Hemmingway, J. D. Phillipson, and R. Verpoorte, *J. Nat. Prod.*, 1981, **44**, 67.
[114] A. Patra, P. K. Mukhopadhyay, and A. K. Mitra, *Indian J. Chem.*, *Sect. B*, 1980, **19**, 561.
[115] C.-H. Chen, T.-M. Chen, and C. Lee, *J. Pharm. Sci.*, 1980, **69**, 1061.
[116] F. R. Stermitz, M. A. Caolo, and J. A. Swinehart, *Phytochemistry*, 1980, **19**, 1469.
[117] G. D. Pandey and K. P. Tiwari, *Heterocycles*, 1980, **14**, 59.
[118] T. R. Govindachari, *J. Indian Chem. Soc.*, 1980, **57**, 353.

N-Benzyl-3,4-dihydroisoquinolinium salts have been shown to react with the anion of methyl methylthiomethyl sulphoxide to give diastereoisomeric adducts of structure (30) that can be cyclized by heating with concentrated hydrochloric acid to dihydroberberines; these can be oxidized to berberines, reduced to tetrahydroberberines, and converted into 13-methyltetrahydroberberines by established procedures. A similar reaction with *N*-benzylisoquinolinium salts affords 3,4-dehydro-analogues of (30) which may be reduced prior to cyclization, so that the method of synthesis is usable in cases where the 3,4-dihydroisoquinoline is not readily accessible by the Bischler–Napieralsky ring-closure. Palmatine iodide (31), (±)-xylopinine (32; $R^1 = R^4 = H$, $R^2 = R^3 = OMe$), (±)-sinactine (32; $R^1R^2 = OCH_2O$, $R^3 = R^4 = H$), and (±)-corydaline (32; $R^1 = R^2 = OMe$, $R^3 = H$, $R^4 = Me$) have been synthesized in this way.[119]

(31)   (32)

Reissert compounds of the type (33) have been cyclized to lactams of the 8-oxodihydroberberine series (34) by sodium hydride in dimethylformamide,[120] though the reactions have not so far been utilized for the synthesis of derivatives of natural bases. 8-Oxotetrahydropalmatine (35; $R^1 = R^2 = R^3 = R^4 = OMe$, $R^5 = H$), 8-oxocanadine (35; $R^1R^2 = OCH_2O$, $R^3 = R^4 = OMe$, $R^5 = H$), 8-oxostylopine (35; $R^1R^2 = R^3R^4 = OCH_2O$, $R^5 = H$) and 8-oxoxylopinine (35; $R^1 = R^2 = R^4 = R^5 = OMe$, $R^3 = H$) have been prepared in yields of 90–100% by the electrolytic reduction of the appropriately substituted 3,4-dihydroisoquinolines and methyl 2-bromomethylbenzoates.[121]

Xylopinine has been synthesized by the condensation of 3,4-dimethoxy-homophthalic anhydride and 6,7-dimethoxy-1-oxotetrahydroisoquinoline, followed by reduction of the resulting lactam.[122] Xylopinine,[123,124] scoulerine,[124] isocoptisine,[124] ψ-epitetrahydroberberine,[124] and govanine[125] have been synthesized by cyclization of the appropriately substituted lactams (36), themselves prepared

[119] Z. Kiparissides, R. H. Fichtner, J. Poplawski, B. C. Nalliah, and D. B. MacLean, *Can. J. Chem.*, 1980, **58**, 2770.

[120] S. Ruchirawat, W. Lertwanawatana, and P. Thepchumrune, *Tetrahedron Lett.*, 1980, **21**, 189.

[121] T. Shono, Y. Usui, T. Mizutani, and H. Hamaguchi, *Tetrahedron Lett.*, 1980, **21**, 3073.

[122] M. A. Haimova, V. I. Ognyanov, and N. M. Mollov, *Synthesis*, 1980, 845.

[123] G. D. Pandey and K. P. Tiwari, *Synth. Commun.*, 1980, **10**, 607.

[124] G. D. Pandey and K. P. Tiwari, *Indian J. Chem., Sect. B*, 1980, **19**, 272.

[125] G. D. Pandey and K. P. Tiwari, *Indian J. Chem., Sect. B*, 1980, **19**, 66.

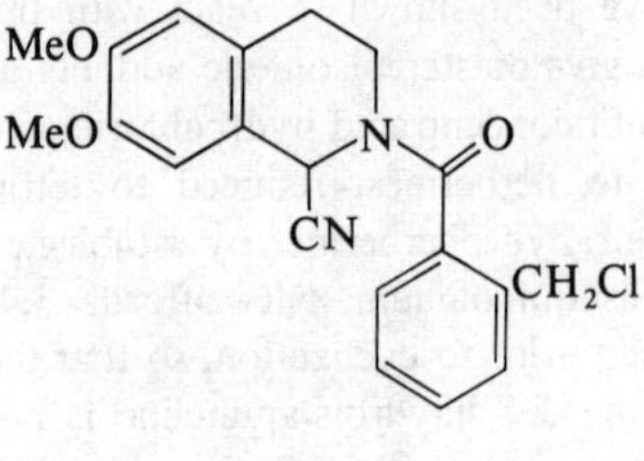

(33)

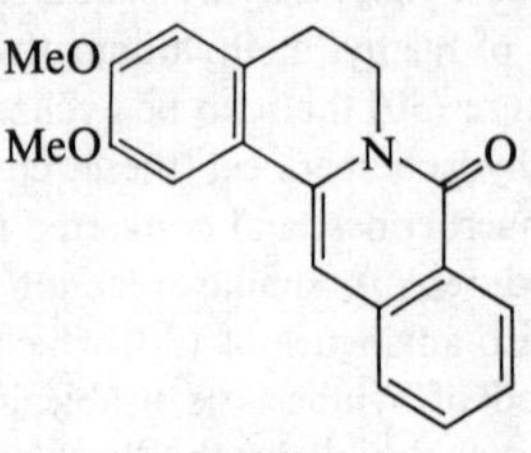

(34)

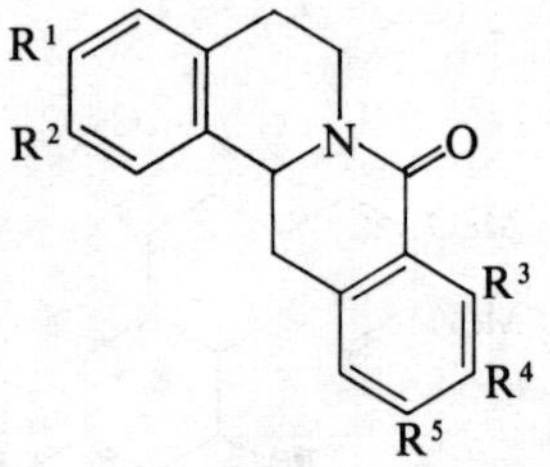

(35)

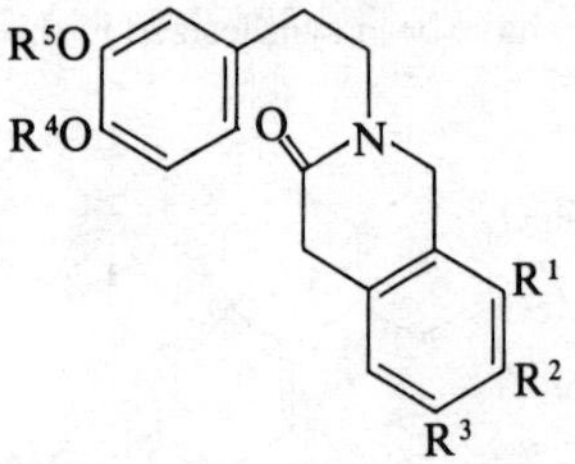

(36)

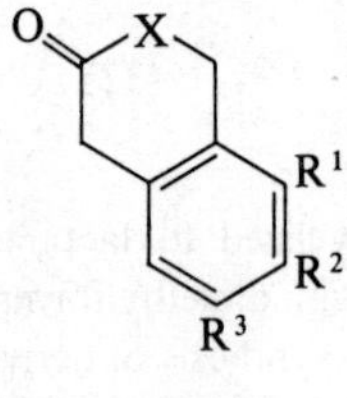

(37)

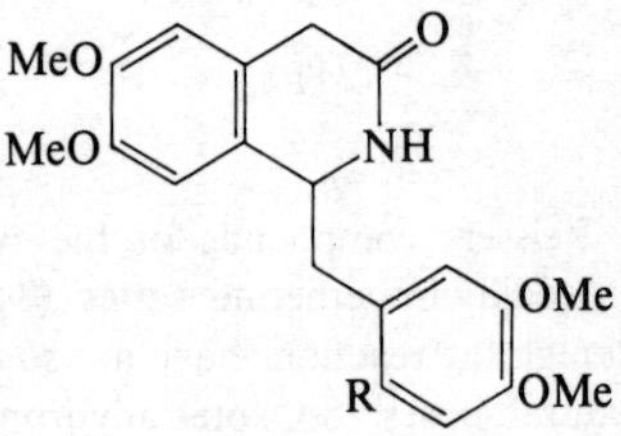

(38)

from the isochromanones (37; X = O) by ring-opening to bromomethyl esters followed by treatment with β-phenylethylamines[124,125] (*cf.* Vol. 11, p. 87), or by alkylation of the lactams (37; X = NH) with β-phenylethyl bromides.[123] Bromination of the benzene ring has been used to direct the formation of an isochromanone ring in the dimethoxy-compounds in the less-favoured form[126] and to control the direction of cyclization of norlaudanosine (38; R = H) with formaldehyde to give tetrahydropalmatine.[127] A review of the use of isochromanones in the synthesis of alkaloids has been published.[128]

Bischler–Napieralsky synthesis of the benzylisoquinoline (39), N-formylation, cyclization, and oxidation has given the pentamethoxy-compound (40), identical with glabrine dimethyl ether and glabrinine methyl ether.[114] Reticuline N-oxide has

[126] G. D. Pandey and K. P. Tiwari, *Curr. Sci.*, 1980, **49**, 216.
[127] G. D. Pandey and K. P. Tiwari, *J. Sci. Res.* (*Bhopal, India*), 1979, **1**, 93.
[128] G. D. Pandey and K. P. Tiwari, *Heterocycles*, 1981, **16**, 449.

been cyclized to a mixture of scoulerine and coreximine by treatment with ferrous sulphate in aqueous methanol.[129]

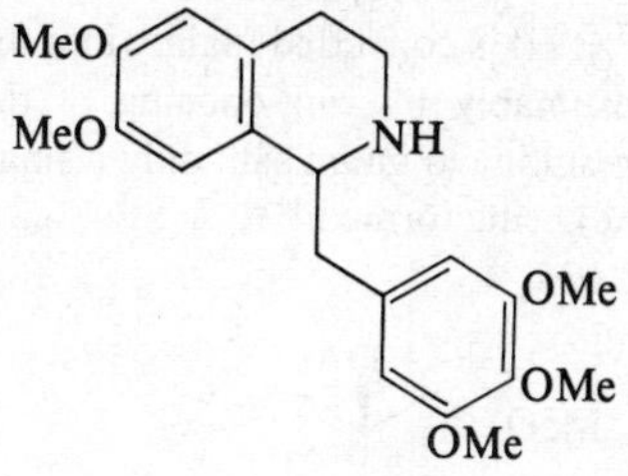

(39)

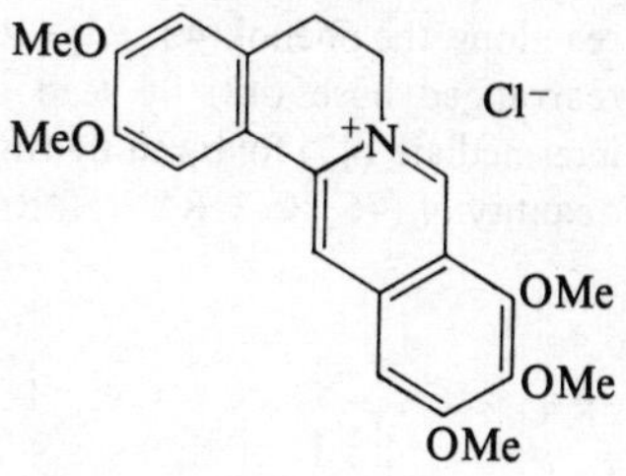

(40)

Treatment of tetrahydroberberine with ethyl chloroformate gives the ring-opened compound (41; R = CO$_2$Et, R^2 = CH$_2$Cl), which can be hydrolysed to the base (41; R^1 = H, R^2 = CH$_2$OH); cyclization of this with formaldehyde gives demethoxymecambridine (42).[130]

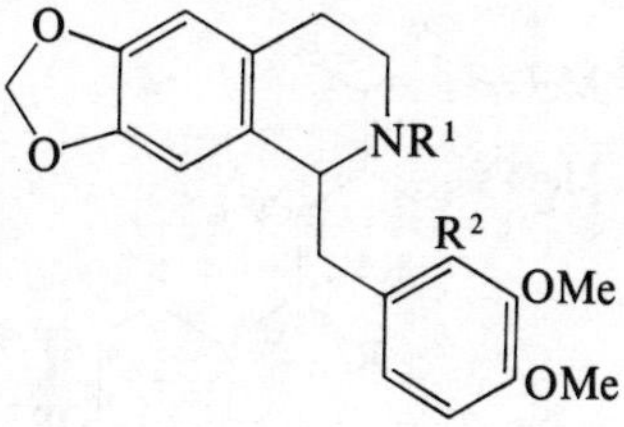

(41)

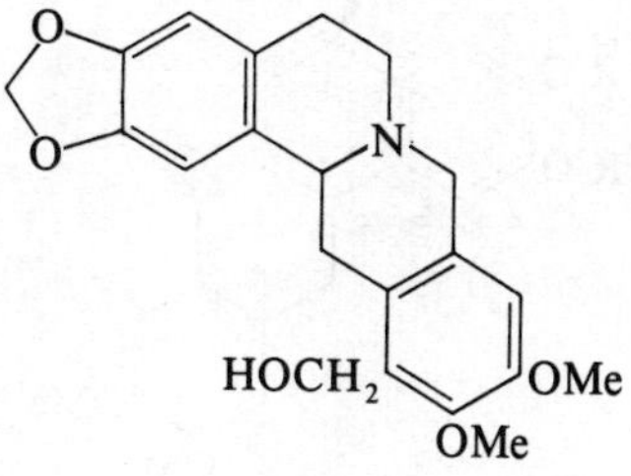

(42)

Fuller details of the oxidation of phenolic tetrahydroberberines with lead tetra-acetate (see Vol. 11, p. 88) have been published. Govanine (43; R^1 = R^3 = R^4 = Me, R^2 = H) reacts with lead tetra-acetate to give the acetoxy-compound (44), which is converted (by acetic anhydride and concentrated sulphuric acid) into acetoxyacetylgovanine (45; R^1 = Me, R^2 = R^3 = Ac); hydrolysis of this with methanolic potassium hydroxide affords both diastereoisomers of methoxygovanine (45; R^1 = R^3 = Me, R^2 = H), with the CHOMe group configured in both senses. The phenol (43; R^1 = H, R^2 = R^3 = R^4 = Me), on oxidation with lead tetra-acetate, gives directly the diacetate (45; R^1 = R^3 = Ac, R^2 = Me), both isomers of the alicyclic CHOAc group being formed. The phenol (43; R^1 = R^2 = R^3 = Me, R^4 = H) gives an analogue of (44) which can be converted, by

---

[129] Sendai Heterocyclic Chemical Research Foundation, Jpn. Kokai Tokkyo Koho 80 139 377 (*Chem. Abstr.*, 1981, **94**, 209 047).

[130] M. Hanaoka, M. Inoue, S. Yasuda, and T. Imanishi, *Heterocycles*, 1980, **14**, 1791.

acetic anhydride and sulphuric acid, into a mixture of the acetate (46; R^1 = Me, R^2 = R^3 = H, R^4 = OAc) and its diastereoisomer (46; R^1 = Me, R^2 = R^4 = H, R^3 = OAc), both of which are hydrolysed by methanolic potash to the single methoxy-compound (46; R^1 = Me, R^2 = R^3 = H, R^4 = OMe). In contrast with these reactions, the phenol (43; R^1 = R^2 = R^4 = Me, R^3 = H) is converted mainly into the rearranged base (48) by lead tetra-acetate, presumably by ring-opening of the intermediate (47) followed by ring-closure of the resulting iminium salt, only a small quantity of (46; R^1 = R^4 = H, R^2 = Me, R^3 = OAc) being formed.[131]

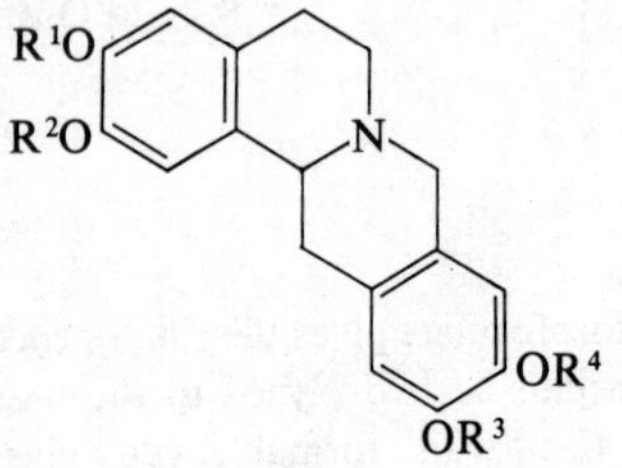

(43)

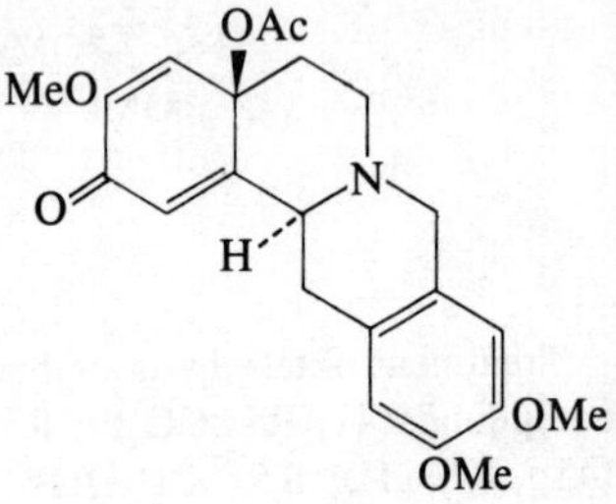

(44)

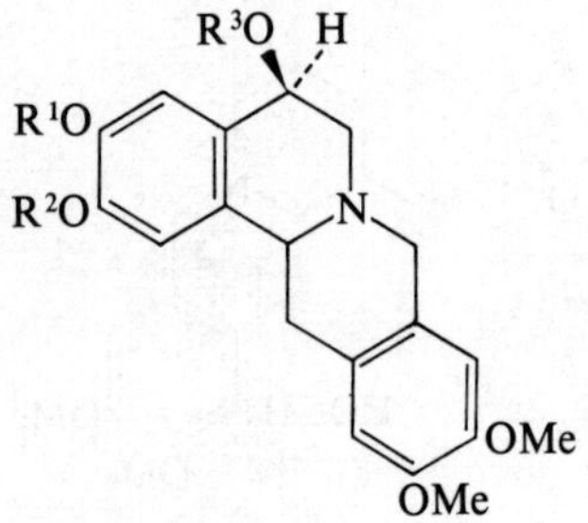

(45)

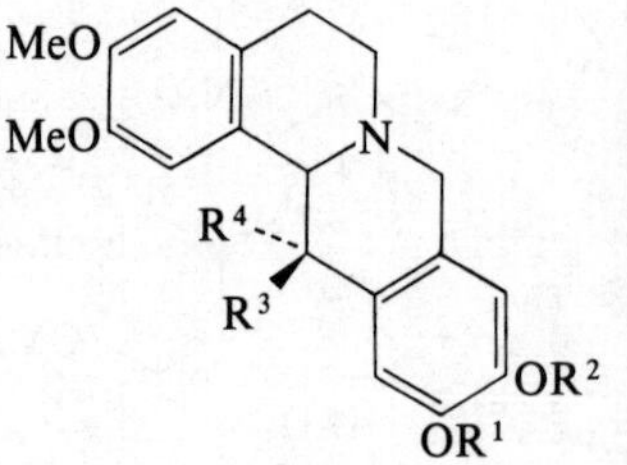

(46)

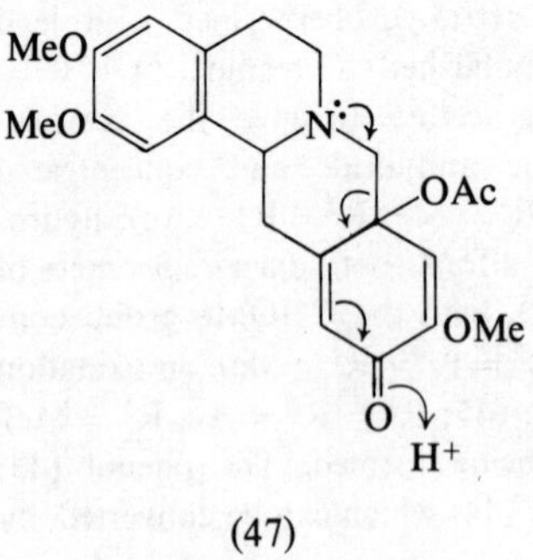

(47)

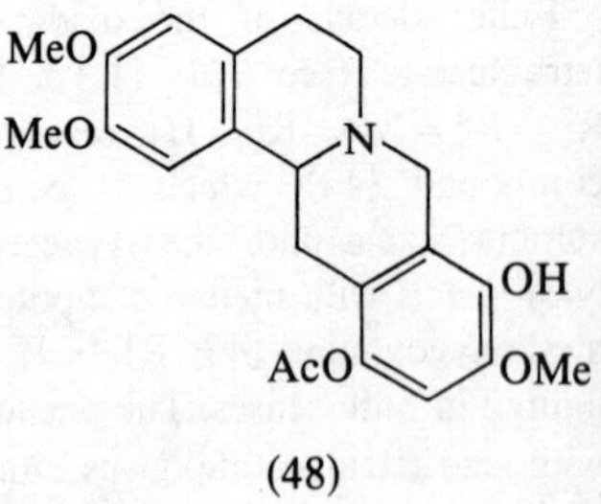

(48)

[131] H. Hara, M. Hosaka, O. Hoshino, and B. Umezawa, *J. Chem. Soc., Perkin Trans. 1*, 1980, 1169.

Oxyberberine (49), when further oxidized by pyridine chlorochromate and then treated with methanol, gives the methoxy-ketone (50; R = Me); the tetramethoxy-analogue of (49) behaves similarly.[132] The carbinolamine (50; R = H) is obtainable, together with the keto-acid (53), by the photolysis of oxidoberberine (51), in the presence of oxygen and rose bengal, to produce the peroxide (52), followed by treatment with pyridine hydrochloride; a tetramethoxy-analogue of oxidoberberine, prepared by the reduction of norcoralyne with zinc and acetic acid, followed by oxidation with a peracid, behaves in the same way.[133]

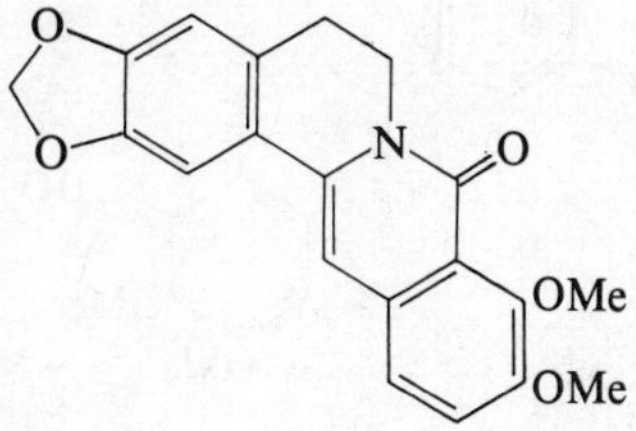

(49)

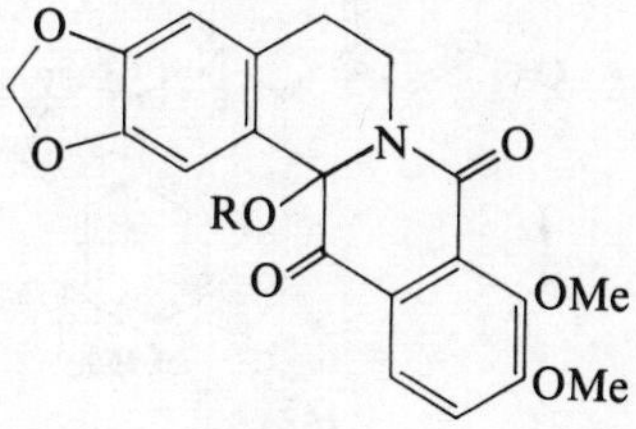

(50)

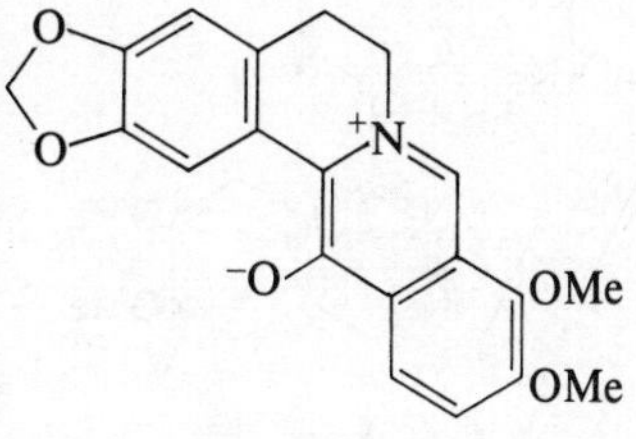

(51)

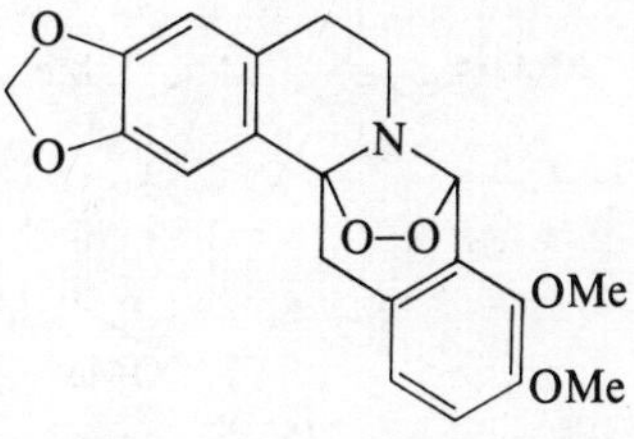

(52)

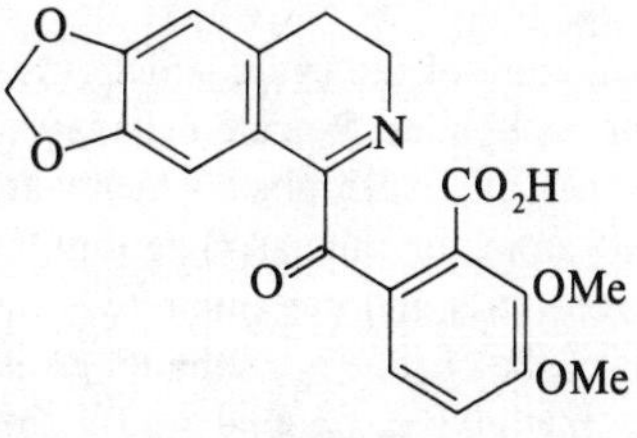

(53)

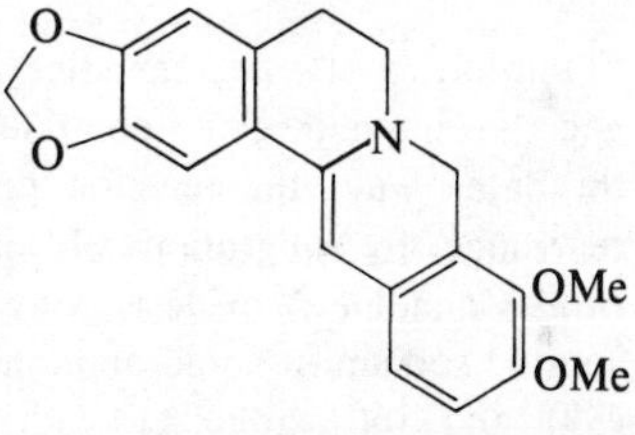

(54)

Dye-catalysed photo-oxidation of dihydroberberine (54) affords oxidoberberine (51), and the peroxide (52), on photolysis at elevated temperature, gives a mixture of the lactone (55) and the aldehyde (56). Dihydrocoralyne (57), on heating with oxygen, in the dark, gives oxidocoralyne, which is a *C*-methylated analogue of

[132] G. Manikumar and M. Shamma, *Heterocycles*, 1980, **14**, 827.
[133] Y. Kondo, J. Imai, and S. Nozoe, *J. Chem. Soc., Perkin Trans. 1*, 1980, 919.

dehydro-(51); dye-catalysed photo-oxidation of oxidocoralyne gives 6'-acetyl-papaveraldine, which gives the phthalazine (58) on treatment with hydrazine.[134]

Coralyne and norcoralyne have been claimed to undergo some change in properties when irradiated with light of wavelength >267 nm, in aqueous solution, these changes being reversed by irradiation with light of wavelength <267 nm; the change is inhibited by DNA and does not occur in non-aqueous media.[135]

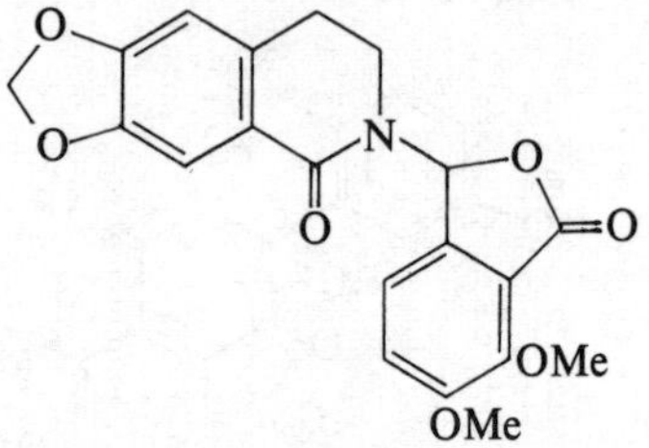

(55)

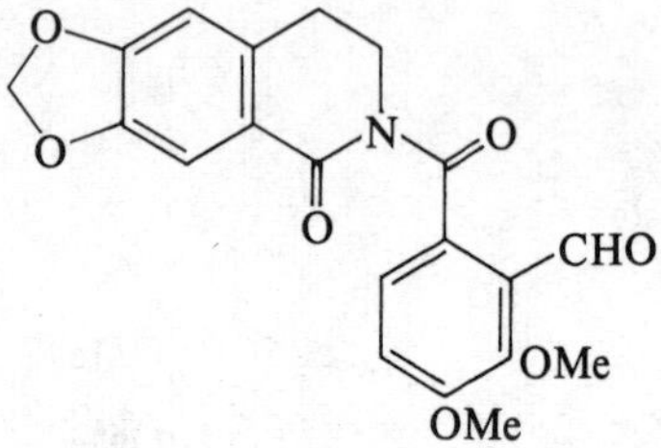

(56)

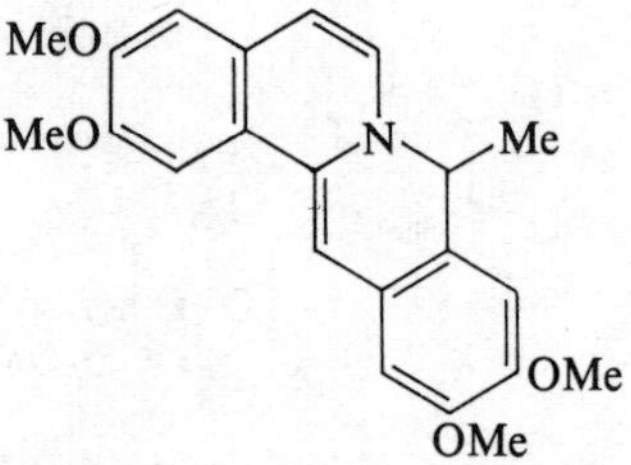

(57)

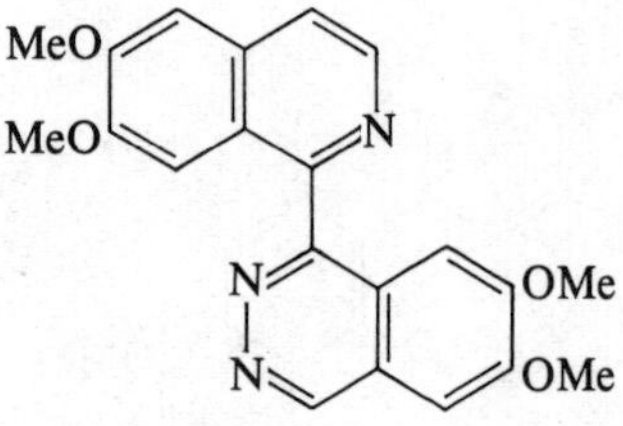

(58)

Photolysis of *trans*-canadine *N*-oxide yields a mixture of the cyclic amide (59) and the ring-contracted compound (60), and *trans*-xylopinine *N*-oxide behaves in the same way; the reaction probably proceeds *via* an oxaziridine, which can rearrange by migration of either a hydrogen atom or an alkyl group.[136] *trans*-Canadine *N*-oxide is converted into the nitrone (61) and the amine (62) by aerated sodium in liquid ammonia, and photolysis of the nitrone in methanol gives (59) and (60), together with the ring-opened methoxy-compound (63); the xylopinine *N*-oxides gave only the analogue of the secondary base (62) under the same conditions.[137]

Berberine reacts with dichlorocarbene in chloroform to give the penta-chloro-compound (64), which is hydrolysed to the aldehyde (65; R = CHO) by aqueous acid; this can be reduced to the alcohol (65; R = CH$_2$OH) by sodium

[134] Y. Kondo, J. Imai, and S. Nozoe, *J. Chem. Soc., Perkin Trans. 1*, 1980, 911.
[135] E. Smekal and J. Kondelka, *Stud. Biophys.*, 1980, **81**, 91.
[136] P. Chinnasamy, R. D. Minard, and M. Shamma, *Tetrahedron*, 1980, **36**, 1515.
[137] K. Iwasa, P. Chinnasamy, and M. Shamma, *J. Org. Chem.*, 1981, **46**, 1378.

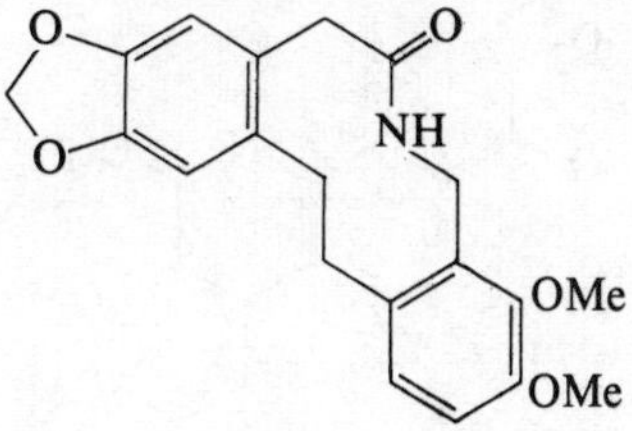

(59)

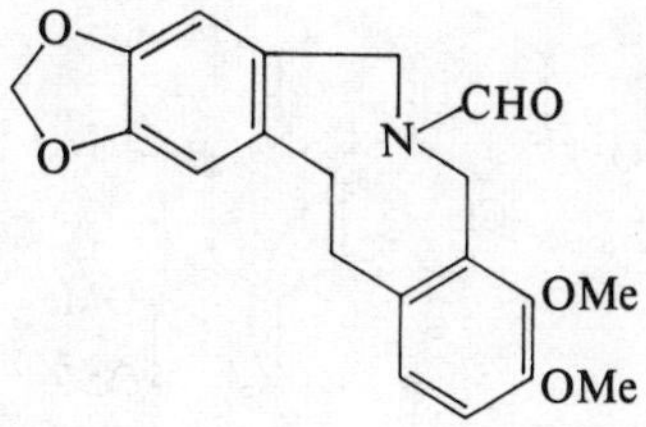

(60)

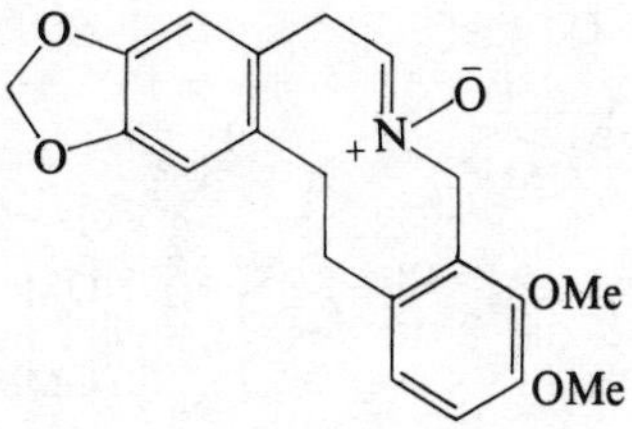

(61)

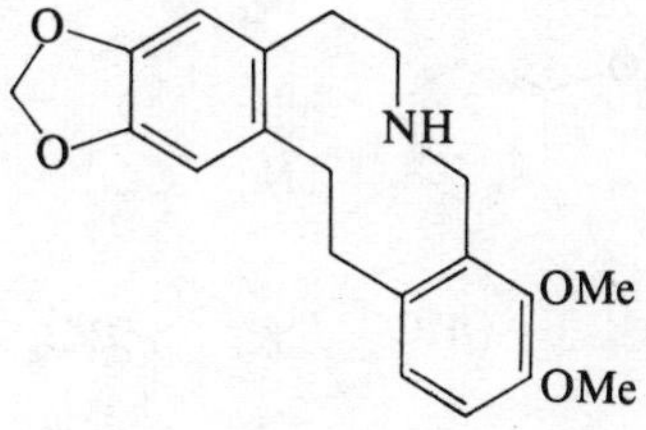

(62)

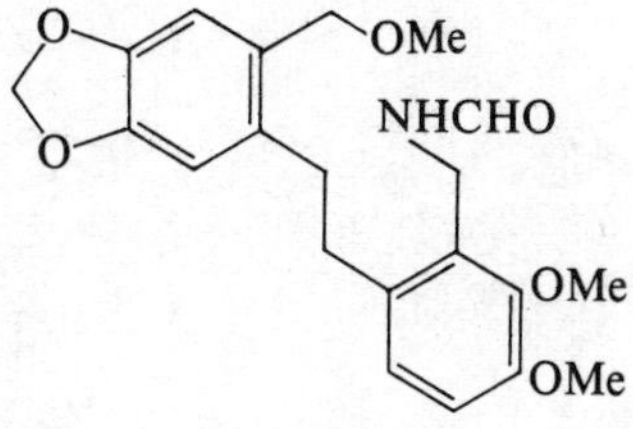

(63)

borohydride or can be dehydrochlorinated to the aldehyde (66) by silver oxide. Oxyberberine (49) reacts with dichlorocarbene to give (67), which can be hydrolysed to the aldehyde (68; R = CHO), reduced either to the 13-methyl compound (68; R = Me) (by zinc and acetic acid) or to the ring-expanded base (69) (by lithium aluminium hydride), or converted into an alternative ring-expanded compound (70) by heating in aqueous pyridine.[138]

The chiroptical properties of the tetrahydroberberine alkaloids in relation to their absolute configurations have been studied.[139] The pharmacokinetics[140] and intercalative binding[141] of berberine, the antipsychotic properties of tetrahydro-

[138] G. Manikumar and M. Shamma, *J. Org. Chem.*, 1981, **46**, 386.
[139] B. Ringdahl, R. P. K. Chan, J. C. Craig, and R. H. F. Manske, *J. Nat. Prod.*, 1981, **44**, 73.
[140] A. Mrozikiewicz, Z. Kowalewski, K. Drost-Karwowska, T. Bobkiewicz, and O. Klimazewska, *Herba Pol.*, 1980, **26**, 123.
[141] M. D. Faddejeva, T. N. Belyaeva, J. P. Novikov, and H. G. Shalabi, *IRCS Med. Sci. Libr. Compend.*, 1980, **8**, 612.

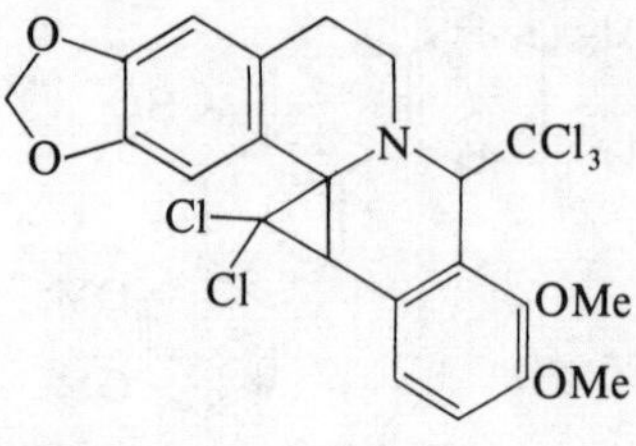

(64)

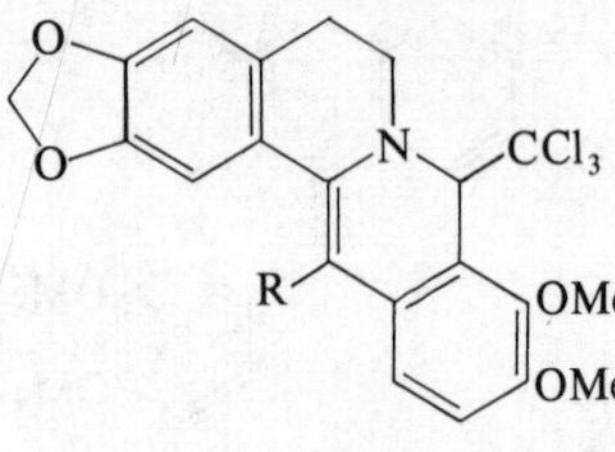

(65)

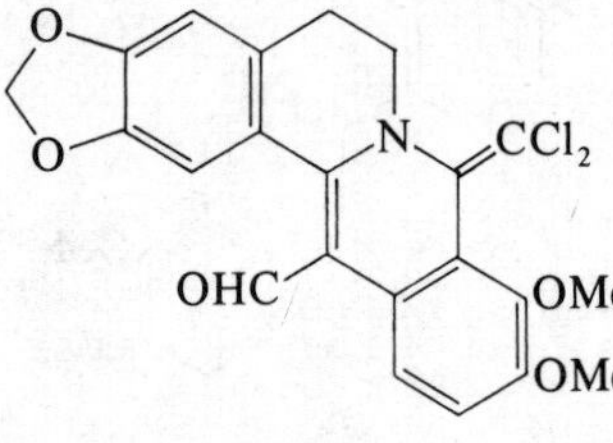

(66)

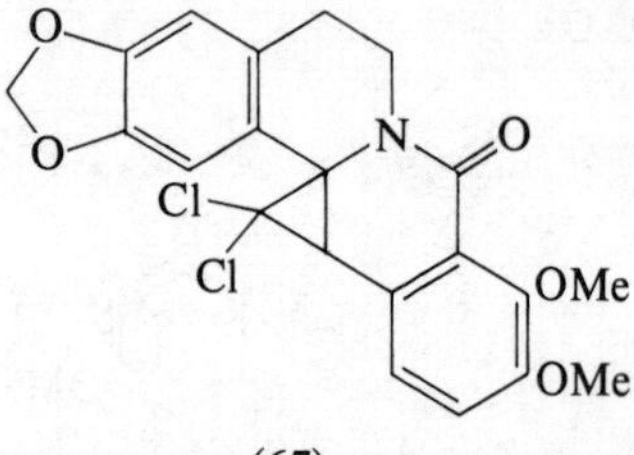

(67)

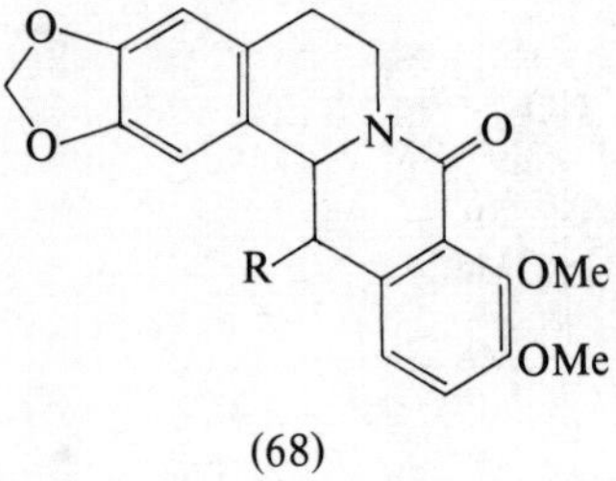

(68)

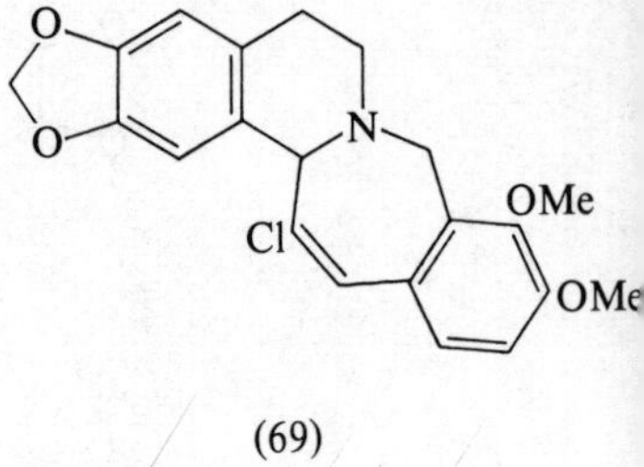

(69)

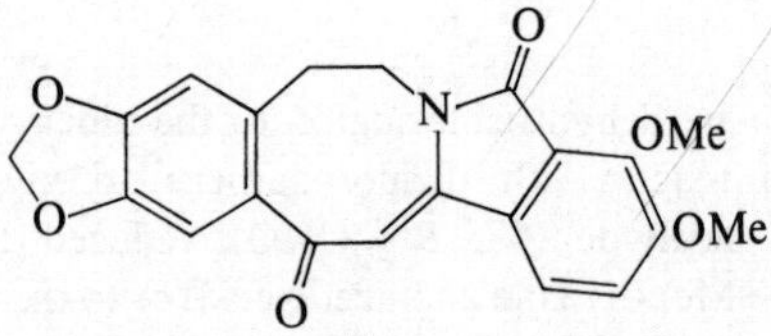

(70)

berberine,[142] and the cataleptic properties of tetrahydropalmatine[143] have also been studied.

Three new alkaloids that can be regarded as 10-aza-berberines have been isolated from *Alangium lamarckii*, namely alangimarine (71; R = H=CH₂), alangimaridine (72), and alamarine [71; R = CH(OH)Me]. Their structures were deduced from

[142] Inst. for Production and Development Science, Jpn. Kokai Tokkyo Koho 80 143 914 (*Chem. Abstr.*, 1981, **94**, 96 320).
[143] G.-Z. Jin, X.-J. Liu, L.-P. Yu, and J. Xu, *Chung-kuo Yao Li Hsueh Pao*, 1980, **1**, 12.

their n.m.r. and high-resolution mass spectra and from the dehydration of alamarine and the dehydrogenation of alangimaridine to alangimarine. Alamarine was isolated as the racemate.[144] The ring-system that is present in alangimaridine has been synthesized (*cf.* Vol. 10, p. 102).

(71)   (72)

7 Secoberberines

The preparation of analogues of hypecorine from 2'-hydroxymethyl-*N*-methyl-papaverinium salts is discussed under benzylisoquinolines. Ring-opening of tetrahydroberberine gives (41; $R^1 = CO_2Et$, $R^2 = CH_2Cl$);[130] on treatment with sodium acetate, the related iodide gives (41; $R^1 = CO_2Et$, $R^2 = CH_2OAc$), which is reducible (using lithium aluminium hydride) to (41; $R^1 = Me$, $R^2 = CH_2OAc$). This last compound can be hydrolysed and oxidized with pyridine chlorochromate to produce canadaline (41; $R^1 = Me$, $R^2 = CHO$).[145]

8 Protopines

Protopine has been isolated from *Bocconia frutescens*,[110] *Fumaria judaica*,[111] *F. schleicheri*,[112] and *Papaver bracteatum*,[146] cryptopine from *F. schleicheri*,[112] and allocryptopine from *B. frutescens*[110] and *Zanthoxylum nitidum*.[147] The protopine ring-system has been prepared from tetrahydrobenzindenoazepines (75) by photo-oxidation to the amides (76) followed by reduction with lithium aluminium hydride and re-oxidation with manganese dioxide.[148-150] The tetrahydro-benzindenoazepines have been prepared from *N*-chloroacetyl-*β*-phenylethylamines (73) by cyclization to the lactam (74) followed by reaction with a benzyl bromide and phosphorus oxychloride. *ψ*-Protopine (77; $R^1R^2 = CH_2$)[148] and fagarine II (77; $R^1 = R^2 = Me$)[149] have been synthesized in this way.

[144] S. C. Pakrashi, B. Achari, E. Ali, P. P. Ghosh Dastidar, and R. R. Sinha, *Tetrahedron Lett.*, 1980, **21**, 2667.
[145] H. Roensch, *Z. Chem.*, 1979, **19**, 447.
[146] H. Meshulam and D. Lavie, *Phytochemistry*, 1980, **19**, 2633.
[147] Z.-X. Huang and Z.-H. Li, *Hua Hsueh Hsueh Pao*, 1980, **38**, 535.
[148] K. Orito and M. Itoh, *Koen Yoshishu-Tennen Yuki Kagobutsu Toronkai 22nd*, 1979, 562.
[149] K. Orito, S. Kudoh, K. Yamada, and M. Itoh, *Heterocycles*, 1980, **14**, 11.
[150] K. Orito, Y. Yurokawa, and M. Itoh, *Tetrahedron*, 1980, **36**, 617.

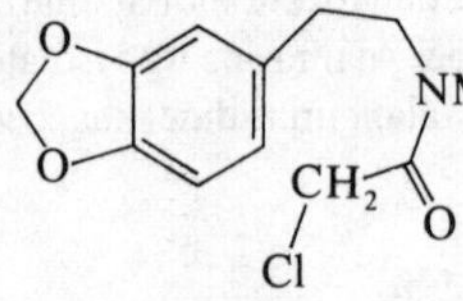

(73)

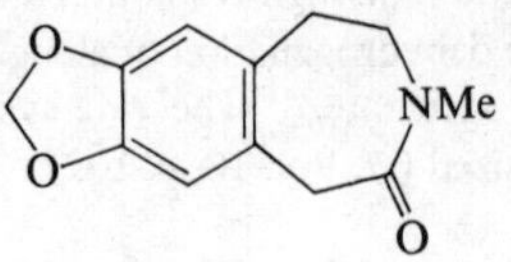

(74)

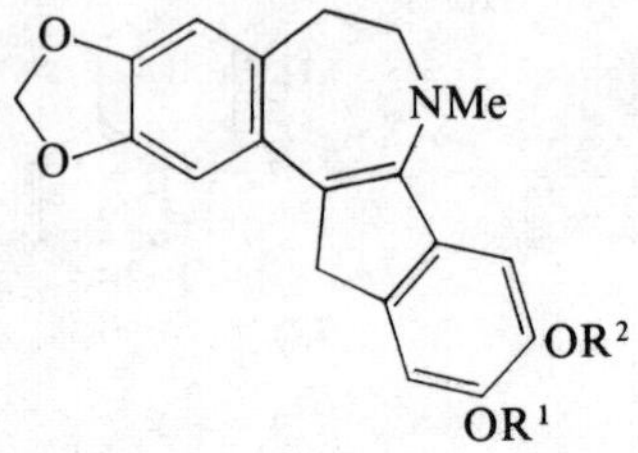

(75)

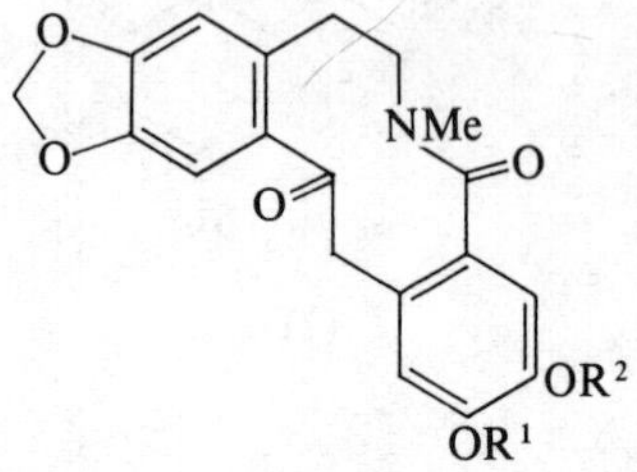

(76)

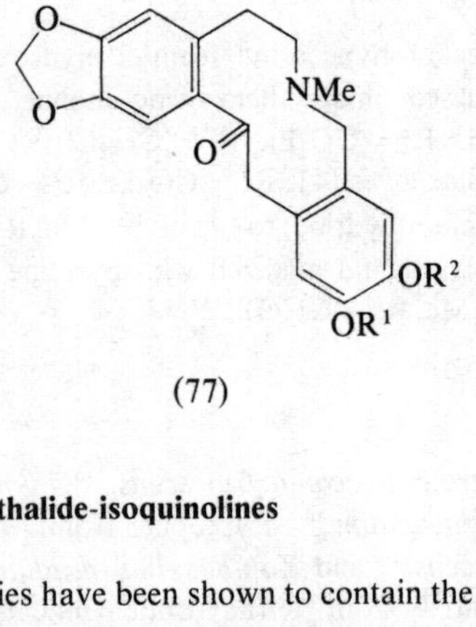

(77)

9 Phthalide-isoquinolines

The following species have been shown to contain the alkaloids stated:

Corydalis pseudo-adunca[151]   corftaline (78; R^1 = Me, R^2 = H)
Fumaria judaica[111]   bicuculline
Fumaria parviflora[2]   fumaramidine
Fumaria schleicheri[112,152]   fumschleicherine (79) and fumaridine
Hydrastis canadensis[153]   β-hydrastine, hydrastidine (78; R^1 = H, R^2 = Me), and
                                    isohydrastidine (78; R^1 = Me, R^2 = H)

Corftaline and isohydrastidine have been assigned the same structure. Fumschleicherine has been assigned the structure (79) on the basis of its spectra and its conversion into fumaramine;[152] fumaridine is identical with hydrastinimide.[112]

[151] I. A. Israilov, T. Irgashev, M. S. Yunusov, and S. Yu. Yunusov, *Khim. Prir. Soedin.*, 1980, 851.
[152] Kh. G. Kiryakov, Z. H. Mardirossian, D. W. Hughes, and D. B. MacLean, *Phytochemistry*, 1980, **19**, 2507.
[153] I. Messana, R. La Bua, and C. Galeffi, *Gazz. Chim. Ital.*, 1980, **110**, 539.

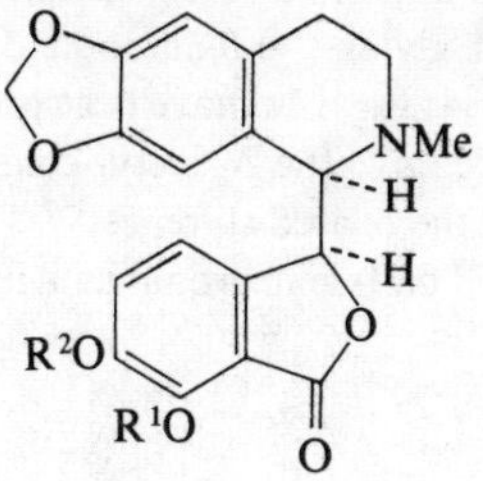

(78) 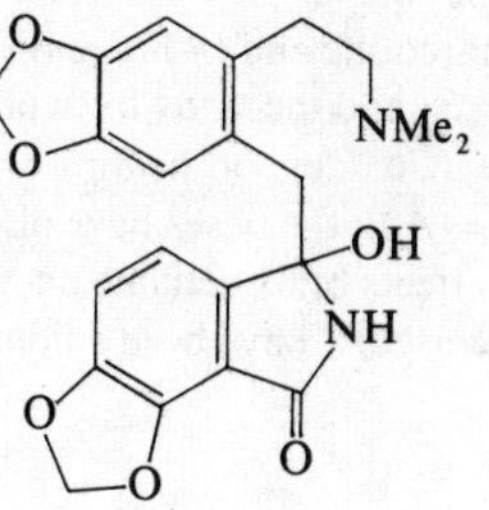   (79)

(−)-β-Hydrastine has been prepared by the hydrogenolysis of the 1-phenyl-1H-5-tetrazolyl ether of (−)-α-narcotoline,[154] and in the racemic form by the reductive cyclization of the quaternary salt of the keto-acid (53) that is obtained from oxidoberberine (51) as described above.[133] The lactone (80), prepared by the dye-sensitized photo-oxidation of oxidonorcoralyne followed by reduction with sodium borohydride and from 6′-acetylpapaveraldine by oxidation with hypo-bromite followed by reduction, has been *N*-methylated and reduced with sodium borohydride to an isomer of cordrastine.[133] Cordrastine itself has been synthesized by the electrolytic reduction of a mixture of the iminium salt (81) and bromomeconine (82), a process that has been shown to be of general applicability to the synthesis of alkaloids of this group.[155]

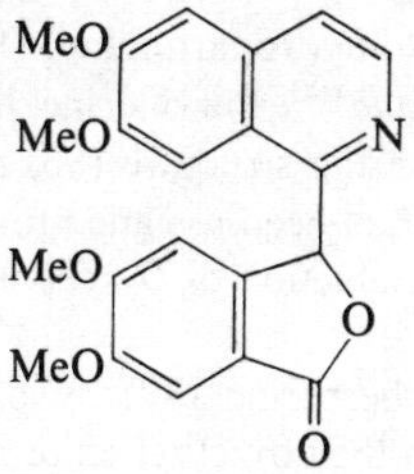

(80) 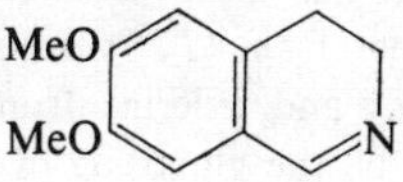   (81)

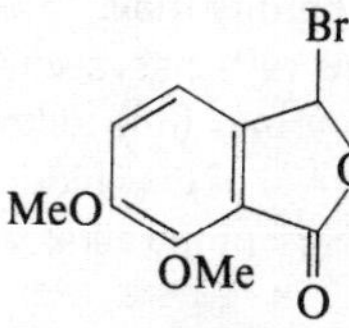

(82)

The reaction between narcotine and cyanogen bromide has been examined. In tetrahydrofuran, the product is the alcohol (83; R = H); in chloroform that contains ethanol, the ether (83; R = Et) is obtained. These products may be hydrolysed and cyclized to narcotine by heating with hydrochloric acid, the

[154] P. Kerekes, G. Gaal, and R. Bognar, *Acta Chim. Acad. Sci. Hung.*, 1980, **103**, 339.
[155] T. Shono, Y. Usui, and H. Hamaguchi, *Tetrahedron Lett.*, 1980, **21**, 1351.

reactions being stereospecific, (−)-α- and (+)-β-narcotine each giving finally (+)-β-narcotine and (+)-α- and (−)-β-narcotine each giving (−)-β-narcotine.[156]

The *cis*- and the *trans*-form of narceine enimide (84; R = Me) have been prepared, and the *N*-oxides, on heating with acetic anhydride, give the *N*-acetyl compounds (84; R = Ac); the bases have also been degraded to the related styrenes.[157]

The effects of bicuculline on myoclonic seizures[158] and on neuronal activity in the motor cortex[159] have been studied.

(83)   (84)

10 Spiro-benzylisoquinolines

Parfumine has been isolated from *Fumaria schleicheri*[112] and *F. judaica*,[111] parviflorine from *F. parviflora*,[2,160] fumaritine (dihydroparfumine) from *F. parviflora*,[2] and lederine from *Corydalis ledebouriana*.[161] Parviflorine has been shown to be the glucoside of parfumine,[160] for which the structure (85) has been determined by X-ray methods.[162] The structure (86) has been assigned to lederine, the *N*-methyl derivative of which is obtained by the reduction of *O*-acetylsibiricine with sodium borohydride.[161]

Fumaricine (90) has been synthesized from the berberine (87) by photolytic transformation into (88) followed by conversion into (89) by ethyl chloroformate, catalytic removal of the chlorine, and reduction with lithium aluminium hydride.[163]

The chemistry of the alkaloids of this group has been reviewed.[164]

[156] P. Kerekes and G. Gaal, *Acta Chim. Acad. Sci. Hung.*, 1980, **103**, 343.
[157] B. Proska and Z. Voticky, *Collect. Czech. Chem. Commun.*, 1980, **45**, 2125.
[158] W. R. Buckett, *Br. J. Pharmacol.*, 1980, **68**, 177P.
[159] J. D. Glass, G. H. Fromm, and A. S. Chattha, *Electroencephalogr. Clin. Neurophysiol.*, 1980, **48**, 16.
[160] S. F. Hussain and M. Shamma, *Tetrahedron Lett.*, 1980, **21**, 1909.
[161] I. A. Israilov, F. M. Melikov, M. S. Yunusov, D. A. Murav'eva, and S. Yu. Yunusov, *Khim. Prir. Soedin.*, 1980, 540.
[162] S. M. Nasirov, L. G. Kuz'mina, Yu, T. Struchkov, I. A. Israilov, M. S. Yunusov, and S. Yu. Yunusov, *Khim. Prir. Soedin.*, 1980, 66.
[163] M. Hanaoka, S. Yasuda, Y. Hirai, K. Nagami, and T. Imanishi, *Heterocycles*, 1980, **14**, 1455.
[164] R. M. Preisner and M. Shamma, *J. Nat. Prod.*, 1980, **43**, 305.

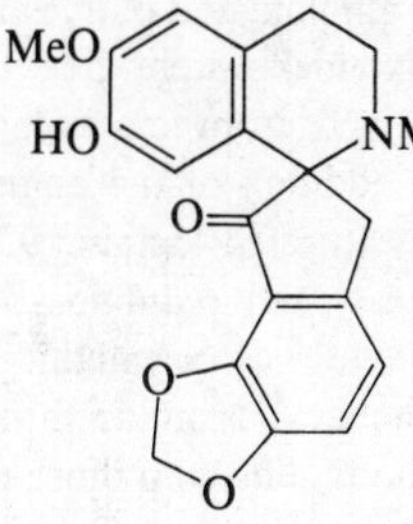

(85)

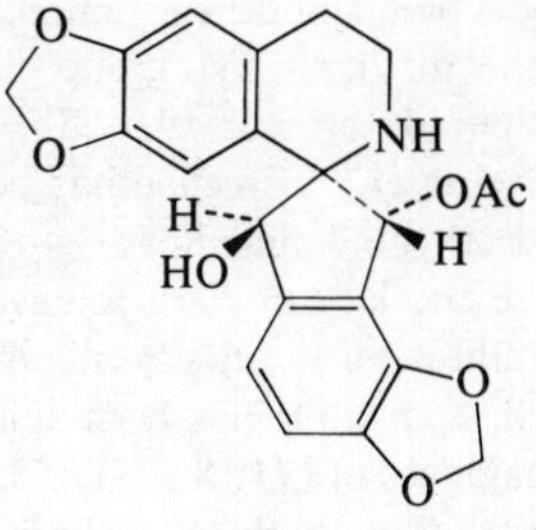

(86)

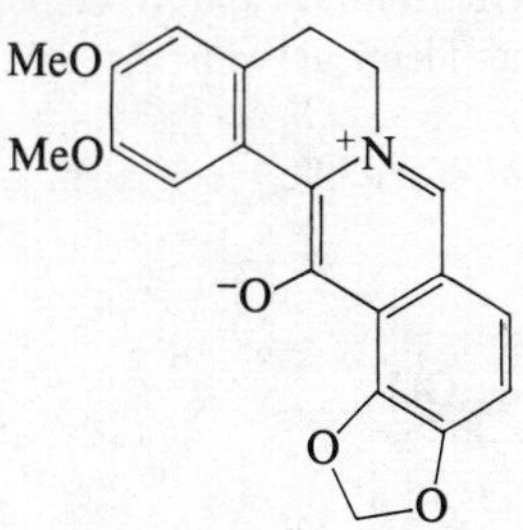

(87)

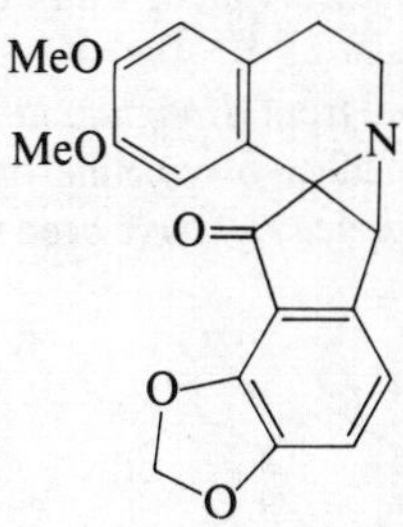

(88)

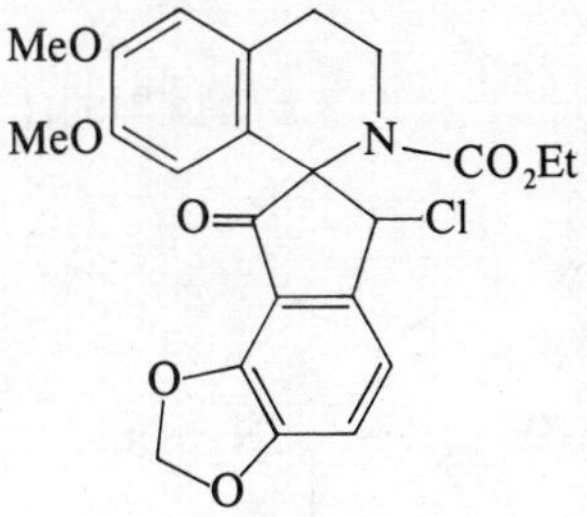

(89)

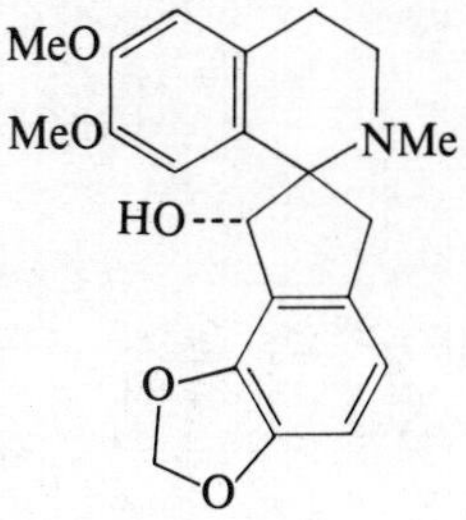

(90)

11 Rhoeadines

Rhoeadine has been isolated from *Bocconia frutescens*,[110] alpinigenine from *Papaver bracteatum*,[146] and rhoeadine, rhoeagenine, glaucamine, glaudine, epi-glaudine, oreodine, and oreogenine from *P. tauricola*.[108]

12 Emetine and Related Bases

(−)-Ankorine (92) has been synthesized from cincholoipon ethyl ester by reaction with 2-benzyloxy-3,4-dimethoxyphenacyl bromide to give (91; X = H$_2$), followed by reduction, oxidation to the lactam by mercuric acetate and EDTA, cyclization, reduction, and debenzylation.[165] A related synthesis of the alkaloid and of its isomer with the hydroxyl group in position 8 in the isoquinoline system has been achieved through (91; X = O), from the lactam derived from cincholoipon ethyl ester.[166] Emetine has been synthesized by Michael addition of the anion of 1-methyl-6,7-dimethoxy-3,4-dihydroisoquinoline to the unsaturated lactam (93) to give the lactam (94), followed by *C*-ethylation, hydrolysis, and reduction.[167] A synthesis of the intermediate (95) that is required for a synthesis of emetine (see Vol. 9, p. 114) has been achieved from 3,4-dimethoxyphenacyl bromide through analogues of (91; X = H$_2$)[168] and (91; X = O),[169] by methods similar to those used for the synthesis of ankorine; omission of the epimerization step affords the intermediate for the synthesis of *cis*-emetine. The benzyl ethers of the isomeric monodemethylated forms of (95) have been converted into 9- and 10-demethyl-tubulosine,[170, 171] the 10-demethyl-compound being identical with the alkaloid obtained from *Alangium lamarckii*.[171]

The effect of emetine on mitochondrial function[172] and the mechanism of its cardiotoxicity[173] have been studied.

(91)

(92)

(93)

(94)

[165] T. Fujii and S. Yoshifuji, *J. Org. Chem.*, 1980, **45**, 1889.
[166] T. Fujii, S. Yoshifuji, and K. Yamada, *Tetrahedron*, 1980, **36**, 965.
[167] T. Kametani, S. A. Surgenor, and K. Fukumoto, *Heterocycles*, 1980, **14**, 303.
[168] T. Fujii and S. Yoshifuji, *Tetrahedron*, 1980, **36**, 1539.
[169] T. Fujii and S. Yoshifuji, *Chem. Pharm. Bull.*, 1979, **27**, 2497.
[170] M. Ohba, H. Hayashi, and T. Fujii, *Heterocycles*, 1980, **14**, 299.
[171] T. Fujii, M. Ohba, A. Popelak, S. C. Pakrashi, and E. Ali, *Heterocycles*, 1980, **14**, 971.
[172] M. A. Dubick and W. C. T. Yang, *J. Pharm. Sci.*, 1981, **70**, 343.
[173] W. C. T. Yang and M. A. Dubick, *Pharmacol. Ther.*, 1980, **10**, 15.

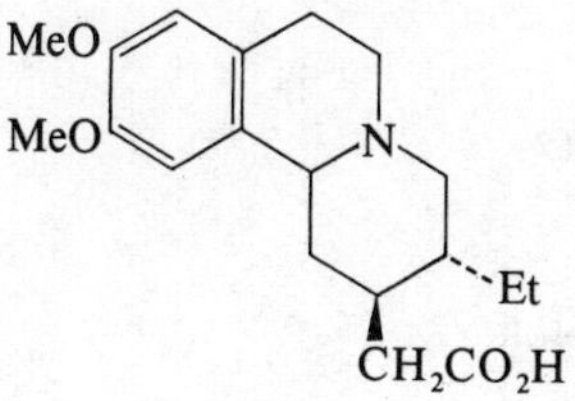

(95)

13 Morphine Alkaloids

Codeine, 14-hydroxycodeine, neopine, and *O*-methylflavinantine have been isolated from *Papaver bracteatum*,[146] flavinantine and amurine from *Meconopsis cambrica*,[113] isosinoacutine from *Stephania elegans*,[174] and a new alkaloid, tridictyophylline, for which the structure (96) was determined by *X*-ray crystallography, from *Triclisia dictyophylla*.[175] Bound morphine, codeine, and thebaine have been found in *P. bracteatum* and *P. somniferum*,[176] and the effect of the period of maturation of the plants on the yield of these three alkaloids from *P. somniferum* has been studied.[177,178] Codeine has been isolated from cell suspension cultures of *P. somniferum* under conditions where no morphine, norcodeine, or thebaine could be detected.[179]

Morphinone methoxymethyl ether has been oxidized by hydrogen peroxide to the 7,8-*β*-epoxide, which has been converted into morphine and heroin 7,8-*β*-epoxides.[180] Neopine has been converted into norneopine through the *N*-nitroso-nor-compound and by treatment with dimethyl azodicarboxylate,[181] into its C-6 epimer, isoneopine, by the reaction of the methanesulphonyl ester with sodium azide, reduction of the azide with zinc and aqueous sodium iodide, and treatment of the resulting amine with nitrous acid,[182,183] and into 6-demethoxythebaine by the action of tetrabutylammonium fluoride on the methanesulphonyl ester. The action of hydrogen peroxide on 6-demethoxythebaine gives 14-hydroxy-*allo*-*ψ*-codeine.[184] Improved conditions have been developed for the conversion of thebaine, northebaine, and *N*-(2-hydroxyethyl)northebaine into morphothebaine and its corresponding analogues, of codeine and *N*-substituted norcodeines into apocodeine and derivatives,[185] and of nalorphine into *N*-allylnorapomorphine.[186]

174 R. L. Khosa, V. K. Lal, A. K. Wahi, and A. B. Ray, *Chem. Ind.* (*London*), 1980, 662.
175 A. L. Spiff, V. Zabel, W. H. Watson, M. A. Zemaitis, A. M. Ateya, D. J. Slatkin, J. E. Knapp, and P. L. Schiff, *J. Nat. Prod.*, 1981, **44**, 160.
176 J. W. Fairbairn and M. J. Steele, *Phytochemistry*, 1980, **19**, 2317.
177 P. J. Hofman and R. C. Menary, *Aust. J. Agric. Res.*, 1980, **31**, 313.
178 J. C. Laughlin, *J. Agric. Sci.*, 1980, **95**, 667.
179 W. H. J. Tam. F. Constabel, and W. G. W. Kurz, *Phytochemistry*, 1980, **19**, 486.
180 N. Miyata, K. Uba, K. Watanabe, and M. Hirobe, *Chem. Pharm. Bull.*, 1980, **28**, 3722.
181 S. Hosztafi, S. Makleit, and R. Bognar, *Acta. Chim. Acad. Sci. Hung.*, 1980, **103**, 371.
182 S. Berenyi and S. Makleit, *Magy, Kem. Foly.*, 1980, **86**, 335.
183 S. Berenyi and S. Makleit, *Acta Chim. Acad. Sci. Hung.*, 1980, **104**, 47.
184 S. Berenyi, S. Makleit, R. Bognar, and A. Tegdes, *Acta Chim. Acad. Sci. Hung.*, 1980, **103**, 365.
185 F. E. Granchelli, C. N. Filer, A. H. Soloway, and J. L. Neumeyer, *J. Org. Chem.*, 1980, **45**, 2275.
186 C. N. Filer, D. Ahern, F. E. Granchelli, J. L. Neumeyer, and S.-J. Law, *J. Org. Chem.*, 1980, **45**, 3465.

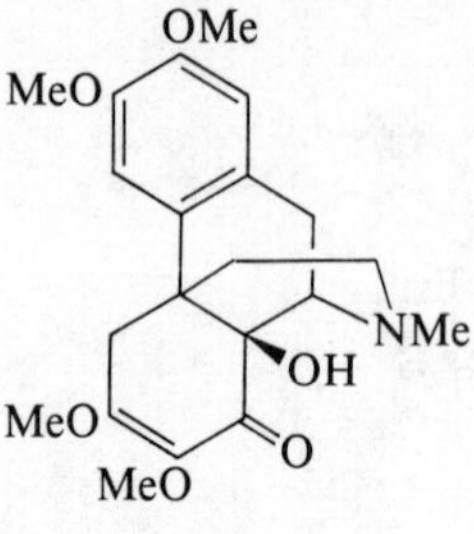

(96)

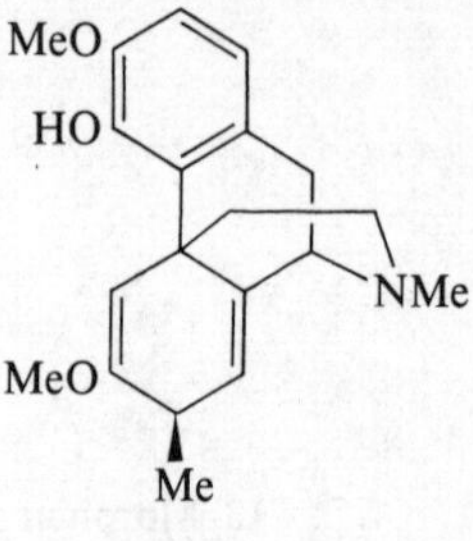

(97)

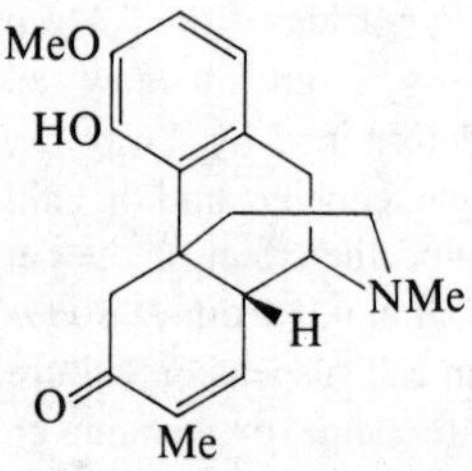

(98)

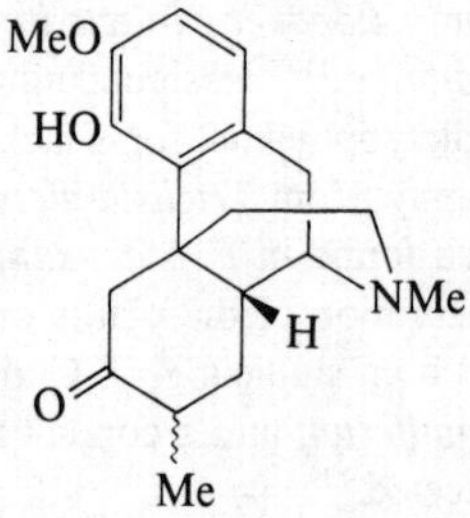

(99)

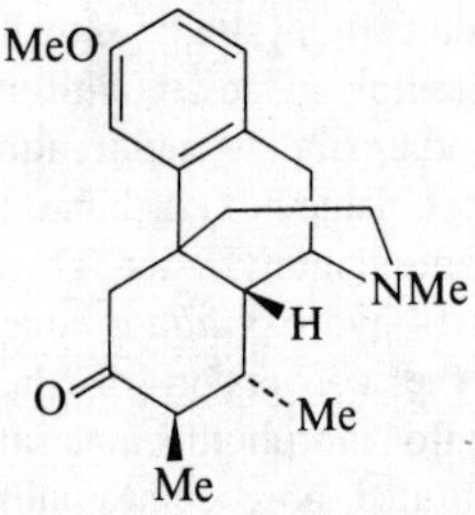

(100)

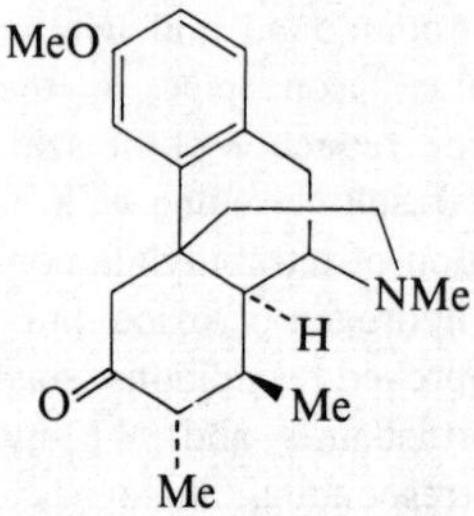

(101)

Thebaine has been shown to react with lithium dimethylcuprate to give 7β-methyldihydro-φ-thebaine (97),[187, 188] acid hydrolysis of which gives 7-methyl-thebainone-A (98) and its B/C-*trans*-isomer, which can be catalytically reduced to 7α- and 7β-dihydrothebainone (99) and their B/C-*trans*-isomers. The 7α- and 7β-dihydrothebainones are also obtained by the action of lithium dimethylcuprate on dihydrothebaine and on dihydrocodeinone enol acetate and subsequent hydrolysis of the products.[187] The phenol (97) can be converted into the

[187] D. L. Leland, J. O. Polazzi, and M. P. Kotick, *J. Org. Chem.*, 1980, **45**, 4026.
[188] D. L. Mites, US P. 4 242 514 (*Chem. Abstr.*, 1981, **94**, 192 532).

4-deoxy-compound by conventional methods, and transformations of this give 4-deoxy-analogues of (98) and (99). Deoxy-(97) itself will react with lithium dimethyl- and diethyl-cuprates to give, after hydrolysis, the ketones (100) and (101) and their 8-ethyl analogues.[189] 14-Methoxycodeinone reacts with lithium dimethylcuprate to give 14-methoxy-7α-methyldihydrocodeinone, which has been converted into the N-cyclobutylmethyl analogue through the nor-compound.[190]

Codeine forms a complex with chromium tricarbonyl, the structure of which has been determined crystallographically. The t-butyldimethylsilyl ether of this complex can be C-methylated by sodium and methyl iodide, and removal of the chromium and silyl residues from the product gives 10β-methylcodeine, which is convertible into 10β-methylmorphine, both of which are active analgesics.[191] Thebaine reacts with tetranitromethane in the presence of dry oxygen to give the peroxide (102), the structure of which was determined by X-ray crystallography,[192] and with 1-chloro-1-nitrosocyclohexane to give 14-hydroxylaminocodeinone, which can be epoxidized at the 7,8-double-bond.[193] A patent covering the production of 14-amino-compounds in the morphine group has been published.[194]

(102)

Derivatives of 14-hydroxydihydrocodeinone have been converted into 6α- and 6β-amino-compounds by condensation with benzylamine and then reduction and hydrogenolysis of the resulting Schiff-bases,[195] and also into their hydrazones.[196] Codeine and ethylmorphine have been nitrated and sulphonated at position 1[197] and fluoro-compounds of structures (103; R = H), (103; R = OH), (104; R = H), and (104; R = OH) have been prepared from dihydrocodeinone and 14-hydroxy-dihydrocodeinone by treatment with diethylaminosulphur trifluoride.[198]

[189] D. L. Leland and M. P. Kotick, J. Med. Chem., 1980, 23, 1427.
[190] R. K. Razdan and A. C. Ghosh, US P. 4 232 028 (Chem. Abstr., 1981, 94, 103 642).
[191] H. B. Arzeno, D. H. R. Barton, S. G. Davies, X. Lusinchi, B. Meunier, and C. Pascard, Nouv. J. Chim., 1980, 4, 369.
[192] R. M. Allen, C. J. Gilmore, G. W. Kirby, and D. McDougall, J. Chem. Soc., Chem. Commun., 1980, 22.
[193] P. Horsewood and G. W. Kirby, J. Chem. Res. (S)., 1980, 401.
[194] R. J. Kobylecki, I. G. Guest, J. W. Lewis, and G. W. Kirby, Ger. (East) P. 135 081 (Chem. Abstr., 1981, 94, 65 925).
[195] L. M. Sayre and P. S. Portoghese, J. Org. Chem., 1980, 45, 3366.
[196] G. W. Pasternak and E. F. Hahn, J. Med. Chem., 1980, 23, 674.
[197] F. Klivenyi and E. Vinkler, Pharmazie, 1979, 34, 668.
[198] G. A. Boswell and R. McK. Henderson, Eur. P. Appl. 9227 (1980) (Chem. Abstr., 1981, 94, 4146).

(103)   (104)

6-Demethylsalutaridine (105; $R^1 = R^2 = H$) can be acetylated to give (105; $R^1 = R^2 = Ac$) and (105; $R^1 = H$, $R^2 = Ac$); of these, the former is hydrolysed to (105; $R^1 = Ac$, $R^2 = H$), which can be methylated to O-acetylsalutaridine (105; $R^1 = Ac$, $R^2 = Me$), which can itself be hydrolysed to salutaridine (105; $R^1 = H$, $R^2 = Me$).[199] A by-product of the cyclization of dihydrosinomenine to (+)-dihydrocodeinone has been identified as the unsaturated ketone (106), which is in tautomeric equilibrium with the saturated ketone (107) in solution; catalytic reduction of the double-bond of (106) has been accomplished.[200]

(105)   (106)

(107)   (108)

[199] G. Horvath and S. Makleit, *Magy, Kem. Foly.*, 1980, **86**, 260.
[200] J. Minamikawa, K. C. Rice, A. E. Jacobson, A. Brossi, T. H. Williams, and J. V. Silverton, *J. Org. Chem.*, 1980, **45**, 1901.

Oxidation of the phenol (108; R = H) with Fremy's salt gives an orthoquinone that is reducible to (108; R = OH); the same sequence of reactions with the bromophenol (108; R = Br) is accompanied by a shift of bromine to position 4.[201]

The acetylation of morphine and of codeine by aspirin[202] and the enzymatic hydrolysis of heroin[203] and the reduction of codeinone[204] have been studied. Dihydronorcodeine and dihydromorphine have been isolated from urine as metabolites of dihydrocodeine.[205] Patents have been published covering the preparation of *N*-(cyclopropylmethyl)normorphine,[206] the conversion of thebaine into codeinone,[207] and the preparation of nalorphine 6-sulphate.[208] Salts of codeine and ethylmorphine with 5-carboxymethyl-2-thio-1,3-thiazan-4-one have been prepared.[209] The circular dichroism[210,211] and fluorescence characteristics[212] of morphine and related bases have been studied.

Descriptions of methods for the estimation or detection of morphine and/or codeine in urine,[213–218] body fluids,[219] blood,[220] blood stains,[221] hair,[222] and opium,[223] for the examination of illicit heroin,[224–226] and for the estimation of dihydromorphinone in plasma[227] have been published, the effect of formaldehyde on the estimation of morphine has been examined,[228] and a bioassay for morphine and naloxone has been described.[229]

[201] C. Olieman, L. Maat, and H. C. Beyerman, *Recl. Trav. Chim. Pays-Bas*, 1980, **99**, 169.

[202] P. Toffel-Nadolny and W. Gielsdorf, *Arch. Kriminol.*, 1980, **166**, 150.

[203] O. Lockridge, N. Mottershaw-Jackson, H. W. Eckerson, and B. N. La Du Bert, *J. Pharmacol. Exp. Ther.*, 1980, **215**, 1.

[204] C. C. Hodges and H. Rapoport, *Phytochemistry*, 1980, **19**, 1681.

[205] W. Gielsdorf, *Fresenius' Z. Anal. Chem.*, 1980, **301**, 434.

[206] D. A. Berkowitz, J. I. DeGraw, H. L. Johnson, J. A. Lawson, and G. H. Loew, US P. 4 218 454 (*Chem. Abstr.*, 1981, **94**, 30 983).

[207] F. Calvo Mondelo, Swiss P. 619 227 (*Chem. Abstr.*, 1981, **94**, 30 984).

[208] H. Yoshimura, K. Ogura, M. Mori, and A. Kawahara, Jpn. Kokai Tokkyo Koho 79 103 889 (*Chem. Abstr.*, 1980, **93**, 8359).

[209] M. O. Oleevskaya, L. I. Petlichnaya, E. L. Besyadetskaya, and B. S. Zimenovskii, *Farm. Zh.* (*Kiev*), 1980, No. 5, p. 51.

[210] J. M. Bowen, T. A. Crone, A. O. Hermann, and N. Purdie, *Anal. Chem.*, 1980, **52**, 2436.

[211] T. A. Crone and N. Purdie, *Anal. Chem.*, 1981, **53**, 17.

[212] W. D. Darwin and E. J. Cone, *J. Pharm. Sci.*, 1980, **69**, 253.

[213] D. F. Mathis and R. D. Budd, *Clin. Toxicol.*, 1980, **16**, 181.

[214] M. Oellerich, *Prakt. Anwend. Enzymimmunoassays. Klin. Chem. Serol.*, 1979, 66.

[215] A. R. Viala, M. Estadieu, A. G. Durand, A. Angeletti, and J. P. Cano, *Trav. Soc. Pharm. Montpellier*, 1980, **40**, 67.

[216] L. Ulrich and P. Ruegsegger, *Arch. Toxicol.*, 1980, **45**, 251.

[217] A. Heyndrickx, P. Demedts, and F. De Clark, *Forensic Toxicol, Proc. Eur. Meet. Int. Assoc. Forensic Toxicol.*, 1979, 119.

[218] M. A. Tahir, A. B. Lateef, and M. Sarwar, *Anal. Lett.*, 1980, **13**, 1635.

[219] J. E. Wallace, S. C. Harris, and M. W. Peek, *Anal. Chem.*, 1980, **52**, 1328.

[220] L. Von Meyer, G. Drasch, and G. Kauert, *Beitr. Gerichtl. Med.*, 1980, **38**, 63.

[221] F. P. Smith, R. C. Shaler, C. R. Mortimer, and L. T. Errichetto, *J. Forensic Sci.*, 1980, **25**, 369.

[222] E. Klug, *Z. Rechtsmed.*, 1980, **84**, 189.

[223] P. V. Vaidya, M. D. Pundlik, and S. K. Meghal, *J. Assoc. Offic. Anal. Chem.*, 1980, **63**, 685.

[224] J. L. Love and L. K. Pannell, *J. Forensic Sci.*, 1980, **25**, 320.

[225] F. Van Vendeloo, J. P. Francke, and R. A. De Zeeuw, *Pharm. Weekbl., Sci. Ed.*, 1980, **2**, No. 5, p. 129.

[226] M. Gloger and H. Neumann, *Arch. Kriminol.*, 1980, **166**, 89.

[227] I. L. Honigberg and J. T. Stewart, *J. Pharm. Sci.*, 1980, **69**, 1171.

[228] W. W. Manders and B. R. Keppler, *Aviat. Space Environ. Med.*, 1980, **51**, 993.

[229] F. Liebman and G. Klinger, *Eur. J. Pharmacol.*, 1980, **62**, 167.

In the synthesis of alkaloids of this group, reticuline has been oxidized to flavinantine and *O*-methylated[230] and *N*-formyl- and *N*-ethoxycarbonyl-nor-reticuline and their 6'-bromo- and 6'-chloro-derivatives have been oxidized to the corresponding derivatives of salutaridine by quaternary ammonium salts of the anions $[I(O_2CCX_3)]^-$, where X is H, F, or Cl.[231] Dihydrothebainone has been synthesized from the isoquinoline (109) by Birch reduction and hydrolysis to produce (110; $R^1 = R^2 = H$), *N*-formylation and bromination [yielding (110; $R^1 = CHO$, $R^2 = Br$)], cyclization, reduction of N—CHO to NMe, and removal of bromine.[232] 1-Acetylamino-*N*-methoxycarbonylnordihydrothebainone has been prepared from (110; $R^1 = CO_2Me$, $R^2 = H$) by coupling with diazotized sulphanilic acid, reduction of the resulting azo-compound to (110; $R^1 = CO_2Me$, $R^2 = NH_2$), acetylation of the amino-group, and then cyclization with boron trifluoride.[233] Using the Birch reduction of suitably substituted isoquinolines and the Grewe cyclization as in the aforementioned syntheses, 4-deoxydihydrothebainone,[234] 2-hydroxy-3-deoxy-*N*-formylnordihydrothebainone, and (from this) 3-deoxydihydromorphine[235] have been prepared.

(109)   (110)

The analgesic,[236-239] local analgesic,[240] hyperalgesic,[241] central,[242] mydriatic,[243] and mutagenic[244] effects of morphine have been studied, as have the effects of the alkaloid on behaviour,[245-273] on the brain,[274-293] on the gastro-intestinal tract,[294-305]

[230] D. S. Bhakuni and A. N. Singh, *Tetrahedron*, 1979, **35**, 2365.

[231] C. Szantay, G. Blasko, M. Barczai-Beke, P. Pechy, and G. Dornyei, *Tetrahedron*, 1980, **36**, 3509.

[232] K. C. Rice, *J. Org. Chem.*, 1980, **45**, 3135.

[233] S. Archer, O. Kenichiro, and D. J. Schepis, *J. Heterocycl. Chem.*, 1981, **18**, 357.

[234] C. Olieman, P. Nagelhout, A. D. De Groot, L. Maat, and H. C. Beyerman, *Recl. Trav. Chim. Pays-Bas*, 1980, **99**, 353.

[235] F. L. Hsu, K. C. Rice, and A. Brossi, *Helv. Chim. Acta*, 1980, **63**, 2042.

[236] A. Puget, *Zentralbl. Veterinaermed., Reihe A*, 1980, **27**, 157.

[237] S. G. Dennis and R. Melzack, *Exp. Neurol.*, 1980, **69**, 260.

[238] R. Poggioli, M. Castelli, S. Genedani, and A. Bertholini, *Riv. Farmacol. Ter.*, 1980, **11**, 11.

[239] S. Kubota, H. Nakano, J. Tamura, T. Koyanagi, C. Sakamoto, K. Harano, and Y. Nomoto, *Fukuoka Igaku Zasshi*, 1980, **71**, 485.

[240] S. H. Ferreira, *Adv. Prostaglandin Thromboxane Res.*, 1980, **8**, 1207.

[241] C. J. Woolf, *Brain Res.*, 1981, **209**, 491.

[242] S.-X. Fu, S.-S. Zhang, Y. Jin, and Y.-S. Li, *Chung-kuo Yao Li Hsueh Pao*, 1980, **1**, 81.

[243] C. Adler, O. Keren, and A. D. Korczyn, *J. Neural. Transm.*, 1980, **48**, 43.

[244] N. Swain, R. K. Das, and M. Paul, *Mutat. Res.*, 1980, **78**, 97.

[245] M. Hirabayashi, F. Iwai, M. Iizuka, T. Mesaki, M. R. Alam, and S. Tadokoro, *Nippon Yakurigaku Zasshi*, 1979, **75**, 683.

[246] M. D. Curley, J. M. Walsh, and L. S. Burch. *Pharmacol. Biochem. Behav.*, 1980, **12**, 413.

[247] J. J. Teal and S. G. Holtzman, *Pharmacol. Biochem. Behav.*, 1980, **12**, 587.

[248] R. A. Howd and G. T. Pryor, *Pharmacol. Biochem. Behav.*, 1980, **12**, 577.

[249] D. S. Kosersky, M. D. Kowolenko, and J. F. Howd, *Pharmacol. Biochem. Behav.*, 1980, **12**, 625.

[250] J. T. Huang, *Res. Commun. Subst. Abuse*, 1980, **1**, 29.

[251] D. J. Leander, *Pharmacol. Biochem. Behav.*, 1980, **12**, 797.

[252] P. A. Caza and L. P. Spear, *Pharmacol. Biochem. Behav.*, 1980, **13**, 45.

[253] R. Langwinski and J. Niedzielski, *Naunyn-Schmiedberg's Arch. Pharmacol.*, 1980, **312**, 225.

[254] J. Panksepp and F. G. DeEskinazi, *J. Comp. Physiol. Psychol.*, 1980, **94**, 650.

[255] D. E. McMillan, W. T. McGivney, and W. C. Hardwick, *J. Pharmacol. Exp. Ther.*, 1980, **215**, 9.

[256] M. Babbini, M. Gaiardi, and M. Bartoletti, *Psychopharmacology*, 1980, **70**, 73.

[257] J. J. Teal and S. G. Holtzman, *J. Pharmacol. Exp. Ther.*, 1980, **215**, 469.

[258] J. E. Sherman, C. Pickman, A. Rice, J. C. Liebeskind, and E. W. Holman, *Pharmacol. Biochem. Behav.*, 1980, **13**, 501.

[259] G. Urca and H. Frenk, *Pharmacol. Biochem. Behav.*, 1980, **13**, 343.

[260] S. G. Beck and J. H. O'Brien, *Physiol. Behav.*, 1980, **25**, 559.

[261] D. M. Grilly, R. F. Genovese, and M. J. Nowak, *Psychopharmacology*, 1980, **70**, 213.

[262] J. D. Levine, D. T. Murphy, D. Siedenwurm, A. Cortez, and H. L. Fields, *Brain Res.*, 1980, **201**, 23.

[263] M. De Ryck, T. Schallert, and P. Teitelbaum, *Brain Res.*, 1980, **201**, 143.

[264] S. Schenk, T. Williams, A. Coupal, and P. Shizgal, *Physiol. Psychol.*, 1980, **8**, 372.

[265] V. A. Lewis and I. C. Liles, *Life Sci.*, 1980, **27**, 2649.

[266] S. Kaakkola, *Acta Pharmacol. Toxicol.*, 1980, **47**, 385.

[267] S. Miksic, G. Sherman, and H. Lal, *Psychopharmacology*, 1981, **72**, 179.

[268] M. Bozarth and R. A. Wise, *Life Sci.*, 1981, **28**, 551.

[269] G. J. Shaefer and S. G. Holtzman, *Pharmacol. Biochem. Behav.*, 1981, **217**, 105.

[270] B. J. Meyerson, *Eur. J. Pharmacol.*, 1981, **69**, 453.

[271] A. L. Bandman, E. T. Vasilenko, and N. N. Pshenkina, *Byull. Eksp. Biol. Med.*, 1981, **91**, 185.

[272] J. M. Nazzaro, T. F. Seeger, and E. L. Gardner, *Pharmacol. Biochem. Behav.*, 1981, **14**, 325.

[273] G. F. Steinfels and G. A. Young, *Pharmacol. Biochem. Behav.*, 1981, **14**, 393.

[274] D. Le Bars, A. H. Dickenson, and J. M. Besson, *Brain Res.*, 1980, **189**, 467.

[275] F. J. Einspahr and M. F. Piercey, *J. Pharmacol. Exp. Ther.*, 1980, **213**, 456.

[276] V. Y. H. Hook, N. M. Lee, and H. H. Loh, *J. Neurochem.*, 1980, **34**, 1274.

[277] M. A. Linseman, *Brain Res. Bull.*, 1980, **5**, 121.

[278] R. Tissot, *Neuropsychobiology*, 1980, **6**, 170.

[279] B. Mukherji, K. Suemaru, N. Sakai, A. K. Ghosh, and H. A. Slooiter, *Biochem. Pharmacol.*, 1980, **29**, 1608.

[280] J. E. Morley, T. Yamata, J. H. Walsh, C. B. Lamers, H. Wong, A. Shulkes, D. A. Damassa, J. Gordon, H. E. Carlson, and J. M. Hershman, *Life Sci.*, 1980, **26**, 2239.

[281] N. Dafny and B. M. Rigor, *J. Neurosci. Res.*, 1980, **5**, 117.

[282] A. G. Phillips and F. G. Le Piane, *Pharmacol. Biochem. Behav.*, 1980, **12**, 965.

[283] J. F. De France, J. C. Stanley, K. M. Taber, J. E. Marchand, and N. Dafny, *Exp. Neurol.*, 1980, **69**, 311.

[284] W. R. Klemm and C. G. Malari, *Prog. Neuro-Psychopharmacol.*, 1980, **4**, 1.

[285] J. W. Phillis, Z. G. Jiang, B. J. Chelack, and P. H. Wu, *Eur. J. Pharmacol.*, 1980, **65**, 97.

[286] T. Miyagama, S. Sakurada, K. Shima, R. Ando, N. Takahashi, and K. Kisara, *Jpn. J. Pharmacol.*, 1980, **30**, 463.

[287] R. Samanin, L. Cervo, C. Rochat, E. Poggesi, and T. Mennini, *Life Sci.*, 1980, **27**, 1141.

[288] F. H. Gage, J. J. Valdes, and R. G. Thompson, *Psychopharmacology*, 1980, **70**, 113.

[289] M. Stanley and S. Wilk, *J. Pharm. Pharmacol.*, 1980, **32**, 567.

[290] T. G. Reigle and J. W. Huff, *Biochem. Pharmacol.*, 1980, **29**, 2249.

[291] T. H. Cicero, E. R. Meyer, S. M. Gabriel, R. D. Bell, and C. E. Wilcox, *Brain Res.*, 1980, **202**, 151.

[292] N. Dafny, J. Marchand, R. McClung, J. Salamy, S. Sands, H. Wachtendorf, and T. F. Burks, *J. Neurosci. Res.*, 1980, **5**, 399.

[293] J. M. Lakoski, J. S. Moorland, and G. F. Gebhart, *Life Sci.*, 1980, **27**, 2639.

[294] R. A. Lefebvre and J. L. Willems, *Arch. Int. Pharmacodyn. Ther.*, 1979, **242**, 310.

[295] G. C. Gillan and D. Pollock, *Br. J. Pharmacol.*, 1980, **68**, 381.

[296] D. Mailman, *Br. J. Pharmacol.*, 1980, **64**, 617.

[297] N. W. Weisbrot, S. E. Sussman, J. J. Stewart, and T. F. Burks, *J. Pharmacol. Exp. Ther.*, 1980, **214**, 333.

[298] Y. Ito, K. Tajima, K. Kitamura, and H. Kuriyama, *Gastrointest. Motil.* [*Int. Symp.*] *7th.*, 1979, 155.

[299] C.-L. Wong, M. B. Roberts, and M.-K. Wai, *Eur. J. Pharmacol.*, 1980, **64**, 289.

[300] N. W. Weisbrot, P. J. Thor, E. M. Copeland, and T. F. Burks, *J. Pharmacol. Exp. Ther.*, 1980, **215**, 515.

[301] M. K. Lee and I. M. Coupar, *Eur. J. Pharmacol.*, 1980, **68**, 501.

[302] A. Tavani, G. Bianchi, P. Ferretti, and L. Manara, *Life Sci.*, 1980, **27**, 2211.

[303] J. S. McKay, B. D. Linaker, and L. A. Turnberg, *Gastroenterology*, 1981, **80**, 279.

[304] J. J. Stewart, *J. Pharmacol. Exp. Ther.*, 1981, **216**, 521.

[305] P. Ferretti, G. Bianchi, A. Tavani, and L. Manara, *Res. Commun. Subst. Abuse*, 1981, **2**, 1.

and on temperature,[306–314] locomotor activity,[315–321] alimentary leucocytes,[322] mitochondrial membranes,[323] intrabiliary pressure,[324] immuno-reactivity,[325,326] peripheral nervous activity,[327] biosynthetic pathways,[328] enzyme systems,[329,330] taste aversion,[331,332] the development of opiate receptors,[333] metabolism of catecholamine,[334–337] neuronal firing,[338–341] food intake,[342–345] the cardio-vascular system,[346–348] respiration,[349–352] androgen binding,[353] foetal metabolism,[354] neuro-muscular[355] and synaptic[356] transmission, intra-ocular pressure,[357] regeneration of

[306] J. N. McDougal, P. R. Marques, and T. F. Burks, *Proc. West. Pharmacol. Soc., 23rd*, 1980, p. 235.
[307] F. C. Tulnay, *Life Sci.*, 1980, **27**, 511.
[308] P. Slater and C. Blundell, *Naunyn-Schmiedeberg's Arch. Pharmacol.*, 1980, **313**, 125.
[309] C. I. Prakash and P. K. Dey, *Indian J. Physiol. Pharmacol.*, 1980, **24**, 205.
[310] J. N. McDougal, P. R. Marques, and T. F. Burks, *Life Sci.*, 1980, **27**, 2679.
[311] J. N. McDougal, P. R. Marques, and T. F. Burks, *Life Sci.*, 1981, **28**, 137.
[312] R. Eikelboom and J. Stewart, *Psychopharmacology*, 1981, **72**, 147.
[313] T. Turski, W. Turski, S. J. Czuczwar, and Z. Kleinrock, *Psychopharmacology*, 1981, **72**, 211.
[314] J. Stewart and R. Eikelboom, *Life Sci.*, 1981, **28**, 1041.
[315] G. E. Martin and N. L. Papp, *Life Sci.*, 1980, **26**, 1731.
[316] P. Slater and C. Blundell, *Naunyn-Schmiedeberg's Arch. Pharmacol.*, 1980, **312**, 219.
[317] E. Alleva, C. Castellano, and A. Oliverio, *Brain Res.*, 1980, **198**, 249.
[318] H. Murakami and T. Segawa, *Jpn. J. Pharmacol.*, 1980, **30**, 565.
[319] K. J. Katz and K. Schmaltz, *Neurosci. Lett.*, 1980, **19**, 85.
[320] M. Sansone and A. Oliverio, *Arch. Int. Pharmacodyn. Ther.*, 1980, **247**, 71.
[321] V. Cuomo, I. Cortese, and G. Racagni, *Pharmacol. Res. Commun.*, 1981, **13**, 87.
[322] J. Luza, *Acta Univ. Palacki. Olomuc., Fac. Med.*, 1979, **89**, 95.
[323] V. V. Chistyakov and G. P. Gegenava, *Biokhimiya*, 1980, **45**, 492.
[324] F. Romo-Salas, J. A. Aldrete, and Y. Franatovic, *Surg., Gynecol. Obstet.*, 1980, **150**, 551.
[325] L. A. Laasberg, E. E. Johnson, and J. Hedley-Whyte, *J. Pharmacol. Exp. Ther.*, 1980, **212**, 496.
[326] M. Gungor, E. Genc, H. Sagduyu, L. Eroglu, and H. Koyuncuoglu, *Experientia*, 1980, **36**, 1309.
[327] Y. Maruyama, K. Shinioji, H. Shimizu, Y. Sato, H. Kuribashi, and R. Kaieda, *Pain*, 1980, **8**, 63.
[328] T. H. Kwong, S. C. Wong, and D. Yeung, *IRCS Med. Sci. Libr. Compend.*, 1980, **8**, 134.
[329] D. Lee, S. C. Wong, and D. Yeung, *IRCS Med. Sci. Libr. Compend.*, 1980, **8**, 148.
[330] S. C. Wong and D. Yeung, *IRCS Med. Sci. Libr. Compend.*, 1980, **8**, 151.
[331] A. L. Riley, W. J. Jacobs, and V. M. Lolordo, *Physiol. Psychol.*, 1978, **6**, 96.
[332] J. E. Gorman, R. N. DeObaldia, R. C. Scott, and L. D. Reid, *Physiol. Psychol.*, 1978, **6**, 101.
[333] D. Tsang and S. C. Ng, *Brain Res.*, 1980, **188**, 199.
[334] R. B. Rastogi, Z. Merali, and R. L. Singhal, *Gen. Pharmacol.*, 1980, **11**, 201.
[335] C. Guaza, A. Torrellas, S. Borrell, and J. Borrell, *Psychopharmacology*, 1980, **68**, 43.
[336] B. L. Roth, M. P. Galloway, and C. J. Coscia, *Brain Res.*, 1980, **197**, 561.
[337] F. Karoum, R. J. Wyatt, and E. Costa, *J. Pharmacol. Exp. Ther.*, 1981, **216**, 321.
[338] D. A. Hosford and H. J. Haigler, *J. Pharmacol. Exp. Ther.*, 1980, **213**, 355.
[339] G. Clarke, D. W. Lincoln, and P. Wood, *J. Physiol.*, 1980, **303**, 59P.
[340] Q. J. Pittman, J. D. Hatton, and F. E. Bloom, *Proc. Natl. Acad. Sci. USA*, 1980, **77**, 5527.
[341] J. D. Wood, *Gastroenterology*, 1980, **79**, 1222.
[342] R. Marks-Kaufman and B. B. Kanarek, *Pharmacol. Biochem. Behav.*, 1980, **12**, 427.
[343] S. Yanaura, T. Suzuki, and T. Kawai, *Jpn. J. Pharmacol.*, 1980, **32**, 145.
[344] J. W. McKearney, *Psychopharmacology*, 1980, **70**, 69.
[345] D. J. Sanger and P. S. McCarthy, *Psychopharmacology*, 1980, **72**, 103.
[346] K. P. Schrank, J. G. Fewel, K. V. Arom, J. K. Trinkle, G. E. Webb, and F. L. Grover, *J. Surg. Res.*, 1980, **28**, 319.
[347] J. Weinberg and B. Altura, *Subst. Alcohol Actions/Misuse*, 1980, **1**, 71.
[348] D. C. Lee, M. O. Lee, and D. H. Clifford, *Am. J. Clin. Med.*, 1980, **24**, 205.
[349] E. Beubler, *Naunyn-Schmiedeberg's Arch. Pharmacol.*, 1980, **311**, 199.
[350] G. E. Isom and M. J. Meldrum, *Proc. West. Pharmacol. Soc., 23rd*, 1980, p. 89.
[351] B. J. Martin, C. W. Zwillich, and J. V. Weil, *Med. Sci. Sports Exercise*, 1980, **12**, 285.
[352] S. Oktay, R. Onur, M. Ilhan, and R. K. Turker, *Eur. J. Pharmacol.*, 1981, **70**, 257.
[353] P. J. Sheridan and M. James, *Int. J. Fertil.*, 1980, **25**, 36.
[354] J. R. Raye, J. W. Dubin, and J. N. Blechner, *Am. J. Obstet. Gynecol.*, 1980, **137**, 505.
[355] M. R. Bennett and N. A. Lavidis, *Br. J. Pharmacol.*, 1980, **69**, 185.
[356] E. Minker, A. Vegh, and M. H. Osman-Taha, *Acta Physiol. Acad. Sci. Hung.*, 1980, **55**, 379.
[357] R. W. Hahnenberger, *Albrecht von Graefes Arch. Klin. Exp. Ophthalmol.*, 1980, **214**, 27.

nerves,[358] trachaeal muscle,[359] reflex latency,[360] transport of ions,[361] vomiting,[362] the formation of triglycerides,[363] the toxicity of nitrogen mustards,[364] the release of acetylcholine,[365] prolactin,[366-368] follicle-stimulating hormone,[369] luteinizing hormone,[369,370] growth hormone,[367,371] thyroid-stimulating hormone,[372] pituitary hormones,[373] vasopressin,[374] insulin,[375] β-endorphin,[376] gonadotrophin,[368,377] bradykinin,[378] noradrenaline,[379] 5-hydroxytryptamine,[380,381] and adenosine,[382] the formation of cyclic AMP,[383] levels of blood sugar[384] and of serum protein,[385] the DNA and RNA content of tissues,[386] the metabolism of phencyclidine,[387] the reduction of naltrexone to 6α-naltrexol,[388] the effects of inhibitors of monoamine oxidases,[389] and in anaesthesia.[390,391] The effects of (+)- and of (−)-morphine on the vas deferens of the rat have been compared[392] and the cross-tolerance between morphine and ethanol[393] and the absence of etorphine–morphine cross-tolerance[394] have been reported.

[358] C. T. Harston, A. Morrow, and R. M. Kostrzewa, *Brain Res. Bull.*, 1980, **5**, 421.

[359] N. Toda and Y. Hatano, *Anesthesiology*, 1980, **53**, 93.

[360] J. D. Devine, D. T. Murphy, D. Seidenwurm, A. Cortez, and H. L. Fields, *Brain Res.*, 1980, **201**, 129.

[361] J. S. McKay, B. D. Linaker, and L. A. Turnberg, *Gastroenterology*, 1981, **80**, 279.

[362] R. A. Lefebvre, J. L. Willems, and M. C. Bogaert, *Eur. J. Pharmacol.*, 1981, **69**, 139.

[363] R. G. Lamb and W. L. Dewey, *J. Pharmacol. Exp. Ther.*, 1981, **216**, 496.

[364] A. Akintonwa, *Clin. Toxicol.*, 1981, **18**, 451.

[365] J. A. Down and J. C. Szerd, *Br. J. Pharmacol.*, 1980, **68**, 47.

[366] S. N. Deyo, R. M. Swift, R. J. Miller, and V. S. Fang, *Endocrinology*, 1980, **106**, 1469.

[367] J. Koenig, M. A. Mayfield, R. J. Coppings, S. M. McCann, and L. Kruhlich, *Brain Res.*, 1980, **197**, 453.

[368] T. Ieiri, H. Suzuki, and S. Shunoda, *Horumon to Rinsho*, 1980, **28**, 1017.

[369] P. W. Sylvester, H. T. Chen, and J. Meites, *Proc. Soc. Exp. Biol. Med.*, 1980, **164**, 207.

[370] B. A. Shainker and T. J. Cicero, *Res. Commun. Subst. Abuse*, 1980, **1**, 225.

[371] T. S. Farris and G. E. Richards, *Life Sci.*, 1980, **27**, 1345.

[372] T. Muraki, T. Nakandate, Y. Tokunaga, and R. Kato, *J. Endocrinol.*, 1980, **86**, 357.

[373] B. Lutz-Bucher and B. Koch, *Eur. J. Pharmacol.*, 1980, **66**, 375.

[374] M. Ukai, K. Okomura, and I. Ishihara, *Kankyo Igaku Kenkyusho Nenpo* (*Nagoya Daigaku*) 1980, **31**, 203.

[375] S. Ryder, G. M. Gollapudi, and D. Ryder, *Horm. Metab. Res.*, 1980, **12**, 412.

[376] B. Hoellt, L. Haarmann, R. Przewlocki, and M. Jerlicz, *Adv. Biochem. Pharmacol.*, 1980, **22**, 399.

[377] T. Muraki, Y. Tokunaga, T. Nakadate, and R. Kato, *Arch. Int. Pharmacodyn. Ther.*, 1980, **246**, 324.

[378] R. Inoki, T. Hayashi, K. Matsumoto, Y. Katani, and M. Oka, *Arch. Int. Pharmacodyn. Ther.*, 1980, **247**, 283.

[379] M. Brodie and E. G. McQueen, *Naunyn-Schmiedeberg's Arch. Pharmacol.*, 1980, **313**, 125.

[380] F. Godefroy, J. Weil-Fugazza, D. Coudert, and J. M. Besson, *Brain Res.*, 1980, **199**, 415.

[381] C. J. Pycock, S. Burns, and R. Morris, *Neurosci. Lett.*, 1981, **22**, 313.

[382] J. W. Phillis, Z. G. Jiang, B. J. Chelack, and P. H. Wu, *Eur. J. Pharmacol.*, 1980, **65**, 97.

[383] E. Beubler and F. Lembeek, *Br. J. Pharmacol.*, 1980, **68**, 513.

[384] E. Ipp, V. Schusdziarra, V. Harris, and R. H. Unger, *Endocrinology*, 1980, **107**, 461.

[385] Z. I. Eltsova and M. V. Komendantova, *Farmakol. Toksikol.* (*Moscow*), 1980, **43**, 422.

[386] N. Swain, B. K. Senapati, R. N. Kar, and R. K. Das, *Indian J. Exp. Biol.*, 1980, **18**, 1188.

[387] T. M. Steeger, R. A. Howd, and G. T. Pryor, *Res. Commun. Subst. Abuse*, 1980, **1**, 131.

[388] S. C. Roerig, T. M. Fujimoto, and R. I. H. Wang, *Drug Metab. Dispos.*, 1980, **8**, 295.

[389] J. Garzon, R. Moratella, and J. Del Rio, *Neuropharmacology*, 1980, **19**, 723.

[390] P. L. Wilkinson, J. V. Tyberg, J. R. Mayers, and A. E. White, *Can. J. Anaesth. Soc.*, 1980, **27**, 230.

[391] I. Susko, D. J. Sturion, A. G. Raiser, and N. L. Pippi, *Rev. Cent. Cienc. Rurais* (*Univ. Fed. St. Maria*), 1980, **10**, 343.

[392] Y. F. Jacquet, *Science*, 1980, **210**, 95.

[393] J. M. Mayer, J. M. Khanna, H. Kalant, and L. Spero, *Eur. J. Pharmacol.*, 1980, **63**, 223.

[394] D. G. Lange, J. M. Fujimoto, C. L. Fuhrman-Lane, and R. I. H. Wang, *Toxicol. Appl. Pharmacol.*, 1980, **54**, 177.

128   *The Alkaloids*

A review of the use of naloxone in mechanisms for the control of pain has been published[395] and the effects of this compound on behaviour,[254,396-415] the brain,[284,416-419] various forms of shock,[420-428] respiration,[349] primary apnoea in neonates,[429] the intake of food[430-433] and of water,[431-437] renal function,[438] the gastro-intestinal tract,[299,305] bodily temperature,[439] memory,[440] electroseizures,[441] epilepsy,[442] immuno-reactivity,[325] ketamine narcosis,[443] reserpine-induced

[395] R. G. Hill, *Neurosci. Lett.*, 1981, **21**, 217.
[396] R. D. E. Sewell, *IRCS Med. Sci. Libr. Compend.*, 1980, **8**, 224.
[397] M. S. Fanselow, *Physiol. Psychol.*, 1979, **7**, 70.
[398] B. M. Myers and M. J. Baum, *Pharmacol. Biochem. Behav.*, 1980, **12**, 365.
[399] G. A. Young, *Life Sci.*, 1980, **26**, 1787.
[400] R. Stretch, C. Pink, and P. Gervaise, *Can. J. Physiol. Pharmacol.*, 1980, **58**, 584.
[401] O. C. Snead and L. J. Bearden, *Neurology*, 1980, **30**, 832.
[402] T. F. Seeger, J. M. Nazzaro, and E. L. Gardner, *Eur. J. Pharmacol.*, 1980, **65**, 435.
[403] L. L. Hernandez and D. A. Powell, *Life Sci.*, 1980, **27**, 863.
[404] T. P. S. Oei, *Pharmacol. Biochem. Behav.*, 1980, **13**, 457.
[405] M. S. Fanselow, R. A. Sigmundi, and R. C. Bolles, *Physiol. Psychol.*, 1980, **8**, 369.
[406] Y. F. Jacquet, *Behav. Brain Res.*, 1980, **1**, 543.
[407] S. Miksic, G. Sherman, and H. Lal, *Psychopharmacology*, 1981, **72**, 179.
[408] S. R. Goldberg, W. H. Morse, and D. M. Goldberg, *J. Pharmacol. Exp. Ther.*, 1981, **216**, 500.
[409] Y. F. Jacquet, *Pharmacol. Biochem. Behav.*, 1980, **13**, 585.
[410] R. D. Spealman, R. T. Kelleher, W. H. Morse, and S. R. Goldberg, *Psychopharmacol. Bull.*, 1981, **17**, 54.
[411] W. Perry, R. U. Esposito, and C. Kornetsky, *Pharmacol. Biochem. Behav.*, 1981, **14**, 247.
[412] C. W. T. Pilcher and S. M. Jones, *Pharmacol. Biochem. Behav.*, 1981, **14**, 299.
[413] S. K. McConnell, M. J. Baum, and T. M. Badger, *Horm. Behav.*, 1981, **15**, 16.
[414] L. J. Botticelli and R. J. Wurtman, *Brain Res.*, 1981, **210**, 479.
[415] G. F. Koob, R. E. Strecker, and F. L. Bloom, *Subst. Alcohol Actions/Misuse*, 1980, **1**, 447.
[416] E. E. Zvartau, *Farmakol. Toksikol. (Moscow)*, 1980, **43**, 313.
[417] C. Hardy, J. Panksepp, J. Rossi, and A. J. Zolovick, *Brain Res.*, 1980, **194**, 293.
[418] F. Baldino, A. L. Beckman, and M. W. Adler, *Brain Res.*, 1980, **196**, 199.
[419] A. A. Artru, P. A. Steen, and J. D. Michenfelder, *Anesthesiology*, 1980, **52**, 517.
[420] A. I. Faden and J. W. Holaday, *J. Pharmacol. Exp. Ther.*, 1980, **212**, 441.
[421] D. G. Reynolds, N. J. Gurll, T. Vargish, R. B. Lechner, A. I. Faden, and J. W. Holaday, *Circ. Shock*, 1980, **7**, 39.
[422] H. F. Janssen and L. O. Lutherer, *Brain Res.*, 1980, **194**, 608.
[423] R. M. Raymond, J. M. Harkema, W. V. Stoffs, and J. E. Emerson, *Surg., Gynecol. Obstet.*, 1981, **152**, 159.
[424] J. W. Holaday and A. I. Faden, *Adv. Biochem. Psychopharmacol.*, 1981, 421.
[425] M. T. Curtis and A. M. Lefer, *Experientia*, 1981, **37**, 403.
[426] J. W. Holaday and A. I. Faden, *Brain Res.*, 1980, **189**, 295.
[427] G. Feurstein, R. Ailam, and F. Bergman, *Eur. J. Pharmacol.*, 1980, **65**, 93.
[428] M. T. Curtis and A. M. Lefer, *Am. J. Physiol.*, 1980, **239**, H416.
[429] V. Chernick, D. L. Madansky, and E. E. Lawson, *Pediatr. Res.*, 1980, **14**, 357.
[430] M. T. Lowry, R. P. Maickel, and G. K. W. Yim, *Life Sci.*, 1980, **26**, 2113.
[431] S. J. Stevens, *Psychopharmacology*, 1980, **71**, 1.
[432] J. A. Foster, M. Morrison, S. J. Dean, M. Hill, and H. Frenk, *Pharmacol. Biochem. Behav.*, 1981, **14**, 419.
[433] J. G. Jones and J. A. Richter, *Life Sci.*, 1981, **28**, 2055.
[434] N. L. Ostrowski, T. L. Foley, M. D. Lind, and L. D. Reid, *Pharmacol. Biochem. Behav.*, 1980, **12**, 431.
[435] D. A. Czech and A. E. Stein, *Pharmacol. Biochem. Behav.*, 1980, **12**, 987.
[436] D. R. Brown and S. G. Holtzman, *Eur. J. Pharmacol.*, 1981, **69**, 331.
[437] M.-F. Wu, M. D. Lind, J. M. Stapleton, and L. D. Reid, *Bull Psychon. Soc.*, 1981, **17**, 101.
[438] B. K. Choi and Y. J. Took, *Taehan Yakrihak Chapchi*, 1980, **16**, 15.
[439] M. T. Linn, A. Chandra, and C. Y. Su, *Neuropharmacology*, 1980, **19**, 435.
[440] I. Izquierdo and M. Grandenz, *Psychopharmacology*, 1980, **67**, 265.
[441] E. W. Snyder, D. E. Shearer, E. C. Beck, and R. E. Dustman, *Psychopharmacology*, 1980, **67**, 211.
[442] E. W. Snyder, R. E. Dustman, and C. Schlehuber, *J. Pharmacol. Exp. Ther.*, 1981, **217**, 299.
[443] B. J. Kraynack, J. Gintantas, L. L. Kraynack, and G. B. Racz, *Proc. West. Pharmacol. Soc.*, *23rd*, 1980, p. 143.

catalepsy,[444] recovery from spinal injury,[445] the development of neuroblastoma,[446] neurones,[447,448] the actions of mescaline,[449] of flunitrazepam,[450] of apomorphine,[451] of amphetamine,[451] and of ethanol,[452,453] the release of prolactin,[454] cortisol,[454] luteinizing hormone,[369,455–458] follicle-stimulating hormone,[369,456] growth hormone,[459,460] vasopressin,[374,461] pituitary and gonadal hormones,[458,462,463] and the levels of corticosteroids in plasma[464] have been investigated. The pharmacokinetics,[465,466] metabolism,[467] binding of receptors,[468] narcotic antagonist activity,[469–475] and hyperalgesic effects[476–478] of naloxone and its use in tests for narcotic addiction[479] have also been studied.

Studies related to other derivatives of the morphine alkaloids include the receptor-binding[480] and narcotic effects[481] of naloxone 6-hydrazone; the effects of

[444] M. M. Namba, R. M. Quock, and M. H. Malone, *Proc. West. Pharmacol. Soc., 23rd*, 1980, p. 285.
[445] A. I. Faden, T. P. Jacobs, and J. W. Holaday, *Science*, 1981, **211**, 493.
[446] I. S. Zagon and P. J. McLaughlin, *Life Sci.*, 1981, **28**, 1095.
[447] J. G. Sinclair, S. S. Mokha, and A. Iggo, *Q. J. Exp. Physiol. Cogn. Med. Sci.*, 1980, **65**, 181.
[448] M. Fitzgerald and C. J. Woolf, *Pain*, 1980, **9**, 293.
[449] R. L. Commisaris, K. E. Moore, and R. H. Rech, *Pharmacol. Biochem. Behav.*, 1980, **13**, 601.
[450] J. E. Schmitz, W. Dick, and P. Lotz, *Anaesthesiol. Intencivmed. (Berlin)*, 1980, **130**, No. 25, p. 66.
[451] P. M. Adams, R. Beauchamp, and C. Alston, *Life Sci.*, 1981, **28**, 629.
[452] W. J. Jeffcoate, P. Platts, M. Ridout, A. G. Hastings, I. MacDonald, and C. Selby, *Pharmacol. Biochem. Behav.*, 1980, **13**, Suppl. 1, p. 145.
[453] K. Blum, A. H. Briggs, L. DeLallo, S. F. A. Elston, and M. Hirst, *Subst. Alcohol Actions/Misuse*, 1980, **1**, 327.
[454] J. Blankstein, F. I. Reyes, J. S. D. White, and C. Faiman, *Proc. Soc. Exp. Biol. Med.*, 1980, **164**, 373.
[455] T. Ieiri, H. T. Chen, and J. Meites, *Life Sci.*, 1980, **26**, 1269.
[456] T. J. Cicero, C. E. Wilcox, R. D. Bell, and E. R. Meyer, *J. Pharmacol. Exp. Ther.*, 1980, **212**, 573.
[457] R. W. Steger, W. E. Sonntag, D. A. Van Vugt, L. J. Forman, and J. Meites, *Life Sci.*, 1980, **27**, 747.
[458] D. P. Owens and T. J. Cicero, *J. Pharmacol. Exp. Ther.*, 1981, **216**, 135.
[459] I. Wakabayashi, M. Kanda, N. Miki, E. Ohmura, R. Demura, and K. Shizume, *Neuroendocrinology*, 1980, **30**, 319.
[460] T. Okajima, T. Motomatsu, K. Kato, and H. Ibayashi, *Life Sci.*, 1980, **27**, 755.
[461] W. Knepel, D. Nutto, H. Anhut, and G. Hertling, *Eur. J. Pharmacol.*, 1980, **65**, 449.
[462] J. H. Mendelson, J. Ellingboe, J. C. Kuehnle, and N. K. Mello, *J. Pharmacol. Exp. Ther.*, 1980, **214**, 503.
[463] T. Barreca, A. Morabini, G. Magnani, A. Sannia, and E. Rolandi, *IRCS Med. Sci. Libr. Compend.*, 1980, **8**, 906.
[464] R. M. Eisenberg, *Life Sci.*, 1980, **26**, 935.
[465] T. A. Moreland, J. E. H. Brice, C. H. M. Walker, and A. C. Parija, *Br. J. Clin. Pharmacol.*, 1980, **9**, 609.
[466] T. A. Moreland, J. E. H. Brice, A. C. Parija, and C. H. M. Walker, *Zentralbl. Pharm. Pharmakother. Laboratoriumsdiagn.*, 1980, **119**, 1140.
[467] S. M. Taylor, R. M. Rodgers, R. K. Lynn, and N. Gerber, *J. Pharmacol. Exp. Ther.*, 1980, **213**, 289.
[468] R. S. Zukin, S. Walczak, and M. H. Makman, *Brain Res.*, 1980, **186**, 238.
[469] M. S. Cohen, A. M. Rudolph, and K. L. Melman, *Dev. Pharmacol. Ther.*, 1980, **1**, 58.
[470] M. K. Menon, L. F. Tseng, H. H. Loh, and W. G. Clark, *Naunyn-Schmiedeberg's Arch. Pharmacol.*, 1980, **312**, 43.
[471] C.-L. Wong, M.-K. Wai, and M. B. Roberts, *Clin. Exp. Pharmacol. Physiol.*, 1980, **7**, 305.
[472] C. Dauthier, J. H. Gaudy, and J. C. Willer, *Ann. Anesthesiol. Fr.*, 1980, **21**, 421.
[473] R. A. Cohen and J. D. Coffman, *Clin. Pharmacol. Ther.*, 1980, **28**, 541.
[474] G. P. Novelli, E. Pieraccioli, V. A. Peduto, and F. Festimanni, *Br. J. Anaesth.*, 1980, **52**, 1107.
[475] T. Hisamitsu, *Showa Igakkai Zasshi*, 1980, **40**, 201.
[476] C. W. T. Pilcher, *Life Sci.*, 1980, **27**, 1905.
[477] C. J. Woolf, *Brain Res.*, 1980, **189**, 593.
[478] C. J. Woolf, *Brain Res.*, 1980, **190**, 578.
[479] B. A. Judson, D. U. Himmelberger and A. Goldstein, *Clin. Pharmacol. Ther.*, 1980, **27**, 492.
[480] G. W. Pasternak, S. R. Childers, and S. H. Snyder, *Science*, 1980, **208**, 514.
[481] G. W. Pasternak, S. R. Childers, and S. H. Snyder, *J. Pharmacol. Exp. Ther.*, 1980, **214**, 455.

naltrexone on behaviour,[408,410,482,483] reserpine-induced catalepsy,[444] intra-ocular pressure,[357] food intake,[484] and the binding of dihydrotestosterone;[456] the effects of nalorphine on the secretion of prolactin[485] and electro-shock;[486] the general pharmacology of nalbuphine,[487] and its effects on respiration;[488] the effects of heroin on behaviour,[489,490] intake of water,[491] and concentrations of gases in the blood;[492] the effects of codeine on the urinary system,[493] and behaviour[494,495] and the comparative pharmacology of the (+)- and (−)-forms;[496] the enzymic N-demethylation of 3-O-ethylmorphine[497,498] and the effects of this base on the oxidation of testosterone;[499] the pharmacokinetics of 6-O-nicotinylcodeine;[500] the binding of dihydromorphine in brain;[501] the use of dihydromorphinone in epidural analgesia;[502] the pharmacology of N-cyclopropylmethyl-8-ethyldihydro-norcodeinone,[503] of 3-(β-aminoethyl) ethers of morphine and morphinone,[504] of 14-hydroxy-, 14-mercapto-, 14-bromo-, 14-chloro-, and 14-nitro-morphines and -morphinones,[505] and of 6-di-(β-chloroethyl)amino-6-deoxy-14-hydroxydihydro-morphine and its N-(cyclopropylmethyl)nor-analogue;[506-508] the physical dependence potential of thebaine;[509] the analgesic properties of 14-phenyl- and 14-(p-fluorophenyl)-hydroxylaminocodeinone;[510] the receptor-binding,[511] cataleptic,[512] and cardiovascular effects[513] of etorphine and its effects on the

[482] N. K. Mello, J. H. Mendelson, J. C. Kuehnle, and M. S. Sellers. *J. Pharmacol. Exp. Ther.*, 1980, **216**, 45.
[483] R. E. Meller, E. B. Keverne, and J. Herbert, *Pharmacol. Biochem. Behav.*, 1980, **13**, 663.
[484] S. Herling, *J. Pharmacol. Exp. Ther.*, 1981, **217**, 105.
[485] E. Rolandi, G. Magnani, A. Sannia, and T. Barreca, *IRCS Med. Sci. Libr. Compend.*, 1980, **8**, 570.
[486] L. Dykstra, *Psychopharmacology*, 1980, **70**, 69.
[487] R. R. Mittler, *Am. J. Hosp. Pharm.*, 1980, **37**, 942.
[488] A. Romagnoli and A. S. Keats, *Clin. Pharmacol. Ther.*, 1980, **27**, 428.
[489] R. Blair, H. Cytryniak, P. Shizgal, and Z. Amit, *Psychopharmacology*, 1980, **59**, 313.
[490] G. J. Gerber, M. A. Bognarth, and R. A. Wise, *Life Sci.*, 1981, **28**, 557.
[491] R. N. Park, M. Hirst, and C. W. Gowdey, *Can. J. Physiol. Pharmacol.*, 1980, **58**, 1295.
[492] I. S. Zagon, P. J. McLaughlin, and W. J. White, *Res. Commun. Subst. Abuse*, 1980, **1**, 235.
[493] S. A. Garcia, C. A. Basilio, and P. M. Dominguez, *Arch. Farmacol. Toxicol.*, 1979, **5**, 139.
[494] F. Hoffmeister, J. Dycka, and K. Raemsch, *J. Pharmacol. Exp. Ther.*, 1980, **214**, 221.
[495] A. M. Young and J. H. Woods, *Psychopharmacology*, 1980, **70**, 263.
[496] T. T. Chau and L. S. Harris, *J. Pharmacol. Exp. Ther.*, 1980, **215**, 668.
[497] M. Kitada, N. Yamaguchi, K. Igarashi, S. Hirose, and H. Kitagawa, *Jpn. J. Pharmacol.*, 1980, **30**, 579.
[498] M. Kitada, S. Omori, T. Igarashi, Y. Kanakubo, and H. Kitagawa, *Biochem. Biophys. Res. Commun.*, 1980, **97**, 1527.
[499] L. N. Pospelova and V. V. Christyakov, *Biokhimiya*, 1981, **46**, 347.
[500] W. Linder and K. Schaupp, *Anal. Lett.*, 1980, **13**, 893.
[501] R. B. Messing, B. J. Vasquez, B. Samaniego, R. A. Jensen, J. L. Martinez, and J. L. McGaugh, *J. Neurochem.*, 1981, **36**, 784.
[502] P. R. Bromage, E. Camporesi, and J. Leslie, *Pain*, 1980, **9**, 145.
[503] J. F. Howes, P. F. Osgood, R. K. Razdan, F. Moreno, A. Castro, and J. Villareal, *NIDA Res. Monogr.*, 1979, **27**, 99.
[504] C. Spyraki, G. Papaioannou, G. Tsatsas, and D. Varonos, *Arzneim.-Forsch.*, 1980, **30**, 467.
[505] P. Osei-Gyimah, S. Archer, M. G. C. Gillan, and H. W. Kosterlitz, *J. Med. Chem.*, 1981, **24**, 212.
[506] T. P. Caruso, D. L. Larson, P. S. Portoghese, and A. E. Takemori, *J. Pharmacol. Exp. Ther.*, 1980, **213**, 539.
[507] T. P. Caruso, D. L. Larson, and P. S. Portoghese, *Life Sci.*, 1980, **27**, 2063.
[508] A. A. Larson and M. J. Armstrong, *Eur. J. Pharmacol.*, 1980, **68**, 25.
[509] World Health Organisation, *Bull. Narc.*, 1980, **32**, 45.
[510] L. S. Schwab, *J. Med. Chem.*, 1980, **23**, 698.
[511] J. A. Glasel, R. F. Venn, and E. A. Barnard, *Biochem. Biophys. Res. Commun.*, 1980, **95**, 263.
[512] B. F. Kania, *Acta Physiol. Pol.*, 1980, **31**, 9.
[513] M. Becker and R. Belinger, *Res. Vet. Sci.*, 1980, **29**, 21.

metabolism of dopamine and serotonin;[514] the cardiovascular effects of diprenorphine;[513] the haemodynamics,[515] pharmacokinetics,[516] and neuro-chemical[517] and behavioural[518] effects of and the radioimmunoassay for[519] buprenorphine; the biopharmacological properties of 6'-ethyl-5'-methylcyclohex-5'-eno[1',2':8,14]dihydrocodeinone;[520] and the effects of *O*-methylflavinantine on guinea-pig ileum.[521]

14 Benzophenanthridines

The following species have been shown to contain the alkaloids stated:

Bocconia frutescens[110]	chelerythrine, chelirubine, and sanguinarine
Corydalis incisa[522]	(+)-corynoline 10-*O*-sulphate
Meconopsis cambrica[113]	sanguinarine
Zanthoxylum monophyllum[116]	chelerythrine
Zanthoxylum nitidum[146]	nitidine, dihydronitidine, oxynitidine, and 6-methoxy-5,6-dihydrochelerythrine
Zanthoxylum williamsii[116]	chelerythrine and nitidine

The chemistry of nitidine and of fagaronine has been reviewed.[523] The absolute configurations of (+)-corynoline[524] and (+)-14-*epi*-corynoline[525] have been determined. After preliminary studies on unsubstituted compounds,[526] (±)-corynoline and related compounds have been synthesized from the enamide (111) by photolytic cyclization to the lactam, followed by dehydrogenation of the C-11–C-12 bond to form the lactam (112; X = O). Treatment of this lactam with performic acid and then with alkali gives a mixture of the 11α,12α- and the 11α,12β-glycols, of general structure (113); reduction of both of these with lithium aluminium hydride, followed by hydrogenolytic removal of the hydroxy-group at C-12, gives (±)-11-*epi*-corynoline. In contrast, reduction of the amide (112; X = O) to the amine (112; X = H₂), followed by conversion into the glycol, gives 91% of the 11β,12α-glycol (113), hydrogenolysis of which yields (±)-corynoline (114). The difference in the behaviours of the lactam and the amine in the oxidation by a peracid has been attributed to participation by the basic nitrogen atom of the amine in the reaction in that case.[527]

[514] S. Algeri, A. Consolazione, and A. Plaznik, *Eur. J. Pharmacol.*, 1980, **68**, 383.
[515] E. Melon, A. Lienhart, and P. Viars, *Anesth., Analg., Reanim.*, 1980, **37**, 121.
[516] J. G. Lloyd-Jones, P. Robinson, R. Henson, S. R. Briggs, and T. Taylor, *Eur. J. Drug Metab. Pharmacokinet.*, 1980, **5**, 233.
[517] P. W. Dettmar, A. Cowan, and D. S. Walter, *Eur. J. Pharmacol.*, 1981, **69**, 147.
[518] S. Kareti, J. E. Morton, and N. Khazan, *Neuropharmacology*, 1980, **19**, 195.
[519] A. J. Bartlett, J. G. Lloyd-Jones, M. J. Rance, I. R. Flockhart, G. J. Dockray, M. R. D. Bennett, and R. A. Moore, *Eur. J. Clin. Pharmacol.*, 1980, **18**, 339.
[520] A. Cowan, *Fed. Proc., Fed. Am. Soc. Exp. Biol.*, 1981, **40**, 1497.
[521] B. K. Noamesi and E. A. Gyang, *Planta Med.*, 1980, **38**, 138.
[522] K. Isawa, N. Takao, G. Nonaka, and I. Nishioka, *Phytochemistry*, 1979, **18**, 1725.
[523] S. D. Phillips and R. N. Castle, *J. Heterocycl. Chem.*, 1981, **18**, 223.
[524] N. Takao, M. Kamigauchi, and K. Isawa, *Tetrahedron*, 1979, **35**, 1977.
[525] N. Takao, M. Kamigauchi, K. Isawa, K. Tomita, T. Fujiwara, and A. Wakahara, *Tetrahedron*, 1979, **35**, 1099.
[526] I. Ninomiya, O. Yamamoto, T. Kiguchi, and T. Naito, *J. Chem. Soc., Perkin Trans. 1*, 1980, 203.
[527] I. Ninomiya, O. Yamamoto, and T. Naito, *J. Chem. Soc., Perkin Trans. 1*, 1980, 212.

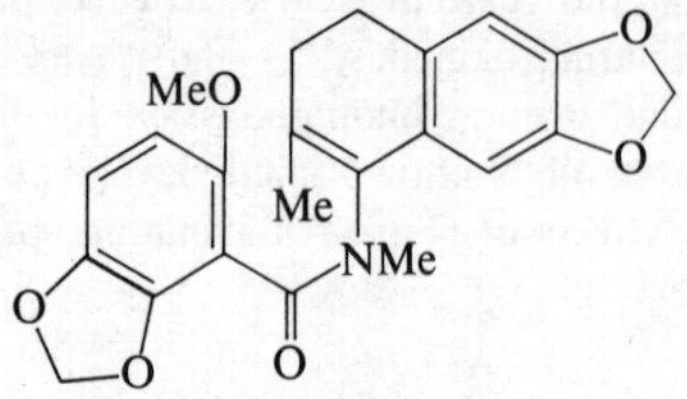

(111)

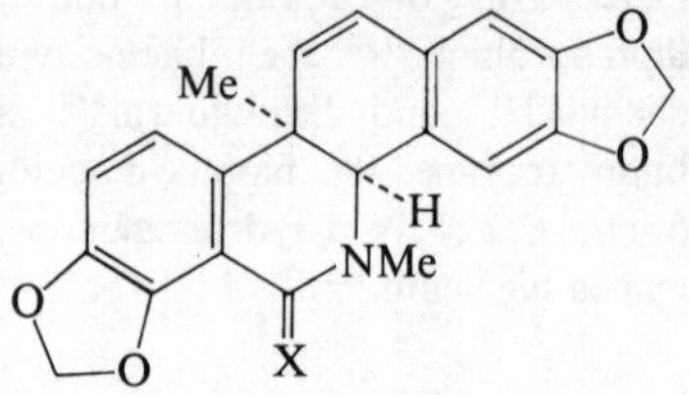

(112)

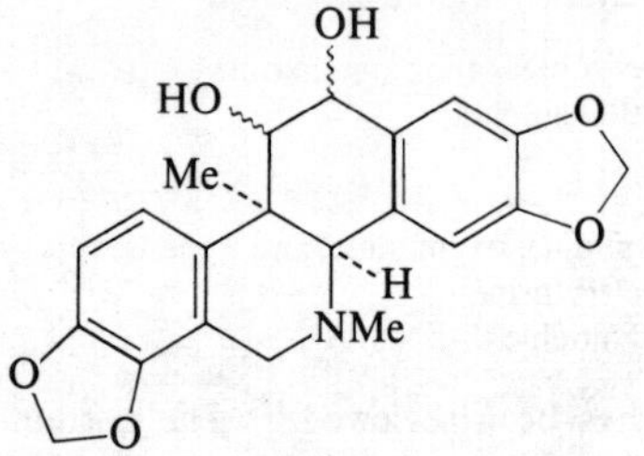

(113)

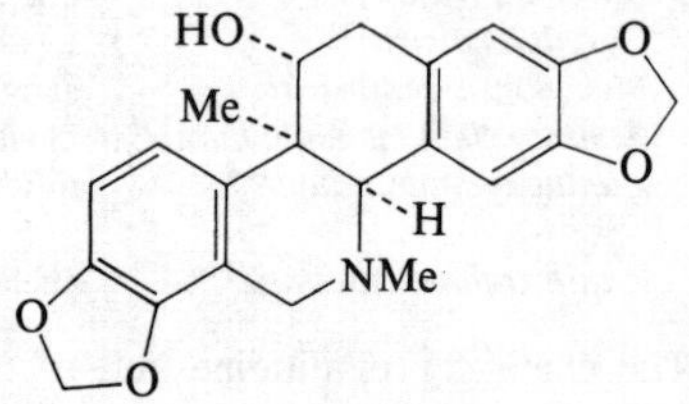

(114)

Following preliminary work,[528] (±)-chelidonine (118) has been synthesized from the acid (115) (obtained from methylenedioxyhomophthalic anhydride and *N*-methyldimethoxybenzalimine) by its conversion into the diazo-ketone (116), cyclization of this (using trifluoroacetic acid) to the keto-lactam (117), and reduction with lithium aluminium hydride.[529, 530]

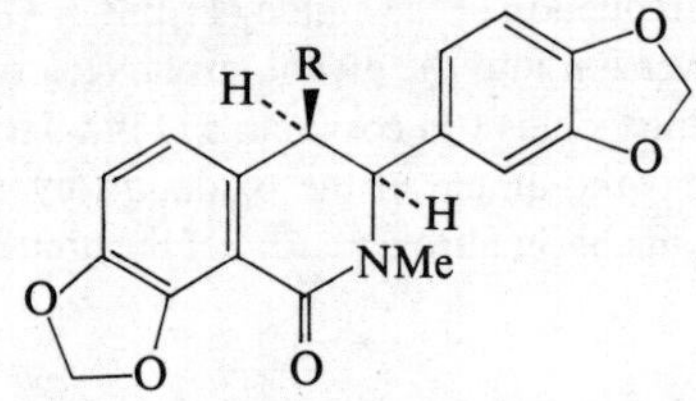

(115) R = CO₂H

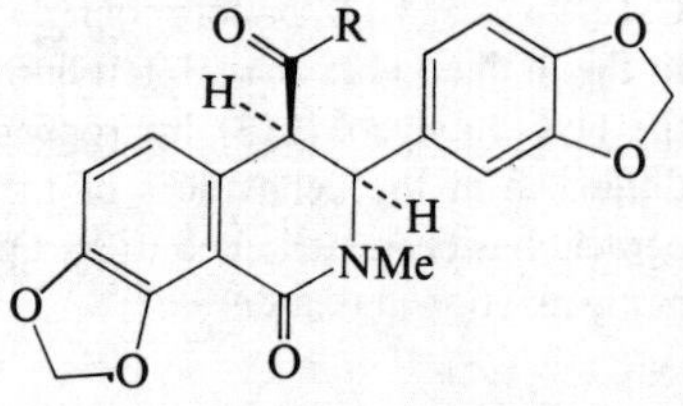

(116) R = CHN₂

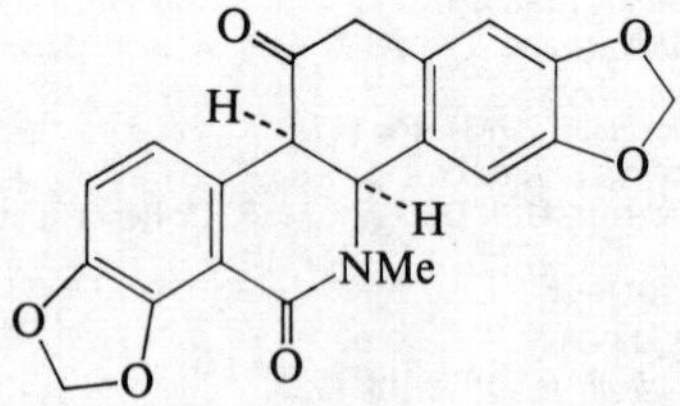

(117)

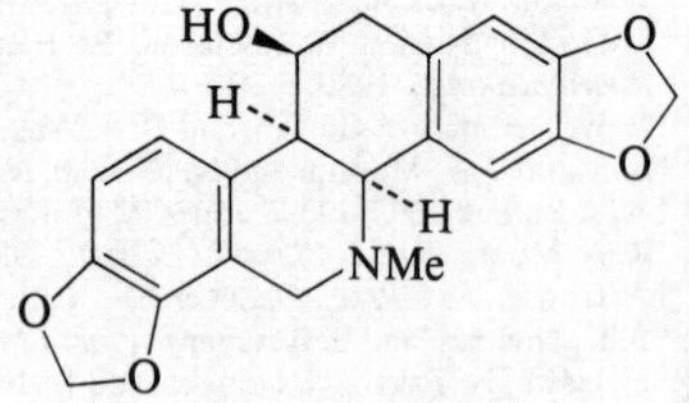

(118)

[528] M. Cushman and D. K. Dikshit, *J. Org. Chem.*, 1980, **45**, 5064.
[529] M. Cushman, T.-C. Choong, J. T. Valko, and M. P. Koleck, *J. Org. Chem.*, 1980, **45**, 5067.
[530] M. Cushman, T.-C. Choong, J. T. Valko, and M. P. Koleck, *Tetrahedron Lett.*, 1980, **21**, 3845.

The fluorescence spectra of chelerythrine, nitidine, and sanguinarine in different solvents have been shown to be dependent on the pH of the solution and to allow the determination of the equilibrium constant for the formation of the pseudo-bases.[531] A salt of chelidonine with 5-carboxymethyl-2-thio-1,3-thiazan-4-one has been prepared.[532] The intercalative binding of sanguinarine[533] and the anti-tumour activities of chelidonine N-oxide[534] and of nitidine[535] have been studied.

15 Colchicine

A review of the chemistry and biological actions of colchicine and its derivatives has been published.[536] The ^{13}C n.m.r. spectra of colchicine and its derivatives have been studied[537-540] and the structure (119) has been determined for epoxy-colchicine (prepared from colchicine by the action of water and sodium peroxide) by X-ray crystallographic methods;[541] other n.m.r. studies have shown that the double-bond in dehydro-7-deacetamidocolchicine is placed as shown in (120), and not in the 6,7-position, as previously supposed.[542] The treatment of deacetyl-colchicine with isocyanates affords the related substituted ureas, which can be converted into N-nitroso-derivatives by treatment with nitrous acid; these have been evaluated as inhibitors of neoplasms and as virucides.[543] N-Demethylation and O-demethylation of N-methylcolchiceinamide by micro-organisms has been achieved.[544]

(119)   (120)

[531] D. Walterova, V. Preininger, F. Grambal, V. Simanek, and F. Santavy, *Heterocycles*, 1980, **14**, 597.
[532] M. O. Oleevskaya, L. I. Petlichnaya, E. I. Besyadetskaya, and B. S. Zimenkovskii, *Farm. Zh. (Kiev)*, 1980, No. 5, p. 51.
[533] M. D. Faddejeva, T. N. Belyaeva, J. P. Navikov, and H. G. Shalabi, *IRCS. Med. Sci. Libr. Compend.*, 1980, **8**, 612.
[534] J. Zbierska and Z. Kowalewski, *Herba Pol.*, 1980, **26**, 61.
[535] Y.-J. Fan, J. Zhou, and M. Li, *Chung-kuo Yao Li Hsueh Pao*, 1981, **2**, 19.
[536] F. Santavy, *Acta Univ. Palacki. Olomuc., Fac. Med.*, 1979, **90**, 15.
[537] C. D. Hufford, H.-G. Capraro, and A. Brossi, *Helv. Chim. Acta*, 1980, **63**, 50.
[538] J. Elguero, R. N. Miller, A. Blade-Font, R. Faure, and E. J. Vincent, *Bull. Soc. Chim. Belg.*, 1980, **89**, 183.
[539] F. G. Kamaev, M. G. Levkovich, N. L. Mukhamed'yarova, M. K. Yusupov, and A. S. Sadykov, *Tezisy Dokl.-Sav. Indiiskii Simp. Khim. Prir. Soedin.*, 5th, 1978, 35.
[540] B. Danieli, G. Palmisano, and G. Severini Ricca, *Gazz. Chim. Ital.*, 1980, **110**, 351.
[541] A. Brossi, M. Roesner, J. V. Silverton, M. A. Iorio, and C. D. Hufford, *Helv. Chim. Acta*, 1980, **63**, 406.
[542] M. Roesner, H.-G. Capraro, A. E. Jacobson, L. Atwell, A. Brossi, M. A. Iorio, T. H. Williams, R. H. Sik, and C. F. Chignell, *J. Med. Chem.*, 1981, **24**, 257.
[543] T.-S. Lin, G. T. Shiau, W. H. Prusoff, and R. E. Harmon, *J. Med. Chem.*, 1980, **23**, 1440.
[544] P. J. Davis, *Antimicrob. Agents Chemother.*, 1981, **19**, 465.

The toxicity[545] of colchicine and the effects of the alkaloid on cell division,[546] leukaemic cells,[547] neuroblastoma cells,[548] and mesophil cells,[549] on chemotaxis of leucocytes[550] and of neutrophils,[551] on the release of lysosomal enzyme from leucocytes,[552] on lymphocytes,[553-555] on dentition,[556,557] on the secretory cycle of the hypothalamus–pituitary system,[558] on the retina,[559] on the transport of tubulin,[560] on the permeability of muscle,[561] on the nervous system,[562-566] on the activity of enzymes,[567-569] on blood flow in mammary glands,[570,571] on the sterilization of flies,[572] on the oxidation of glucose,[573] on the synthesis of DNA,[574] on the dynamics of aqueous humour,[575] and on intracellular transport[576-578] and behaviour have been studied.[579] The binding of colchicine to proteins,[580,581] the action of colchamine on leukaemic spleen cells,[582] and the pharmacology of colchiceinamide[583] have also been studied.

[545] F. R. Quinn, Z. Nieman, and J. A. Beisler, *J. Med. Chem.*, 1981, **24**, 636.
[546] A. R. Hardham and B. E. S. Gunning, *Protoplasma*, 1980, **102**, 31.
[547] F. R. Quinn and J. A. Biesler, *J. Med. Chem.*, 1981, **24**, 251.
[548] R. L. Koerker, *Toxicol. Appl. Pharmacol.*, 1980, **53**, 458.
[549] J. H. Dodds, Z. *Pflanzenphysiol.*, 1980, **99**, 283.
[550] M. Ehrenfeld, M. Levy, M. Bar Eli, R. Gallily, and M. Eliakim, *Br. J. Clin. Pharmacol.*, 1980, **10**, 297.
[551] C. C. Daughaday, A. N. Bohrer, and I. Spillberg, *Experientia*, 1981, **37**, 199.
[552] D. Mikulikova and K. Truavsky, *Cas. Lek. Cesk.*, 1980, **119**, 1398.
[553] D. R. Forsdyke, *Proc. Int. Leucocyte Cult. Conf., 13th*, 1979, 355.
[554] C. E. Rudd and J. G. Kaplan, *Proc. Int. Leucocyte Cult. Conf., 13th*, 1979, 364.
[555] K. A. Rogers and D. L. Brown, *Proc. Int. Leucocyte Cult. Conf., 13th*, 1979, 368.
[556] T. de O. Nogueira, T. Stene, and H. S. Koppang, *Scand. J. Dent. Res.*, 1980, **88**, 15.
[557] M. Chiba, K. Takizawa, and S. Ohshima, *Arch. Oral Biol.*, 1980, **25**, 115.
[558] B. Gajkowska and J. Borowicz, *Acta Med. Pol.*, 1979, **20**, 387.
[559] H. Pasantes-Morales, R. Salceda, and A. M. Lopez-Colone, *J. Neurochem.*, 1980, **34**, 172.
[560] Y. Komiya and M. Kurokawa, *Brain Res.*, 1980, **190**, 505.
[561] G. L. A. Graff, C. Gueuning, and C. Dictus-Vermeulen, *Arch. Int. Physiol. Biochem.*, 1980, **88**, 303.
[562] G. M. Sole, *J. Anat.*, 1980, **130**, 777.
[563] M. Morelli, M. L. Porceddu, and G. DiChiara, *Pharmacol. Res. Commun.*, 1980, **12**, 719.
[564] T. A. Murakami, *Shinkei Kenkyu no Shinpo*, 1980, **24**, 948.
[565] R. Schafer and P. D. Reagan, *J. Neurobiol.*, 1981, **12**, 155.
[566] G. Baux, M. Simmoneau, and L. Tauc, *J. Neurobiol.*, 1981, **12**, 75.
[567] I. H. Fraser, S. Ratnam, J. M. Collins, and S. Mookerjea, *J. Biol. Chem.*, 1980, **225**, 6617.
[568] M. M. Mitranic, J. M. Boggs, and M. A. Moscarello, *Biochim. Biophys. Acta*, 1981, **672**, 57.
[569] D. Rachmilewitz and F. Karmeli, *Eur. J. Pharmacol.*, 1980, **67**, 235.
[570] G. Burvenich and G. Peeters, *Z. Tierphysiol., Tierernaehr. Futtermittelkd.*, 1980, **44**, 211.
[571] A. J. Henderson and M. Peaker, *Q. J. Exp. Physiol. Cogn. Med. Sci.*, 1980, **65**, 367.
[572] G. S. Shukla and R. N. Singh, *Bioresearch (Ujjain, India)*, 1979, **3**, 27.
[573] J. W. Rosenthal, *Gen. Pharmacol.*, 1981, **12**, 47.
[574] M. Friedkin, A. Legg, and E. Rozengurt, *Exp. Cell Res.*, 1980, **129**, 23.
[575] R. Ritch, A. Mulbreg, C. Rosen, G. Chubak, K. Pokorny, and M. E. Yablonski, *Exp. Eye Res.*, 1981, **32**, 143.
[576] M. Mori, M. Oyamada, A. Kaneko, and T. Onoue, *Igaku no Ayumi*, 1980, **113**, 829.
[577] J. Blok, L. A. Ginsel, A. A. Mulder-Stapel, J. J. M. Onderwater, and W. T. Daems, *Cell Tissue Res.*, 1981, **215**, 1.
[578] B. M. Mullock, R. S. Jones, J. Peppard, and R. H. Hinton, *FEBS Lett.*, 1980, **120**, 278.
[579] B. Siegfried, J. Fischer, and J. Bures, *Neuroscience*, 1980, **5**, 529.
[580] Z. Trnavska and K. Trnavsky, *Acta Univ. Palacki, Olomuc., Fac. Med.*, 1980, **93**, 47.
[581] S. D. M. Watts, *Biochim. Biophys. Acta*, 1981, **667**, 59.
[582] V. N. Dobrokhotov and V. I. Vasil'eva, *Byull. Eksp. Biol. Med.*, 1980, **90**, 357.
[583] D.-H. Li, S.-K. Zhang, X.-G. Hao, K.-S. Ma, X.-R. Tan, Z.-L. Wang, and N.-K. Li, *Chin. Med. J.*, 1980, **93**, 188.

9

Aporphinoid Alkaloids

BY M. SHAMMA AND H. GUINAUDEAU

1 Introduction

The year under review has witnessed the isolation and structural elucidation of no less than nine new aporphines, as well as of five aporphine–benzylisoquinoline dimers. Two of these five aporphine–benzylisoquinolines, namely kalashine and khyberine, are the first such dimers known to be substituted at C-11 of the aporphine moiety. Thallium(III) trifluoroacetate is an oxidizing agent which effects the cyclization of tetrahydrobenzylisoquinolines to aporphines in satisfactory yields.[1] A listing of aporphinoid alkaloids has appeared.[2]

2 Proaporphines

The new proaporphine iso-oridine (1), isomeric with the known oridine (2), has been isolated from *Papaver oreophilum*.[3] The supporting structural data include ^{13}C n.m.r. spectroscopy. *N*-Acetylstepharine (3) has been found for the first time as a natural compound in *Stephania sasakii*.[4]

(1) $R^1 = H$, $R^2 = Me$
(2) $R^1 = Me$, $R^2 = H$

(3)

[1] E. C. Taylor, J. C. Andrade, G. J. H. Rall, and A. McKillop, *J. Am. Chem. Soc.*, 1980, **102**, 6513.
[2] I. A. Israilov, S. U. Karimova, M. S. Yunusov, and S. Yu. Yunusov, *Khim. Prir. Soedin.*, 1980, 279 [*Chem. Natural Compd. (Engl. Transl.)*, 1980, **16**, 197].
[3] F. Věžník, P. Sedmera, V. Preininger, V. Šimánek, and J. Slavík, *Phytochemistry*, 1981, **20**, 247.
[4] J. Kunitomo, Y, Murakami, M. Oshikata, T. Shingu, S.-T. Lu, I.-S. Chen, and M. Akasu, *J. Pharm. Soc. Jpn.*, 1981, **101**, 431.

Known proaporphines that have been re-isolated from plants are tabulated in Table 1.[5-10]

Table 1 *Proaporphines that have recently been re-isolated, and their natural sources*

Alkaloid	Source	Ref.
Glaziovine	*Uvaria chamae*	5
	Litsea laurifolia	6
Pronuciferine	*Uvaria chamae*	5
	Papaver lacerum	7
	P. tauricola	8
	P. fugax	8
	Meconopsis cambrica	9
Mecambrine	*Meconopsis cambrica*	9
	Papaver lacerum	7
	P. rhoeas	10
N-Methylcrotonosine	*Meconopsis cambrica*	9
Oridine	*Papaver oreophilum*	3
Roehybrine	*Roemeria hybrida*	10

3 Aporphines

A detailed study of the alkaloidal content of the Formosan *Stephania sasakii* (Menispermaceae) has led to four new aporphines: stesakine (4), dehydrostesakine (5), dehydrocrebanine (6),[11] and dehydrophanostenine (7).[4] Additionally, dehydro-stephanine (8) has been isolated from *S. kwangsiensis*.[12]

(4) (5) R = OH (6) R = OMe (8) R = H (7)

A total of five other new aporphines have also been found and characterized, including isocorytuberine (9), dehydrocorydine (10), and corydine *N*-oxide (11), all from *Glaucium fimbrilligerum*,[13] laetine (12), from *Litsea laeta*,[14] and anaxagoreine (13), from *Anaxagorea dolichocarpa* and *A. prinoides*.[15]

Table 2[16-42] includes a listing of aporphines that have recently been re-isolated.

The reaction of reticuline *N*-oxide (14) with cuprous chloride in methanol, followed by treatment with sodium hydrosulphite, furnished corytuberine (15) in good yield.[43] The structure of hernagine (16) was confirmed by the synthesis of N-methylhernagine (17) *via* alkylation of a Reissert intermediate and Pschorr cyclization.[28]

In a modification of a known sequence, base-catalysed condensation of hydrastininium iodide (18) with 6-methoxy-2-nitrotoluene (19) provided (20) and (21). Catalytic hydrogenation of (20), followed by Pschorr cyclization, afforded stephanine (22).[44]

Oxidation of the tetrahydroisoquinoline (23) by thallium(III) trifluoroacetate in methanol leads to ocoteine (24) in 46% yield. When the same oxidizing reagent, in trifluoroacetic acid, was used in connection with the acetyl amide (25), N-acetyl-3-methoxynornantenine (26) and the corresponding dehydroaporphine

[5] M. Leboeuf and A. Cavé, *Plant. Med. Phytother.*, 1980, **14**, 143.

[6] M. Leboeuf, A. Cavé, J. Provost, and P. Forgacs, *Plant. Med. Phytother.*, 1979, **13**, 262.

[7] G. Sariyar and J. D. Phillipson, *J. Nat. Prod.*, 1981, **44**, 239.

[8] J. D. Phillipson, O. O. Thomas, A. I. Gray, and G. Sariyar, *Planta Med.*, 1981, **41**, 105.

[9] S. R. Hemingway, J. D. Phillipson, and R. Verpoorte, *J. Nat. Prod.*, 1981, **44**, 67.

[10] J. D. Phillipson, A. I. Gray, A. A. R. Askari, and A. A. Khalil, *J. Nat. Prod.*, 1981, **44**, 296.

[11] J. Kunitomo, Y. Murakami, M. Oshikata, T. Shingu, M. Akasu, S.-T. Lu, and I.-S. Chen, *Phytochemistry*, 1980, **19**, 2735.

[12] Z.-D. Min and S.-M. Zhong, *Yao Hsueh Hsueh Pao*, 1980, **15**, 532.

[13] S. U. Karimova, I. A. Israilov, M. S. Yunusov, and S. Yu. Yunusov, *Khim. Prir. Soedin.*, 1980, 224 [*Chem. Natural Compd. (Engl. Transl.)*, 1980, **16**, 177].

[14] R. C. Rastogi and N. Borthakur, *Phytochemistry*, 1980, **19**, 998.

[15] R. Hocquemiller, S. Rasamizafy, C. Moretti, H. Jacquemin, and A. Cavé, *Planta Med.*, 1981, **41**, 48.

[16] S. El-Masry, M. G. El-Ghazooly, A. A. Omar, S. Khafagy, and J. D. Phillipson, *Planta Med.*, 1980, **41**, 61.

[17] M. A. Manushakyan, I. A. Israilov, V. A. Mnatsakanyan, M. S. Yunusov, and S. Yu. Yunusov, *Khim. Prir. Soedin.*, 1980, 849.

[18] G. B. Lockwood, *Phytochemistry*, 1981, **20**, 1463.

[19] P. W. LeQuesne, J. E. Larrahondu, and R. F. Raffauf, *J. Nat. Prod.*, 1980, **43**, 353.

[20] L. Castedo, J. M. Saá, R. Suau, C. Villaverde, and P. Potier, *An. Quim., Ser. C*, 1980, **76**, 171.

[21] L. A. Djakoure, D. Kone, and L. L. Douzoua, *Ann. Univ. Abidjan, Ser. C*, 1978, **14**, 49.

[22] W. N. Wu, J. L. Beal and R. W. Doskotch, *J. Nat. Prod.*, 1980, **43**, 372.

[23] S. F. Hussain, A. Amin, and M. Shamma, *J. Chem. Soc. Pak.*, 1980, **2**, 157.

[24] S. F. Hussain, R. D. Minard, A. J. Freyer, and M. Shamma, *J. Nat. Prod.*, 1981, **44**, 169.

[25] I. R. C. Bick, T. Sevenet, W. Sinchai, B. Skelton, and A. H. White, *Aust. J. Chem.*, 1981, **34**, 195.

[26] N. K. Saxena and D. S. Bhakuni, *J. Indian Chem. Soc.*, 1979, **56**, 1020.

[27] S.-T. Lu, T.-L. Su, and C.-Y. Duh, *T'ai-wan Yao Hsueh Tsa Chih*, 1979, **31**, 23.

[28] J. Bruneton, *C. R. Hebd. Seances Acad. Sci., Ser. C*, 1980, **291**, 187.

[29] N. W. Wu, J. L. Beal, and R. W. Doskotch, *J. Nat. Prod.*, 1980, **43**, 567.

[30] S. V. Bhat, H. Dornauer, and N. J. Descuza, *J. Nat. Prod.*, 1980, **43**, 588.

[31] C. H. Chen, T. M. Chen, and C. Lee, *J. Pharm. Sci.*, 1980, **69**, 1061.

[32] R. T. Boulware and F. R. Stermitz, *J. Nat. Prod.*, 1981, **44**, 200.

[33] J. A. Swinehart and F. R. Stermitz, *Phytochemistry*, 1980, **19**, 1219.

[34] D. Dwuma-Badu, J. S. K. Ayim, O. Rexford, A. M. Ateya, D. J. Slatkin, J. E. Knapp, and P. L. Schiff, Jr., *Phytochemistry*, 1980, **19**, 1564.

[35] P. Pachaly and C. Schneider, *Arch. Pharm. (Weinheim, Ger.)*, 1981, **314**, 251.

[36] L. Slavíková and J. Slavik, *Collect. Czech. Chem. Commun.*, 1980, **45**, 761.

[37] R. Verpoorte, J. Siwon, M. E. M. Tieken, and A. B. Svendsen, *J. Nat. Prod.*, 1981, **44**, 221.

[38] B. Podolesov and Z. Zdravkovski, *Acta Pharm. Jugosl.*, 1980, **30**, 161.

[39] R. H. Burnell, A. Chapelle, and P. H. Bird, *J. Nat. Prod.*, 1981, **44**, 238.

[40] J. Siwon, R. Verpoorte, T. van Beck, H. Meerburg, and A. B. Svendsen, *Phytochemistry*, 1981, **20**, 323.

[41] D. A. Murav'eva, I. A. Israilov, and F. M. Melikov, *Farmatsiya (Moscow)*, 1981, **30**, 25.

[42] W. N. Wu, W. T. Liao, Z. F. Mahmoud, J. L. Beal, and R. W. Doskotch, *J. Nat. Prod.*, 1980, **43**, 472.

[43] T. Kametani and M. Ihara, *J. Chem. Soc., Perkin Trans. 1*, 1980, 629.

[44] V. Sharma and R. S. Kapil, *Indian J. Chem., Sect. B*, 1981, **20**, 70.

(9)   (10)   (11)

(12)   (13)

Table 2 *Aporphines that have recently been re-isolated, and their natural sources*

Alkaloid	Source	Ref.
Liridinine	*Papaver armeniacum*	8
	P. tauricola	8
Asimilobine	*Uvaria chamae*	5
	Anaxagorea prinoides	15
	A. dolichocarpa	15
N-Methylasimilobine	*Papaver lacerum*	7
	P. rhoeas	16
	P. urbanianum	17
Nuciferine	*Papaver tauricola*	8
Roemerine	*Stephania kwangsiensis*	12
	Papaver tauricola	8
	P. lacerum	7
Dehydroroemerine	*Stephania sasakii*	4
	S. kwangsiensis	12
Isothebaine	*Papaver bracteatum*	18
Laurelliptine	*Nectandra rigida*	19
	Litsea laurifolia	6
	L. triflora	20
	Monodora tenuifolia	21
Isoboldine	*Uvaria chamae*	5
	Thalictrum alpinum	22
	Glaucium fimbrilligerum	13
	Machilus duthei	23
	Litsea triflora	20
	L. laurifolia	6
	Fumaria parviflora	24
	Cryptocarya longifolia	25

Table 2—continued

Alkaloid	Source	Ref.
Laurifoline	*Cocculus laurifolius*	26
Thaliporphine	*Uvaria chamae*	5
	Thalictrum alpinum	22
Laurolitsine	*Litsea laurifolia*	6
	L. kawakamii	27
	L. akoensis	27
	Machilus duthei	23
Boldine	*Litsea laurifolia*	6
	Machilus duthei	23
Predicentrine	*Litsea triflora*	20
Laurotetanine	*Machilus duthei*	23
	Cryptocarya longifolia	25
	Litsea laurifolia	6
	Hernandia cordigera	28
N-Methyl-laurotetanine	*Cryptocarya longifolia*	25
	Litsea laurifolia	6
	L. triflora	20
	Thalictrum revolutum	29
Glaucine	*Litsea laeta*	14
	L. triflora	20
	Uvaria chamae	5
Nornantenine	*Hernandia cordigera*	28
Nantenine	*Papaver tauricola*	8
Actinodaphnine	*Litsea laurifolia*	6
N-Methylactinodaphnine	*Litsea kawakamii*	27
	L. laurifolia	6
Nordicentrine	*Litsea salicifolia*	14
Magnoflorine	*Pachygone ovata*	30
	Thalictrum fauriei	31
	Cocculus laurifolius	26
	Zanthoxylum microcarpum	32
	Z. culantrillo	33
	Kolobopetalum auriculatum	34
	Tinospora cordifolia	35
	Papaver rupifragum	36
	Anamirta cocculus	37
	Meconopsis cambrica	9
	Aristolochia macedonica	38
	Croton turumiquirensis	39
	Pycnarrhena longifolia	40
Norcorydine	*Glaucium fimbrilligerum*	13
Corydine	*Glaucium fimbrilligerum*	13
	Dicentra spectabilis	41
	Litsea triflora	20
N-Methylcorydine	*Kolobopetalum auriculatum*	34
N-Methyl-lindcarpine	*Glaucium fimbrilligerum*	13
Norisocorydine	*Hernandia cordigera*	28
	Cryptocarya longifolia	25
	Glaucium fimbrilligerum	13

Table 2—continued

Alkaloid	Source	Ref.
Isocorydine	*Glaucium fimbrilligerum*	13
	Papaver rhoeas	16
	P. fugax	8
	Litsea triflora	20
	Hernandia cordigera	28
	Stephania sasakii	4
	S. kwangsiensis	12
N-Methylisocorydine	*Cocculus laurifolius*	26
	Zanthoxylum coriaceum	33
	Z. culantrillo	33
Nandigerine	*Hernandia cordigera*	28
	Litsea laurifolia	6
N-Methylnandigerine	*Litsea laurifolia*	6
Ovigerine	*Hernandia cordigera*	28
Thalphenine	*Thalictrum minus* race B	42

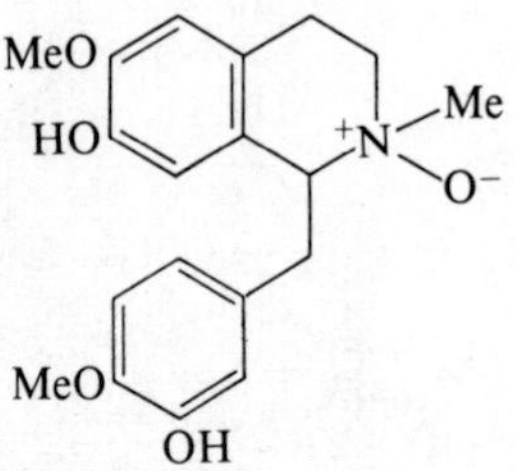

(14)

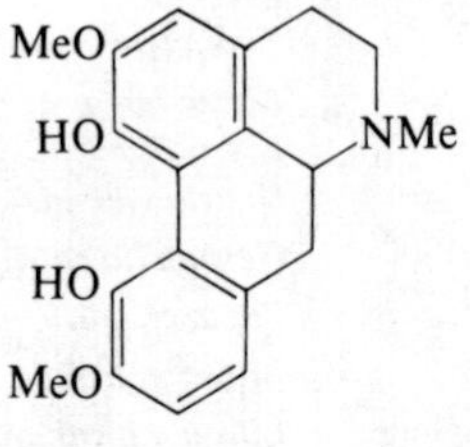

(15)

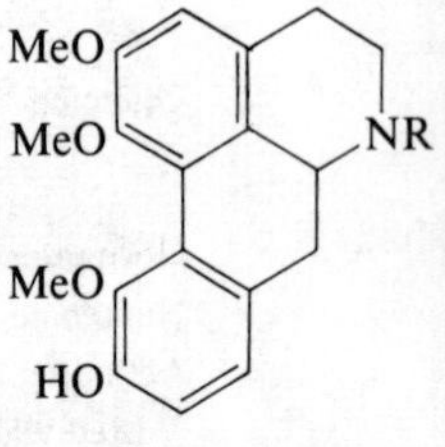

(16) R = H
(17) R = Me

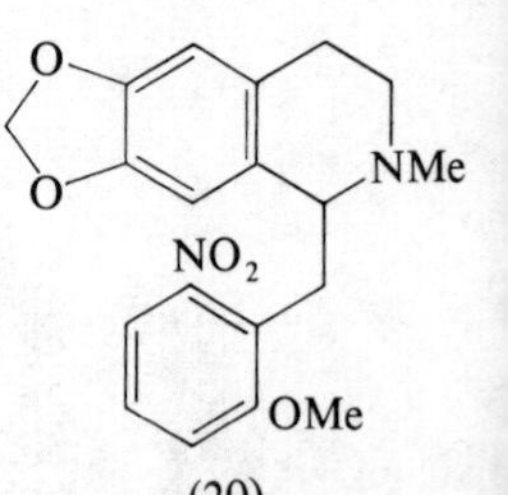

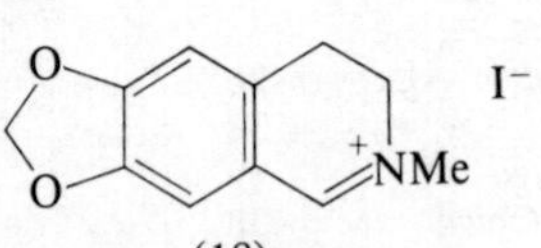

(18)

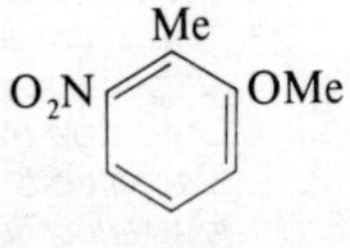

(19)

(20)

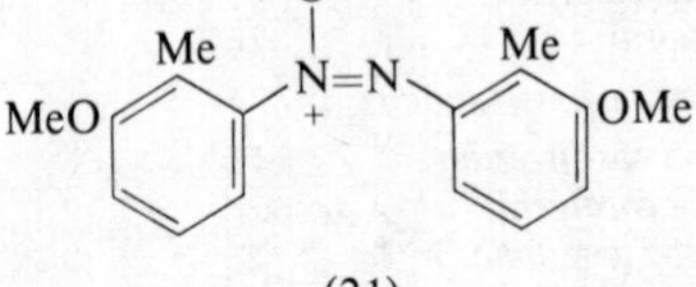

(21)

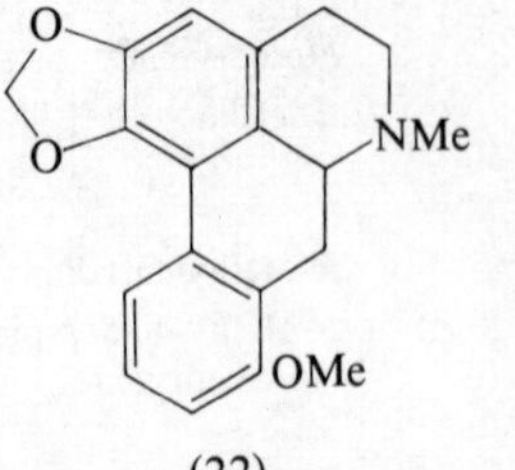

(22)

(27) were obtained in 40% and 31% yields, respectively. In like fashion, the oxidation of the tetrahydrobenzylisoquinoline (28) by thallium(III) trifluoroacetate in trifluoroacetic acid led to neolitsine (29) in an impressive yield of 68%.[1]

On the other hand, the oxidation of the tetrahydrobenzylisoquinoline (23) by thallium(III) acetate resulted in cyclization as well as in acetoxylation, so that the aporphine (30) was obtained in 35% yield.[1]

In a continuing study of the synthesis of aporphines *via* o-quinol acetates, 1,2-diacetoxyaporphines of the type (32) were obtained in good yields by treatment of solutions of the o-quinol acetates (31), in acetonitrile, with concentrated sulphuric acid in acetic anhydride.[45]

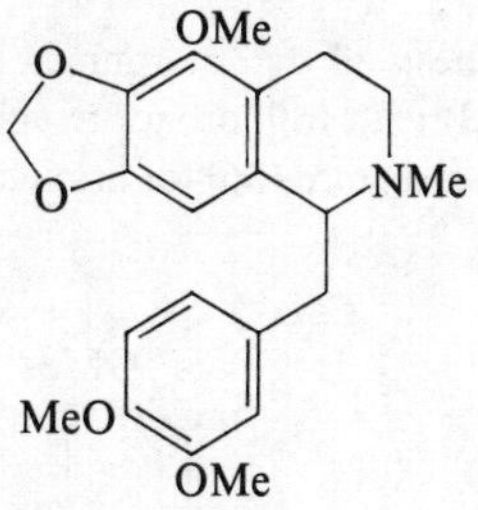

(23)

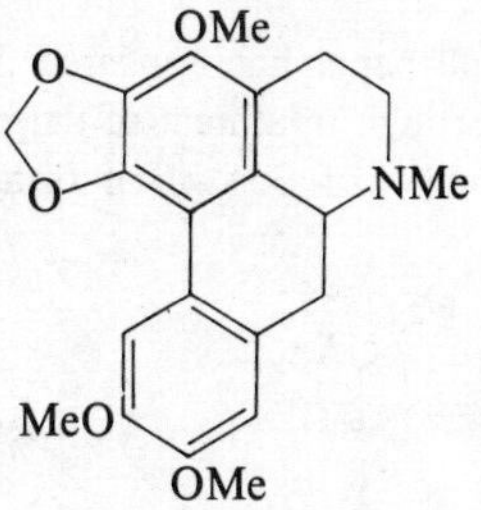

(24)

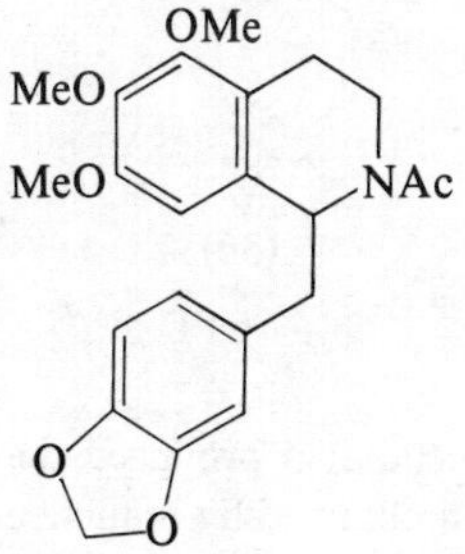

(25)

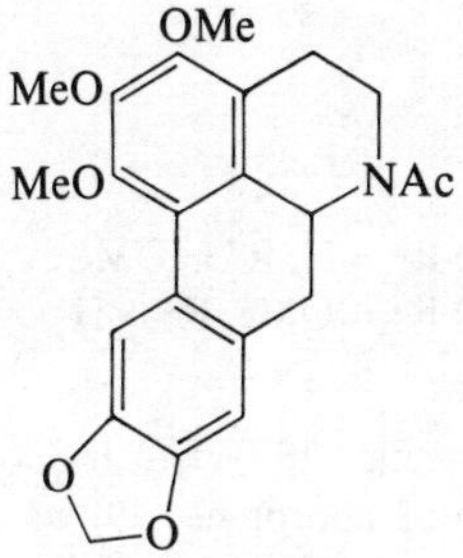

(26)

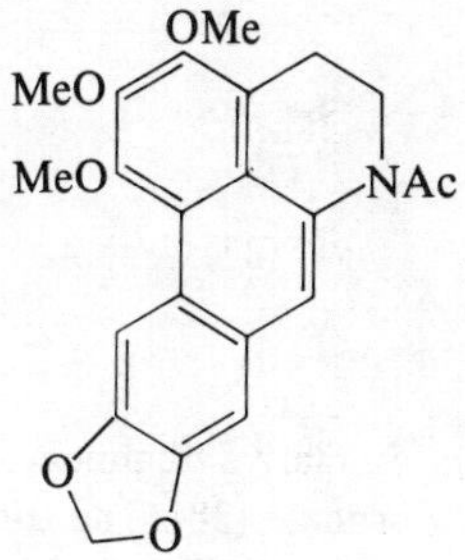

(27)

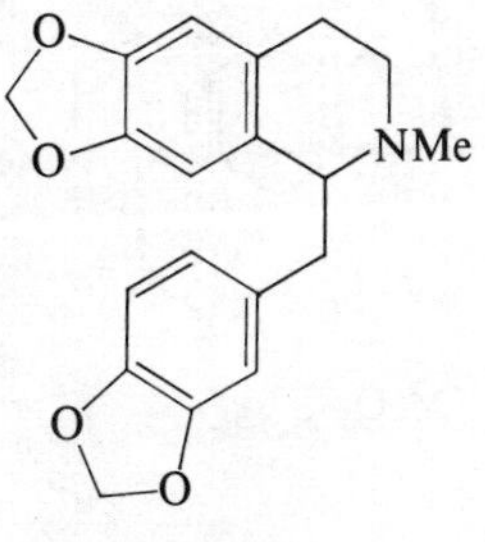

(28)

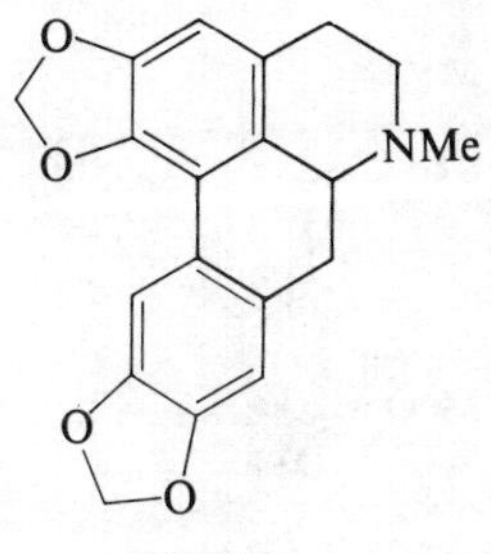

(29)

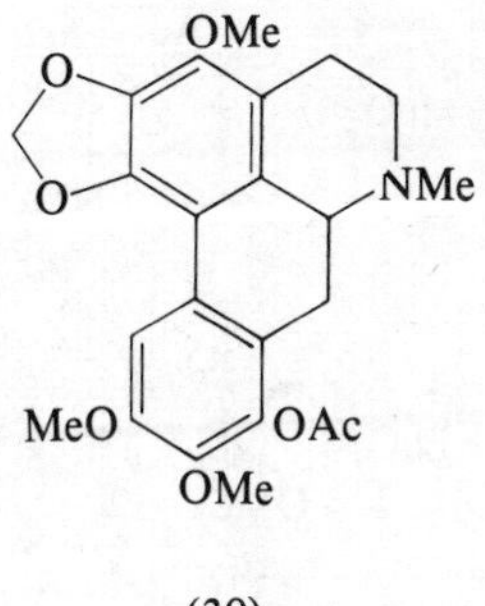

(30)

[45] O. Hoshino, M. Ohtani, and B. Umezawa, *Heterocycles*, 1981, **16**, 793.

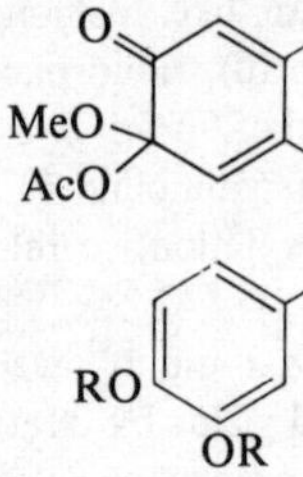

(31)

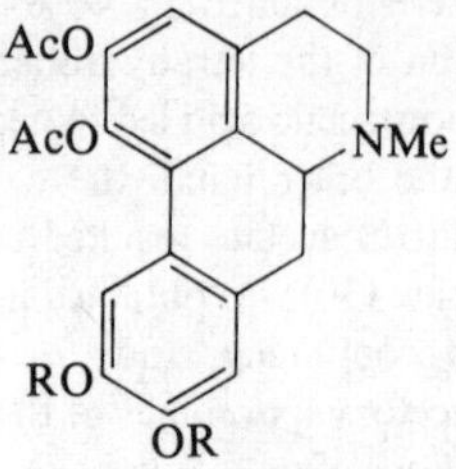

(32)

A full paper has appeared describing the synthesis of the aporphines (34) and (35) through treatment of the *p*-quinol acetate (33) with trifluoroacetic acid.[46] The non-identity of (35) with natural lirinine confirmed structure (36) for lirinine.

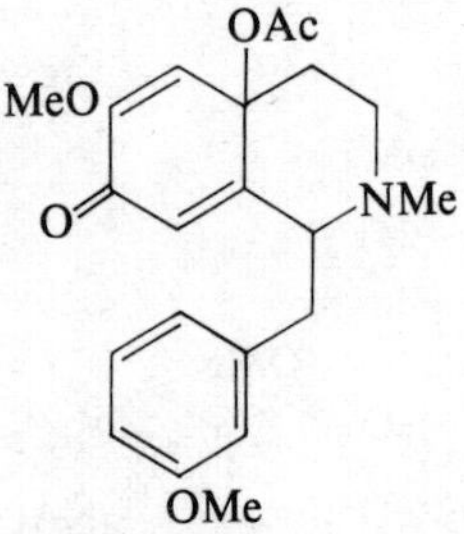

(33)

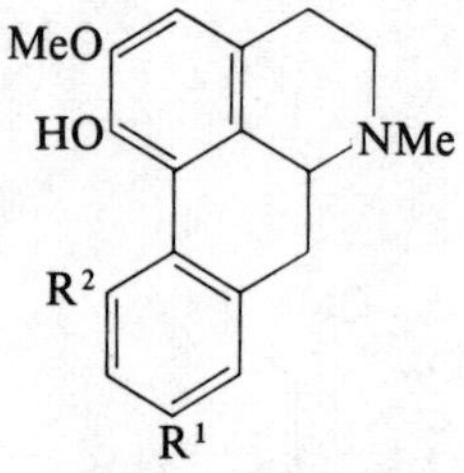

(34) R¹ = H, R² = OMe
(35) R¹ = OMe, R² = H

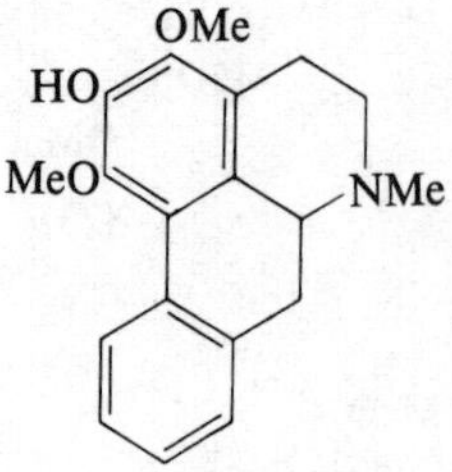

(36)

Similarly, treatment of the acetate (37) with trifluoroacetic acid produced the aporphine (38). The trioxygenated aporphine (39) was then obtained by reductive dephenoxylation of (38) with sodium in liquid ammonia.[46]

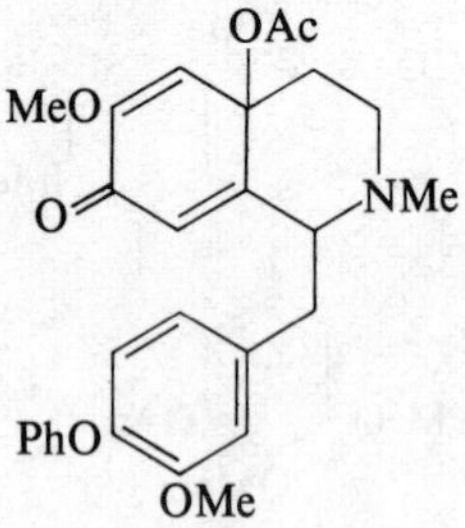

(37)

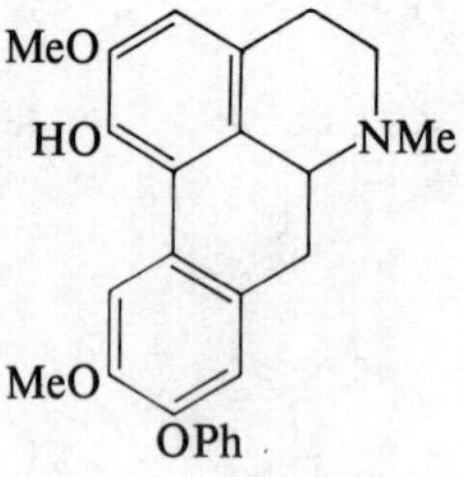

(38)

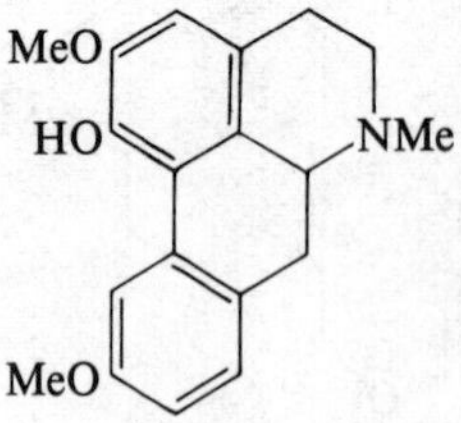

(39)

[46] H. Hara, O. Hoshino, T. Ishige, and B. Umezawa, *Chem. Pharm. Bull.*, 1981, **29**, 1083.

The oxidation of the nor-reticuline derivatives (40a) and (40b), using a variety of tetraethylammonium diacyloxyiodates as oxidizing agents, supplied good yields of the norisoboldines (41a) and (41b).[46a] This marks the first recorded use of diacyliodates for the synthesis of aporphines.

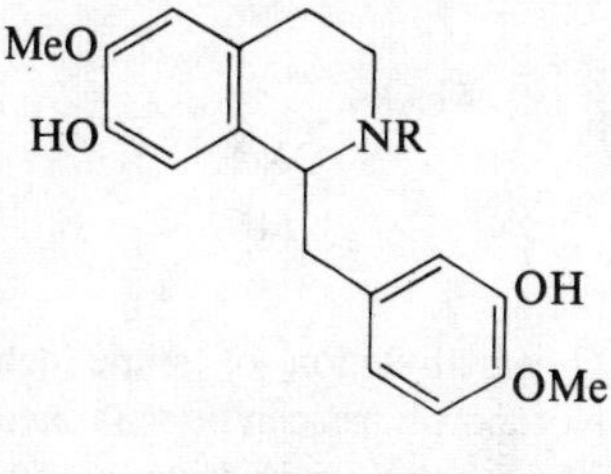

(40) a; R = CO$_2$Et
b; R = CHO

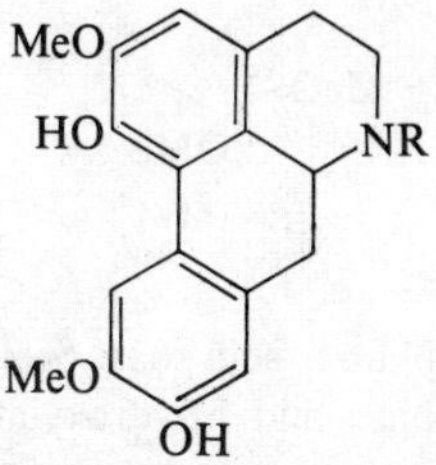

(41) a; R = CO$_2$Et
b; R = CHO

A method has been developed for the preparation of *erythro-* and *threo-*7-hydroxylated aporphines. Reduction of the *N*-metho-oxoaporphinium salt (42) with potassium borohydride resulted in a 60% yield of *erythro-*(43), whereas reduction with Adams catalyst followed by further treatment with borohydride gave *erythro-*(43) along with a minor amount of the *threo-*compound (44).[47]

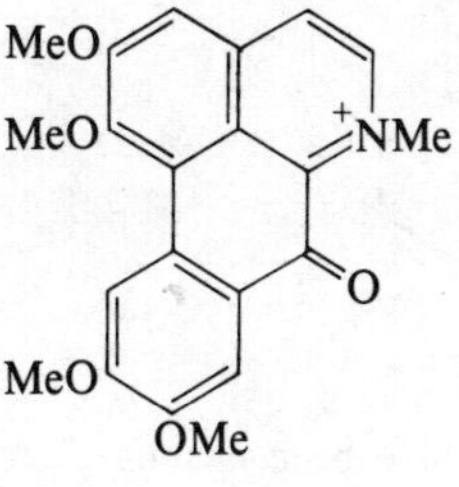

(42)

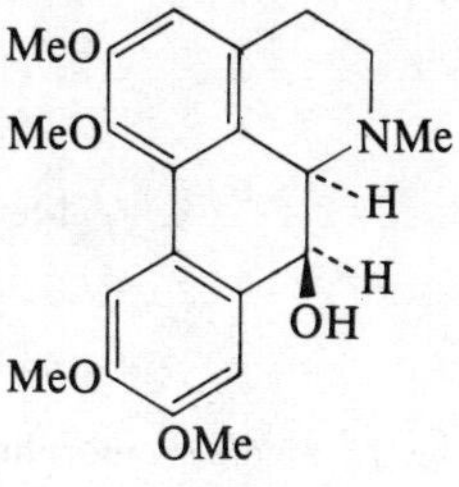

(43)

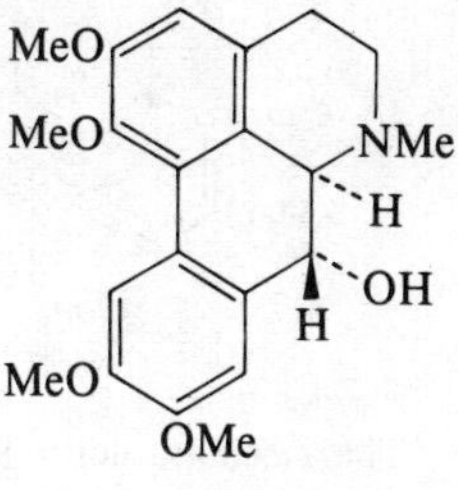

(44)

The photo-oxidation of aporphines to dehydroaporphines has been described.[48] A stereospecific and quantitative oxidation of (+)-glaucine (45) to dehydroglaucine (46) was achieved with *Fusarium solani*. Using this same micro-organism, (−)-glaucine was not metabolized, while (±)-glaucine was oxidized to the extent of 50%.[49]

[46a] C. Szantay, G. Blaskó, M. Barczai-Beke, P. Pechy, and G. Dörnyei, *Tetrahedron Lett.*, 1980, **21**, 3509.

[47] S. Chackalamannil and D. R. Dalton, *Tetrahedron Lett.*, 1980, **21**, 2029.

[48] L. Castedo, T. Iglesias, A. Puga, J. M. Saá, and R. Suau, *Heterocycles*, 1981, **15**, 915.

[49] P. J. Davis and J. P. Rosazza, *Bioorg. Chem.*, 1981, **10**, 97.

(45)   (46)

Sulphuric acid has been used for the *O*-demethylation of some dehydro-aporphines and 7-hydroxylated aporphines.[50] Noraporphines can be separated from aporphines by trifluoroacetylation, using trifluoroacetic anhydride in pyridine. Aporphines, however, tend to be converted by this reagent into the corresponding phenanthrenes.[51]

It has been shown that (+)-magnoflorine (47) and (+)-*NN*-dimethyl-lindcarpine (48) are easily distinguishable by analysis of their ¹H n.m.r. spectra in different solvents. On this basis, the quaternary aporphine found in *Caltha leptosepala* is clearly (+)-magnoflorine.[52]

(47)   (48)

The circular dichroism curves of nineteen aporphines have been examined. A generalization worth bearing in mind is that the terminal absorption near 216 nm is positive in the case of the *R* configuration, and negative for the *S*.[53] A study of natural (−)-crebanine confirmed the *R* absolute configuration, instead of the *S* configuration which is usual for aporphines that are disubstituted in ring D.

A method for the preparation of (−)-apomorphine (50) and (−)-*N*-(n-propyl)norapomorphine (51) from thebaine (49) has been described. The sequence has also been extended to the transformation of (+)-bulbocapnine (52) into (+)-apomorphine.[55]

[50] L. Castedo, A. Roderiguez de Lera, J. M. Saá, R. Suau, and C. Villaverde, *Heterocycles*, 1980, **14**, 1135.

[51] M. Leboeuf, F. Bévalot, and A. Cavé, *Planta Med.*, 1980, **38**, 33.

[52] F. R. Stermitz, L. Castedo, and D. Dominguez, *J. Nat. Prod.*, 1980, **43**, 140.

[53] B. Ringdahl, R. P. K. Chan, J. C. Craig, M. P. Cava, and M. Shamma, *J. Nat. Prod.*, 1981, **44**, 80.

[54] K. Pharadai, B. Tantisewie, S. F. Hussain, and M. Shamma, *Heterocycles*, 1981, **15**, 1067.

[55] V. J. Ram and J. L. Neumeyer, *J. Org. Chem.*, 1981, **46**, 2830.

(49)　　　　　(50) R = Me　　　　　(52)
　　　　　　　(51) R = Pr^n

The conversion of (−)-apomorphine into (+)-apomorphine has been achieved.[56] The O-dealkylation of 10,11-dimethoxyaporphine, using sodium thioethoxide in dimethylformamide, has been reported.[57] The reactions of the enamine dehydronuciferine with dimethyl acetylenedicarboxylate, methyl propiolate, methyl acrylate, and diethyl azodicarboxylate have been investigated, and have resulted in the preparation of a novel series of 7-substituted aporphines.[58]

The synthesis of a number of radiolabelled apomorphine derivatives has been described.[59–62] A full paper has appeared on the preparation of (−)-2,10,11-trihydroxy-N-(n-propyl)noraporphine, a dopaminergic compound with anticonvulsant activity.[63] The potential dopamine-inhibiting properties of (−)-N-(2-chloroethyl)norapomorphine and related compounds have been assessed. N-Chloroethylation in the aporphine series can abolish dopamine agonist action, and can confer a long-lasting dopamine antagonist potential.[64]

(53)　　　　　(54) R = H (*rac.*)
　　　　　　　(55) R = Me

[56] P. J. Davis, S. Seyhan, W. Soine, and R. V. Smith, *J. Pharm. Sci.*, 1980, **69**, 1056.
[57] J. C. Kim, *Taehan Hwahakhoe Chi*, 1980, **24**, 266.
[58] M. D. Menachery, J. M. Saá, and M. P. Cava, *J. Org. Chem.*, 1981, **46**, 2584.
[59] C. N. Filer, D. G. Ahern, F. E. Granchelli, J. L. Neumeyer, and L. J. Law, *J. Org. Chem.*, 1980, **45**, 3465.
[60] C. N. Filer and D. G. Ahern, *J. Org. Chem.*, 1980, **45**, 3918.
[61] W. H. Soine, J. E. Hudson, B. A. Shoulders, and R. V. Smith, *J. Pharm. Sci.*, 1980, **69**, 1040.
[62] W. H. Soine, P. Salgo, and R. V. Smith, *Radiopharmacy*, 1979, **16**, 597.
[63] J. L. Neumeyer, S. J. Law, B. Meldrum, G. Anlezaku, and K. J. Watling, *J. Med. Chem.*, 1981, **24**, 898.
[64] B. Costall, D. H. Fortune, F. E. Granchelli, S. J. Law, R. J. Naylor, J. L. Neumeyer, and V. Nohria, *J. Pharm. Pharmacol.*, 1980, **32**, 571; J. L. Neumeyer, S. J. Law, R. J. Baldessarini, and N. S. Kula, *J. Med. Chem.*, 1980, **23**. 594.

The preparation and anti-tumour activity of a series of aporphine nitrogen mustards have been reported.[65]

The biosynthesis of magnoflorine (47) and laurifoline (53) has been studied, and the specific incorporation [of (±)-nor-reticuline (54)] and (+)-(S)-reticuline (55) has been demonstrated.[66] This work is described in greater detail elsewhere in this volume.

4 Dimeric Aporphinoids

A study of three *Berberis* species, namely *B. orthobotrys*, *B. calliobotrys*, and *B. zabeliana*, has yielded four new aporphine dimers, *i.e.* chitraline (56), 1-*O*-methylpakistanine (57),[67, 68] kalashine (58),[68, 69] and khyberine (59).[70] Kalashine and khyberine are the first aporphine–benzylisoquinoline dimers incorporating a diaryl ether terminal at C-11 of the aporphine moiety.

A careful study of the dienone–phenol rearrangement *in vitro* of pakistanamine (60), which is the sole proaporphine–benzylisoquinoline alkaloid presently known,

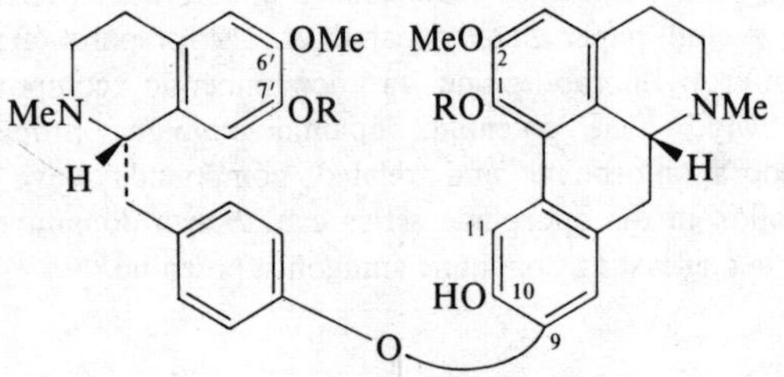

(56) R = H
(57) R = Me

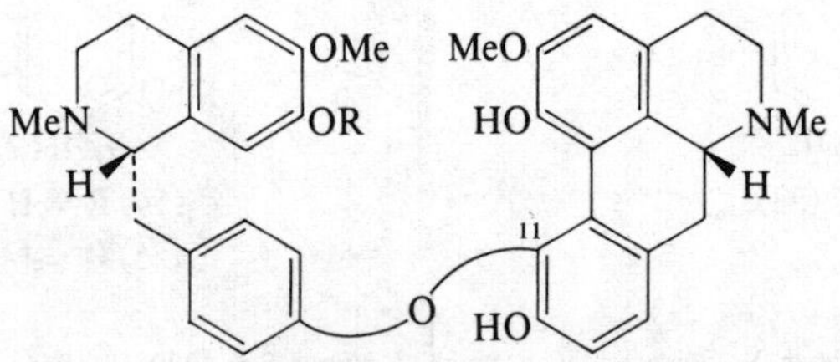

(58) R = Me
(59) R = H

[65] J. L. Neumeyer, F. E. Granchelli, C. N. Filer, A. H. Soloway, and S. J. Law, *J. Med. Chem.*, 1980, **23**, 1008.

[66] D. S. Bhakuni, S. Jain, and R. S. Singh, *Tetrahedron*, 1980, **36**, 2525.

[67] S. F. Hussain, L. Khan, and M. Shamma, *Heterocycles*, 1981, **15**, 191.

[68] S. F. Hussain, L. Khan, K. K. Sadozai, and M. Shamma, *J. Nat. Prod.*, 1981, **44**, 274.

[69] S. F. Hussain and M. Shamma, *Tetrahedron Lett.*, 1980, **21**, 3315.

[70] S. F. Hussain, M. T. Siddiqui, and M. Shamma, *Tetrahedron Lett.*, 1980, **21**, 4573.

has shown that migration of an aryl group to the less-hindered side of the dienone system is the heavily favoured (but not the exclusive) process.[70] Studies of natural alkaloidal mixtures from *Berberis* spp. have indicated that dimers with the diaryl ether linkage at C-11 of the aporphine moiety instead of at C-9 are very minor components of the mixture.[67-70] It may thus be that the dienone–phenol rearrangement of a proaporphine that occurs in Nature is simply an acid-catalysed process which is not mediated by an enzyme, and that the migrating aryl group prefers to move, albeit not exclusively, to the less-hindered side of the dienone. The stereochemistry of the proaporphine may also play a role in the specific direction of the aryl migration.

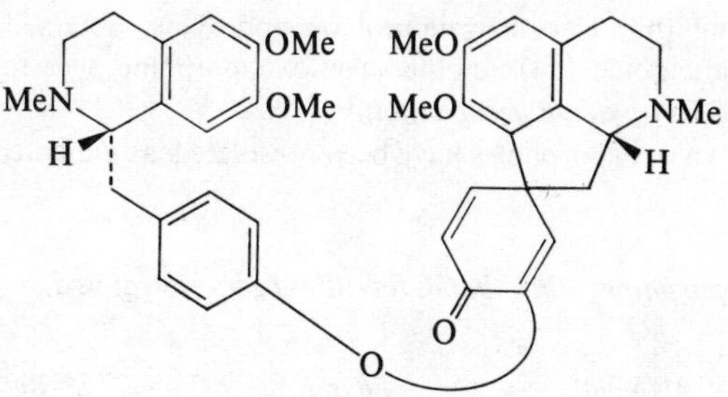

(60)

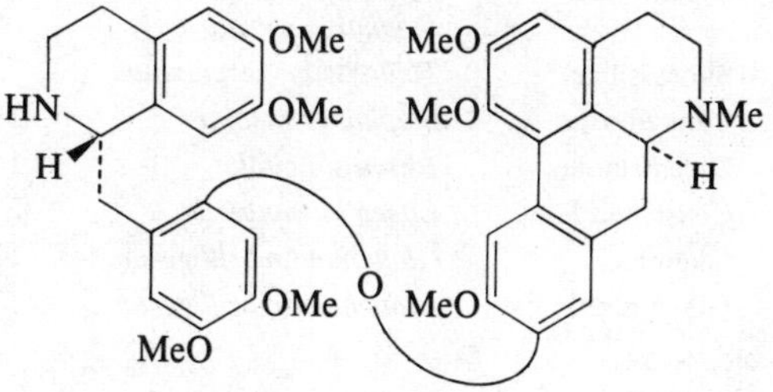

(61)

Northalicarpine (61), which is present in *Thalictrum revolutum*, is the first known noraporphine–benzylisoquinoline dimer.[29] *Berberis orthobotris* has also furnished two known dimeric alkaloids, pakistanamine (60) and pakistanine.[67] Thalicarpine has been re-isolated from *Thalictrum alpinum*.[42] Cancer patients showed no objective response when treated with this alkaloid.[71]

5 Oxoaporphines

An interesting and unusual aporphinoid obtained from *Telitoxicum peruvianum* (Menispermaceae) is the reddish-brown telazoline, which possesses two nitrogen atoms. The tentative structure (62) has been advanced for this compound.[73]

---

[71] J. T. Leimert, M. P. Corder, T. M. Elliott, and J. M. Lovett, *Cancer Treat. Rep.*, 1980, **64**, 1389.

(62)   (63)   (64)

Oxocrebanine (63) is a new natural oxoaporphine, obtained from *Stephania sasakii*.[4] Oxoanolobine (64), another new oxoaporphine, was found in *Guatteria melosma*[72] and *Telitoxicum peruvianum*.[73]

Several known oxoaporphines have been re-isolated, as indicated in Table 3.

Table 3 *Oxoaporphines that have recently been re-isolated, and their natural sources*

Alkaloid	Source	Ref.
Liriodenine	*Stephania sasakii*	11
Lysicamine	*Telitoxicum peruvianum*	73
	Stephania sasakii	4
Subsessiline	*Telitoxicum peruvianum*	73
Lanuginosine	*Stephania sasakii*	11
Dicentrinone	*Litsea salicifolia*	14
Cassameridine	*Litsea kawakamii*	27
Glaunine	*Glaucium fimbrilligerum*	13
Glaunidine	*Glaucium fimbrilligerum*	13

Conclusive support for the identity of glauvine with corunnine (65) has been provided. Reduction of so-called glauvine (65) with zinc in acetic acid did not give norlirioferine (66), as previously claimed, but instead supplied the known thaliporphine (69). Authentic norlirioferine and its *N,O*-diacetyl derivative (67) were then obtained *via* reduction of oxolirioferine (68), which was synthesized by two independent routes.[74]

Oxidation of noraporphines and aporphines with Fremy's salt gives oxoaporphines and oxoaporphinium salts, respectively.[75] Aerial oxidation of dehydro-aporphines in the presence of alkali provides the corresponding oxoaporphines, 4,5-dioxoaporphines, and *N*-methylaristolactams.[76] An analysis of the ^{13}C n.m.r.

[72] C. H. Phoebe, Jr., P. L. Schiff, Jr., J. E. Knapp, and D. J. Slatkin, *Heterocycles*, 1980, **14**, 1977.

[73] M. D. Menachery and M. P. Cava, *J. Nat. Prod.*, 1981, **44**, 320.

[74] L. Castedo, J. M. Saá, R. Suau, and C. Villaverde, *Heterocycles*, 1980, **14**, 1131.

[75] L. Castedo, A. Puga, J. M. Saá, and R. Suau, *Tetrahedron Lett.*, 1981, **22**, 2236.

[76] J. Kunitomo, Y. Murakami, and M. Akasu, *J. Pharm. Soc. Jpn.*, 1980, **100**, 337.

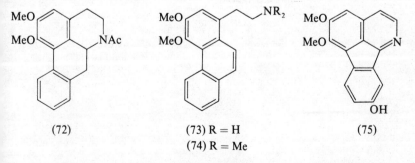

(65)

(66) R = H
(67) R = Ac

(68)

(69)

(70)

(71)

spectra of several oxoaporphines has confirmed the structures of oxo-*O*-methyl-pukateine and *O*-methylmoschatoline.[77]

Liriodenine (70) and other oxoaporphines were evaluated for their antibacterial and antifungal activity against several micro-organisms. Only (70) showed significant activity.[78]

6 4,5-Dioxoaporphines

The new alkaloid 4,5-dioxodehydrocrebanine (71) has been obtained from *Stephania sasakii*[11] and cepharadione A has been re-isolated from that same plant.[4]

(72)

(73) R = H
(74) R = Me

(75)

[77] A. Marsaioli, A. F. Magalhaes, E. A. Ruveda, and F. de A. M. Reis, *Phytochemistry*, 1980, **19**, 995.
[78] C. D. Hufford, A. S. Sharma, and B. O. Oguntimein, *J. Pharm. Sci.*, 1980, **69**, 1180.

7 Phenanthrenes

When *N*-acetylnornuciferine (72) was heated with hydrochloric acid in methanol, bisnoratherosperminine (73) was obtained. This compound was then *N*-methylated to atherosperminine (74).[78]

8 Azafluoranthenes

A new azafluoranthene is telitoxine (75), found together with the known nor-rufescine in *Telitoxicum peruvianum*.[73]

10

Amaryllidaceae Alkaloids

BY M. F. GRUNDON

1 Isolation and Structural Studies

An investigation of the alkaloid content of *Crinum augustum* resulted in the isolation of lycorine (8), buphanisine (1; $R^1 = R^2 = H$, $R^3 = Me$), and crinamine (2) and the identification of six new alkaloids.[1] Augustine, one of the latter group, was shown by a thorough study of its ^1H n.m.r., ^{13}C n.m.r., and mass spectra to be an epoxide with relative structure (3).[2] Four of the new alkaloids were separated into two pairs of compounds and were shown by ^1H n.m.r. and mass spectroscopy to be

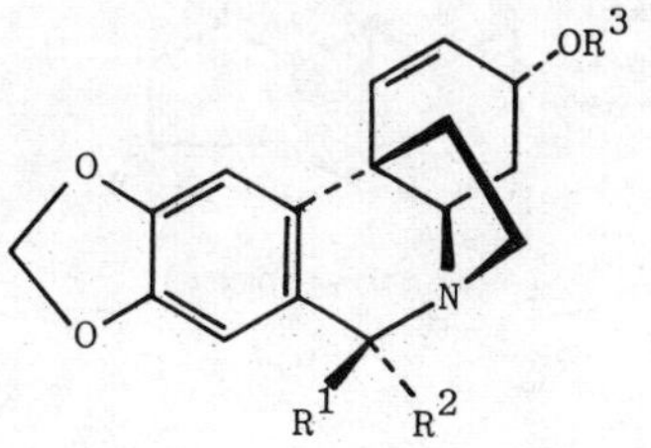

(1)

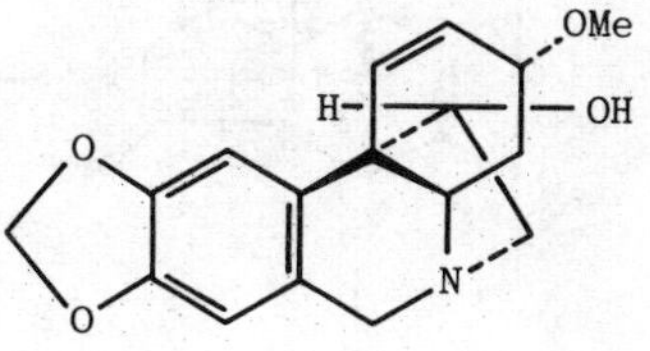

(2)

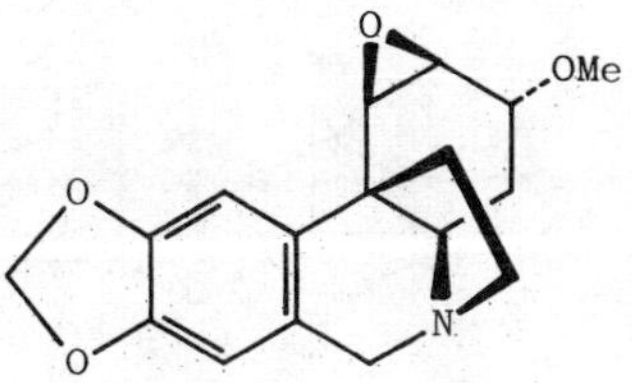

(3)

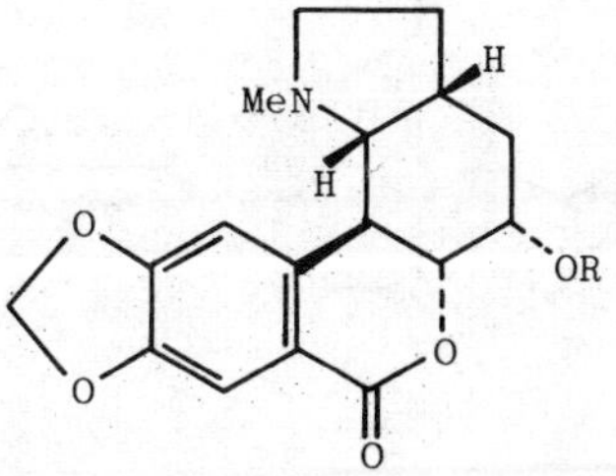

(4)

[1] A. A. Ali, H. Kating, A. W. Frahm, A. M. El-Moghazi, and M. A. Ramadan, *Phytochemistry*, 1981, **20**, 1121; A. A. Ali, H. Kating, and A. W. Frahm, *ibid.*, p. 1731.
[2] A. W. Frahm, A. A. Ali, and H. Kating, *Phytochemistry*, 1981, **20**, 1735.

6-α- and 6-β-hydroxybuphanisine (1; $R^1 = H$, $R^2 = OH$, $R^3 = Me$) and (1; $R^1 = OH$, $R^2 = H$, $R^3 = Me$), respectively, and 6-α- and 6-β-hydroxycrinine (1; $R^1 = R^3 = H$, $R^2 = OH$) and (1; $R^1 = OH$, $R^2 = R^3 = H$), respectively. Within each pair, epimerization at C-6 to an equilibrium mixture of stereoisomers occurred readily and the constituents could not be separated by chromatographic methods. The constitution of the remaining new alkaloid was not established; it has a molecular formula of $C_{17}H_{19}NO_4$, contains an *N*-methyl group but no olefinic protons, and may represent a new type of carbon skeleton in the Amaryllidaceae alkaloids.

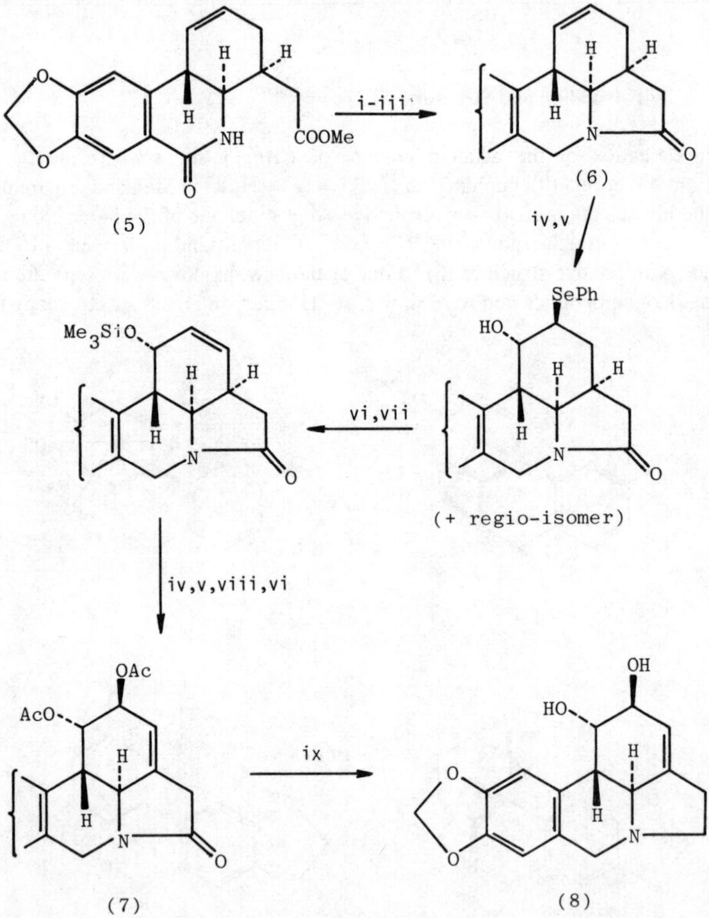

Reagents: i, Meerwein reagent, CH_2Cl_2; ii, $NaBH_4$, $SnCl_2$, $MeO(CH_2)_2OMe$, at −60 °C; iii, K_2CO_3, MeOH; iv, m-$ClC_6H_4CO_3H$, CH_2Cl_2; v, Ph_2Se_2, $NaBH_4$, EtOH; vi, $NaIO_4$; vii, trimethyl-silylimidazole, MeCN; viii, acetylation; ix, $LiAlH_4$.

Scheme 1

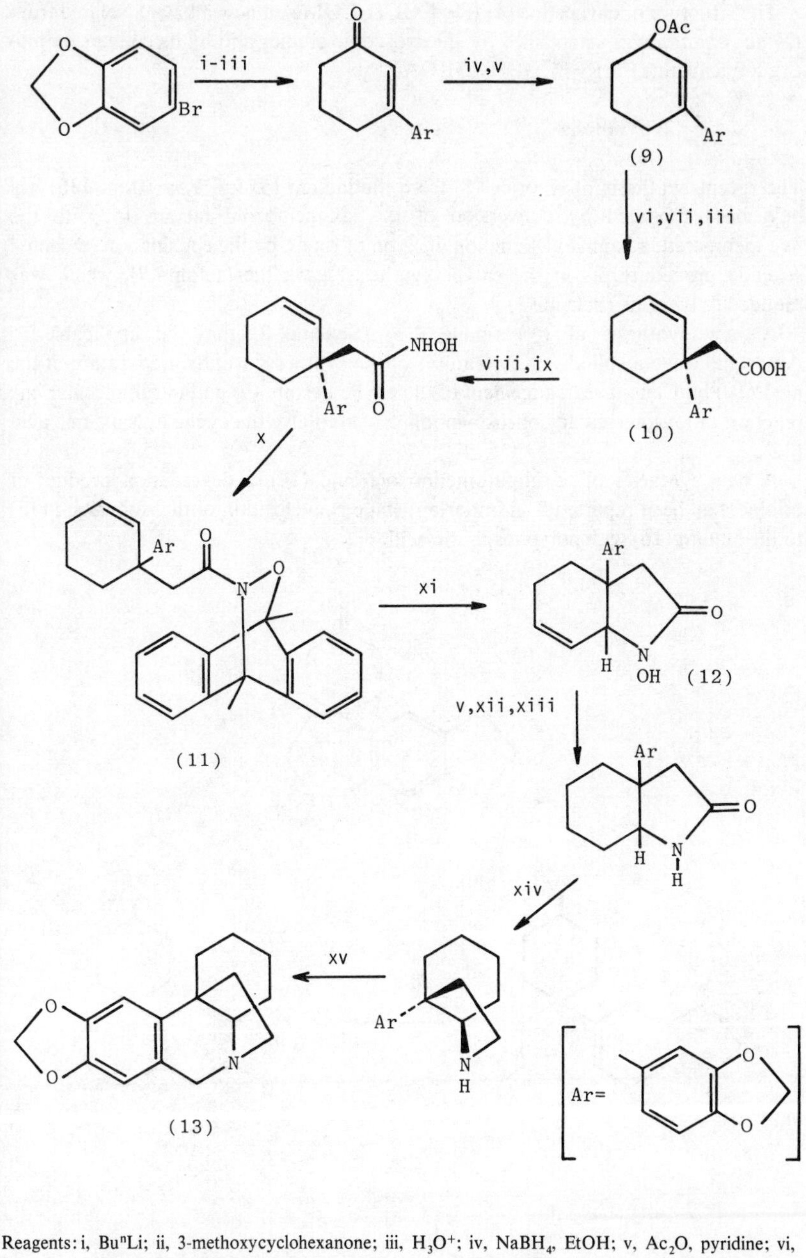

Reagents: i, Bu^{n}Li; ii, 3-methoxycyclohexanone; iii, H$_3$O$^+$; iv, NaBH$_4$, EtOH; v, Ac$_2$O, pyridine; vi, LDA, THF; vii, Bu^{t}Me$_2$SiCl; viii, SOCl$_2$; ix, NH$_2$OH · HCl; x, Pr$^n_4$N$^+$ IO$_4^-$, 9,10-dimethyl-anthracene; xi, PhMe, reflux; xii, H$_2$, Pd/C, EtOAc; xiii, TiCl$_3$, MeOH–H$_2$O, Na$_2$CO$_3$; xiv, LiAlH$_4$, THF, reflux; xv, *NN*-dimethylmethyleneammonium iodide, THF, at 40 °C.

Scheme 2

The structure of clivacetine (4; R = COCH$_2$COMe), a new alkaloid isolated from *Clivia miniata*, was established by spectroscopic studies and by its conversion into *O*-acetylclivatine [4; R = COCH$_2$CH(OAc)Me].[3]

2 Synthesis

The recent synthesis of lycorine (8) from the lactam (5) (*cf.* Vol. 10, p. 136) has now been modified by conversion of the six-membered lactam (5) into the five-membered lactam (6).[4] Functionalization of ring C by the epoxidation–diphenyl selenide procedure, as in the earlier synthesis, gave the lactam (7), which was reduced to lycorine (Scheme 1).

A new synthesis of ($\pm$)-crinane (13) (Scheme 2) may be applicable to Amaryllidaceae alkaloids.[5] Key features of the synthesis are the preparation of the acid (10) by Claisen rearrangement of the allylic acetate (9) and intramolecular ene reaction of a protected acylnitroso-enophile (11) to give the cyclic hydroxamic acid (12).

A new synthesis of tetrahydrometinoxocrinine (14), a degradation product of crinine, has been reported;[6] an important stage is cyclization of the isocyanate (15) to the lactam (16) with polyphosphoric acid.

(14)

(15)   (16)

[3] S. Kobayashi, H. Ishikawa, E. Sasakawa, M. Kihara, T. Shingu, and A. Kato, *Chem. Pharm. Bull.*, 1980, **28**, 1827.
[4] T. Sano, N. Kashiwaba, J. Toda, Y. Tsuda, and H. Irie, *Heterocycles*, 1980, **14**, 1097.
[5] G. E. Keck and R. R. Webb, *J. Am. Chem. Soc.*, 1981, **103**, 3173.
[6] I. H. Sanchez and M. T. Mendoza, *Tetrahedron Lett.*, 1980, **21**, 3651.

11

Erythrina and Related Alkaloids

BY A. S. CHAWLA AND A. H. JACKSON

This report embodies the work published on the isolation, structure determination, and synthesis of *Erythrina, Cocculus, Cephalotaxus*, and other related alkaloids.

1 Isolation and Structure Determination

Erysodine (1a) and erysotrine (1b) have been isolated from the bark of *Erythrina blakei*.[1] From the leaves of *Cocculus laurifolius*, Bhakuni and Jain[2] have reported the isolation of four new alkaloids, *i.e.* cocculitinine (2a), cocculidinone (3), cocculimine (4a), and coccudienone (5), in addition to isococculidine (4b), isococculine (4c), coccuvine (6a),[3] coccuvinine (6b), coccoline (6c), coccolinine (6d), and cocculitine (2b), which had been described previously (*cf.* Vol. 8, p. 144; Vol. 9, p. 145; Vol. 11, p. 138). Their structures (including stereochemistry) were assigned by chemical transformations and spectral studies. From the same plant, Ju-ichi *et al.*[4] isolated two new erythrinan alkaloids, named erythlaurine (7a) and erythramide (7b), and likewise assigned their structures by chemical and spectral studies. These two new alkaloids both contain a C_1 unit that is directly attached to the aromatic ring; this is a relatively novel feature, only observed once previously in the *Cocculus* alkaloids [in erythroculine (7c)[5]]. There has also appeared a review on

(1) a; R = H
   b; R = Me

(2) a; R = H
   b; R = Me

(3)

[1] H. Singh, A. S. Chawla, V. K. Kapoor, and J. Kumar, *Planta Med.*, 1981, **41**, 101.
[2] D. S. Bhakuni and S. Jain, *Tetrahedron*, 1980, **36**, 3107.
[3] A. N. Singh, H. Pande, and D. S. Bhakuni, *Experientia*, 1976, **32**, 1368.
[4] M. Ju-ichi, Y. Fujitani, T. Shingu, and H. Furukawa, *Heterocycles*, 1981, **16**, 555.
[5] Y. Inubushi, H. Furukawa, and M. Ju-ichi, *Tetrahedron Lett.*, 1969, 153; *Chem. Pharm. Bull.*, 1970, **18**, 1951.

classification, structural characteristics, structure elucidation, and biosynthesis of erythrinan alkaloids in the genus *Cocculus*.[6] The isolation of cephalotaxine, 11-hydroxycephalotaxine, drupacine, demethylcephalotaxinone, 3-*epi*-wilsonine, and wilsonine from *Cephalotaxus sinensis* has also been reported.[7]

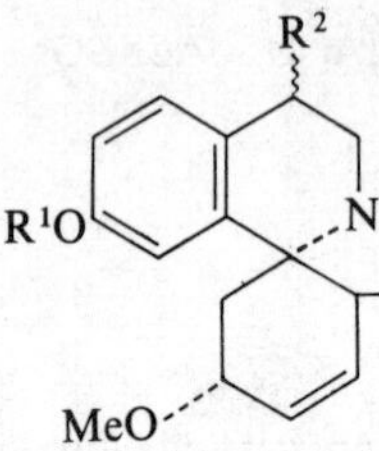

(4) a; $R^1 = H$, $R^2 = OMe$
    b; $R^1 = Me$, $R^2 = H$
    c; $R^1 = R^2 = H$

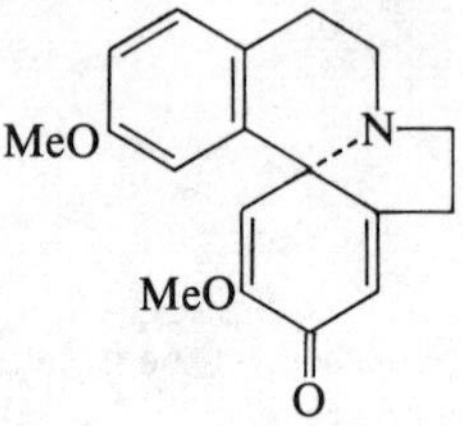

(5)

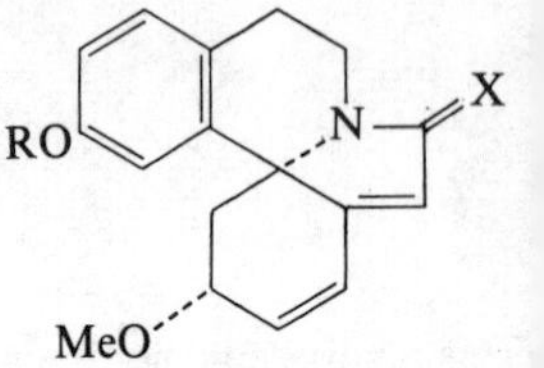

(6) a; $R = H$, $X = H_2$
    b; $R = Me$, $X = H_2$
    c; $R = Me$, $X = O$
    d; $R = H$, $X = O$

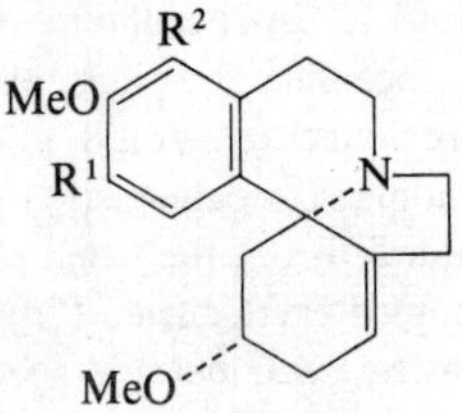

(7) a; $R^1 = CO_2Me$, $R^2 = OH$
    b; $R^1 = CONH_2$, $R^2 = H$
    c; $R^1 = CO_2Me$, $R^2 = H$

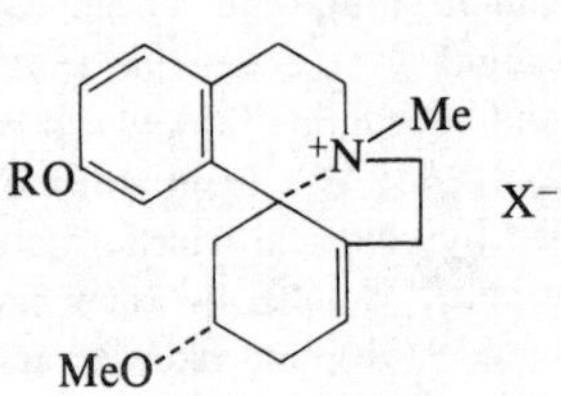

(8) a; $R = H$, $X = OH$
    b; $R = Me$, $X = I$

A new quaternary alkaloid, pachygonine (8a), was isolated from the roots of *Pachygone ovata* (Menispermaceae) and the structure established on the basis of spectroscopic evidence and its methylation to give cocculidine methiodide (8b).[8] This is the first reported occurrence of an alkaloid with the erythrinan skeleton outside the genera *Erythrina* and *Cocculus*, and moreover it is also uncommon in being quaternary.[9, 10] Pharmacological studies with pachygonine showed that it is a negative chromotropic agent.

[6] M. Ju-ichi, *Mukogawa Joshi Daigaku Kiyo, Yakugaku Hen*, 1979, **27**, 13 (*Chem. Abstr.*, 1981, **94**, 188 602).
[7] *Chih Wu Hsueh Pao*, 1980, **22**, 156 (*Chem. Abstr.*, 1980, **93**, 235 142).
[8] S. V. Bhat, H. Dornauer, and N. J. De Souza, *J. Nat. Prod.*, 1980, **43**, 588.
[9] J. S. Glasby, 'Encyclopaedia of the Alkaloids', Vol. 1—3. Plenum Press, New York, 1976–7.
[10] S. W. Pelletier, 'Chemistry of the Alkaloids', Van Nostrand Reinhold, New York, 1970, p. 173.

2 Synthesis

In the *Erythrina* series, Tsuda and co-workers[11] have developed new methods for the synthesis of spiro-type compounds that are related to *Erythrina* alkaloids, either by intramolecular nucleophilic cyclization of dioxopyrrolines or by cycloaddition of activated butadienes to dioxopyrrolines. The conversion of 2-(ethoxycarbonyl)-cycloalkanones (9) into the spiro-type compounds (12), in high yield, *via* 3,3-disubstituted dioxopyrrolines (11) is shown in Scheme 1.[12] They also reported

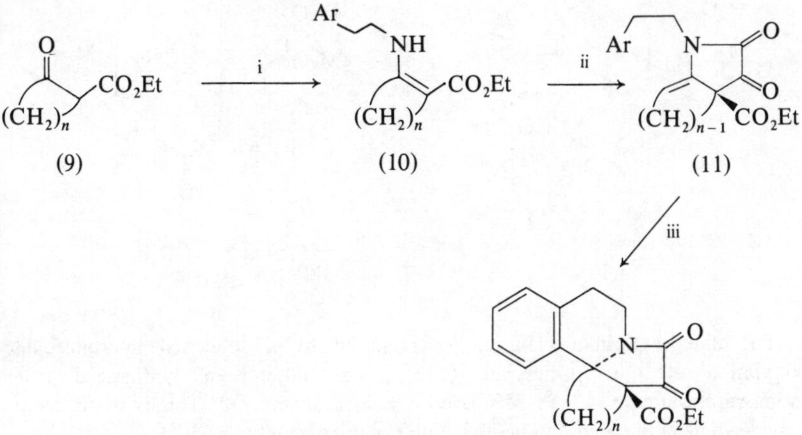

Reagents: i, $ArCH_2CH_2NH_2$; ii, $(COCl)_2$; iii, PPA or an appropriate Lewis acid

Scheme 1

that the thermal cycloaddition of butadiene to isoquinolinopyrrolinedione (13) proceeded regio- and stereo-selectively to give 1,4-cyclo-adducts which, on hydrogenation over palladium–carbon, gave a tetracyclic product (14); this was identical with the material obtained by heating 2-ethoxycarbonylcyclohexanone (9; $n = 6$) with homoveratrylamine followed by treatment with oxalyl chloride and anhydrous phosphoric acid. Tsuda *et al.*[12] have also shown the wide applicability of this method by synthesizing D-nor- and D-homo-erythrinans and other variants in which ring A was heteroaromatic rather than benzenoid. Phenolic erythrinans can be synthesized without protection of hydroxyl groups. The dioxopyrroline derivative (15) has been cyclized[13] to the corresponding erythrinan (16a) and the latter converted into (16b) by decarbethoxylation with $MgCl_2$ in DMSO.[14]

[11] Y. Tsuda, Y. Sakai, N. Kashiwaba, T. Sano, J. Toda, and K. Isobe, *Heterocycles*, 1981, **16**, 189.
[12] Y. Tsuda, Y. Sakai, M. Kaneko, Y. Ishiguro, K. Isobe, J. Taga, and T. Sano, *Heterocycles*, 1981, **15**, 431.
[13] Y. Tsuda, Y. Sakai, and T. Sano, *Heterocycles*, 1981, **15**, 1097.
[14] Y. Tsuda and Y. Sakai, *Synthesis*, 1981, 119.

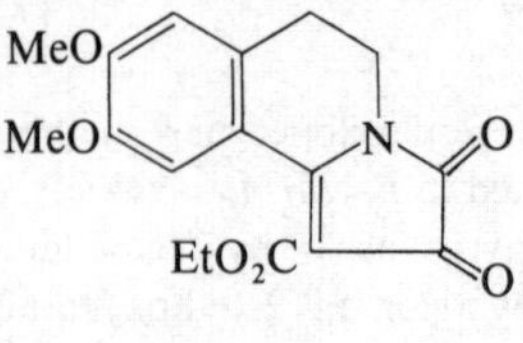

(13)

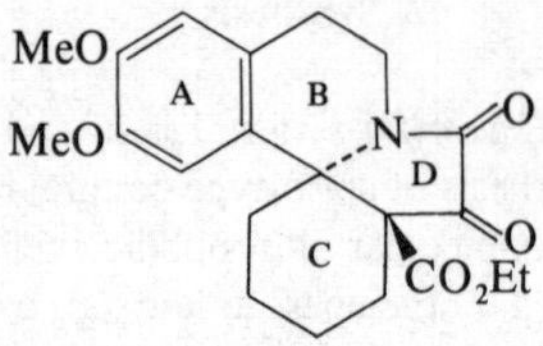

(14)

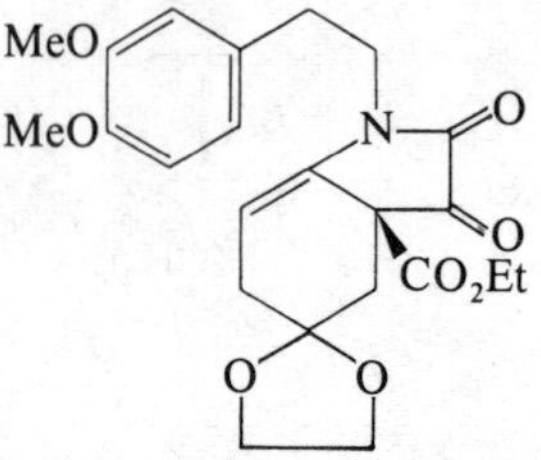

(15)

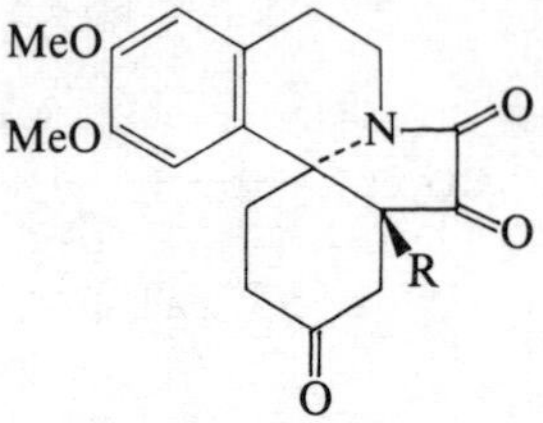

(16) a; R = CO$_2$Et
     b; R = H

The dioxoerythrinan (19), earlier prepared by a concerted intermolecular alkylation of the 7β-mesylate (18b),[15] has now been synthesized from homoveratrylamine (17) in 35% overall yield (Scheme 2).[16] This is useful as an intermediate in the preparation of alkaloids of the dienoid type.[15,16]

The dibenzazonine (21), related to a biosynthetic precursor of *Erythrina* alkaloids, was obtained in 36% yield by intramolecular oxidative coupling of tetramethoxytrifluoroacetamide (20) by thallium(III) trifluoroacetate in trifluoroacetic acid at 25 °C.[17] The trifluoroacetyl group was readily removed by alkaline hydrolysis, and the ¹H n.m.r. spectrum of the free base confirmed that coupling had taken place *para–para* (with respect to the 3-methoxy-group of each aromatic ring).

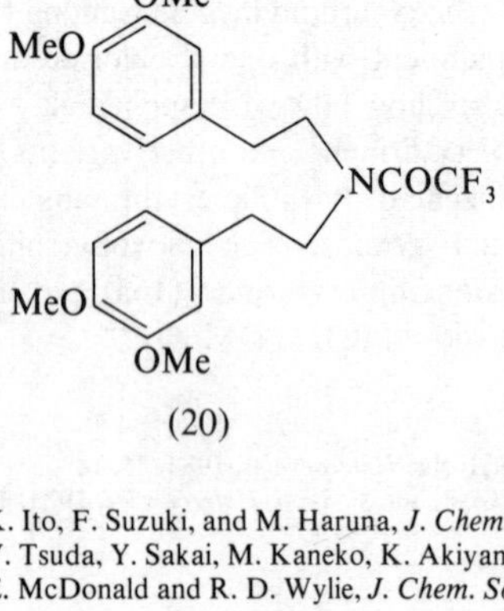

(20)

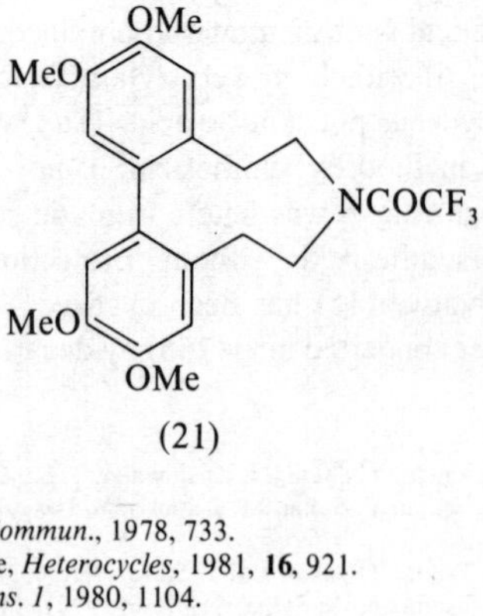

(21)

[15] K. Ito, F. Suzuki, and M. Haruna, *J. Chem. Soc., Chem. Commun.*, 1978, 733.
[16] Y. Tsuda, Y. Sakai, M. Kaneko, K. Akiyama, and K. Isobe, *Heterocycles*, 1981, **16**, 921.
[17] E. McDonald and R. D. Wylie, *J. Chem. Soc., Perkin Trans. 1*, 1980, 1104.

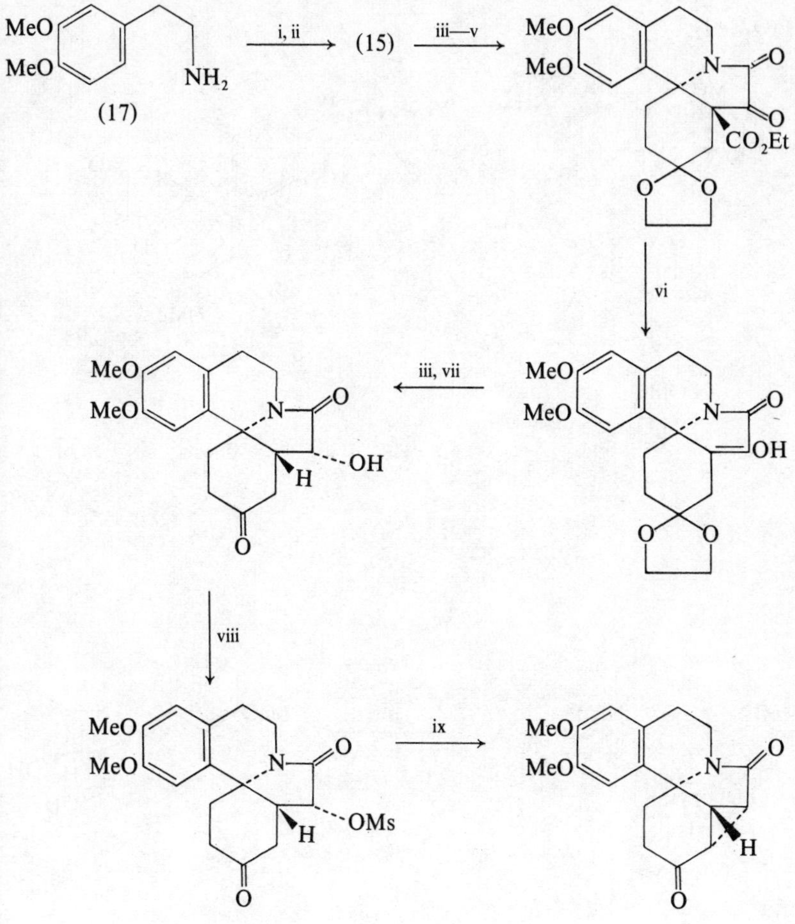

(18) a; 7α-mesylate
b; 7β-mesylate

(19)

Reagents: i,

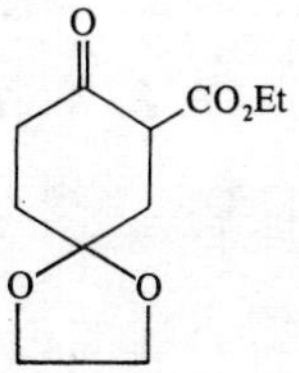

; ii, (COCl)$_2$; iii, NaBH$_4$; iv, BF$_3$·Et$_2$O; v, Collins oxidation;

vi, MgCl$_2$, HMPA; vii, HCl, acetone; viii, MsCl, pyridine; ix, KOH, MeOH

Scheme 2

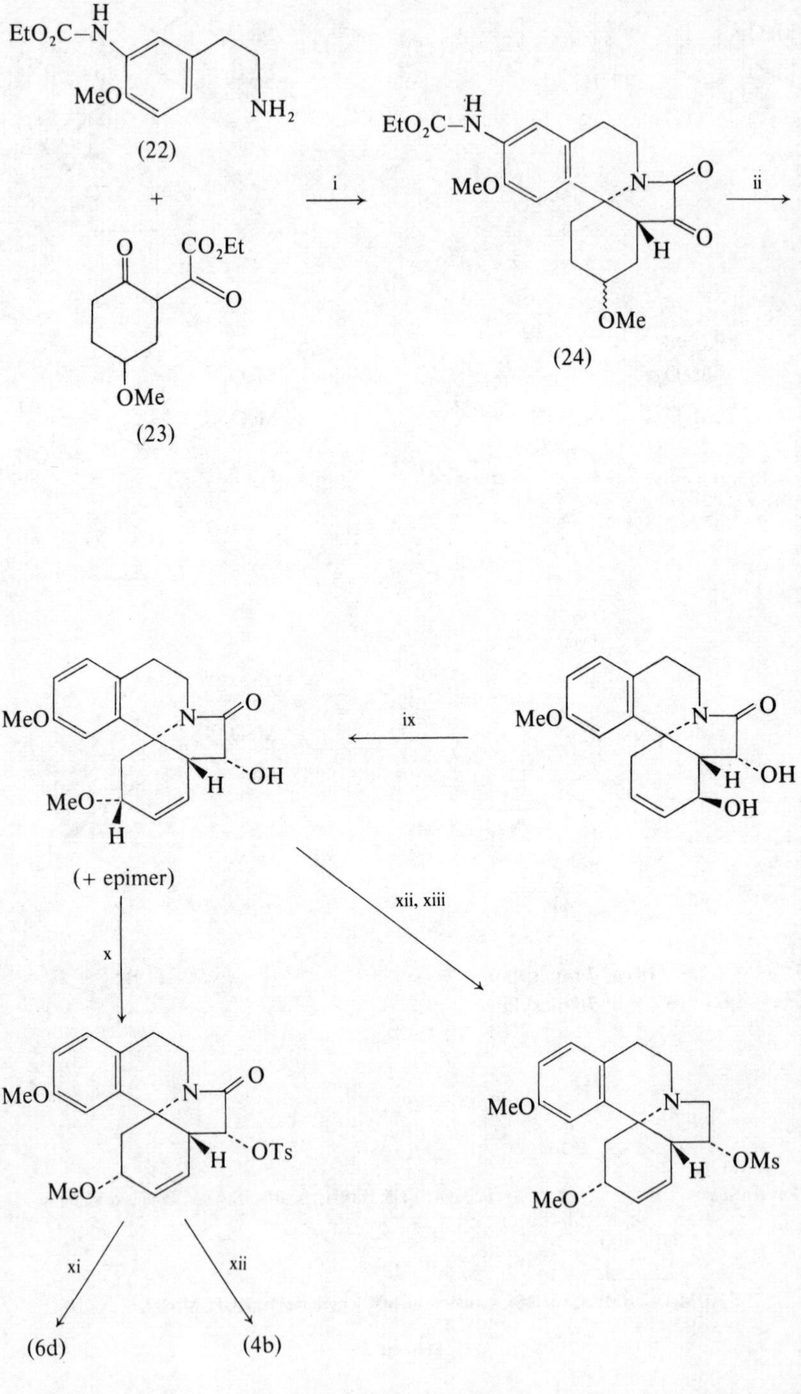
EtO2C–N(H)
MeO
NH2
(22)
+
CO2Et
O
OMe
(23)
i
EtO2C–N(H)
MeO
N
O
O
H
OMe
(24)
ii
MeO
N
O
OH
H
MeO
H
(+ epimer)
ix
MeO
N
O
H
OH
OH
x
xii, xiii
MeO
N
O
OTs
H
MeO
xi
xii
(6d)
(4b)
MeO
N
H
OMs
MeO

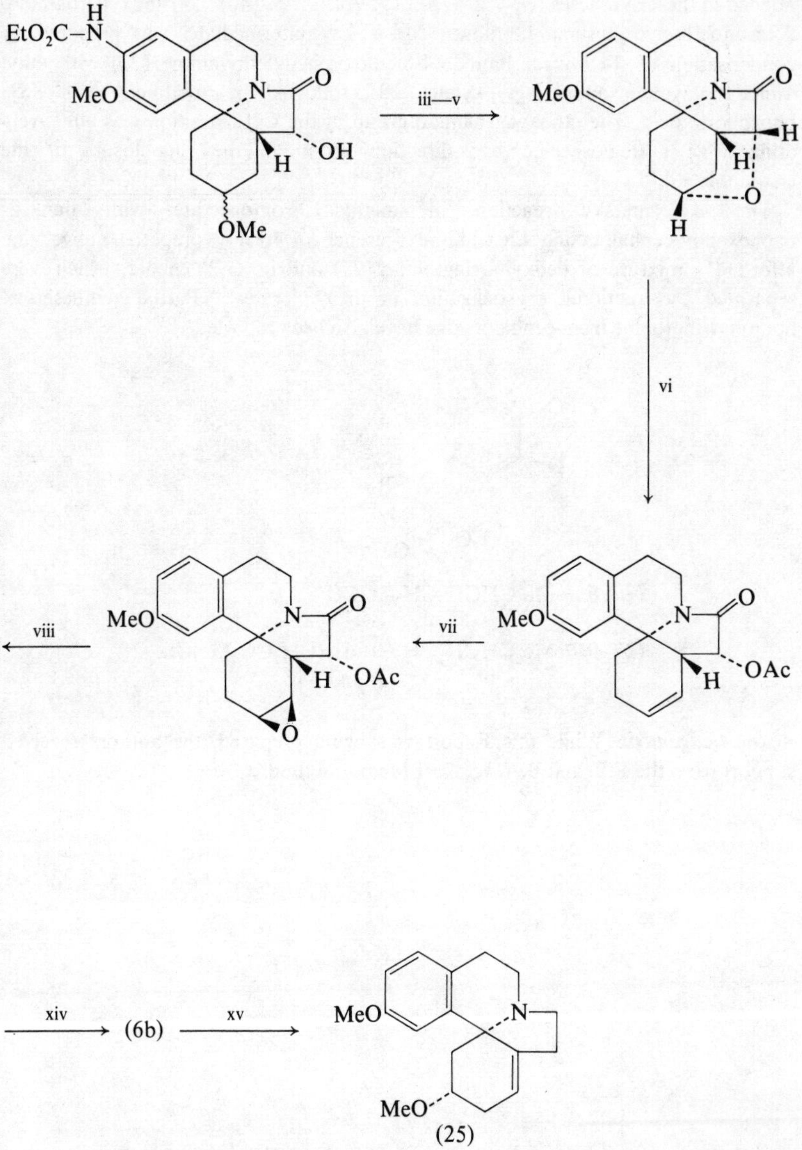

Reagents: i, 85% H_3PO_4; ii, $NaBH_4$; iii, conc. H_2SO_4; iv, 20% HCl; v, $NaNO_2$, HCl, H_2PO_3; vi, *p*-TsOH, Ac_2O; vii, *m*-ClC$_6$H$_4$CO$_3$H; viii, $NaBH_4$, PhSeSePh, 30% H_2O_2, THF; ix, 6M-HCl; x, *p*-TsCl, pyridine; xi, DBU; xii, $LiAlH_4$; xiii, MsCl, pyridine; xiv, KOH in $MeOCH_2CH_2OH$; xv, H_2, Pd/C, MeOH

Scheme 3

Total syntheses of 'abnormal' erythrinan alkaloids, *e.g.* ($\pm$)-coccolinine (6d), ($\pm$)-isococculidine (4b), ($\pm$)-coccuvinine (6b), and ($\pm$)-cocculidine (25), have now been described (Scheme 3).[18] Mondon's glyoxylic ester synthesis was successfully applied in these syntheses (*cf.* Vol. 1, p. 147; Vol. 11, p. 140). 16-Ethoxycarbamido-2,15-dimethoxyerythrinan-7,8-dione (24), a key intermediate, was prepared by condensation of 3-ethoxycarbamido-4-methoxyphenylethylamine (22) with ethyl 4-methoxycyclohexanone-2-glyoxylate (23), followed by treatment with 85% phosphoric acid. The ethoxycarbamido-group at the C-16 position was effectively utilized as a regiospecific *para*-directing group in the ring-closure of the isoquinoline.

The Reformatsky reaction of methyl bromoacetate with the 2-oxohexanoylcephalotaxine (26) in the presence of freshly prepared active zinc afforded a mixture of deoxyharringtonine (27) and its C-2' epimer, which were separated by fractional crystallization of their picrates.[19] Partial syntheses of homoharringtonine from cephalotaxine have also been reported.[20,21]

(26) R = Me$_2$CHCH$_2$CH$_2$COCO–

(27) R = Me$_2$CHCH$_2$CH$_2$C(OH)(CH$_2$CO$_2$Me)CO–
(*R* and *S*)

Acknowledgement. While this Report was being prepared, the authors received support from the F.B. and B.A. Krukoff Memorial fund.

[18] M. Ju-ichi, Y. Fujitani, and Y. Ando, *Chem. Pharm. Bull.*, 1981, **29**, 396.
[19] H. Wenkui, L. Yulin, and P. Xinfu, *Sci. Sin.*, 1980, **23**, 835.
[20] Z.-Z. Zhao, Y.-G. Xi, H.-F. Zhao, J.-Y. Hou, J.-Y. Zhang, and Z.-H. Wang, *Yao Hsueh Hsueh Pao*, 1980, **15**, 46 (*Chem. Abstr.*, 1981, **94**, 103 627).
[21] Y.-K. Wang, Y.-L. Li, H.-F. Pan, C.-P. Li, and W.-K. Huang, *K'o Hsueh Tung Pao*, 1980, **25**, 576 (*Chem. Abstr.*, 1981, **94**, 103 628).

12
Indole Alkaloids

BY J. E. SAXTON

1 General

Hesse's excellent general introduction to alkaloid chemistry naturally contains numerous references to indole alkaloids, and in particular contains a detailed discussion of the structure elucidation of villalstonine, by way of illustration of modern experimental methods used in alkaloid chemistry.[1] Indole alkaloids also receive mention in a brief review of hypotensive principles from plants,[2a] in a review of synthesis *via* photocyclization of enamides,[2b] and in another on the use of lactones in alkaloid synthesis.[2c] Kametani has also reviewed (unfortunately, in Japanese) the synthesis, by his group, of alkaloids by 'retro mass spectral synthesis'.[2d] The proceedings of the 1979 meeting of the Phytochemical Society of Europe, which were entirely devoted to indole alkaloids, are now available.[3]

2 Simple Alkaloids

Non-tryptamines.—The structure (1) of a new bromo-indole derivative, isolated from an Australian sponge of the genus *Iotrocha*, has been elucidated and confirmed by synthesis.[4] The indole-3-acetic ester of *myo*-inositol occurs in rice kernels (*Oryza sativa*);[5a] this is the first reported occurrence of this ester in a plant other than maize (*Zea mays*).

Carbazomycins A and B, isolated from an unidentified *Streptomyces* species, prove to be the first known antibiotics that contain a carbazole skeleton. Carbazomycin B is 4-hydroxy-3-methoxy-1,2-dimethylcarbazole and carbazomycin A is its *O*-methyl ether.[5b]

The structures of several carbazoles and pyranocarbazoles have been confirmed by synthesis; these include lansine [2-hydroxy-3-formyl-6-methoxycarbazole – a

[1] M. Hesse, 'Alkaloid Chemistry', Wiley–Interscience, New York, 1981.
[2] (a) S. Funayama and H. Hikino, *Heterocycles*, 1981, **15**, 1239; (b) I. Ninomiya and T. Naito, *ibid.*, p. 1433; (c) G. D. Pandey and K. P. Tiwari, *ibid.*, 1981, **16**, 449; (d) T. Kametani, *Yakugaku Zasshi*, 1981, **101**, 1.
[3] 'Indole and Biogenetically Related Alkaloids', ed. J. D. Phillipson and M. H. Zenk, Academic Press, New York, 1980.
[4] G. Dellar, P. Djura, and M. V. Sargent, *J. Chem. Soc., Perkin Trans. 1*, 1981, 1679.
[5] (a) P. J. Hall, *Phytochemistry*, 1980, **19**, 2121; (b) M. Kaneda, K. Sakano, S. Nakamura, Y. Kushi, and Y. Iitaka, *Heterocycles*, 1981, **15**, 993.

new alkaloid of *Clausena lansium* (Lour.) Skeels] and 6-methoxyheptaphylline,[6] hyellazole,[7] mupamine,[8a] murrayacinine,[8b] and koenigicine.[8c] Mahanimbinol (2)[9a] and mahanimboline (3)[9b] are two new prenylated carbazoles from the stem wood and root bark, respectively, of *Murraya koenigii* Spreng.

Details of the structure elucidation of melosatins A (4) and B (5)[10] have now been published;[11a] melosatin C, a new metabolite from *Melochia tomentosa* L., is the closely related 4-(5-phenylpentyl)isatin derivative (6).

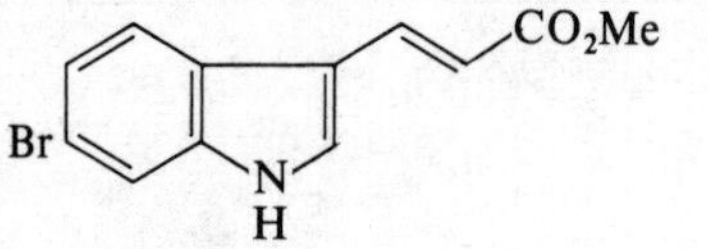

(1)

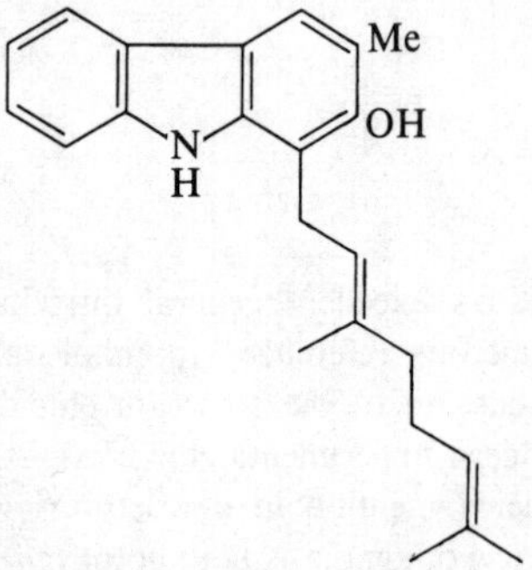

Mahanimbinol (2)

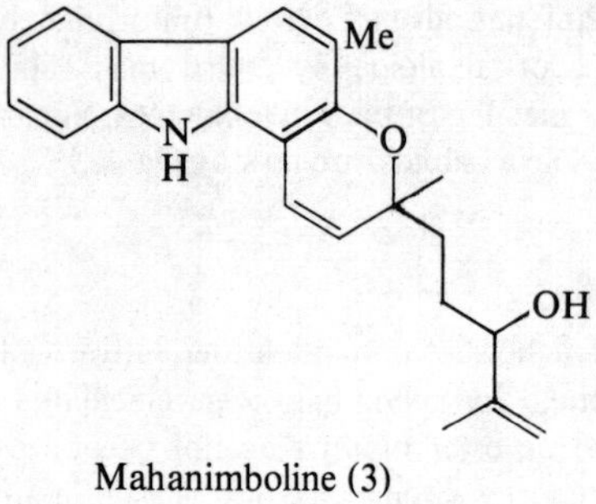

Mahanimboline (3)

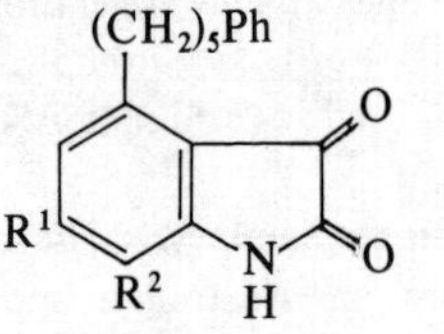

(4) $R^1 = R^2 = OMe$
(5) $R^1 = R^2 = H$
(6) $R^1 = H, R^2 = OMe$

[6] R. B. Sharma and R. S. Kapil, *Chem. Ind.* (*London*), 1980, 158; D. Prakash, K. Raj, R. S. Kapil, and S. P. Popli, *Indian J. Chem., Sect. B*, 1980, **19**, 1075.

[7] S. Kano, E. Sugino, and S. Hibino, *J. Chem. Soc., Chem. Commun.*, 1980, 1241.

[8] (a) I. Mester, M. K. Choudhury, and J. Reisch, *Liebigs Ann. Chem.*, 1980, 241; (b) S. Roy and D. P. Chakraborty, *J. Indian Chem. Soc.*, 1980, **57**, 759; (c) R. B. Sharma, R. Seth-Verma, and R. S. Kapil, *Experientia*, 1980, **36**, 815.

[9] (a) A. V. Rama Rao, K. S. Bhide, and R. B. Mujumdar, *Chem. Ind.* (*London*), 1980, 697; (b) S. Roy, S. Gosh, and D. P. Chakraborty, *ibid.*, 1979, 669.

[10] J. E. Saxton, in 'The Alkaloids', ed. M. F. Grundon (Specialist Periodical Reports), The Chemical Society, London, 1979, Vol. 9, p. 152.

Three new, closely related indolosesquiterpene derivatives have been isolated from the stems of *Polyathia suaveolens* Engl. et Diels, a Nigerian tree which affords a decoction used in native medicine for the treatment of blackwater fever and stomach disorders.[11b] Mainly on the basis of their n.m.r. spectra and inter-conversions, polyavolensinol has been attributed the structure (7), polyavolensin is the corresponding acetate (8), and polyavolensinone is the related ketone; a completely unequivocal proof of these structures, however, has not been provided.

Polyavolensinol (7) R = H
Polyavolensin (8) R = Ac

(9)

Full details of the syntheses of gliotoxin, dehydrogliotoxin, and hyalodendrin, by Kishi and his collaborators, have now been published.[12a]

Bisdethiobis(methylthio)dehydrogliotoxin (9) is one of four new metabolites isolated from *Gliocladium deliquescens*.[12b]

The synthesis[13] of the pentacyclic intermediate (10) constitutes a second formal synthesis of (±)-tryptoquivaline G, since (10) has already been converted into tryptoquivaline G by Büchi *et al.*[14a] The most noteworthy stages in the synthesis of (10) (Scheme 1) involved the oxidation of the indolepropionic acid derivative (11) by means of thallium trinitrate, which gave the oxindole-lactone (12), and the oxidative removal of the unwanted methyl group in (12) by means of selenium dioxide.[13]

Asterriquinone, the pigment isolated earlier from *Aspergillus terreus*, proves to be one of eleven closely related pigments, based on a bisindolyl-*p*-benzoquinone skeleton substituted at positions 1, 2 or 7 of the indole rings by isoprene or

[11] (a) G. J. Kapadia, Y. N. Shukla, S. P. Basak, E. A. Sokoloski, and H. M. Fales, *Tetrahedron*, 1980, **36**, 2441; (b) D. A. Okorie, *ibid.*, p. 2005.

[12] (a) T. Fukuyama, S.-I. Nakatsuka, and Y. Kishi, *Tetrahedron*, 1981, **37**, 2045; (b) J. R. Hanson and M. A. O'Leary, *J. Chem. Soc., Perkin Trans. 1*, 1981, 218.

[13] T. Ohnuma, Y. Kimura, H. Kasuya, and Y. Ban. *Heterocycles*, 1981, **16**, 173.

[14] J. E. Saxton, in 'The Alkaloids', ed. M. F. Grundon (Specialist Periodical Reports), The Royal Society of Chemistry, London, 1981, Vol. 11, (a) p. 152; (b) p. 158; (c) p. 168; (d) p. 173; (e) p. 185; (f) p. 195; (g) p. 192.

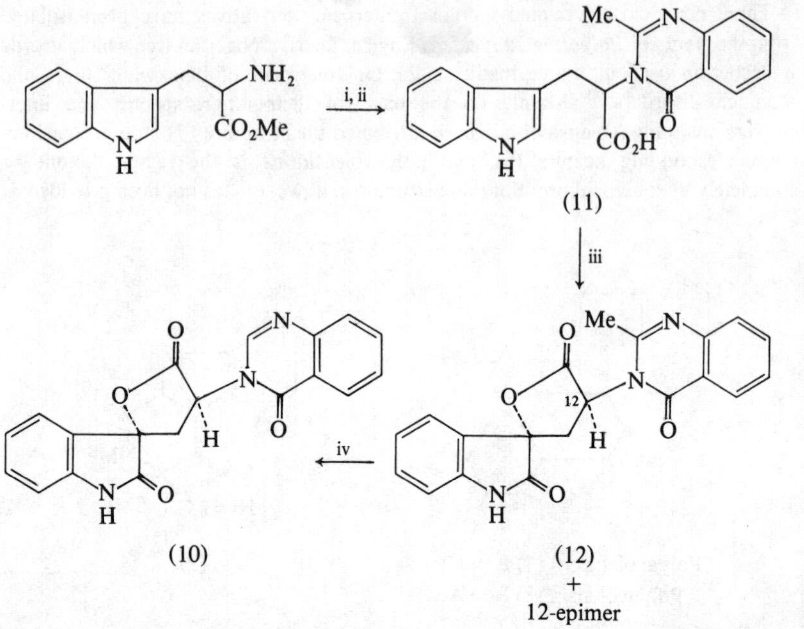

Reagents: i, *N*-Acetylanthranilic acid, POCl₃; ii, LiOH; iii, Tl(NO₃)₃, H₂O, MeCN, DMF; iv, SeO₂, AcOH, heat

Scheme 1

reversed isoprene units;[15a,b] the structures of these eleven metabolites are illustrated in (13a)—(13k).[15a] Eight colourless indolic metabolites were also isolated;[15c] two of these were shown to be the quinols corresponding to asterriquinones A-1 (13a) and C-1 (13i), three others were dimethyl ethers of the quinols corresponding to A-1 (13a), B-3 (13g), and C-2 (13j), and the remaining three were cyclic derivatives (14a)—(14c) of the quinol corresponding to asterriquinone D.

The fungal tremorgens aflatrem and paspalinine have been found in *A. flavus* cultures, together with a non-tremorgenic metabolite, dihydroxyaflavinine (15), whose structure was elucidated by the *X*-ray method.[15d]

Six mycotoxins, penitrems A—F [(16a)—(16f)], derived from tryptophan (with loss of the side-chain) and a diterpene unit (with loss of a carbon atom), have been isolated from *Penicillium crustosum*;[15e] of these, penitrem A was first isolated[15f] in 1968 from *P. cyclopium*, and penitrems B and C in 1971[15g] from a micro-organism originally presumed to be *P. palitans*, but now classified as *P. crustosum*. The occurrence of penitrem A in *P. cyclopium* has recently been confirmed, and the presence of penitrem B was also demonstrated.[15h]

[15] (a) K. Arai, K. Masuda, N. Kiriyama, K. Nitta, Y. Yamamoto, and S. Shimizu, *Chem. Pharm. Bull.*, 1981, **29**, 961; (b) K. Arai, S. Shimizu, Y. Taguchi, and Y. Yamamoto, *ibid.*, p. 991; (c) K. Arai, S. Shimizu, and Y. Yamamoto, *ibid.*, p. 1005; (d) R. J. Cole, J. W. Dorner, J. P. Springer, and R. H. Cox, *J. Agric. Food Chem.*, 1981, **29**, 293; (e) A. E. de Jesus, P. S. Steyn, F. R. Van Heerden, R. Vleggaar, P. L. Wessels, and W. E. Hull, *J. Chem. Soc., Chem. Commun.*, 1981, 289; (f) B. J. Wilson, C. H. Wilson, and A. W. Hayes, *Nature* (*London*), 1968, **220**, 77; (g) C. T. Hou, A. Ciegler, and C. W. Hesseltine, *Can. J. Microbiol.*, 1971, **17**, 599; (h) R. F. Vesonder, L. Tjarks, W. Rohwedder, and D. O. Kieswetter, *Experientia*, 1980, **36**, 1308.

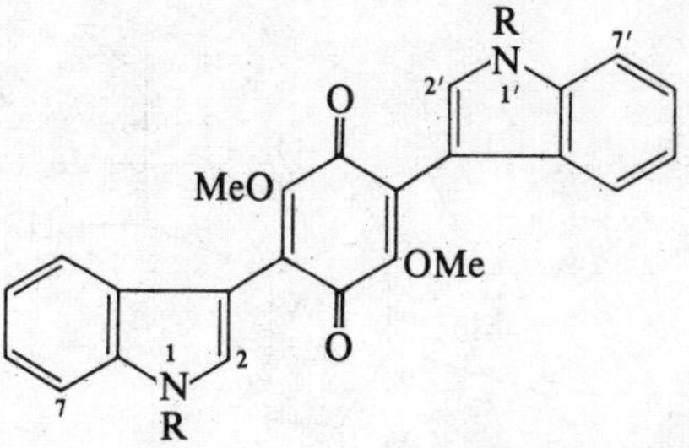

	Positions of prenyl groups	
Asterriquinone A-1 (13a)	1 & 1'	—
A-2 (13b)	2	7 & 7'
A-3 (13c)	1 & 2'	—
A-4 (13d)	2 & 2'	7
B-1 (13e)	2	7'
B-2 (13f)	2	7
B-3 (13g)	1	—
B-4 (13h)	2 & 2'	—
C-1 (13i)	2	—
C-2 (13j)	—	7
D (13k)	—	—

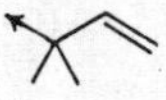

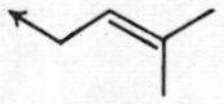

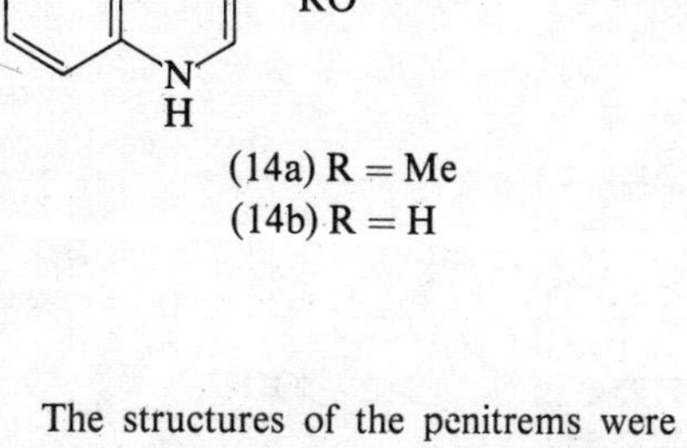

(14a) R = Me
(14b) R = H

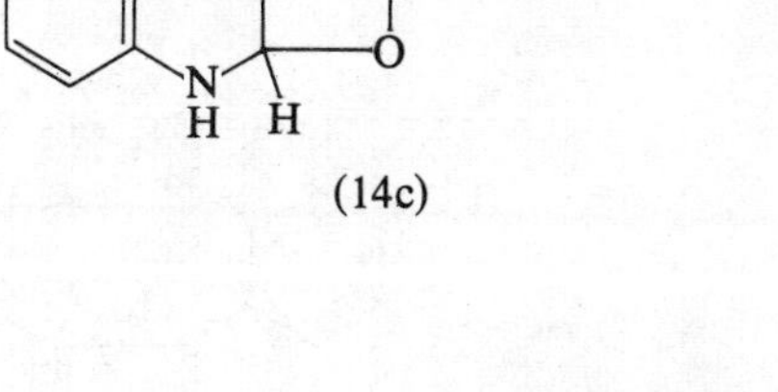

(14c)

The structures of the penitrems were elucidated[15e] mainly by extensive n.m.r. studies, both on the natural material and on ^{13}C-enriched material isolated after administration of appropriate ^{13}C-labelled precursors (tryptophan and acetate) to the cultures. Structure (16a) incorporates the proposed relative stereochemistry of penitrem A, but no deductions concerning the stereochemistry of the other metabolites were made. Penitrem A is a powerful toxin, and causes sustained tremors, disco-ordination, and convulsions in laboratory and farm animals.

Dihydroxyaflavinine (15)

	R¹	R²
Penitrem A (16a)	Cl	OH
B (16b)	H	H
E (16e)	H	OH
F (16f)	Cl	H

Penitrem C (16c)   R = Cl
D (16d)   R = H

Non-isoprenoid Tryptamines.—Psilocybin has been isolated from the fruiting bodies of *Psilocybe argentipes* K. Yokoyama, a mushroom which causes hallucinogenic intoxication when ingested;[16a] it has also been detected in three other mushrooms, namely *Gymnopilos liquiritae* (Fr.) Karst., *Psathyrella candolleana* (Fr.) A. H. Smith, and *Agrocybe farinacea* Hongo.[16a]

The conformations of bufotenine and psilocin in solution have been studied by 360 MHz ^{1}H n.m.r. spectroscopy.[16b] Bufotenine adopts preferentially a staggered conformation in deuteriochloroform, whereas psilocin exists mainly in the *gauche* conformation, presumably owing to weak hydrogen-bonding of the 4-hydroxy-group with the basic nitrogen atom.

The β-carboline alkaloids reported to April 1979 are the subject of a brief review.[17a] New alkaloids include strychnocarpine (N_b-methyl-1,2,3,4-tetrahydro-1-oxo-β-carboline), which is a constituent of the stem bark of *Strychnos elaeocarpa* Gilg. ex Leeuwenberg,[17b] N_a-acetyl-3,4-dihydro-β-carboline, which occurs in the roots of *Adhatoda vasica* Nees,[17c] and komarovine (17) and its 3,4-dihydro-derivative komarovidine (18), from the aerial parts of *Nitraria komarovii*.[17d] The structures of these last two alkaloids have been confirmed by synthesis from tryptamine and quinoline-8-carboxylic acid.[17e] Pyridindolol (19), a specific inhibitor of bovine liver β-galactosidase, from *Streptomyces alboverticillatus*, was first isolated[18a] in 1975, and has recently been synthesized.[18b] Although in this synthesis the tetrahydro-β-carboline ring was closed by Pictet–Spengler condensation of tryptophan methyl ester with a D-glyceraldehyde derivative, the final product was racemic. A recent formal synthesis[18c] makes use of the masked glyceraldehyde derivative (20), which, on condensation with tryptophan, affords the intermediate (21); this has already[18b] been converted into ($\pm$)-pyridindolol.

Cook's stereospecific synthesis[19a] of *trans*-1,3-disubstituted tetrahydro-β-carbolines by condensation of N_b-benzyltryptophan methyl ester with aldehydes, coupled with a simple ^{13}C n.m.r. method for differentiating *cis*- and *trans*-tetrahydro-β-carbolines,[19b] should afford a promising starting point for the synthesis of complex alkaloids. Angenot and his collaborators have also discussed the ^{13}C n.m.r. spectra of a number of β-carboline and 3,4-dihydro-β-carboline derivatives.[19c]

[16] (a) Y. Koike, K. Wada, G. Kusano, S. Nozoe, and K. Yokoyama, *J. Nat. Prod.*, 1981, **44**, 362; (b) G. P. Migliaccio, T.-L. N. Shieh, S. R. Byrn, B. A. Hathaway, and D. E. Nichols, *J. Med. Chem.*, 1981, **24**, 206.

[17] (a) J. R. F. Allen and B. R. Holmstedt, *Phytochemistry*, 1980, **19**, 1573; (b) W. Rolfsen, A. M. Bresky, M. Andersson, and J. Strömbom, *Acta Pharm. Suec.*, 1980, **17**, 333; (c) M. P. Jain, S. K. Koul, K. L. Dhar, and C. K. Atal, *Phytochemistry*, 1980, **19**, 1880; (d) T. S. Tulyaganov, A. A. Ibragimov, and S. Yu. Yunusov, *Khim. Prir. Soedin.*, 1980, 732; (e) *ibid.*, 1981, 192.

[18] (a) T. Aoyagi, M. Kumagai, T. Hazato, M. Hamada, T. Takeuchi, and H. Umezawa, *J. Antibiot.*, 1975, **28**, 555; (b) D. Soerens, J. Sandrin, F. Ungemach, P. Mokry, G. S. Wu, E. Yamanaka, L. Hutchins, M. DiPierro, and J. M. Cook, *J. Org. Chem.*, 1979, **44**, 535; (c) H. Bieräugel, R. Plemp, and U. K. Pandit, *Heterocycles*, 1980, **14**, 947.

[19] (a) F. Ungemach, M. DiPierro, R. Weber, and J. M. Cook, *J. Org. Chem.*, 1981, **46**, 164; (b) F. Ungemach, D. Soerens, R. Weber, M. DiPierro, O. Campos, P. Mokry, J. M. Cook, and J. V. Silverton, *J. Am. Chem. Soc.*, 1980, **102**, 6976; (c) C. A. Coune, L. J. G. Angenot, and J. Denoël, *Phytochemistry*, 1980, **19**, 2009.

Komarovine (17)
Komarovidine (18) 3,4-dihydro

Pyridindolol (19) R = CH$_2$OH

(20)

(21)

1-Acetyl-3-methoxycarbonyl-β-carboline, the alkaloid of *Vestia lycioides*, has been synthesized by two routes;[20a,b] the second involves the benzylic oxidation–dehydrogenation of 1-ethyl-3-methoxycarbonyl-1,2,3,4-tetrahydro-β-carboline by means of selenium dioxide in dioxan. A by-product in this oxidation was 1-acetyl-β-carboline, which occurs in *Ailanthus malabarica*. An extension of this reaction led to a simple two-stage synthesis of canthin-6-one (22) from N_b-benzyltryptamine and α-ketoglutaric acid (Scheme 2).[20b]

The root bark of *Ailanthus altissima* Swingle contains the known alkaloids canthin-6-one (22), 1-methoxycanthin-6-one, canthin-6-one N-oxide, and 1-acetyl-4-methoxy-β-carboline, together with three new alkaloids, identified as 1-methoxy-canthin-6-one N-oxide, 1-(2-hydroxyethyl)-4-methoxy-β-carboline, and 1-(1,2-dihydroxyethyl)-4-methoxy-β-carboline.[21]

An extension of Kametani's earlier synthesis has afforded a neat synthesis of rutaecarpine (24) and hortiacine (10-methoxyrutaecarpine).[22a] In this modification, the presence of a trifluoromethyl group in (23) (instead of hydrogen, as in Kametani's synthesis) increases the electrophilicity of the protonated form, and also provides a useful leaving group for the final stage of the synthesis; a dehydro-genation step is therefore unnecessary (Scheme 3). Rutaecarpine has also been synthesized.[22b] 11-Methoxyrutaecarpine has been simply synthesized by conden-sation of 7-methoxy-1-oxo-1,2-dihydro-β-carboline with methyl anthranilate and phosphorus oxychloride.[22c]

[20] (a) I. Razmilić, M. Castillo, and J. T. López, *J. Heterocycl. Chem.*, 1980, **17**, 595; (b) O. Campos, M. DiPierro, M. Cain, R. Mantei, A. Gawish, and J. M. Cook, *Heterocycles*, 1980, **14**, 975.

[21] T. Ohmoto, K. Koike, and Y. Sakamoto, *Chem. Pharm. Bull.*, 1981, **29**, 390.

[22] (a) J. Bergman and S. Bergman, *Heterocycles*, 1981, **16**, 347; (b) H. Moehle, C. Kamper, and R. Schmid, *Arch. Pharm.* (*Weinheim, Ger.*), 1980, **313**, 990; (c) Atta-ur-Rahman and M. Ghazala, *Synthesis*, 1980, 372.

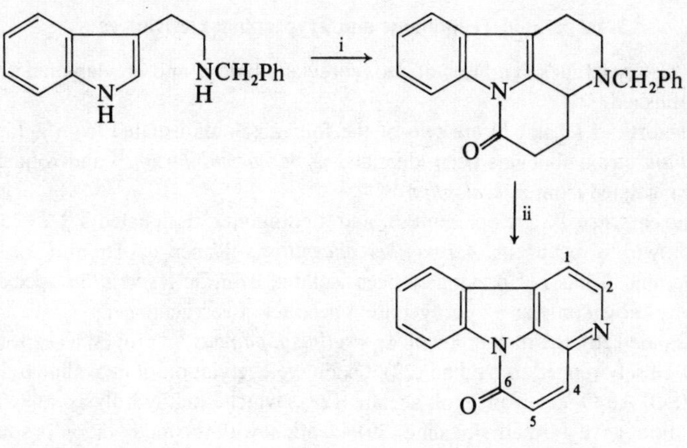

Canthin-6-one (22)

Reagents: i, $HO_2CCH_2CH_2COCO_2H$, TsOH, PhMe, heat for 7 days; ii, SeO_2, heat for 3 days

Scheme 2

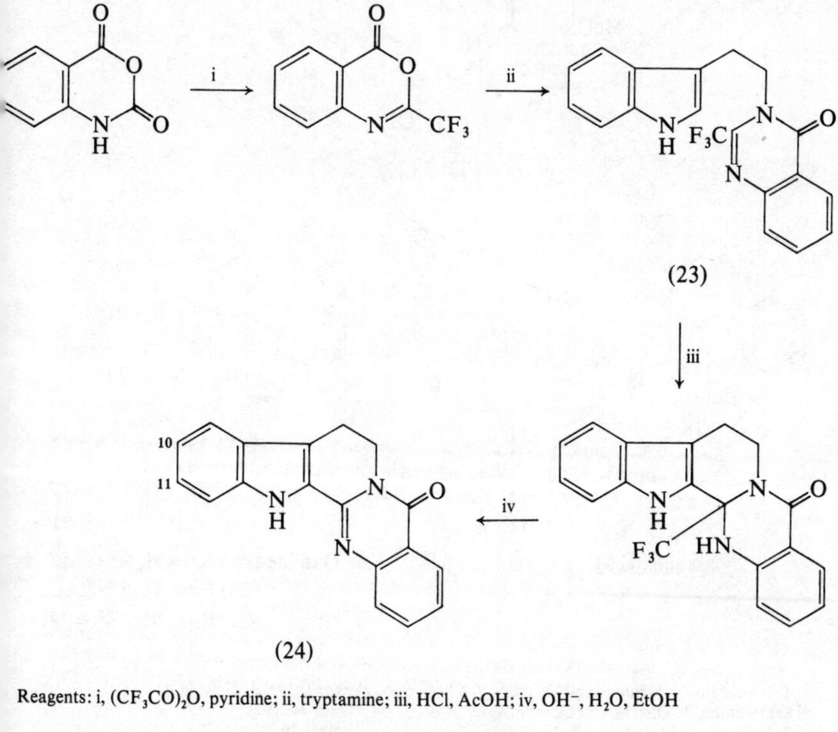

(23)

(24)

Reagents: i, $(CF_3CO)_2O$, pyridine; ii, tryptamine; iii, HCl, AcOH; iv, OH^-, H_2O, EtOH

Scheme 3

3 Isoprenoid Tryptamine and Tryptophan Derivatives

Details of Kametani's synthesis of deoxybrevianamide E and brevianamide E have been published.[23]

Roquefortines C and D are two of the four alkaloids isolated from a Japanese *Penicillium* strain that has been identified as *P. corymbiferum*,[24] and roquefortine has been isolated from *P. cyclopium*.[15h]

Fumitremorgen B, cyclopiamine B, and a compound designated TR 2 (25) have been shown to occur in *Aspergillus caespitosus* Raper et Thom.;[25] of these, cyclopiamine B has not previously been isolated from an *Aspergillus* species, and TR 2 was known only as a hydrogenation product of verruculogen.

Neoxaline (26), the metabolite of *Aspergillus japonicus*,[14b] proves, as expected, to be very closely related to oxaline (27). Oxidative acetylation of neoxaline by means of DMSO–Ac$_2$O gave an enol acetate (28), which, on hydrolysis followed by methylation, gave 14-methyloxaline (29), identical with the methylation product (by CH$_2$N$_2$) of oxaline.[26]

Paraherquamide, a toxic metabolite of *Penicillium paraherquei*, has the structure

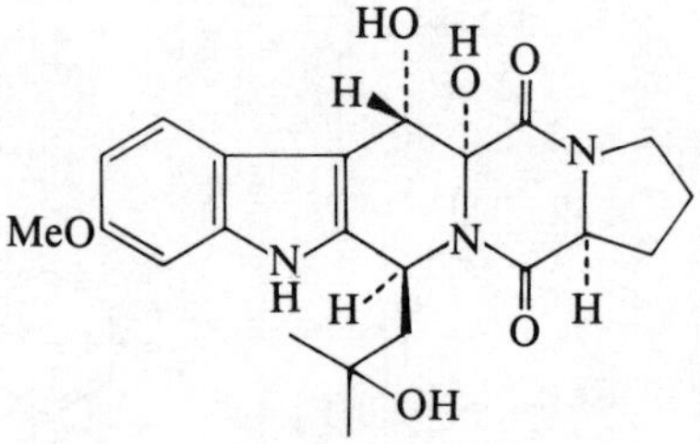

TR 2 (25)

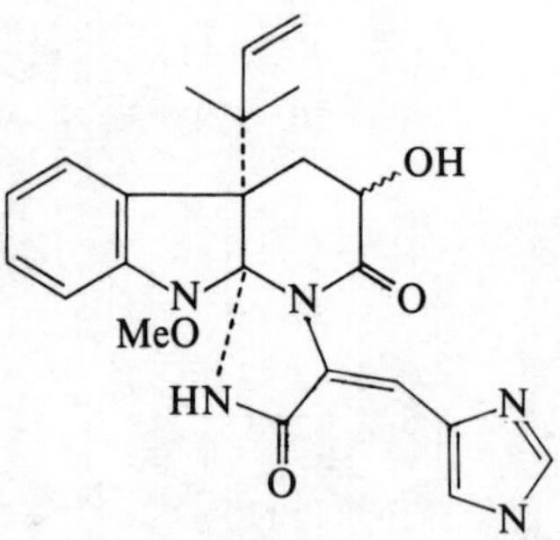

Neoxaline (26)

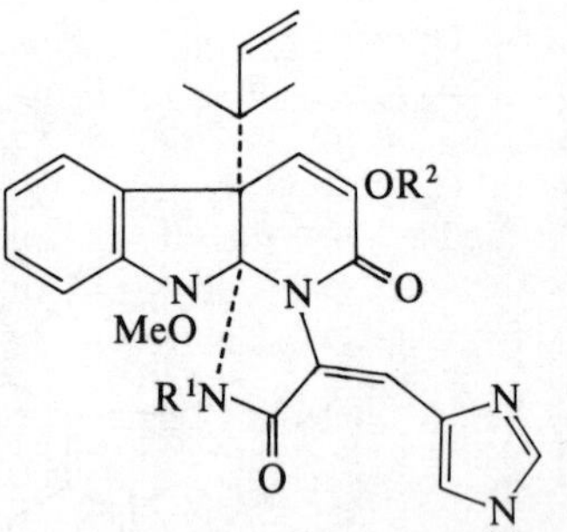

Oxaline (27) R^1 = H, R^2 = Me
(28) R^1 = H, R^2 = Ac
(29) R^1 = Me, R^2 = Me

[23] T. Kametani, N. Kanaya, and M. Ihara, *J. Chem. Soc., Perkin Trans. 1*, 1981, 959.
[24] S. Ohmomo, T. Ohashi, and M. Abe, *Agric. Biol. Chem.*, 1980, **44**, 1929.
[25] P. S. Steyn, R. Vleggaar, and C. J. Rabie, *Phytochemistry*, 1981, **20**, 538.
[26] Y. Konda, M. Onda, A. Hirano, and S. Omura, *Chem. Pharm. Bull.*, 1980, **28**, 2987.

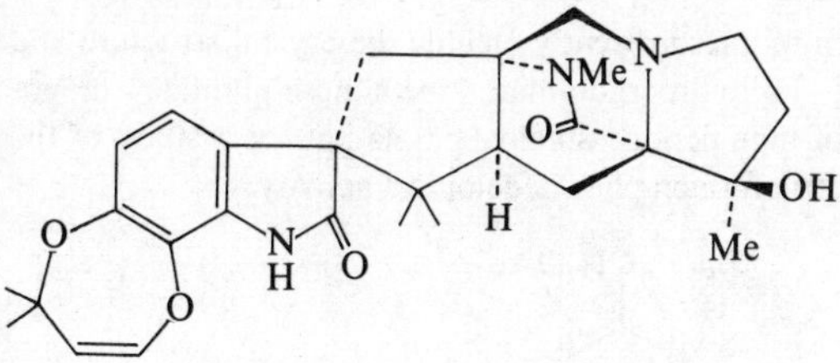

Paraherquamide (30)

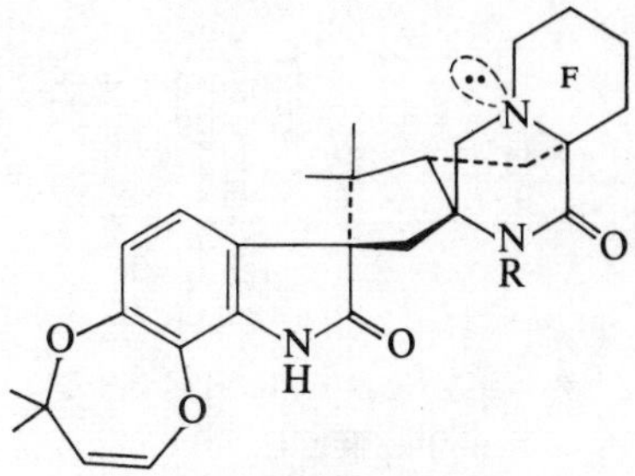

Marcfortine A; R = Me
Marcfortine B (31) R = H

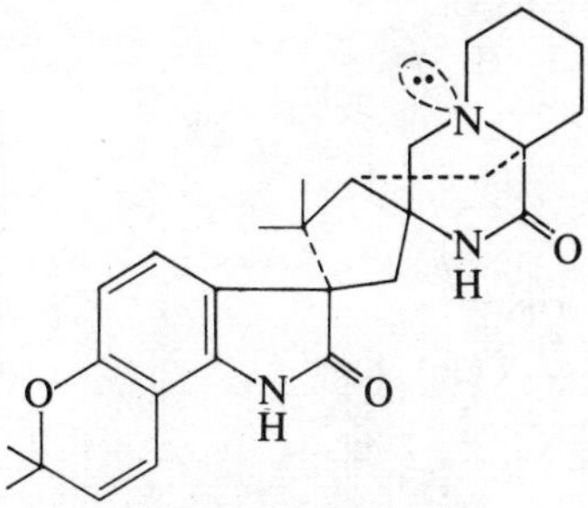

Marcfortine C (32)

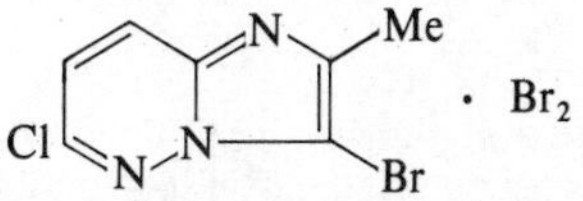

(33)

(30), according to *X*-ray structure analysis.[27] Marcfortine B (31) proves to be desmethylmarcfortine A; these two metabolites differ structurally from para-herquamide (30) only in the presence of a pipecolic acid unit (ring F) instead of 2-hydroxy-2-methylproline.[28] Marcfortine C (32) differs from marcfortine B only in the direct attachment of an isopentenyl unit to ring A, rather than *via* oxygen, as in (31); the structure of (32) was also determined by the *X*-ray method.[28]

Ergot Alkaloids.—The presence of ergot alkaloids in contaminated winter wheat in northern Scotland has been established.[29] The major alkaloid (46% of total) is ergotamine, but ergometrine, ergocryptine, ergocornine, and ergocristine were also detected. The total level of alkaloids was considered to be sufficient to constitute a potential hazard in animal foodstuffs. It was also noted that the alkaloid composition is different from that of Spanish and East European ergots.

[27] M. Yamazaki, E. Okuyama, M. Kobayashi, and H. Inoue, *Tetrahedron Lett.*, 1981, **22**, 135.
[28] T. Prangé, M. A. Billion, M. Vuilhorgne, C. Pascard, and J. Polonsky, *Tetrahedron Lett.*, 1981, **22**, 1977.
[29] B. G. Osborne and R. D. Watson, *J. Agric. Sci.*, 1980, **95**, 239.

X-Ray studies that have been reported recently include the crystal structure and absolute configuration of (−)-dihydroergotamine methanesulphonate mono-hydrate[30a] and the structures of nine peptide alkaloids[30b] as part of a study of the influence of conformation and structure on pharmacological activity.

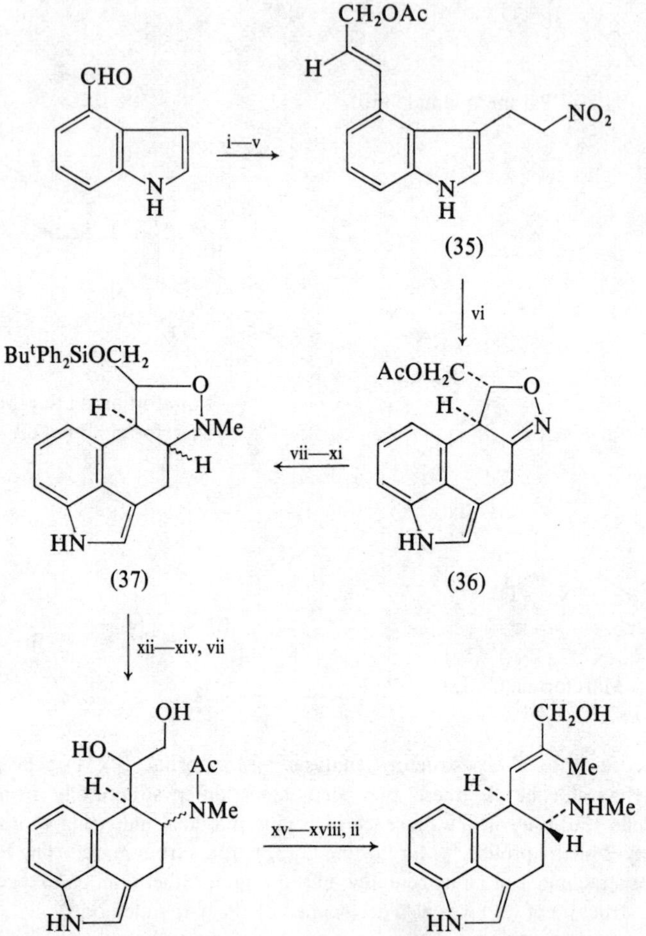

Chanoclavine I (34)

Reagents: i, $(EtO)_2POCHCO_2Et$; ii, AlH_3; iii, acetylation; iv, $Me_2\overset{+}{N}=CH_2$ Cl^-, CH_2Cl_2; v, $MeNO_2$, THF, $MeO_2CC\equiv CCO_2Me$; vi, PhNCO, NEt_3; vii, K_2CO_3, EtOH, H_2O; viii, Bu^tPh_2SiCl; ix, *N*-acetylimidazole; x, Meerwein reagent; xi, $NaBH_4$, EtOH; xii, Bu_4N^+ F^-; xiii, H_2, Pd/C; xiv, Ac_2O, py; xv, HIO_4; xvi, $Ph_3P=C(Me)CO_2Et$; xvii, Et_3O^+ BF_4^-, Na_2CO_3, CH_2Cl_2; xviii, 3% AcOH–H_2O

Scheme 4

[30] (a) H. Hebert, *Acta Crystallogr., Sect. B*, 1979, **35**, 2978; (b) H. P. Weber, *Adv. Biochem. Psychopharmacol.*, 1980, **23**, 25.

The bromination of peptide alkaloids at position 2 can be smoothly achieved by means of the 3-bromo-6-chloro-2-methylimidazo[1,2-*b*]pyridazine–bromine complex (33) in methylene chloride solution; in particular, the clinically important 2-bromo-α-ergocryptine can be prepared by this method in 75% yield.[31]

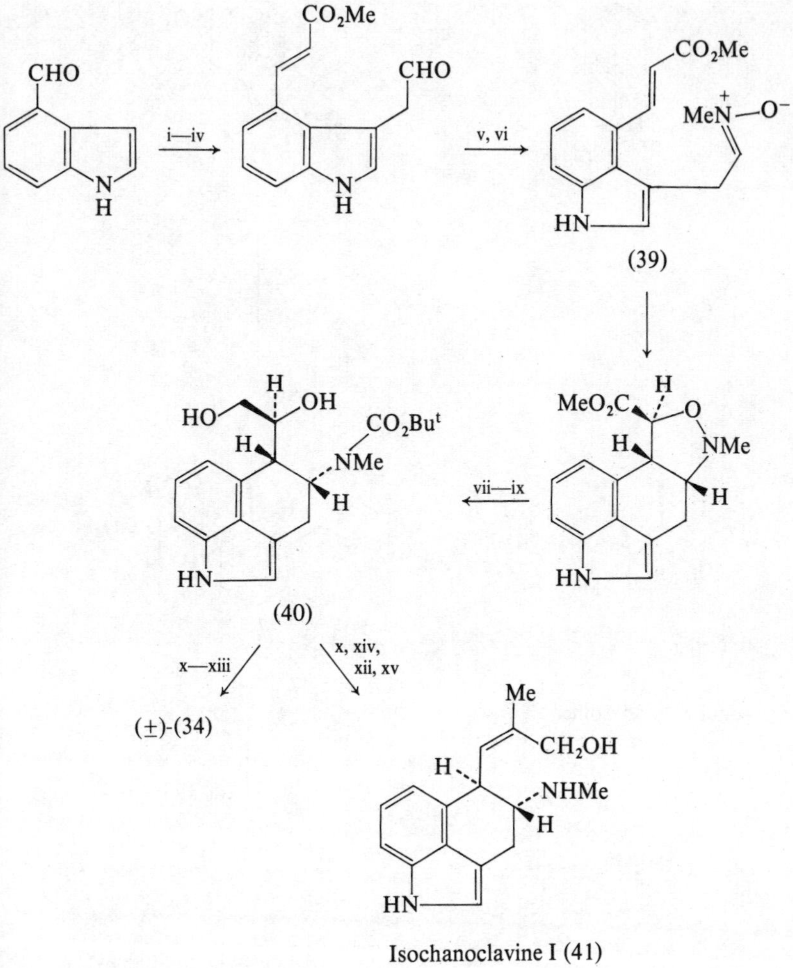

Isochanoclavine I (41)

Reagents: i, (MeO)$_2$POCH$_2$CO$_2$Me, NaH, THF; ii, Me$_2$NH, CH$_2$O, H$_2$O, AcOH; iii, MeI, KCN, Me$_2$CHOH; iv, Ni, NaH$_2$PO$_2$, pyridine, AcOH, H$_2$O; v, MeNHOH, PhH, MeOH; vi, PhH, MeOH, remove H$_2$O azeotropically; vii, LiAlH$_4$, THF; viii, H$_2$, Ni, MeOH; ix, (Bu^{t}OCO)$_2$O, NaOH, H$_2$O, THF; x, NaIO$_4$, MeOH, H$_2$O; xi, Ph$_3$P=C(Me)CO$_2$Me, CH$_2$Cl$_2$; xii, CF$_3$CO$_2$H, CHCl$_3$; xiii, Bu$^i_2$AlH, THF; xiv, (EtO)$_2$POCHMeCO$_2$Me, NaH, THF; xv, LiAlH$_4$, Et$_2$O

Scheme 5

[31] B. Stanovnik, M. Tišler, M. Jurgec, and R. Ručman, *Heterocycles*, 1981, **16**, 741.

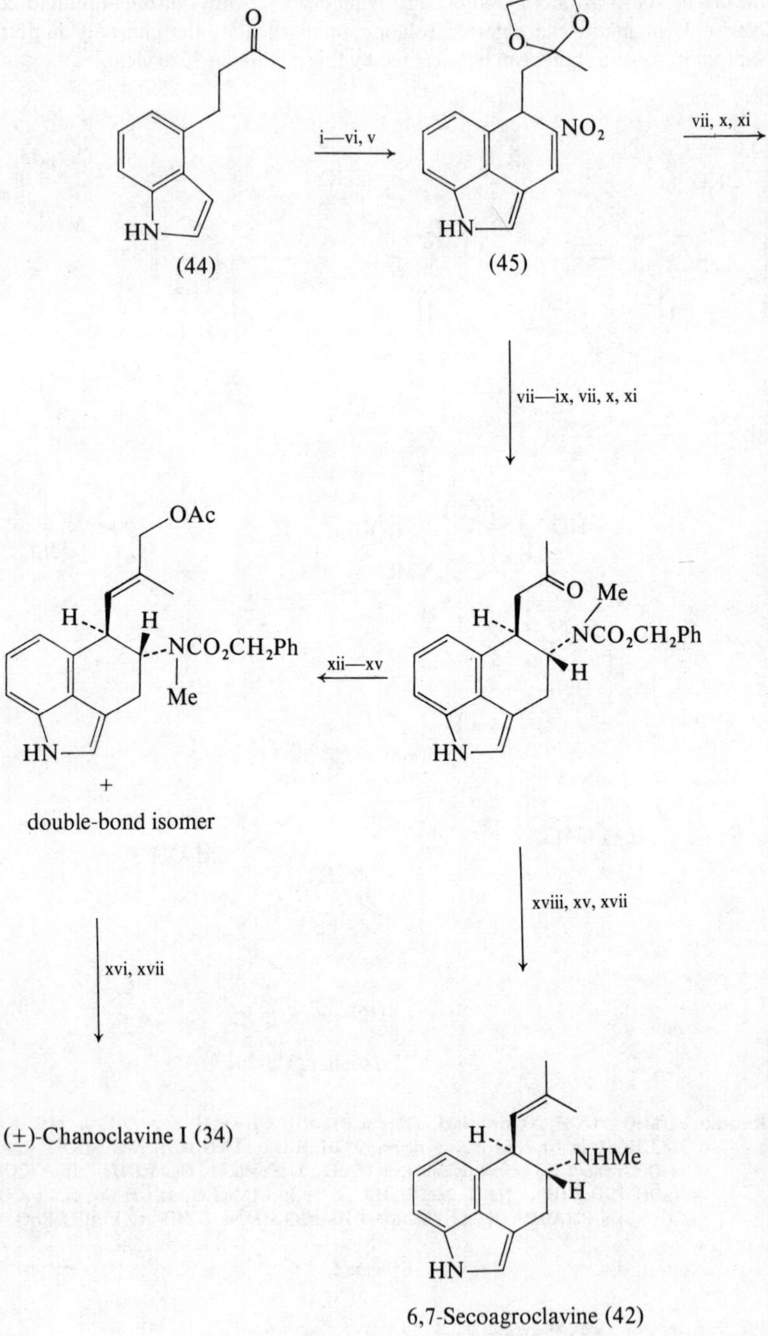
i—vi, v
NO2
vii, x, xi
(44)
(45)
vii—ix, vii, x, xi
OAc
H
H
NCO2CH2Ph
Me
xii—xv
O
Me
NCO2CH2Ph
H
H
HN
+
double-bond isomer
xvi, xvii
xviii, xv, xvii
(±)-Chanoclavine I (34)
NHMe
6,7-Secoagroclavine (42)

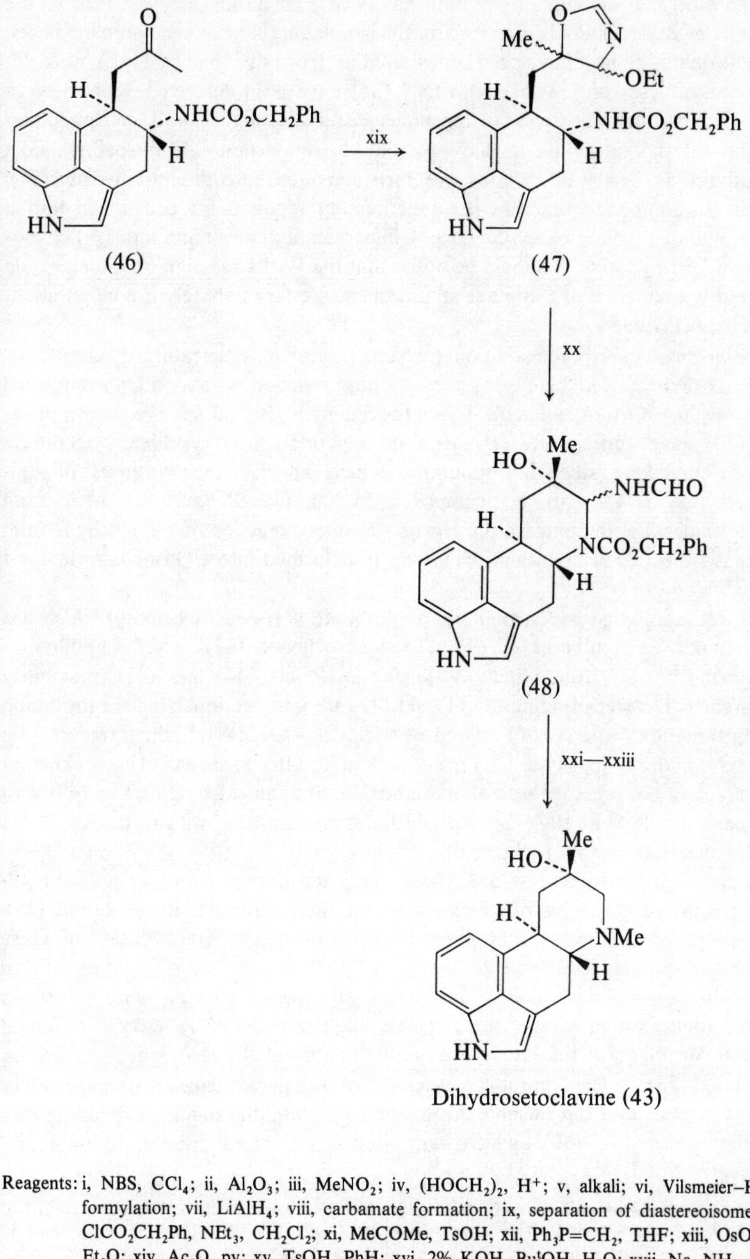

Dihydrosetoclavine (43)

Reagents: i, NBS, CCl$_4$; ii, Al$_2$O$_3$; iii, MeNO$_2$; iv, (HOCH$_2$)$_2$, H$^+$; v, alkali; vi, Vilsmeier–Haack formylation; vii, LiAlH$_4$; viii, carbamate formation; ix, separation of diastereoisomers; x, ClCO$_2$CH$_2$Ph, NEt$_3$, CH$_2$Cl$_2$; xi, MeCOMe, TsOH; xii, Ph$_3$P=CH$_2$, THF; xiii, OsO$_4$, py, Et$_2$O; xiv, Ac$_2$O, py; xv, TsOH, PhH; xvi, 2% KOH, Bu^{t}OH, H$_2$O; xvii, Na, NH$_3$, THF; xviii, MeMgI; xix, TsCH$_2$NC, TlOEt, EtOH, DME; xx, TsOH, DME, H$_2$O; xxi, H$_2$, Pd/C, MeOH; xxii, 5% KOH, MeOH, H$_2$O; xxiii, H$_2$, Pd/C, CH$_2$O, H$_2$O, MeOH

Scheme 6

A considerable amount of attention has been paid during the past year to the synthesis of ergot alkaloids, particularly the simpler ergoline or seco-ergoline bases. Reviews on this subject have been contributed by Horwell[32a] and by Kozikowski.[32b]

Three syntheses of (±)-chanoclavine I (34) have been described. Kozikowski's route (Scheme 4)[33a] involves an ingenious intramolecular [3 + 2] cycloaddition reaction on the nitrile oxide derived from the important 3-(2-nitroethyl)indole derivative (35). The product (36) was then converted into the intermediate (37), a strategic compound that has the functionality required for conversion into a wide range of ergoline bases. Scheme 4 illustrates its conversion into (±)-chano-clavine I.[33a] In passing, it should be noted that the Wittig reaction on the aldehyde obtained by oxidation of (38) gave an unsaturated ester of entirely E configuration, as in chanoclavine I.

An intramolecular cycloaddition reaction is also a vital feature of Oppolzer's synthesis (Scheme 5).[33b] Here the cycloaddition reaction occurs on an unsaturated nitrone ester (39) (not isolated). Again, the aldehyde derived from oxidation of the diol (40) gave entirely the (E)-olefin on reaction with crystalline α-methoxy-carbonylethylidenetriphenylphosphorane, which allowed the synthesis of (±)-chanoclavine I (34) to be completed in an overall yield of 14% from indole-4-aldehyde. In contrast, the Horner–Emmons reaction on the aldehyde from (40) gave the (Z)-olefin, which was then transformed into (±)-isochanoclavine I (41).[33b]

The route adopted by Natsume and Muratake[33c] is one (Scheme 6) which has also resulted in syntheses of (±)-6,7-secoagroclavine (42)[33d] and (±)-dihydro-setoclavine[33c] (43) from common starting materials (44) and (45), previously prepared.[33e] The steps leading to (34) and (42) are unexceptional, but the formation of ring D in the synthesis of (±)-dihydrosetoclavine (43) involved the formation of a masked α-hydroxy-aldehyde (47) from the ketone (46) by means of p-tosylmethyl isocyanide in the presence of thallium ethoxide. Subsequent treatment of (47) with acid caused ring D to close and opened the oxazoline ring, with formation of the desired intermediate (48) (Scheme 6).

Independently, Somei *et al.*[34] have reported three syntheses of (±)-6,7-secoagroclavine (42), each of which starts with the unsaturated indole-ketone (49), itself prepared in three stages from 5-nitroisoquinoline. The shortest of these syntheses is outlined in Scheme 7.

Rugulovasines A (50) and B (51) have been isolated from *Penicillium biforme* Thom., found on mouldy canned pears, and from *P. rubrum* Stoll, a fungal contaminant of corn; the latter fungus also contained the chlororugulovasines A (52) and B (53).[35a] The first total synthesis[35b] of the rugulovasines is summarized in Scheme 8. Starting from racemic dibenzoyltryptophan, the sole initial product was (±)-rugulovasine A (50), which was isomerized to a mixture of racemic

[32] (a) D. C. Horwell, *Tetrahedron*, 1980, **36**, 3123; (b) A. P. Kozikowski, *Heterocycles*, 1981, **16**, 267.

[33] (a) A. P. Kozikowski and H. Ishida, *J. Am. Chem. Soc.*, 1980, **102**, 4265; (b) W. Oppolzer and J. I. Grayson, *Helv. Chim. Acta*, 1980, **63**, 1706; (c) M. Natsume and H. Muratake, *Heterocycles*, 1981, **16**, 375; (d) M. Natsume and H. Muratake, *ibid.*, 1980, **14**, 1101; (e) M. Natsume and H. Muratake, *Tetrahedron Lett.*, 1979, 3477; *Heterocycles*, 1980, **14**, 445.

[34] M. Somei, F. Yamada, Y. Karasawa, and C. Kaneko, *Chem. Lett.*, 1981, 615.

[35] (a) J. W. Dorner, R. J. Cole, R. Hill, D. Wicklow, and R. H. Cox, *Appl. Environ. Microbiol.*, 1980, **40**, 685; (b) J. Rebek and Y. K. Shue, *J. Am. Chem. Soc.*, 1980, **102**, 5426.

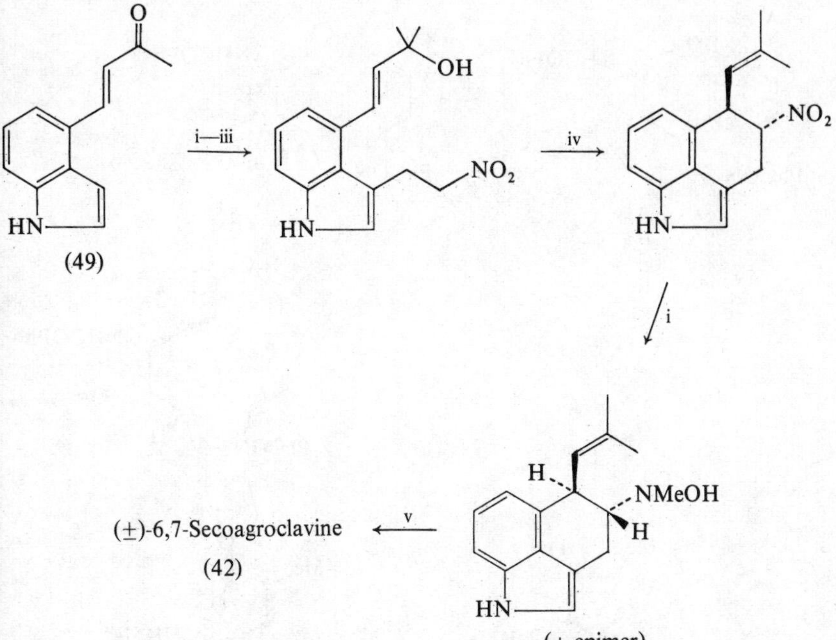

Reagents: i, MeMgI; ii, $Me_2\overset{+}{N}=CH_2\ Cl^-$; iii, $MeNO_2$, Bu_3^nP; iv, $ZnCl_2$, NEt_3; v, $TiCl_3$, NH_4OAc, H_2O

Scheme 7

rugulovasines A (50) and B (51) in methanol solution at room temperature. Subsequently, repetition of the synthesis, starting from L-tryptophan, gave (−)-N-benzoylrugulovasine A of very high (>98%) optical purity.[35b]

An improved synthesis of Uhle's ketone (54), a useful intermediate in ergoline synthesis, has been described,[36] in which the closure of the ketone ring is accomplished by the Dieckmann reaction; this route is also applicable to the homologous ketone.

Details of Ramage's synthesis of (±)-lysergic acid have now been published.[37a] A new synthesis by Oppolzer et al.[37b] (Scheme 9) relies on an intramolecular Diels–Alder cycloaddition on the diene derived by thermolysis of the oxime-ether-ester (55) for the formation of rings C and D. Initial experiments led to dimers of the required diene; however, the desired cycloaddition product (56) was ultimately obtained by thermolysis of (55) at high temperature, under argon, and at very high dilution. Methylation of (56), followed by reductive removal of the N-methoxy-group, and isomerization of the double-bond to the more stable 9,10-position, then gave (±)-lysergic acid (57).[37b]

Two routes to the synthesis of the rather unstable peptide alkaloid (58) have been

[36] G. S. Ponticello, J. J. Baldwin, P. K. Lumma, and D. E. McLure, *J. Org. Chem.*, 1980, **45**, 4236.

[37] (a) R. Ramage, V. W. Armstrong, and S. Coulton, *Tetrahedron*, 1981, **37**, Suppl., p. 157; (b) W. Oppolzer, E. Francotte, and K. Bättig, *Helv. Chim. Acta*, 1981, **64**, 478.

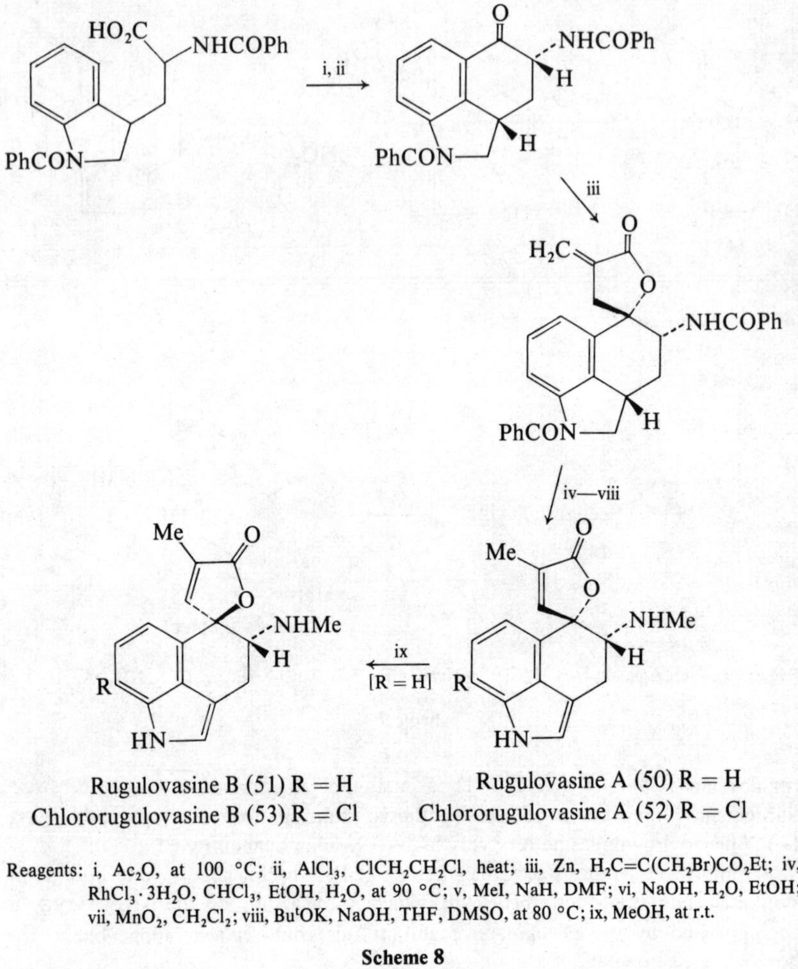

Rugulovasine B (51) R = H
Chlororugulovasine B (53) R = Cl

Rugulovasine A (50) R = H
Chlororugulovasine A (52) R = Cl

Reagents: i, Ac_2O, at 100 °C; ii, $AlCl_3$, $ClCH_2CH_2Cl$, heat; iii, Zn, $H_2C{=}C(CH_2Br)CO_2Et$; iv, $RhCl_3 \cdot 3H_2O$, $CHCl_3$, EtOH, H_2O, at 90 °C; v, MeI, NaH, DMF; vi, NaOH, H_2O, EtOH; vii, MnO_2, CH_2Cl_2; viii, Bu^tOK, NaOH, THF, DMSO, at 80 °C; ix, MeOH, at r.t.

Scheme 8

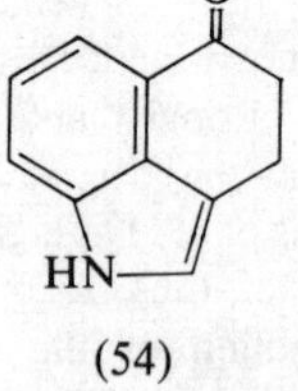

(54)

reported.[38a] A method has been developed for the removal of the pyrrole ring from ergoline derivatives.[38b] The resulting depyrrolo-ergolines, however, exhibited little or no dopaminergic activity.

[38] (*a*) G. Losse and K. D. Wehrstedt, *Z. Chem.*, 1981, **21**, 148; (*b*) N. J. Bach, E. C. Kornfeld, J. A. Clemens, and E. B. Smalstig, *J. Med. Chem.*, 1980, **23**, 812.

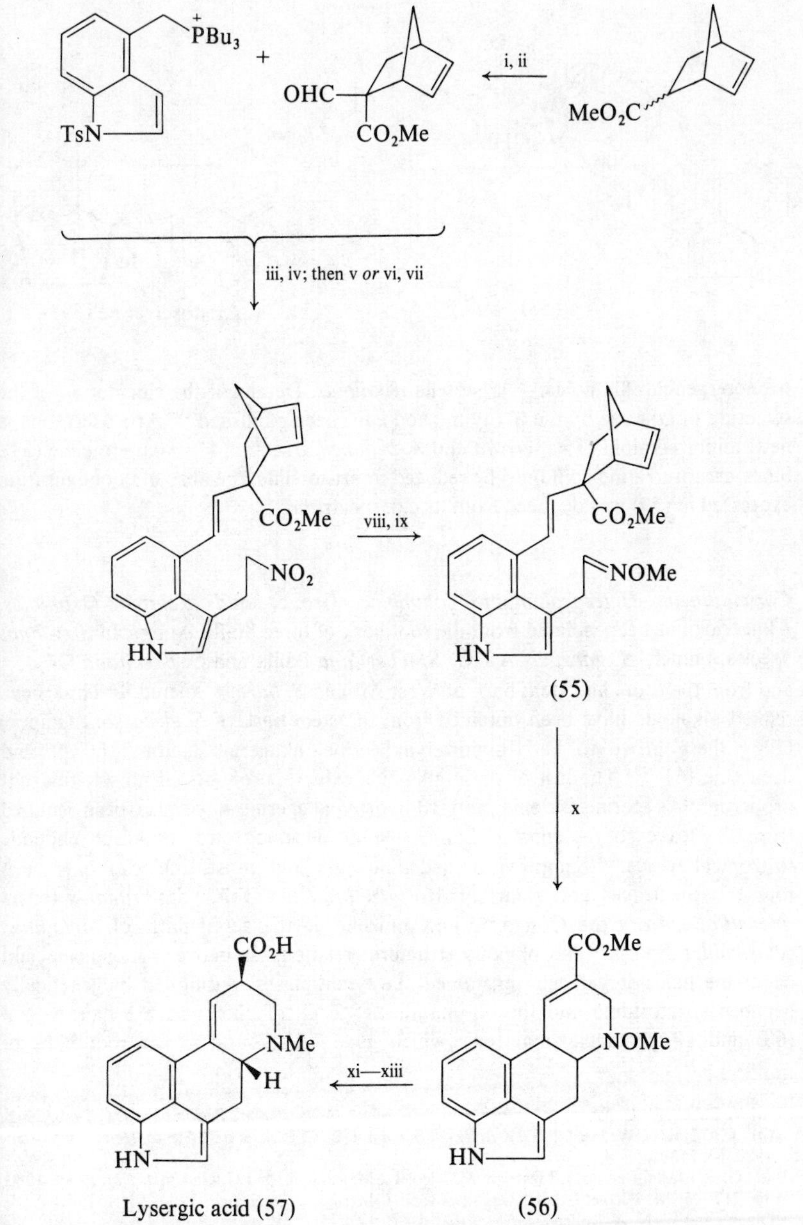

Reagents: i, LiNPr$_2^i$, THF, at −75 °C; ii, HCO$_2$Me, THF, at −75 °C; iii, NaH, DMSO; iv, NaOH, MeOH; v, H$_2$C=CHNO$_2$, PhMe; vi, CH$_2$O, H$_2$O, Me$_2$NH; vii, MeNO$_2$, DMAD; viii, NaOMe, MeOH; ix, TiCl$_3$, NH$_4$OAc, NH$_2$OMe, MeOH, H$_2$O; x, at 200 °C, C$_6$H$_3$Cl$_3$; xi, MeOSO$_2$F, CH$_2$Cl$_2$; xii, Al/Hg, THF, H$_2$O; xiii, 0.5 M-KOH, EtOH, H$_2$O

Scheme 9

(58)   Aristoserratine (59)

Monoterpenoid Alkaloids.—Aristotelia *Alkaloids*. Details of the elucidation of the structure of aristone by the *X*-ray method have been published.[39a] Aristoserratine, a new, minor alkaloid of *A. serrata* and *A. peduncularis*,[39b] is 15-oxoaristoteline (59). Since aristoserratine could not be reduced to aristoteline, the absolute configuration expressed in (59) was deduced from its c.d. spectrum.

Corynantheine–Heteroyohimbine–Yohimbine Group, and Related Oxindoles. Akagerine has been isolated from the root bark of three South American *Strychnos* species, namely *S. gardneri* A.DC, *S. jobertiana* Baill., and *S. parvifolia* DC., [40a] and from the roots and stem bark of West African *S. barteri* Solered.[40b] Three new, related alkaloids have been obtained from the stem bark of *S. decussata* (Pappe) Gilg.; these prove to be 10-hydroxyakagerine, akagerine lactone (60),[40c] and decussine (61).[40d] The last of these alkaloids exhibits pronounced muscle-relaxant properties. Akagerine, kribine, and 10-hydroxyakagerine have also been isolated from the leaves of *S. spinosa* Lam., a small deciduous tree or shrub endemic to tropical Africa.[40e] Sempervirine and a new alkaloid, mostueine (62), have been found in the leaves and stems of *Mostuea brunonis* Didr. var. *brunonis* forma *angustifolia*, from the Congo,[40f] and antirhine in the aerial parts of *Melodinus celastroides* Baill.[40g] The obvious structural relationship between mostueine and decussine has not yet been examined. Two new bases, positioned biogenetically between strictosidine and the corynantheine group of alkaloids, are naucleidinal (63) and 19-*epi*-naucleidinal (64), which have been found in the root bark of

[39] (a) V. Zabel, W. H. Watson, M. Bittner, and M. Silva, *J. Chem. Soc., Perkin Trans. 1*, 1980, 2842; (b) M. A. Hai, N. W. Preston, R. Kyburz, E. Schöpp, I. R. C. Bick, and M. Hesse, *Helv. Chim. Acta*, 1980, **63**, 2130.

[40] (a) G. B. Marini-Bettòlo, I. Messana, M. Nicoletti, M. Patamia, and C. Galeffi, *J. Nat. Prod.*, 1980, **43**, 717; (b) M. Nicoletti, J. U. Okuakwa, and I. Messana, *Fitoterapia*, 1980, **51**, 131; (c) A. A. Olaniyi and W. N. A. Rolfsen, *J. Nat. Prod.*, 1980, **43**, 595; (d) W. N. A. Rolfsen, A. A. Olaniyi, and F. Sandberg, *Acta Pharm. Suec.*, 1980, **17**, 105; A. Kvick, *Acta Crystallogr., Sect. B*, 1981, **37**, 1304; (e) J. U. Oguakwa, C. Galeffi, M. Nicoletti, I. Messana, M. Patamia, and G. B. Marini-Bettòlo, *Gazz. Chim. Ital.*, 1980, **110**, 97; (f) M. Onanga and F. Khuong-Huu, *C. R. Hebd. Seances Acad. Sci., Ser. C*, 1980, **291**, 191; (g) S. Baassou, H. Mehri, A. Rabaron, M. Plat, and T. Sévenet, *Ann. Pharm. Fr.*, 1981, **39**, 167.

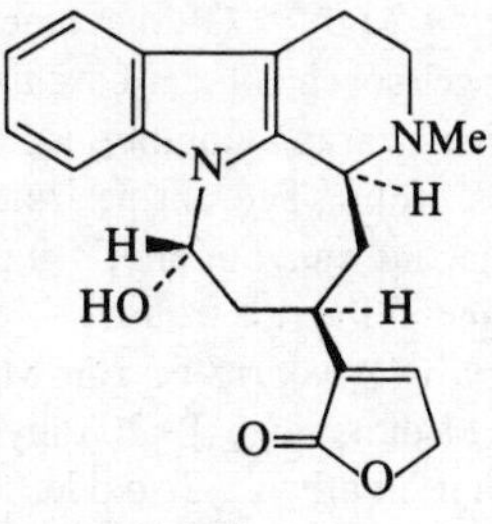

(60)

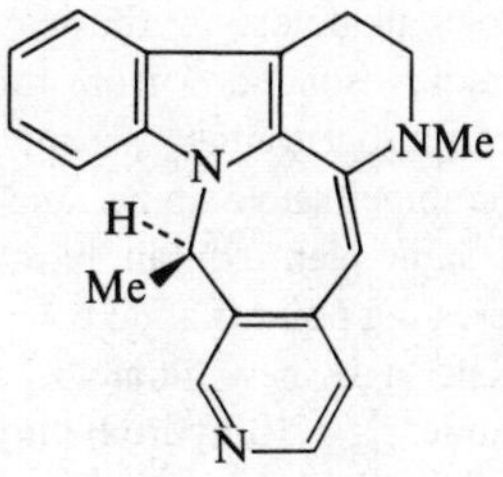

Decussine (61)

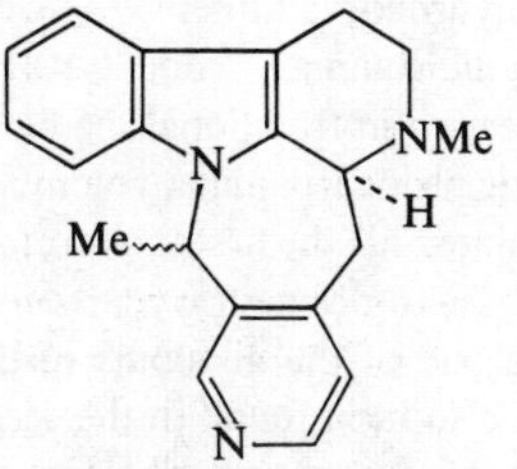

Mostueine (62)

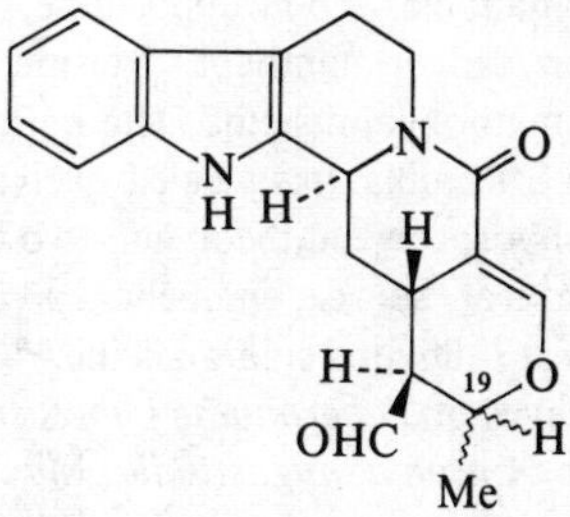

19-H

Naucleidinal (63)   β
19-*epi*-Naucleidinal (64)   α

Nauclea latifolia.[41a] Isodolichantoside and 16-*epi*-diploceline are new epimers of known bases which have been reported[41b] to occur in the root bark of *Strychnos gossweileri*.

Ajmalicine and vallesiachotamine are amongst the alkaloids produced by cultures of cell lines 943,[42a] 953,[42b] and 200 GW of *Catharanthus roseus*;[42a] the first two of these cell lines also produced yohimbine and isositsirikine, while cell lines 953 and 200 GW also produced strictosidine lactam.

A number of new extractions of intact plants, or specific organs of plants, have also been recorded. Geissoschizine occurs, together with aricine, pleiocarpamine, 10-methoxypleiocarpamine, 19,20-dehydroadirubine acetate, and sixteen other

[41] (a) F. Hotellier, P. Delaveau, and J. L. Pousset, *Phytochemistry*, 1980, **19**, 1884; (b) C. A. Coune and L. J. G. Angenot, *Herba Hung.*, 1980, **19**, 189 (*Chem. Abstr.*, 1980, **93**, 217 906).
[42] (a) J. P. Kutney, L. S. L. Choi, P. Kolodziejczyk, S. K. Sleigh, K. L. Stuart, B. R. Worth, W. G. W. Kurz, K. B. Chatson, and F. Constabel, *Phytochemistry*, 1980, **19**, 2589; *Heterocycles*, 1980, **14**, 765; (b) W. G. W. Kurz, K. B. Chatson, F. Constabel, J. P. Kutney, L. S. L. Choi, P. Kolodziejczyk, S. K. Sleigh, K. L. Stuart, and B. R. Worth, *Helv. Chim. Acta*, 1980, **63**, 1891.

indole alkaloids, in the leaves of *Rauwolfia oreogiton* Mgf.[43a] Of the nineteen alkaloids isolated[43b] from the leaves of *R. vomitoria*, geissoschizol, geissoschizine, tetrahydroalstonine, aricine, isoreserpiline, reserpiline, carapanaubine, isocarapanaubine, rauvoxinine, and rauvoxine belong to this group. The unripe fruits of this same plant contain tetrahydroalstonine, yohimbine, and rauvomitine;[43c] it is an interesting point that no trace of alkaloids could be found in the ripe fruits.

The eight new alkaloids of the trunk bark of *Ochrosia moorei* F. von Muell. include[43d] 10-hydroxydihydrocorynantheol, 10,11-dimethoxy-19,20-dihydro-(16S,20R)-sitsirikine, 10-methoxy-(3S,4R)-dihydrocorynantheol N-oxide, 10-methoxy-(3S,4S)-dihydrocorynantheol N-oxide, (3R,4R)-reserpiline N-oxide, (3R,4S)-reserpiline N-oxide, 11-methoxypicraphylline, and (3S,7S)-ochropposinine oxindole (65). Amongst the twenty known alkaloids also isolated are tetrahydroalstonine, aricine, reserpinine, ochropposine, isoreserpiline, reserpiline, isoreserpiline pseudoindoxyl, 10,11-dimethoxypicraphylline, rauvoxine, isocarapanaubine, ochropposinine, 10-methoxy-18,19-dihydrocorynantheol, 18,19-dihydrocorynantheol, desmethoxycarbonyldihydrogambirtannine, and 10,11-dimethoxyajmalicine. The bark of *Hunteria zeylanica*, used in traditional medicine in Sri Lanka, has yielded twelve alkaloids, which include pleiocarpamine, yohimbol, dihydrocorynantheol, and two bases that have been isolated for the first time from a natural source, namely epiyohimbol and 17-hydroxy-16-desmethoxycarbonyl-16,17-dihydro-epiajmalicine.[43e] Tetrahydroalstonine is one of the alkaloids of the leaves of *Tabernanthe pubescens*, from Zaïre;[43f] it has also been found in the seeds of *Amsonia angustifolia* Michx.,[43g] and is also one of the major alkaloids of *Uncaria attenuata*, from southern Thailand,[43h,i] in which it occurs with rauniticine (which has not previously been isolated from any *Uncaria* species) and 14-β-hydroxy-3-isorauniticine (66), which is the first known heteroyohimbine base to contain a hydroxy-group in this position. A specimen of *U. salaccensis* Bakh. f. (now included in *U. attenuata* Korth.) from northern Thailand contained 3-isoajmalicine, 19-*epi*-3-isoajmalicine, and the related normal oxindole alkaloids mitraphylline and uncarine B.[43i]

(65)   (66)

[43] (a) B. A. Akinloye and W. E. Court, *Phytochemistry*, 1980, **19**, 2741; (b) M. M. Amer and W. E. Court, *ibid.*, p. 1833; (c) M. M. Iwu, *Planta Med.*, 1980, Suppl. p. 13; (d) A. Ahond, H. Fernandez, M. Julia-Moore, C. Poupat, V. Sánchez, P. Potier, S. K. Kan, and T. Sévenet, *J. Nat. Prod.*, 1981, **44**, 193; (e) L. S. R. Arambewela and F. Khuong-Huu, *Phytochemistry*, 1981, **20**, 349; (f) T. Mulamba, C. Delaude, L. Le Men-Olivier, and J. Lévy, *J. Nat. Prod.*, 1981, **44**, 184; (g) H. Tomczyk and W. Kisiel, *Pol. J. Chem.*, 1980, **54**, 2397; (h) D. Ponglux, T. Supavita, R. Verpoorte, and J. D. Phillipson, *Phytochemistry*, 1980, **19**, 2013; (i) J. D. Phillipson, *J. Pharm. Pharmacol.*, 1980, **32**, Suppl., 73P.

Recent extractions[44a] of the leaves of *Cinchona ledgeriana* have resulted in the isolation of quinamine (previously observed), 3-*epi*-quinamine, aricine, and a new alkaloid which may prove to be stereoisomeric with quinamine. 11-Hydroxy-pleiocarpamine occurs in *Vinca erecta*,[44b] and herbacine and herbaine in *V. herbacea*.[44c] Yohimbine appears to be the major alkaloid of the trunk bark of *Pausinystalia macroceras*, in which it occurs with four other alkaloids of this group.[44d] Pleiocarpamine occurs in association with nine alkaloids of the aspidospermine–eburnamine group in the stem and root bark of *Hunteria elliottii* (Stapf.) Pichon.[44e]

A new weak base, neonor-reserpine, from *Rauwolfia vomitoria*, is regarded as a stereoisomer of psuedoreserpine;[45] since the new base appears to have the *epiallo* stereochemistry, it must differ from pseudoreserpine in the stereochemistry at one or more of the positions C-16, C-17, and C-18.

Two new oxindole alkaloids from the leaves of *Hamelia patens* Jacq. prove to be the carboxylic acids corresponding to pteropodine and isopteropodine; they have been named maruquine and isomaruquine.[46]

Deoxytubulosine has been extracted from the seeds of *Alangium lamarckii*.[47]

In connection with structure elucidation in this area, the compilations of ^{1}H n.m.r. data for geissoschizine, ajmalicine, 19-*epi*-ajmalicine, tetrahydroalstonine, and rauniticine[48a] and of ^{13}C n.m.r. data for diploceline[19c] should prove useful. The effect of stereoisomerism on the mass-spectrometric behaviour of some corynantheine and yohimbine alkaloids has also been studied.[48b]

A considerable amount of attention has been devoted to the total or partial synthesis of alkaloids in this group during the past year. The application of 3-halogeno-indolenine derivatives of alkaloids in synthesis is the subject of a review.[49]

The first synthesis[50] of akagerine (67) makes use of the tetracyclic lactam (68a), previously synthesized. Formation of the (*E*)-dilactam (69) was observed from both (68a) and its (*Z*)-isomer. Preferential opening of the N_b,21 lactam function in (69) was followed by reduction stages to the diol (70), which was then oxidized preferentially at the allylic alcohol function to give akagerine (Scheme 10).

Takano *et al.* have reported the first synthesis of (±)-antirhine (71) (Scheme 11),[51] in which the problem of generating the desired, less stable (*anti*) stereochemistry at C-3 and C-15 was overcome by preparing the non-tryptamine fragment (72) from (±)-trinorcamphor (73) *via* a sequence of stereospecific reactions. Condensation of (72) with tryptamine, followed by cyclization and

[44] (*a*) M. Zeches, B. Richard, P. Thepenier, L. Le Men-Olivier, and J. Le Men, *Phytochemistry*, 1980, **19**, 2451; (*b*) M. M. Khalmirzaev, M. R. Yagudaev, and S. Yu. Yunusov, *Khim. Prir. Soedin.*, 1980, 426; (*c*) G. V. Chkhikvadze, V. Yu. Vachnadze, and K. S. Mujiri, *ibid.*, p. 850; (*d*) M. Leboeuf, A. Cavé, P. Mangeney, and A. Bouquet, *Planta Med.*, 1981, **41**, 374; (*e*) J. Vercauteren, J. Kerharo, A. M. Morfaux, G. Massiot, L. Le Men-Olivier, and J. Le Men, *Phytochemistry*, 1980, **19**, 1959.
[45] A. Malik, S. Siddiqui, and W. Voelter, *Z. Naturforsch., Teil. B*, 1980, **35**, 920.
[46] J. B. del Castillo, M. T. M. Ferrero, J. L. M. Ramon, F. R. Luis, and P. V. Bueno, *An. Quim.*, 1980, **76**, 294.
[47] B. Achari, E. Ali, P. P. G. Dastidar, R. R. Sinha, and S. C. Pakrashi, *Planta Med.*, 1980, Suppl. p. 5.
[48] (*a*) G. Hoefle, P. Heinstein, J. Stöckigt, and M. H. Zenk, *Planta Med.*, 1980, **40**, 120; (*b*) J. Tamás, G. Czira, and G. Bujtás, *Adv. Mass Spectrom., Sect. A*, 1980, **8**, 733.
[49] M. Ikeda and Y. Tamura, *Heterocycles*, 1980, **14**, 867.
[50] W. Benson and E. Winterfeldt, *Heterocycles*, 1981, **15**, 935.
[51] S. Takano, M. Takahashi, and K. Ogasawara, *J. Am. Chem. Soc.*, 1980, **102**, 4282.

reduction stages, then gave a tetrahydro-β-carboline derivative (74) with a *cis* C/D ring junction (inferred from the n.m.r. spectrum and the absence of Bohlmann bands in the i.r. spectrum). Finally, the vinyl group was introduced by preferential

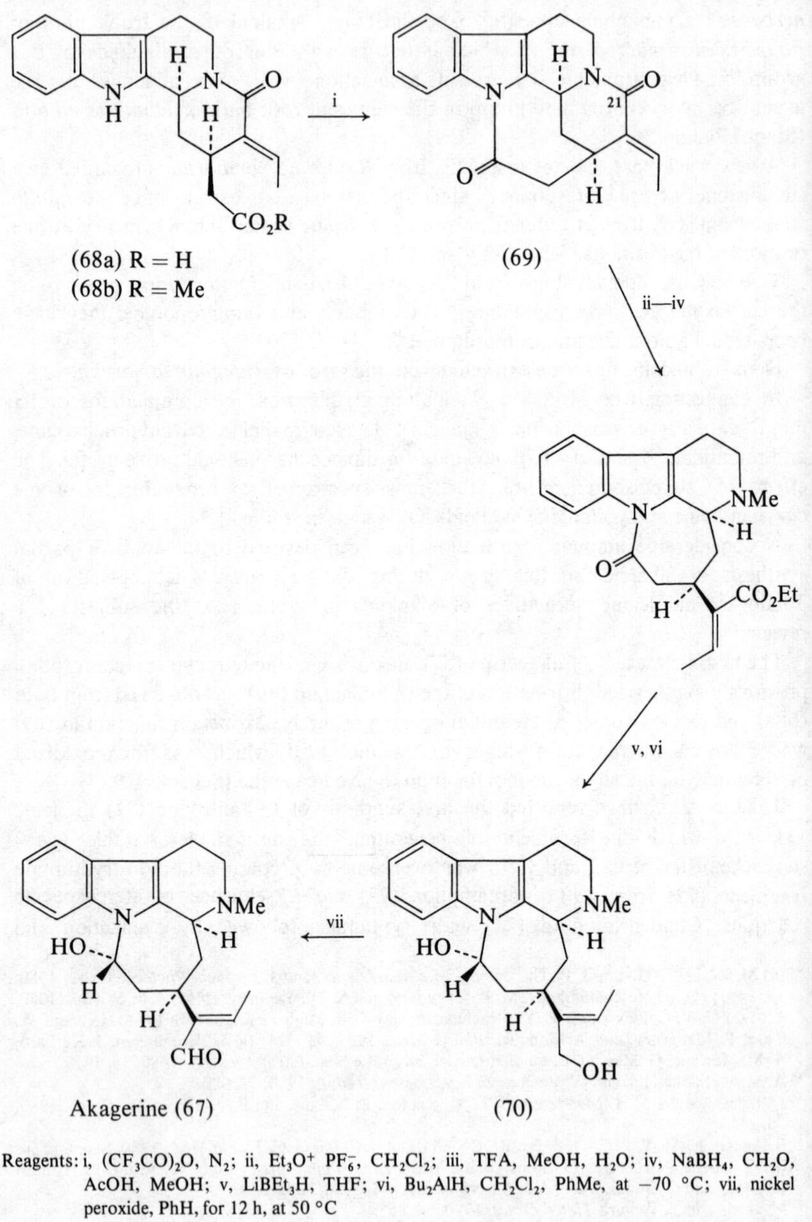

(68a) R = H
(68b) R = Me

(69)

(70)

Akagerine (67)

Reagents: i, (CF$_3$CO)$_2$O, N$_2$; ii, Et$_3$O$^+$ PF$_6^-$, CH$_2$Cl$_2$; iii, TFA, MeOH, H$_2$O; iv, NaBH$_4$, CH$_2$O, AcOH, MeOH; v, LiBEt$_3$H, THF; vi, Bu$_2$AlH, CH$_2$Cl$_2$, PhMe, at -70 °C; vii, nickel peroxide, PhH, for 12 h, at 50 °C

Scheme 10

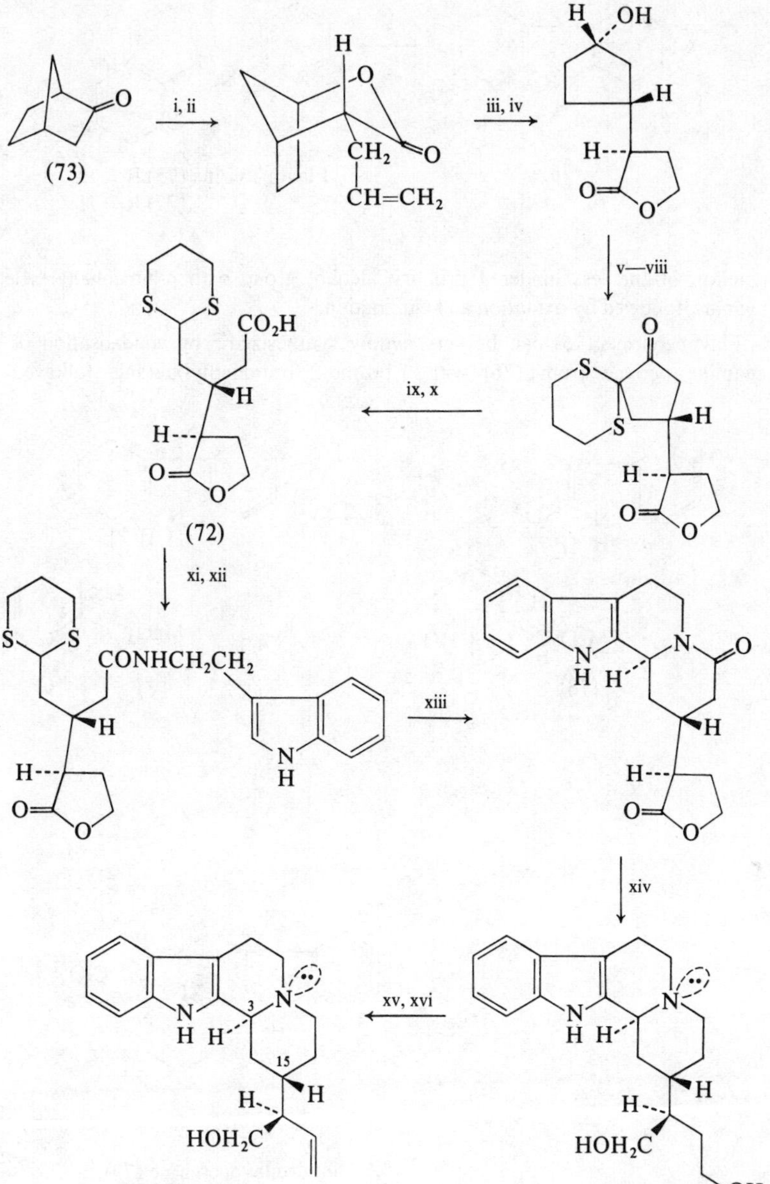

Reagents: i, Baeyer–Villiger oxidation; ii, H₂C=CHCH₂Br, LiNPr$_2^i$, HMPA, THF; iii, O₃, MeOH, at
−78 °C; iv, NaBH₄, at −78 °C → r.t.; v, Jones' reagent; vi, pyrrolidine, PhH; vii,
CH₂(CH₂STs)₂, NEt₃; viii, HCl, H₂O; ix, KOH, Bu^{t}OH, at 60 °C; x, H⁺; xi, ClCO₂Et,
NEt₃, CH₂Cl₂; xii, tryptamine, CH₂Cl₂; xiii, MeI, MeCN, H₂O; xiv, LiAlH₄; xv,
o-O₂NC₆H₄SeCN, Bu$_3^n$P, THF; xvi, *m*-ClC₆H₄CO₃H, CH₂Cl₂

Scheme 11

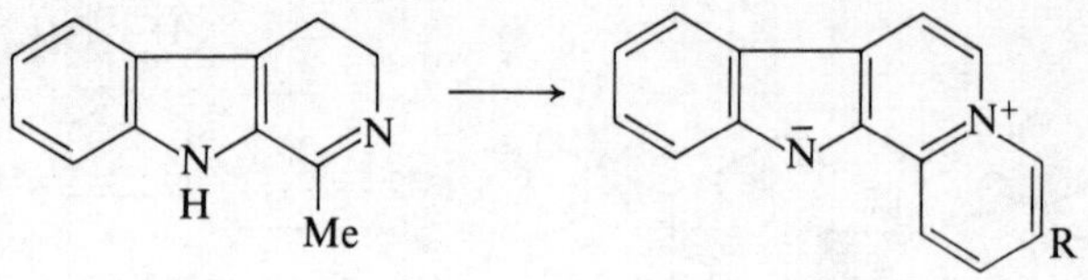

(76)   Flavopereirine (75) R = Et
                                                        (77) R = H

reaction of the less hindered primary alcohol group with *o*-nitrophenyl seleno-cyanate, followed by oxidation and elimination.

Flavopereirine (75) can be very simply synthesized[52] by condensation of the enamine derived from (76) with 1-bromo-2-(bromoethyl)butane, followed by

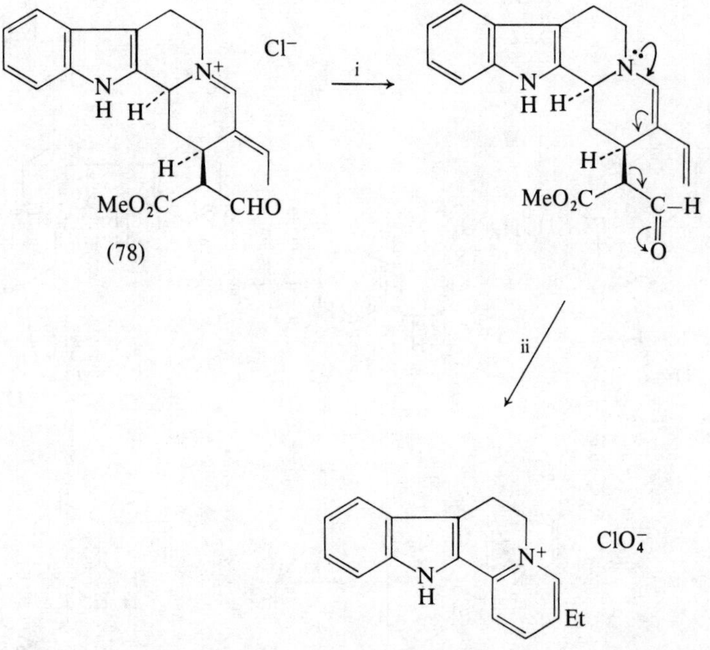

(78)

5,6-Dihydroflavopereirine (79)

Reagents: i, KOH, MeOH; ii, NaClO$_4$, H$_2$O, MeOH

Scheme 12

[52] B. Danieli, G. Lesma, and G. Palmisano, *J. Chem. Soc., Chem. Commun.*, 1980, 860.

dehydrogenation; the simpler base (77) can be efficiently synthesized by an analogous route.

N_b,21-Dehydrogeissoschizine (78) has earlier been proposed to be a precursor of the flavopereirine group of alkaloids. A method has now been developed for the analogous removal *in vitro* of the three-carbon unit attached to C-15, by a retro-Mannich reaction.[53] Treatment of N_b,21-dehydrogeissoschizine chloride with base presumably released the dienamine base, which then lost the β-aldehydo-ester unit attached to C-15. Migration of a double-bond then gave 5,6-dihydro-flavopereirine (79), which was isolated as its perchlorate on addition of sodium perchlorate to the reaction medium (Scheme 12).

The dilactam (69), obtainable as described above (see Scheme 10) from (68a) or its (Z)-isomer, has been used in an improved synthesis of geissoschizine (80).[54a] Methanolysis of (69) gave a quantitative yield of the tetracyclic ester (68b), which, on Borch reduction and formylation, gave geissoschizine (80).

In further extensions of their versatile approach to the yohimbinoid and heteroyohimbinoid alkaloids, Wenkert and his collaborators have developed three further syntheses of ($\pm$)-geissoschizine and a new synthesis of hirsutine.[54b] One of the syntheses of ($\pm$)-geissoschizine, and the synthesis of hirsutine, are outlined in Scheme 13. The intermediate keto-ester (81), previously prepared,[54c] was converted by the Borch reaction into a mixture of three products (82a)—(82c), of which the desired unsaturated ester (82a) was obtained in 26% yield. Reduction of (82a) gave ($\pm$)-3-isogeissoschizine (83), which was then converted into its dimethyl acetal, epimerized, and hydrolysed, to give ($\pm$)-geissoschizine (80). Alternatively, the reaction of (81) with Meerwein's reagent, followed by hydrogenation, gave the ester (84), which on reduction and methylation gave ($\pm$)-hirsutine (85).[54b]

The approach by Harley-Mason and his collaborators[55a] to the synthesis of geissoschizine involved closure of the 15,20 bond in the formation of the tetracyclic framework, rather than the 2,3 or N_b,21 bond, as in all the other reported syntheses (Scheme 14). Alkylation of the tetrahydro-β-carboline ester (86) by means of *erythro*-2-bromo-3-methoxybutyroyl chloride gave a mixture of C-3 epimers (87), *only one* of which cyclized, to give the tetracyclic ester (88). Hydrolysis and decarboxylation, followed by reduction, gave a mixture of primary alcohols (89), epimeric only at C-20, which were then converted into the unsaturated nitrile (90), having the desired *E* configuration. Fortunately, both epimers of (89) gave (90), and no material of *Z* configuration was obtained. The synthesis was then completed by removal of the lactam carbonyl group *via* reduction of the oxazoline derivative, followed by acid-catalysed methanolysis. The product, ($\pm$)-methyl 3-isogeisso-schizoate (91), has previously[55b] been converted into ($\pm$)-geissoschizine.

[53] C. Kan Fan and H.-P. Husson, *Tetrahedron Lett.*, 1980, **21**, 4265.

[54] (a) W. Benson and E. Winterfeldt, *Angew. Chem., Int. Ed. Engl.*, 1979, **18**, 862; (b) E. Wenkert, Y. D. Vankar, and J. S. Yadav, *J. Am. Chem. Soc.*, 1980, **102**, 7972; (c) E. Wenkert, C. J. Chang, H. P. S. Chawla, D. W. Cochran, E. W. Hagaman, J. C. King, and K. Orito, *ibid.*, 1976, **98**, 3645.

[55] (a) B. J. Banks, M. J. Calverley, P. D. Edwards, and J. Harley-Mason, *Tetrahedron Lett.*, 1981, **22**, 1631; (b) K. Yamada, K. Aoki, T. Kato, D. Uemura, and E. E. van Tamelen, *J. Chem. Soc., Chem. Commun.*, 1974, 908.

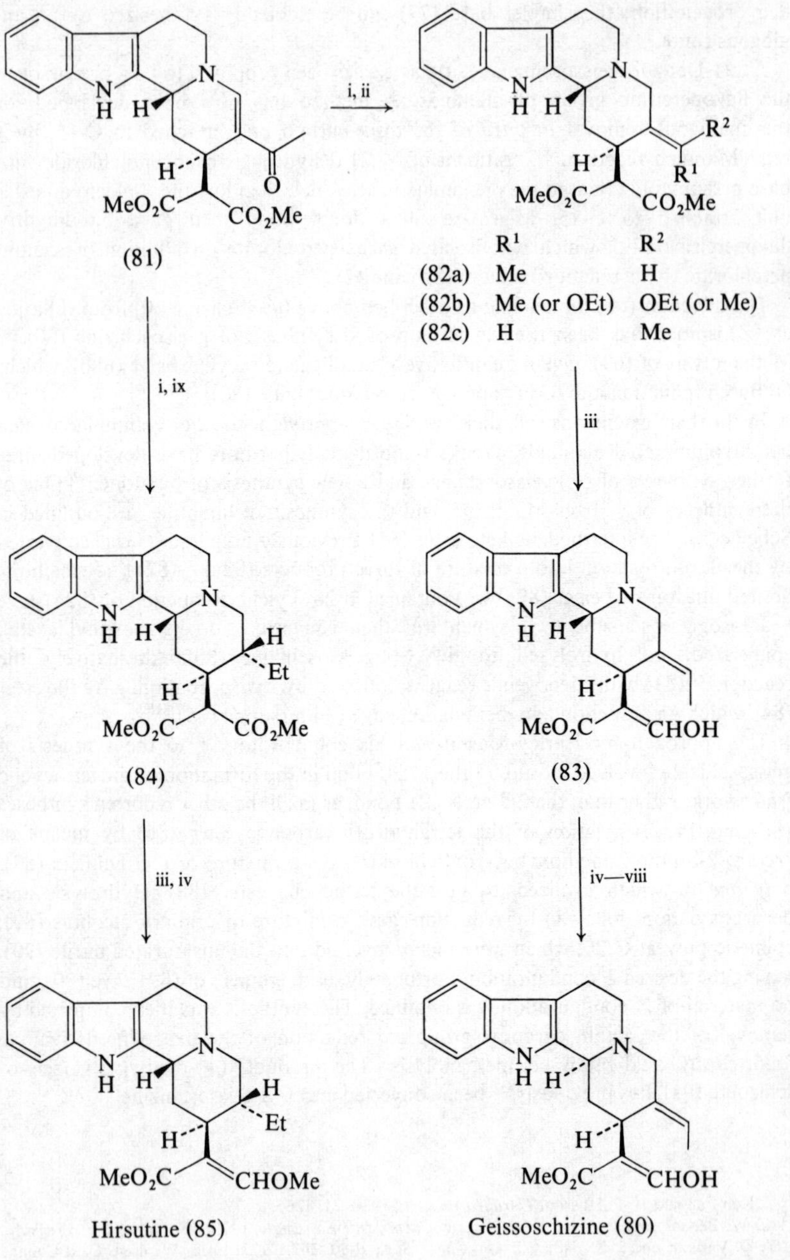

Reagents: i, Et$_3$O$^+$ BF$_4^-$, CH$_2$Cl$_2$; ii, NaBH$_4$, MeOH; iii, Bu$_2^i$AlH, CH$_2$Cl$_2$, at -78 °C; iv, HCl, MeOH; v, m-ClC$_6$H$_4$CO$_3$H; vi, (CF$_3$CO)$_2$O, CH$_2$Cl$_2$, at -78 °C $\rightarrow$ 0 °C; vii, NaBH$_4$, THF; viii, 4.5 M-HCl, MeCOMe; ix, H$_2$, Pd/C

Scheme 13

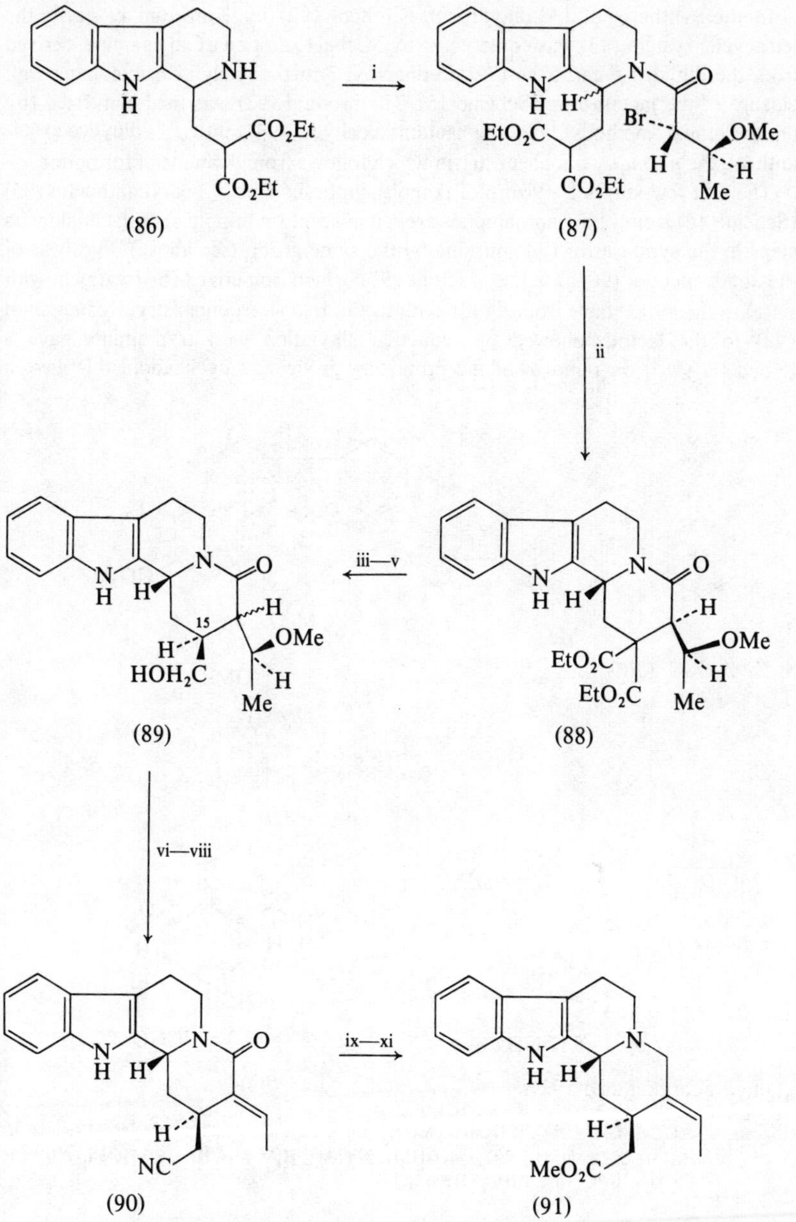

Reagents: i, (±)-*erythro*-MeCH(OMe)CH(Br)COCl, pyridine; ii, NaH, THF; iii, NaOH, H₂O, EtOH; iv, DMF, at 100 °C; v, NaBH₄, MeOH, heat; vi, MsCl, pyridine; vii, NaCN, DMF; viii, NaOMe, MeOH; ix, NH₂CMe₂CH₂OH, ZnCl₂, at 140 °C; x, Bu$_2^i$AlH, THF, at −20 °C; xi, MeOH, H⁺

Scheme 14

In the synthesis of (±)-dihydrocorynantheol (92) by Kametani *et al.*,[56a] the tetracyclic system (93) was constructed by Michael addition of an enamine, derived from the dihydro-β-carboline (76), to dimethyl 3-methoxyallylidenemalonate, with closure of the lactam ring (Scheme 15). The product (93) was then converted (by unexceptional methods), *via* the lactam aldehyde (94), into (±)-dihydrocorynantheol, the *normal* stereochemistry in which follows from its mode of formation.

The first few stages [→(96)] of Takano's synthesis[56b] of (±)-corynantheidol (95) (Scheme 16) from (±)-trinorcamphor are reminiscent (in principle) of the analogous steps in the synthesis of (±)-antirhine by the same group (see above). Pyrolysis of the amide-alcohol (96) gave the δ-lactone (97), which isomerized (by treatment with base) to the more stable isomer (98), with the desired stereochemistry. Reduction of (98) to the lactol, followed by reductive alkylation with tryptamine, gave a thioacetal (99); on removal of the protecting group and cyclization, this gave a

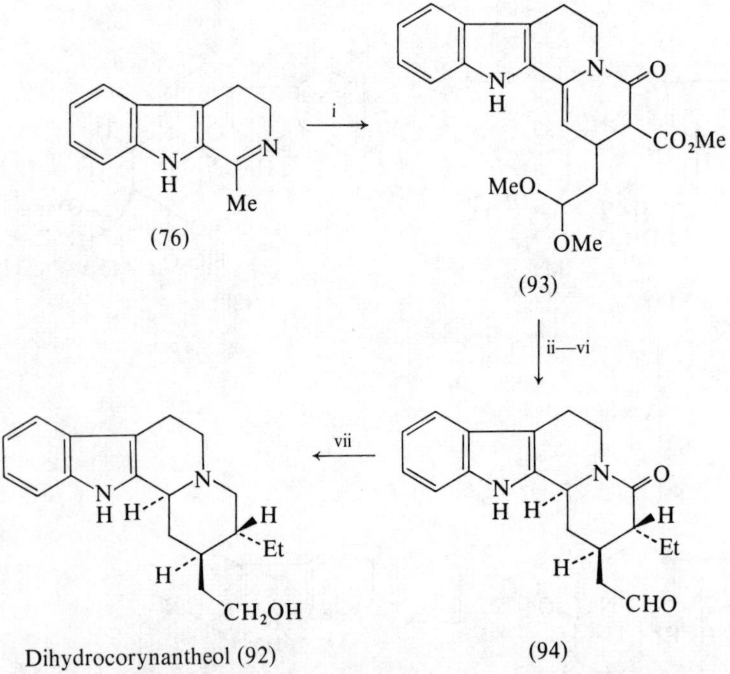

Reagents: i, MeOCH=CH–CH=C(CO₂Me)₂, MeOH, for 3 days at r.t., then heat for 24 hours; ii, NaH, EtI, DMF; iii, H₂, PtO₂, MeOH; iv, NaOMe, H₂O, MeOH; v, DMSO, at 160 °C; vi, TsOH, at 0 °C; vii, LiAlH₄, THF, Et₂O

Scheme 15

⁵⁶ (*a*) T. Kametani, N. Kanaya, H. Hino, S. P. Huang, and M. Ihara, *Heterocycles*, 1980, **14**, 1771; (*b*) S. Takano, K. Masuda, and K. Ogasawara, *J. Chem. Soc., Chem. Commun.*, 1980, 887; (*c*) S. Takano, M. Sato, and K. Ogasawara, *Heterocycles*, 1981, **16**, 799; (*d*) E. E. van Tamelen and J. B. Hester, *J. Am. Chem. Soc.*, 1969, **91**, 7342.

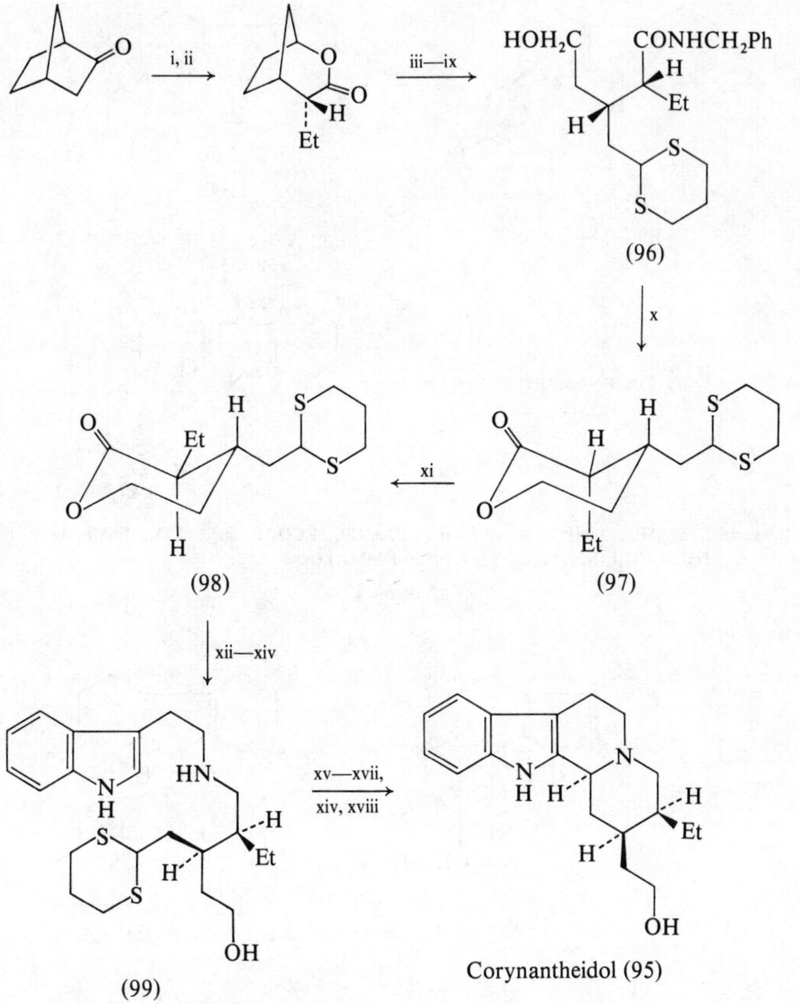

Reagents: i, *m*-ClC₆H₄CO₃H; ii, EtBr, LiNPr₂ⁱ, HMPA, THF; iii, PhCH₂NH₂, at 180 °C; iv, Jones' reagent; v, pyrrolidine, PhH; vi, CH₂(CH₂STs)₂, NEt₃; vii, Bu'OH, KOH; viii, ClCO₂Et, NEt₃; ix, reduction; x, PhCH₂Ph, at 200 °C; xi, Bu'OH, THF; xii, Bu₂ⁱAlH, PhMe, at −25 °C; viii, tryptamine, MeOH, heat; xiv, NaBH₄; xv, MeI, MeCN, H₂O; xvi, acetylation; xvii, Hg(OAc)₂; xviii, alkaline hydrolysis

Scheme 16

mixture of C-3 epimers, in which 3-*epi*-corynantheidol was predominant. The remaining stages were thus required to invert the configuration at position 3.

The formal synthesis⁵⁶ᶜ of dihydrocorynantheine (Scheme 17), also by Takano's group, relies on the stereospecific addition of malonate to the unsaturated lactam (100) for the generation of the correct stereochemistry at positions 15 and 20. The

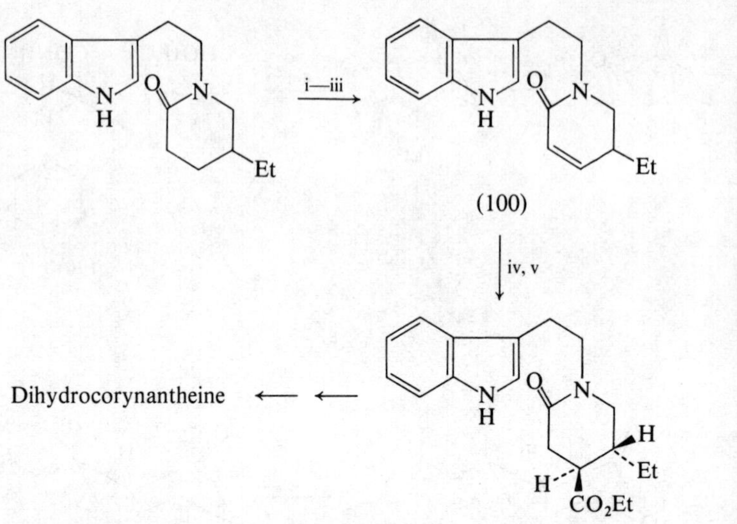

(100)

Dihydrocorynantheine ← ←

Reagents: i, PhSSPh, LiNPr$_2^i$, at -78 °C; ii, m-ClC$_6$H$_4$CO$_3$H; iii, CaCO$_3$, PhMe, heat; iv, NaCH(CO$_2$Et)$_2$; v, MgCl$_2 \cdot$ 6H$_2$O, moist DMSO, heat

Scheme 17

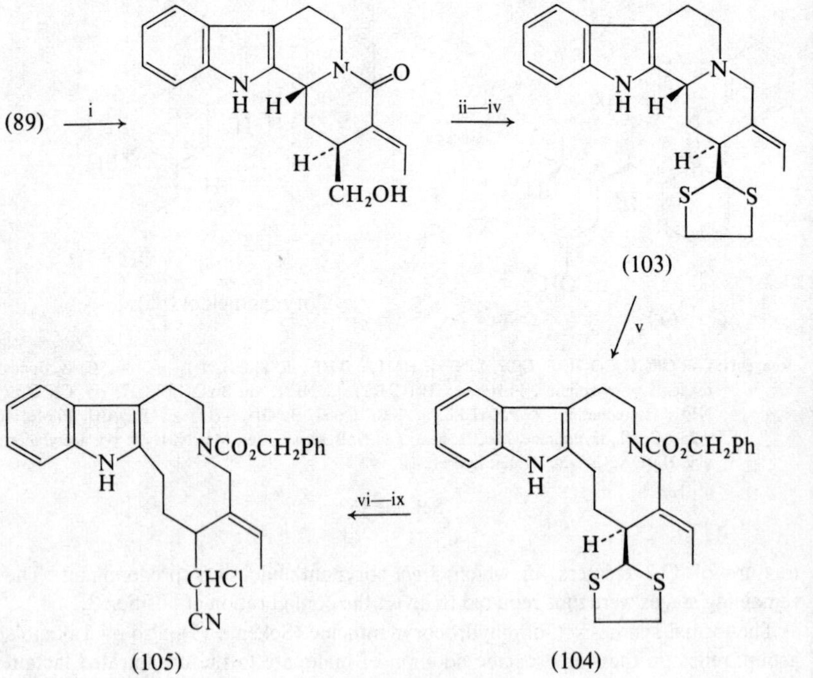

(89)

(103)

(105)

(104)

Scheme 18 (*part*)

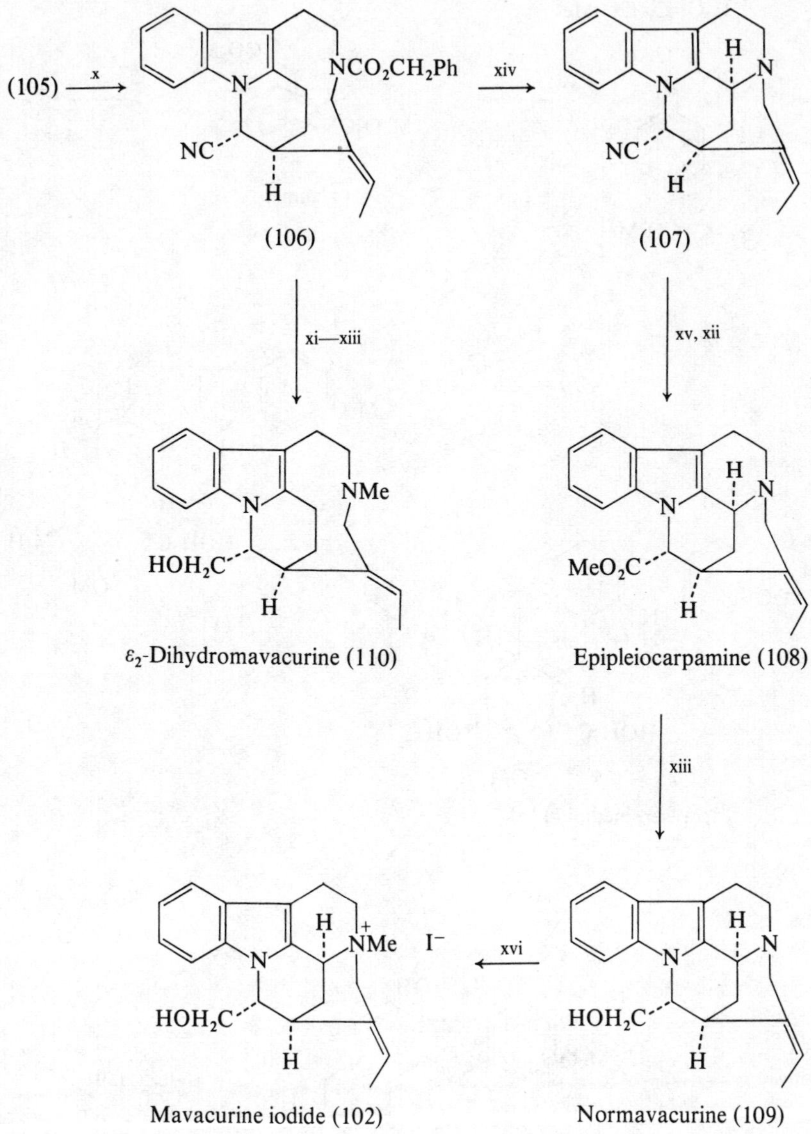

(105) $\xrightarrow{x}$ (106) $\xrightarrow{xiv}$ (107)

(106) $\xrightarrow{xi-xiii}$ ε_2-Dihydromavacurine (110)

(107) $\xrightarrow{xv, xii}$ Epipleiocarpamine (108)

Epipleiocarpamine (108) $\xrightarrow{xiii}$ Normavacurine (109)

Normavacurine (109) $\xrightarrow{xvi}$ Mavacurine iodide (102)

Reagents: i, NaOEt, EtOH; ii, DMSO, DCC, H_3PO_4; iii, $(CH_2SH)_2$, $BF_3 \cdot Et_2O$; iv, Bu_2^iAlH; v, $ClCO_2CH_2Ph$, $NaBH_3CN$, THF, at $-70\,^\circ C \rightarrow$ r.t.; vi, MeI, MeCN, H_2O, MeCOMe, $CaCO_3$; vii, HCN, H_2O, THF; viii, $MeSO_2Cl$, pyridine, CH_2Cl_2; ix, LiCl, DMF, at $110\,^\circ C$; x, NaH, DMSO; xi, EtOH, KOH; xii, BF_3, MeOH; xiii, $LiAlH_4$; xiv, 1-chlorobenzotriazole, NEt_3, CH_2Cl_2; xv, KOH, EtOH, H_2O; xvi, MeI, MeOH

Scheme 18 (*continued*)

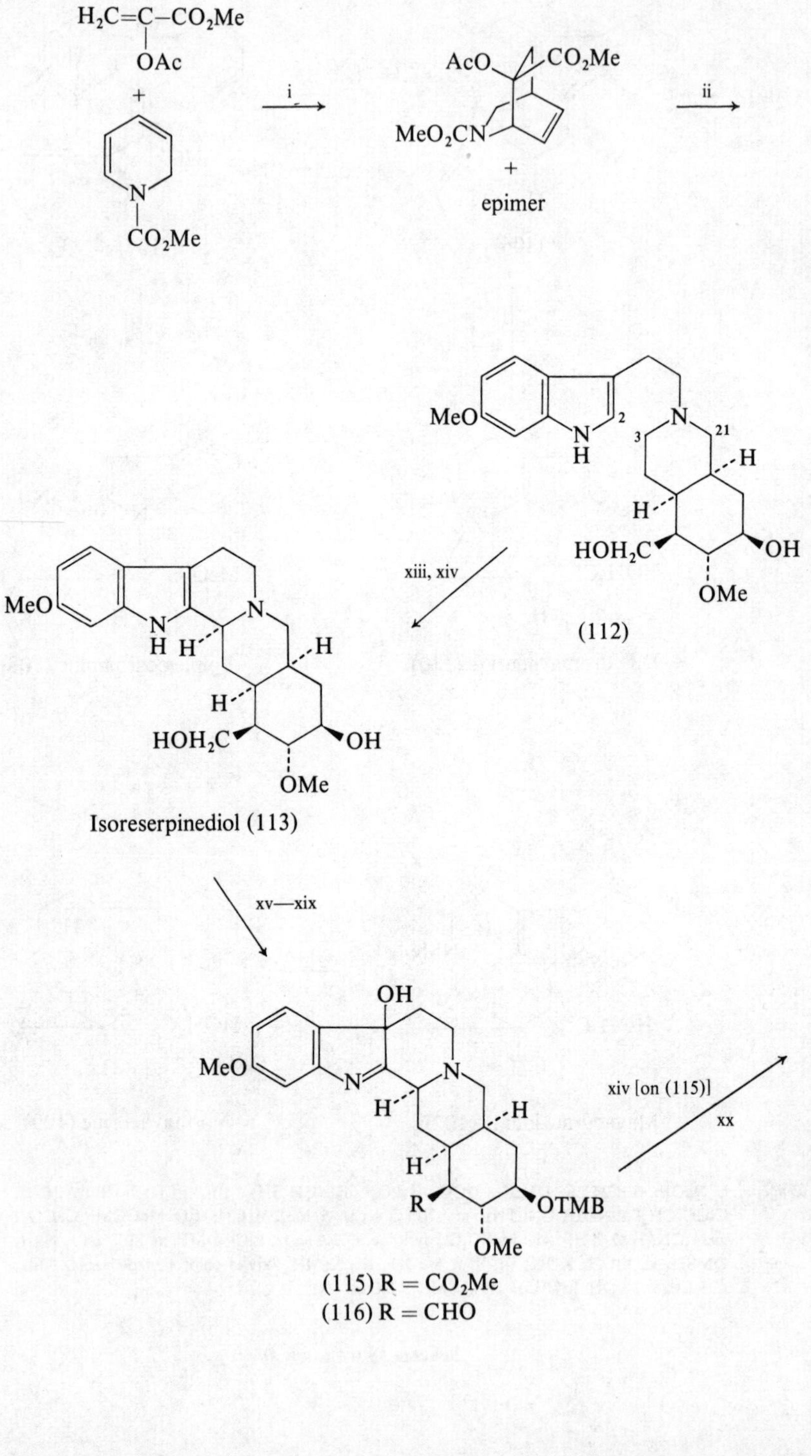

Isoreserpinediol (113)

(112)

(115) R = CO₂Me
(116) R = CHO

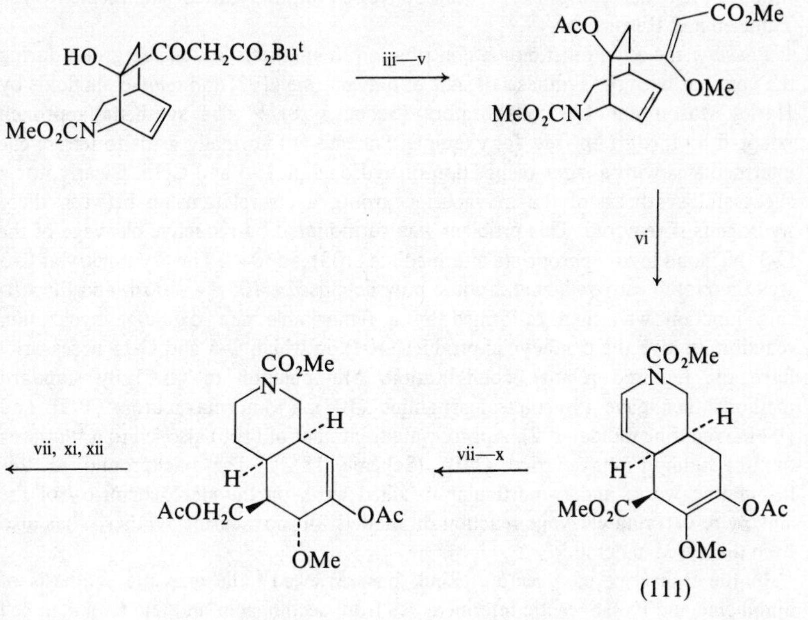

3-Isoreserpine (114)

Reagents: i, PhMe, hydroquinone, at 120 °C, for 56 h; ii, LiCH₂CO₂Buᵗ, THF, at −78 °C → r.t.; iii, Ac₂O, NEt₃, DMAP; iv, CF₃CO₂H; v, CH₂N₂, MeOH, Et₂O; vi, C₆H₄Me₂, at 243 °C; vii, H₂, Pd/C, EtOAc; viii, LiAlH₄, Et₂O, at 0 °C, for 2 minutes, then EtOAc, and basic work-up; ix, LiN(SiMe₃)₂, at −10 °C; x, AcCl, at −78 °C; xi, Me₃SiI; xii, 6-methoxytryptophyl bromide, K₂CO₃, MeOH, heat; xiii, Hg(OAc)₂; xiv, NaBH₄, (MeO)₃C₆H₂COCl; xvi, 0.3M-KOH, MeOH, at 25 °C; xvii, DMSO, DCC, H₃PO₄; xviii, Me₂C(OH)CN, NEt₃; xix, oxalyl chloride, DMSO, then MeOH; xx, H⁺, H₂O

Scheme 19

product (101) had previously[56d] been converted into dihydrocorynantheine by van Tamelen and Hester.

Possibly the most outstanding contribution to synthesis in this sub-group during the year has been the synthesis of salts of mavacurine (102) and related alkaloids by Harley-Mason and his collaborators (Scheme 18).[57a] The synthetic approach adopted by this group (see, for example, Scheme 14) normally leads to tetracyclic intermediates with a *trans* disposition of hydrogen at C-3 and C-15. Clearly, for a successful synthesis of the mavacurine group, a *cis* relationship between these hydrogens is required. This problem was surmounted by reductive cleavage of the C-3–N_b bond in an appropriate intermediate [(103) → (104)]. The asymmetry at C-3 was therefore destroyed, ring E could now be closed [(105) → (106)], and the c/D ring junction was then re-formed by a remarkable new oxidative cyclization reaction, to give the pentacyclic product (107), in which C-3 and C-15 necessarily have the required relative configuration. Manipulation of (107) by standard methods then gave (±)-epipleiocarpamine (108), (±)-normavacurine (109), and (±)-mavacurine iodide (102). Appropriate treatment of (106) also led to a synthesis of (±)-ε_2-dihydromavacurine (110) (Scheme 18).[57a] The background to this fascinating work, and in particular to allied work on the stereochemistry of the solvolytic c/D ring-cleavage reaction on models for mavacurine synthesis, has also been discussed in detail.[57b]

In the transcript of a lecture, Zenk has reviewed[58] the enzymic synthesis of ajmalicine and its biogenetic intermediates from secologanin and tryptamine in cell clones of *Catharanthus roseus*.

Several groups of workers have reported briefly on new or modified syntheses of the yohimbane ring system.[59] The last of these communications[59e] describes syntheses of 14-isogambirtannine and related compounds.

A new synthesis of reserpine (Scheme 19)[60] makes use of a very neat synthesis of *cis*-hydroisoquinoline derivatives, *e.g.* (111), by means of a Diels–Alder/Cope rearrangement sequence. Manipulation of (111) by unexceptional methods then gives (112), which possesses the required stereochemistry in ring E. Oxidative cyclization of (112) affords 3-isoreserpinediol (113) but, unfortunately, some 'inside' isomer, originating from the cyclization of C-2 with C-21, is also obtained. The synthesis also loses some elegance in the multi-stage conversion of 3-isoreserpinediol into 3-isoreserpine (114), since, in the Swern oxidation of the C-16 aldehyde cyanhydrin by means of DMSO with oxalyl chloride as activator, the over-oxidized products (115) and (116) were obtained. However, reduction of (115) gave 3-isoreserpine (114), which has previously been converted into reserpine by four different methods.

The conclusion reached earlier[14c] that the O-desmethyltubulosine that occurs in

[57] (a) M. J. Calverley, B. J. Banks, and J. Harley-Mason, *Tetrahedron Lett.*, 1981, **22**, 1635; (b) M. J. Calverley, J. Harley-Mason, S. A. Quarrie, and P. D. Edwards, *Tetrahedron*, 1981, **37**, 1547.

[58] M. H. Zenk, *J. Nat. Prod.*, 1980, **43**, 438.

[59] (a) G. D. Pandey and K. P. Tiwari, *Synth. Commun.*, 1980, **10**, 523; (b) H. Bieraügel, H. C. Hiemstra, and U. K. Pandit, *Heterocycles*, 1981, **16**, 239; (c) Atta-ur-Rahman and M. Ghazala, *ibid.*, p. 261; (d) P. Rosenmund, W. Trommer, D. Dorn-Zachertz, and E. Ewerdwalbesloh, *Liebigs Ann. Chem.*, 1979, 1643; (e) M. Haimova, E. Alexandrova, E. Stanoeva, and C. Thal, *C. R. Hebd. Seances Acad. Sci., Ser. C*, 1980, **291**, 303.

[60] P. A. Wender, J. M. Schaus, and A. W. White, *J. Am. Chem. Soc.*, 1980, **102**, 6157.

Alangium lamarckii is 10-desmethyltubulosine, rather than the 9-desmethyl isomer (as originally proposed), has now been confirmed by the total synthesis of (±)-10-desmethyltubulosine by a conventional sequence of reactions.[61]

Sarpagine–Ajmaline–Picraline Group. Normacusine B, perakine, raucaffrinoline, peraksine, akuammiline, desacetylakuammiline, desacetyldesformoakuammiline, and picrinine have been found in the leaves of *Rauwolfia vomitoria*,[43b] and ajmaline 17-*O*-3,4,5-trimethoxybenzoate in the unripe fruits.[43c] Leaves of *R. oreogiton* contain[43a] 10-methoxyakuammine, akuammiline, desacetylakuammiline, 5-hydroxymethylakuammiline, 1,2β-dihydroakuammiline, desacetyl-1,2β-dihydro-akuammiline, picraline, picrinine, volkensine, quaternine, peraksine, normacusine B, nortetraphyllicine, and a base which is suspected to be (117). *Hazunta modesta* var.

(117)

modesta subvar. *montana*, from Madagascar, has been thoroughly studied.[62] Tabernaemontanine, dregamine, methuenine, and silicine are present in the leaves, stem bark, and root bark, vobasine and pericyclivine are also present in the leaves, and 6-oxosilicine and its 16-epimer in the stem bark and root bark. The presence of isocorymine and vobasine in the leaves and bark of *Hunteria zeylanica* has been established,[43e] and 10-methoxyvincamidine has been found in *Vinca minor*.[63a] Ajmalinol, a new alkaloid from *Rauwolfia vomitoria*, has been formulated as 11-hydroxyajmaline.[63b] Akuammidine has been isolated, together with eleven other alkaloids, from the stems of *Tabernaemontana olivacea* Müll. Arg.[63c] Pericyclivine has been found in leaves of *Catharanthus roseus*, and periformyline in the leaves of *C. trichophyllus* (Bak.) Pichon.[63d]

The importance of unequivocally identifying botanical material, and naming it correctly, complete with the authority, is well illustrated in a study of four distinct *Ervatamia* species, all of which have at some time been named *E. orientalis*, with consequent inevitable confusion.[64] The leaves of *E. orientalis* (R. Brown) Domin (≡*Tabernaemontana orientalis* R. Brown), from New Guinea, unlike the other

[61] T. Fujii, M. Ohba, A. Popelak, S. C. Pakrashi, and E. Ali, *Heterocycles*, 1980, **14**, 971.

[62] A. M. Bui, B. C. Das, and P. Potier, *Phytochemistry*, 1980, **19**, 1473.

[63] (*a*) J. Trojánek, Z. Koblicova, P. M. McCurry, J. Dadok, and L. Pijewska, *Acta Univ. Palacki, Olomuc., Fac. Med.*, 1980, **93**, 111 (*Chem. Abstr.*, 1981, **94**, 44045); (*b*) S. Siddiqui and A. Malik, *J. Chem. Soc. Pak.*, 1979, **1**, 1 (*Chem. Abstr.*, 1980, **92**, 111204); (*c*) H. Achenbach and B. Raffelsberger, *Z. Naturforsch., Teil. B*, 1980, **35**, 885; (*d*) S. Mukhopadhyay and G. A. Cordell, *J. Nat. Prod.*, 1981, **44**, 335.

[64] L. Allorge, P. Boiteau, J. Bruneton, T. Sévenet, and A. Cavé, *J. Nat. Prod.*, 1980, **43**, 514.

three species examined, contain mainly bases of the voacangine group (*q.v.*), and no trace of the ervatamine group could be found. *Ervatamia daemeliana* Domin, from Queensland, contained akuammidine, but no alkaloids of the ervatamine group, while *E. obtusiuscula* Mgf. (≡ *E. orientalis* Turrill, ≡ *Tabernaemontana orientalis* Seeman), from the New Hebrides, contained tabernaemontanine and dregamine, together with small amounts of vobasine; again, the ervatamine bases were absent. The fourth species, *E. lifuana* Boiteau (≡ *E. orientalis* Guillaumin), from the Loyalty Islands, contains tabernaemontanine and dregamine along with *iboga* bases as major alkaloids, together with smaller quantities of vobasine, and very small amounts of ervatamine and 20-*epi*-ervatamine.[64]

The presence of vincamajine in *Vinca herbacea* has been confirmed.[44c]

The stem bark of *Alstonia lanceolata* V. Heurck et Müll. Arg. has so far yielded thirteen alkaloids, including 10,11-dimethoxy-N_a-methyldesacetylpicraline 3,4,5-trimethoxybenzoate (118), 10,11-dimethoxy-N_a-methylpicraline (119), 10,11-dimethoxy-N_a-methyldesacetylpicraline (120), picraline, pseudoakuammigine (121), and two new alkaloids, lanceomigine and its N_b-oxide.[65a] Lanceomigine also

Pseudoakuammigine (121) R = H
17-Hydroxy-
pseudoakuammigine (122) R OH

	R^1	R^2
(118)	MeO—C₆H₂(OMe)₂—CO	CO_2Me
(119)	Ac	CO_2Me
(120)	H	CO_2Me

Lanceomigine (123)

(124)

[65] (*a*) J. Vercauteren, G. Massiot, T. Sévenet, B. Richard, V. Lobjois, L. Le Men-Olivier, and J. Lévy, *Phytochemistry*, 1981, **20**, 1411; (*b*) J. Vercauteren, G. Massiot, L. Le Men-Olivier, and J. Lévy, *Tetrahedron Lett.*, 1981, **22**, 2871.

occurs in *Hunteria zeylanica* and, together with 17-hydroxypseudoakuammigine (122), in the seeds of *H. congolana*.[65b] The structure of 17-hydroxypseudo-akuammigine was deduced[65a] from a comparison of its ^{13}C n.m.r. spectrum with that of pseudoakuammigine, and confirmed by the reduction of (122), by means of triethylsilane and trifluoroacetic acid, to pseudoakuammigine.

Lanceomigine (123) is an isomer of 17-hydroxypseudoakuammigine, and, in similar fashion, can be reduced to a desoxy-derivative (124), whose structure was established by X-ray crystal structure analysis.[65b] If it is assumed that the additional oxygen atom in lanceomigine (123) is attached to C-17, this new alkaloid must be the alternative carbinolamine-ether-hemiacetal to 17-hydroxypseudoakuammigine (122), a conclusion that was confirmed by the quantitative conversion of (122) into lanceomigine (123) by treatment with aqueous (3M) hydrochloric acid.[65] Finally, it should be noted that lanceomigine is not regarded as an artifact, since it can be isolated from *A. lanceolata* in the total absence of acid.[65b]

Raucubaine (125), an alkaloid of *Rauwolfia salicifolia* Griseb., which is endemic to Cuba, has been described[66a,b] as having a novel ring system, but it surely contains the same ring system as quaternoline (126),[66c] and the two alkaloids may even be identical.

Raucubaine (125)

Quaternoline (126)

Koumine (127)

Lanciferine (128) R = CH$_2$OCOCH=CHPh
(129) R = H

[66] (a) J. P. Kutney, J. Trotter, R. A. Pauptit, B. R. Worth, and P. Sierra, *Heterocycles*, 1980, **14**, 1309; (b) R. A. Pauptit and J. Trotter, *Can. J. Chem.*, 1981, **59**, 1007; (c) S. Mamatas-Kalamaras, T. Sévenet, C. Thal, and P. Potier, *Phytochemistry*, 1975, **14**, 1849.

Koumine (127),[67a] an alkaloid of *Gelsemium elegans* Benth. first isolated[67b] in 1931, is included in this group because it contains a 5,16 bond as well as the novel 7,20 bond; its structure was elucidated by the *X*-ray method.

The structure of lanciferine (128), from *Alstonia boulindaensis* Boiteau, has been confirmed[68] by comparison of the spectra of descinnamoyldesformolanciferine (129) with those of the base (132), prepared from desacetyldesformopicraline (130), as shown in Scheme 20. Formylation of (130) gave an *NN*-diformyl derivative (131), with concomitant fission of the N_b,5 bond and lactol formation. Oxidation to the lactone, selective removal of the N_a-formyl group (presumably assisted by the lactone function), and reduction *via* the thioamide gave the desired N_b-methyl derivative (132). Unfortunately, the vital epoxidation of the 19,20 double-bond to give the lanciferine derivative (129) could not be achieved; hence spectrographic comparison of (129) and (132) was undertaken. This confirmed the structure and stereochemistry of (129), and therefore of lanciferine (128), with the exception of the configuration at C-19.

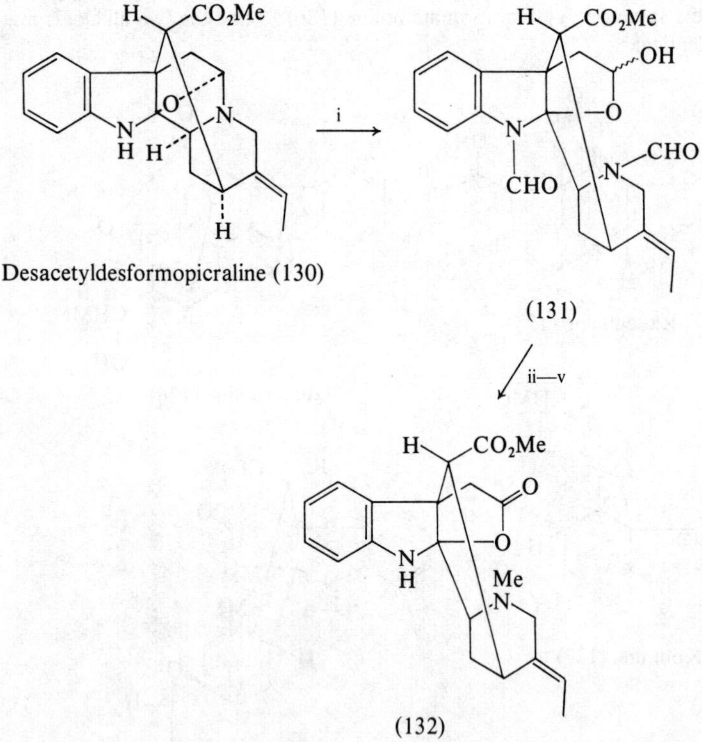

Desacetyldesformopicraline (130)

(131)

ii—v

(132)

Reagents: i, HCO$_2$H, Ac$_2$O; ii, Jones' reagent; iii, MeOH, NaOH; iv, P$_2$S$_5$, PhH; v, NaBH$_4$, MeOH

Scheme 20

[67] (a) F. Khuong-Huu, A. Chiaroni, and C. Riche, *Tetrahedron Lett.*, 1981, **22**, 733; (b) T. Q. Chou, *Chin. J. Physiol.*, 1931, **5**, 345 (*Chem. Abstr.*, 1932, **26**, p. 806).
[68] G. Lewin and J. Poisson, *Bull. Soc. Chim. Fr.*, Part 2, 1980, 400.

The remaining uncertainties relating to the stereochemistry of isocorymine, erinine, erinicine, and eripine have been resolved by a detailed examination of their n.m.r. spectra.[69] Isocorymine, for which the stereochemistry at C-16 (and therefore at C-3) and C-17 was previously unknown, as well as the geometry about the 19,20 double-bond, has now been shown to have the complete stereochemistry expressed in (133). Similarly, erinine is (134) and erinicine is a 19,20-dihydroerinine, of structure (135). The pentacyclic base eripine must have the same configuration at

	R^1	R^2
Isocorymine (133)	OH	H_2
Erinine (134)	H	O

Erinicine (135)

C-16, and the *E* configuration at the double-bond, in view of the earlier correlation with erinine.

The absolute configuration of (+)-macroline, obtained by degradation of the dimeric alkaloid villalstonine, follows[70a] from its transformation into a mixture of the tetracyclic base (138) and its C-20 epimer. Base (138) was also obtained by degradation of ajmaline (139), as outlined in Scheme 21; the C-20 epimer of (138) was similarly obtained from isoajmaline, which is epimeric with ajmaline at both C-20 and C-21.

The partial synthesis[70b] of macroline (136) from normacusine B (140) was inspired by its postulated biosynthesis from a sarpagine-type precursor. Nor-macusine B (140), prepared by a previously published route from perivine, was protected at the primary alcohol group and then methylated on N_a. Direct epoxidation of the product (141) failed; however, osmylation gave the desired diol together with the related oxindole obtained by simultaneous oxidation of the indole double-bond, followed by rearrangement. Conversion of these diols into the related epoxides gave a mixture of (142) and (143), from which the desired epoxide (142) could be separated satisfactorily by fractional crystallization.

[69] F. Heatley, D. I. Bishop, and J. A. Joule, *J. Chem. Soc., Perkin Trans. 2*, 1981, 725.
[70] (a) G. Neukomm, E. Kletzhändler, and M. Hesse, *Helv. Chim. Acta*, 1981, **64**, 90; (b) R. W. Esmond and P. W. Le Quesne, *J. Am. Chem. Soc.*, 1980, **102**, 7117.

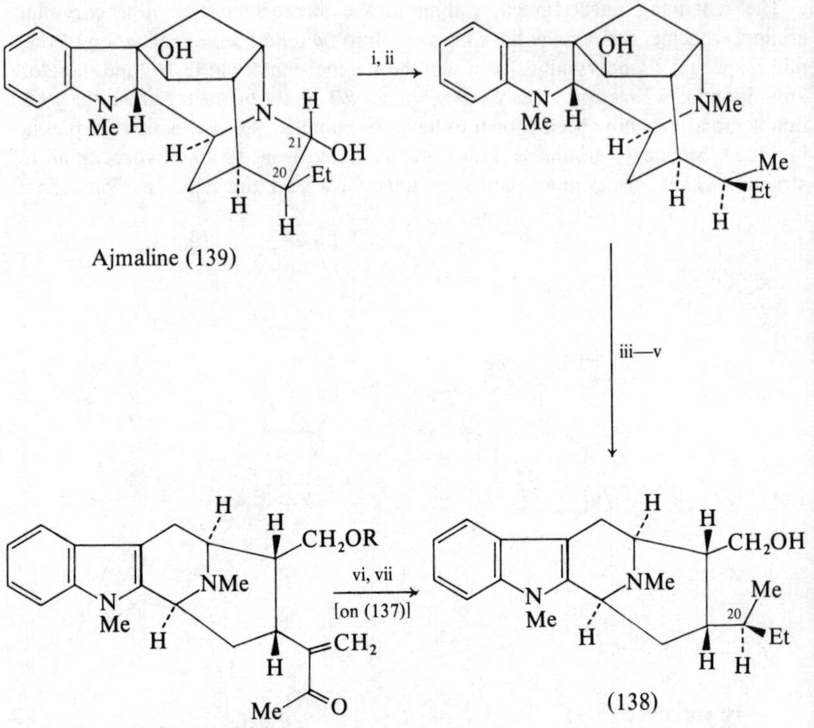

Ajmaline (139)

(+)-Macroline (136) R = H
(137 R = Ac

(138)

Reagents: i, NH_2NH_2; ii, MeI; iii, $Pb(OAc)_4$; iv, NaOMe; v, $NaBH_4$; vi, H_2, PtO_2, MeOH; vii, NH_2NH_2, KOH, $(HOCH_2CH_2)_2O$

Scheme 21

Isomerization then gave a β-amino-ketone (144), whose methosulphate readily suffered β-elimination and loss of the protecting group, with formation of the unsaturated amino-ketone identified as macroline (136) (Scheme 22).[70b]

The oxidation of dregamine (145) by trifluoroperacetic acid also results in rearrangement, the product being the oxindole analogue (146).[71]

An X-ray diffraction study[72] of the methanol solvate of ervitsine has confirmed the structure and stereochemistry previously deduced.

A description[73] of the conversion of gardnerine into ochropine duplicates one published earlier.[14d]

[71] C. Ouannès and C. Thal, *Tetrahedron Lett.*, 1981, **22**, 951.
[72] C. Riche, *Acta Crystallogr.*, Sect. B, 1979, **35**, 2738.
[73] S. Sakai, Y. Yamamoto, and S. Hasegawa, *Chem. Pharm. Bull.*, 1980, **28**, 3454.

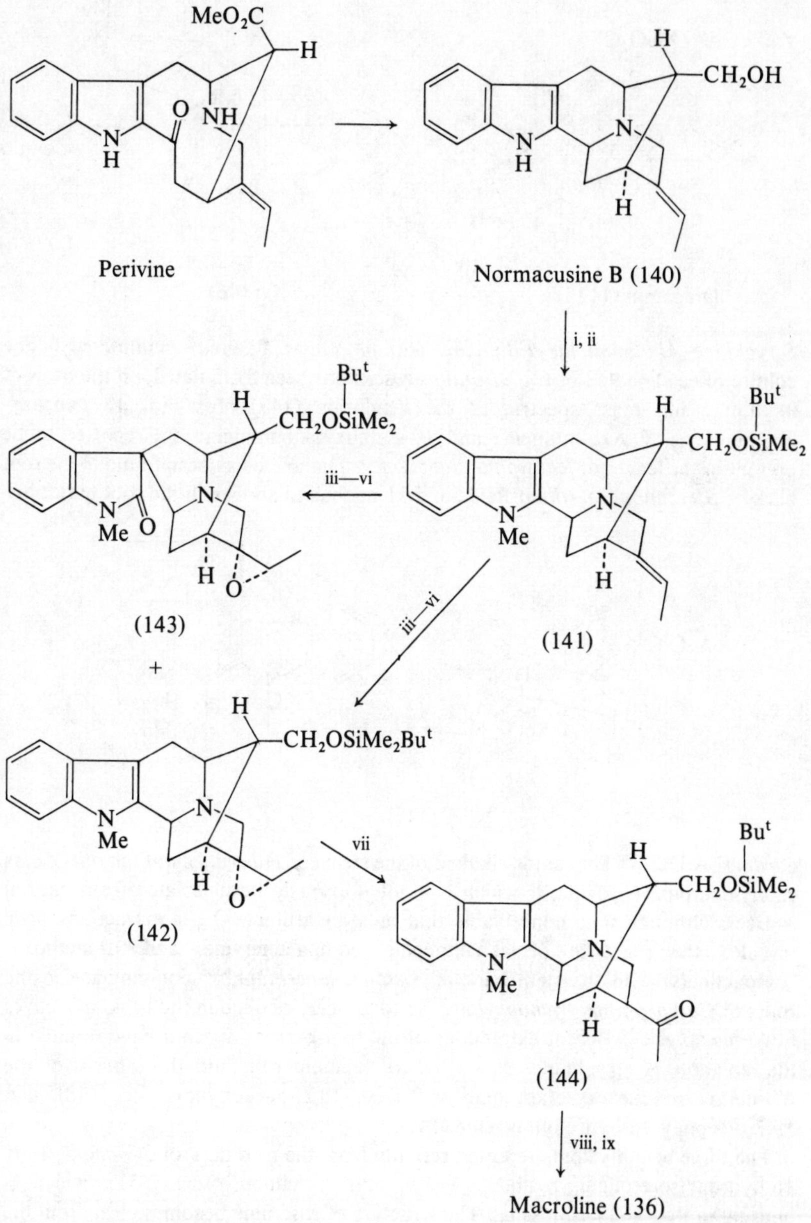

Perivine → Normacusine B (140)

i, ii

(143) + (142)

iii—vi

(141)

iii—vi

vii

(144)

viii, ix

Macroline (136)

Reagents: i, Bu^{t}Me$_2$SiCl, imidazole, DMF; ii, KH, NH$_3$, THF, MeI; iii, OsO$_4$, THF, pyridine; iv, Na$_2$S$_2$O$_5$; v, TsCl, pyridine; vi, NaH, THF; vii, MgBr$_2$, Et$_2$O, PhH, heat; viii, Me$_2$SO$_4$, K$_2$CO$_3$, PhH, heat; ix, Bu$_4$N$^+$ F$^-$, THF, H$_2$O

Scheme 22

Dregamine (145)   (146)

Strychnine–Akuammicine–Ellipticine Group. A new alkaloid encountered in the culture of cell line 943 of *Catharanthus roseus* has been formulated, on the basis of its n.m.r. and mass spectra, as the β-xyloside (147) of 9- (or 12-)hydroxy-akuammicine.[42a] Akuammicine and 10-methoxyakuammicine are reported to be present in the leaves of *Rauwolfia oreogiton*.[43a] Diaboline has been found in the root bark of *Strychnos jobertiana* Baill., and 11-methoxydiaboline in the root bark of *S.*

(147)

gardneri A.DC.[40a] The major alkaloid of the stems of *Tabernaemontana olivacea* is condylocarpine N_b-oxide,[63c] which has not previously been isolated from natural sources, although it is conceivable that it is an artifact. Other extractions have revealed the presence of akuammicine, compactinervine, and 10-methoxy-compactinervine in the stem bark of *Alstonia lanceolata*,[65a] akuammicine in the roots of *Catharanthus trichophyllus*,[63d] and *O*-acetylretuline in the bark and leaves of *S. henningsii*.[74] Recent extractions of the root bark of *S. icaja* have resulted in the isolation of strychnine, N_b-methylstrychninium salt, and three bases of the N_b-methyl *sec*-pseudo-series, namely icajine, 19,20-α-epoxynovacine (148), and 19,20-α-epoxy-15-hydroxynovacine (149).[75]

The three new alkaloids reported recently from the root bark of *S. variabilis* are 16-hydroxyisoretulinal, rosibiline (150), and strychnopivotine (151), which is unusual in that C-17 is missing. The structure of rosibiline becomes clear from the formation of desacetylretuline on treatment with acid, and is confirmed by the preparation of rosibiline by the reaction of desacetylretuline with formaldehyde in the presence of acetic acid.[75]

[74] L. Angenot and M. Tits, *Planta Med.*, 1981, **41**, 240.
[75] K. Kambu, C. Coune, and L. Angenot, *Planta Med.*, 1979, **37**, 161.

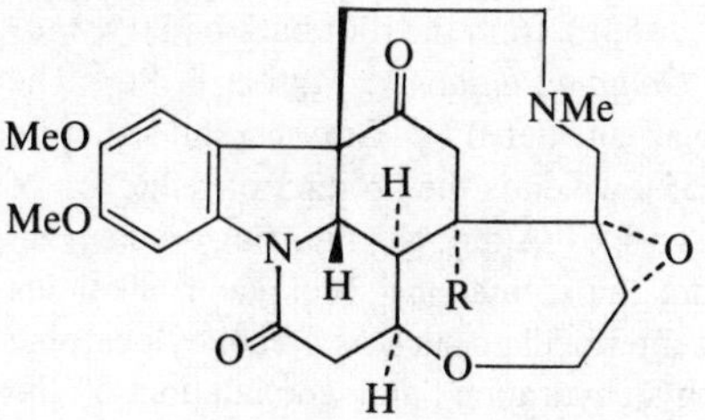

(148) R = H
(149) R = OH

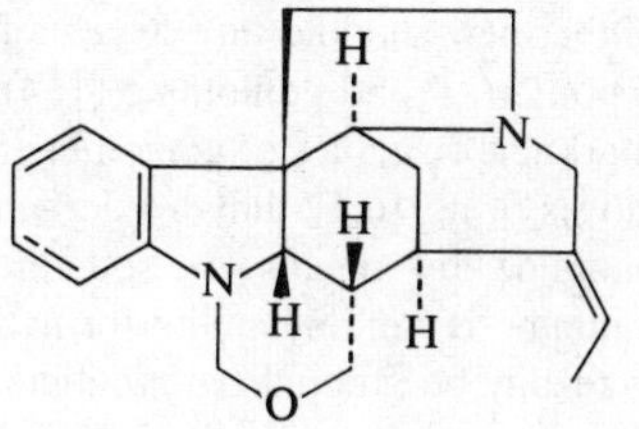

Rosibiline (150)

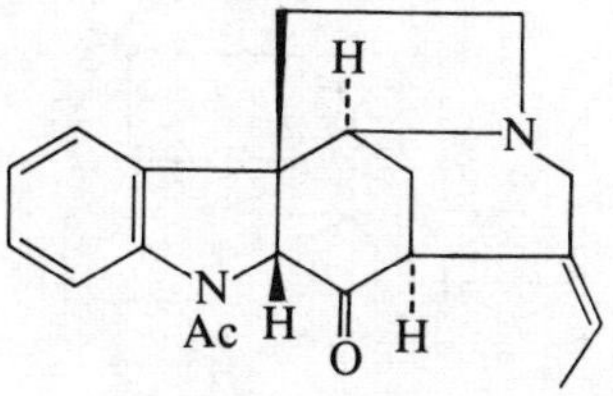

Strychnopivotine (151)

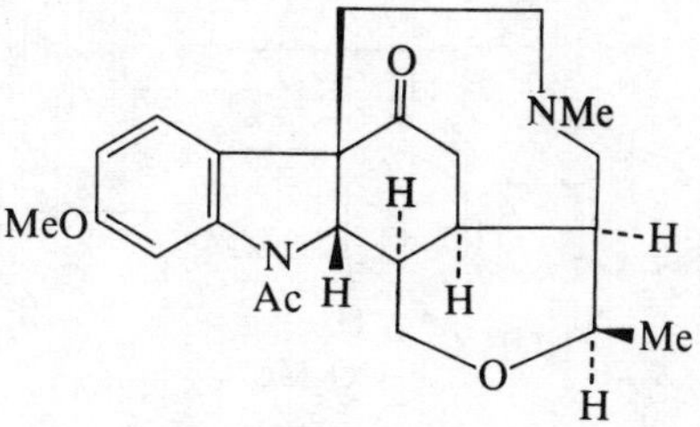

11-Methoxystrychnofendlerine (152)

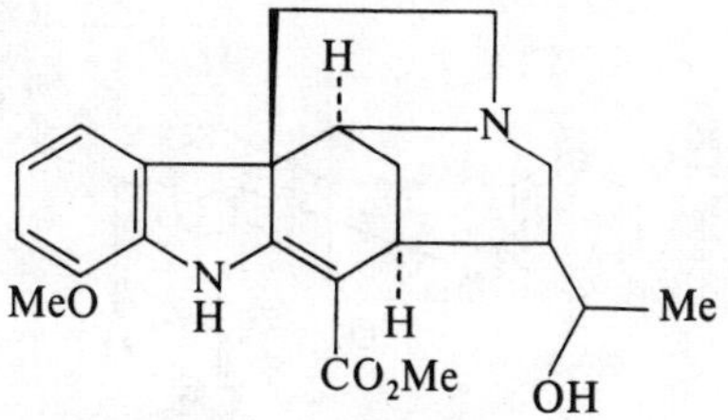

Scholarine (153a)

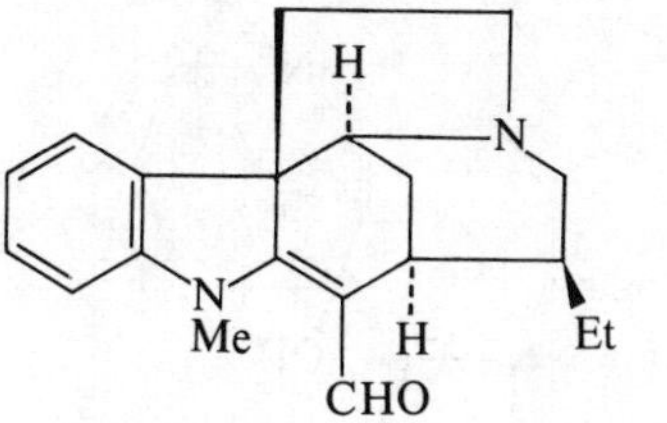

Strychnofluorine (153b)

Details of the isolation and structure determination of strychnofendlerine, 12-hydroxy-11-methoxystrychnofendlerine, N_a-acetylstrychnosplendine, and N_a-acetyl-12-hydroxy-11-methoxystrychnosplendine, from the stem bark of *S. fendleri* Sprague et Sandwith, have been published, together with those of the isolation of four minor alkaloids.[76] These were identified as diaboline, spermostrychnine, and henningsamine; the fourth, 11-methoxystrychnofendlerine (152), is new. Another new alkaloid, scholarine, from the leaves of *Alstonia scholaris* R. Br., proves to be 12-hydroxyechitamidine (153a);[77] unlike most β-anilino-acrylate alkaloids, which exhibit very high optical rotatory power, this one is optically inactive, and hence is only the second racemic alkaloid of this group to be encountered, the other one being (±)-akuammicine.

[76] C. Galeffi and G. B. Marini-Bettòlo, *Gazz. Chim. Ital.*, 1980, **110**, 81.
[77] A. Banerji and A. K. Siddhanta, *Phytochemistry*, 1981, **20**, 540.

Other new alkaloids are strychnofluorine (153b), from the root bark of *Strychnos gossweileri*,[41b] and goniomine (154), from *Gonioma malagasy* Mgf. et P. Bt.[78] The remarkable structure of goniomine (154) was elucidated by *X*-ray crystal-structure analysis of its 16,17-dihydro-derivative (155), which is the product of reduction of goniomine by means of sodium borohydride. Although the ring system in goniomine is unfamiliar, its formation from the tryptamine-secologanin alkaloids can readily be rationalized *via* a biogenetic intermediate such as precondylocarpine (156). Decarboxylative loss of C-22 with dehydration, and epoxidation of the ethylidene group in (156), are unexceptional stages which would lead to (157). Hydrolysis of the imine group, followed by re-cyclization by attack of N_a on the epoxide function, then leads to goniomine (154).

The leaves and stems of *Mostuea brunonis* Didr. var *brunonis* forma *augustifolia*

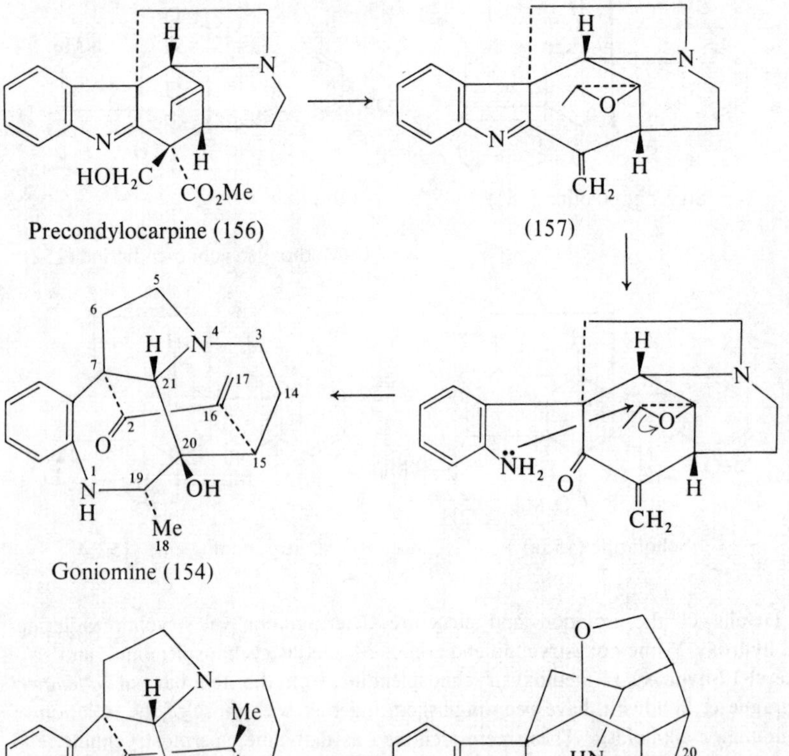

Precondylocarpine (156)   (157)

Goniomine (154)

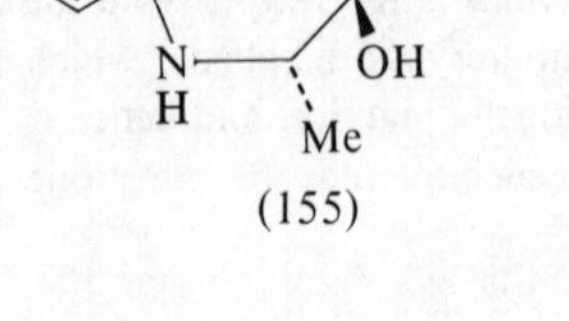

(155)

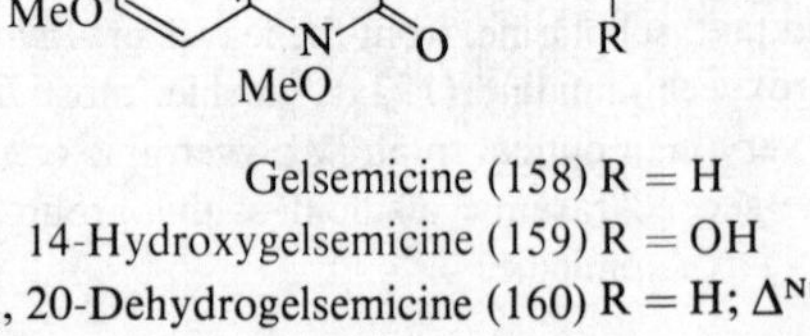

Gelsemicine (158) R = H
14-Hydroxygelsemicine (159) R = OH
N_b, 20-Dehydrogelsemicine (160) R = H; $\Delta^{N_b, 20}$

[78] A. Chiaroni, L. Randriambola, C. Riche, and H.-P. Husson, *J. Am. Chem. Soc.*, 1980, **102**, 5920.

contain[40f] gelsemicine (158), 14-hydroxygelsemicine (159), and a new base, N_b,20-dehydrogelsemicine (160).

The alkaloids of *Hazunta modesta* var. *modesta* subvar. *montana* are a mixture of sarpagine, *strychnos*, pyridocarbazole, aspidospermane, and *iboga* types. Those that belong to this group that were found in this species include vallesamine in the leaves, apparicine in the leaves, stems, and roots, and 3,14-dihydroellipticine in the stems and roots.[62] Another alkaloid from the ellipticine group encountered recently is 17-oxoellipticine (161), a constituent of the bark of *Strychnos dinklagei* Gilg.[79] The structure of (161) becomes clear from an examination of its spectra, particularly its n.m.r. spectrum, and is confirmed by its preparation from ellipticine by oxidation with Jones' reagent.

(161)

Ellipticine is the major alkaloid of the trunk bark of *Ochrosia moorei* F. von Muell., in which it occurs in association with 27 other alkaloids, including 10-methoxyellipticine, ellipticine N_b-oxide, and 3,14-dihydroellipticine.[43d] A point of chemotaxonomic interest is that the genera *Ochrosia* and *Neisosperma* can be distinguished on the basis of their ellipticine/10-methoxyellipticine content; one or both of these alkaloids are present in *Ochrosia* species, but absent from *Neisosperma* species.

Ellipticine, olivacine, and a new alkaloid (162), based on an angular pyridocarbazole skeleton, have been found in the bark of *Aspidosperma gilbertii* A. P. Duarte.[80] The structure of 5-ethyl-2-methyl-11*H*-pyrido[3,4-*a*]carbazolium hydroxide (162) was deduced from its u.v. and n.m.r. spectra, in conjunction with biogenetic arguments. Of the possible structures consistent with its spectra, only (162) is plausible on biogenetic grounds, as shown in Scheme 23, in which (162) is considered to arise *via* a minor variant on the pathway to ellipticine and olivacine, starting essentially from stemmadenine (163). Structure (162) was confirmed by a very brief, direct synthesis (Scheme 23).[80]

In recent years, considerable attention has been paid to the ^{13}C n.m.r. spectra of strychnine and brucine. Two recent papers[81a,b] were prompted by errors and inconsistencies in earlier contributions on this topic; it is to be hoped that the assignments now made are completely reliable.

In the proton n.m.r. spectrum of apparicine (164), spin–lattice relaxation times and selected steady-state and transient n.O.e. enhancements have been studied.[82] In

[79] S. Michel, F. Tillequin, and M. Koch, *Tetrahedron Lett.*, 1980, **21**, 4027.
[80] E. C. Miranda, C. H. Brieskorn, and S. Blechert, *Chem. Ber.*, 1980, **113**, 3245.
[81] (*a*) R. Verpoorte, *J. Pharm. Sci.*, 1980, **69**, 865; (*b*) G. E. Martin, *ibid.*, 1981, **70**, 81.
[82] F. Heatley, L. Akhter, and R. T. Brown, *J. Chem. Soc., Perkin Trans. 2*, 1980, 919.

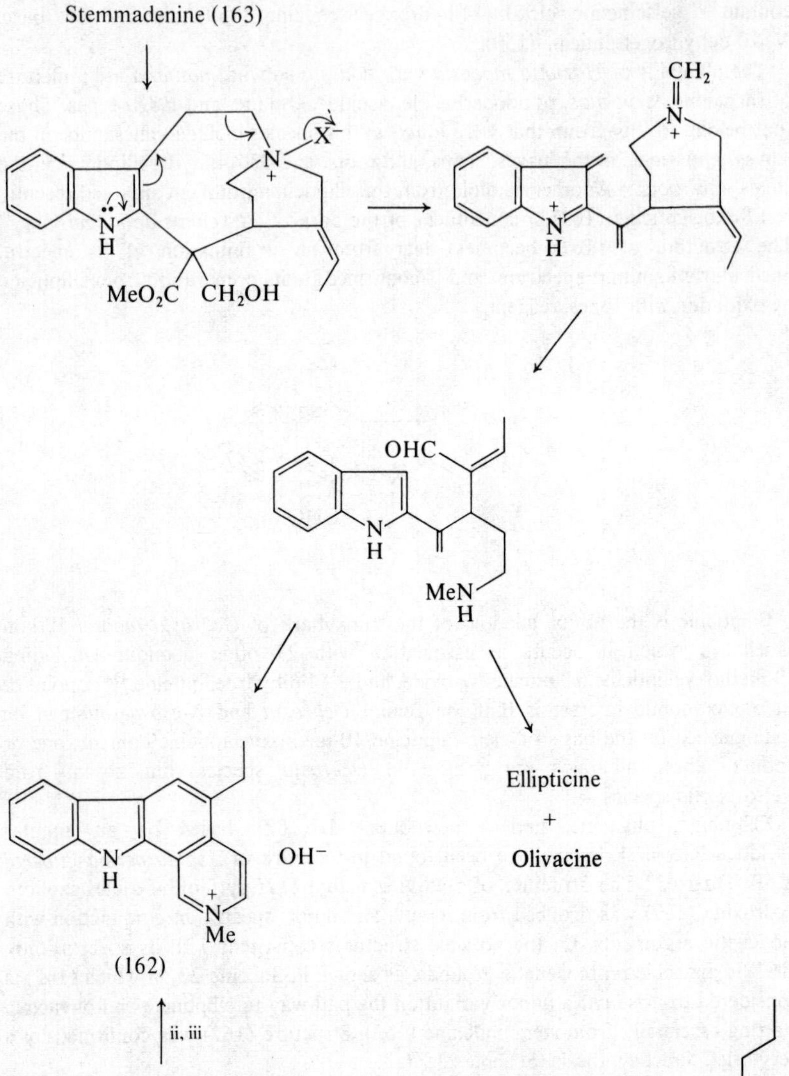

Stemmadenine (163)

Ellipticine
+
Olivacine

(162)

ii, iii

I⁻

Reagents: i, Piperidine; ii, MeOH, $h\nu$; iii, Al$_2$O$_3$

Scheme 23

conjunction with spin–spin coupling constants, the results permit assignment of the molecular conformation; in summary, the piperidine ring is a boat, and the eight-membered ring is also boat-shaped, but strained so that the 16,17 double-bond is inclined ~30° with respect to the indole plane, the latter bisecting the angle subtended by the hydrogen atoms attached to C-6.

The sulphonation of strychnine by means of sulphur dioxide and manganese dioxide in the presence of water (the Leuchs sulphonation) has again been examined.[83] The major product is the zwitterionic 15-sulphonic acid; minor products have been identified as the zwitterionic 19- and 21-sulphonic acids.

Two nine-stage procedures have been developed[84a] for the conversion of strychnine into vomicine (165); these serve to confirm the structure, on which little work appears to have been done since the structure determination by Bailey and Robinson.[84b]

Apparicine (164)   Vomicine (165)

The microbiological transformation of ellipticine by means of *Aspergillus alliaceus* has been investigated;[85] the principal product is 10-hydroxyellipticine,* which is the same as that observed in the metabolism of ellipticine in mammalian liver, but some 11-hydroxyellipticine is also obtained.

Besselièvre and Husson have published full details of their syntheses of ellipticine, olivacine, and guatambuine.[86] A new approach to the synthesis of ellipticine[87] involves construction of an appropriately substituted anilino-isoquinoline, *e.g.* (166); closure of the five-membered ring then constitutes the last stage (Scheme 24). Other synthetic work in this area includes syntheses of 16-desmethyl-16-methoxyellipticine[88] and 11-hydroxy- and 11-methoxy-ellipticines.[89]

A new synthesis[90] of olivacine (167) (Scheme 25) proceeds *via* an intermediate

---

* The numbering used throughout this review is the biogenetic numbering; this may differ from the numbering used in the original literature.

[83] J. T. Edward, P. G. Farrell, S. A. Samad, R. Wojtowski, and S. C. Wong, *Can. J. Chem.*, 1980, **58**, 2380.

[84] (*a*) P. Rosenmund, M. P. Schmitt, and H. Franke, *Liebigs Ann. Chem.*, 1980, 895; (*b*) A. S. Bailey and Sir Robert Robinson, *Nature* (*London*), 1948, **161**, 433.

[85] M. M. Chien and J. P. Rosazza, *Drug Metab. Dispos.*, 1979, **7**, 211; *Appl. Environ. Microbiol.*, 1980, **40**, 741.

[86] R. Besselièvre and H.-P. Husson, *Tetrahedron*, 1981, **37**, Suppl. 1, p. 241.

[87] R. B. Miller and T. Moock, *Tetrahedron Lett.*, 1980, **21**, 3319.

[88] Y. Oikawa, M. Tanaka, H. Hirasawa, and O. Yonemitsu, *Heterocycles*, 1981, **15**, 207.

[89] D. Dolman and M. Sainsbury, *Tetrahedron Lett.*, 1981, **22**, 2119.

[90] T. Naito, N. Iida, and I. Ninomiya, *J. Chem. Soc., Chem. Commun.*, 1981, 44.

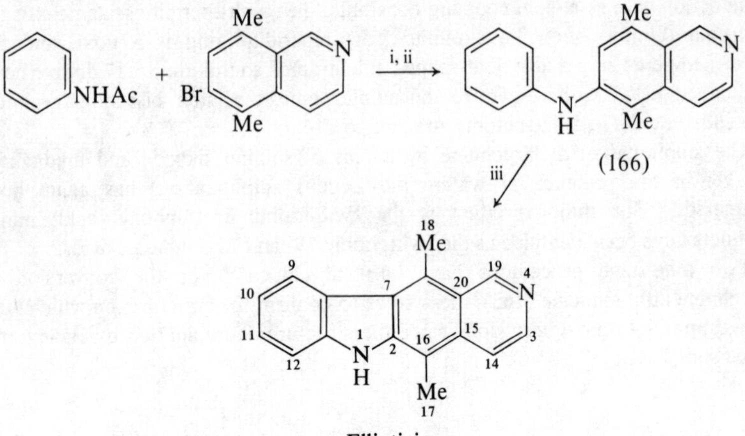

Ellipticine

Reagents: i, Cu, K$_2$CO$_3$, heat; ii, 3 M-HCl; iii, Pd(OAc)$_2$, CF$_3$CO$_2$H, AcOH

Scheme 24

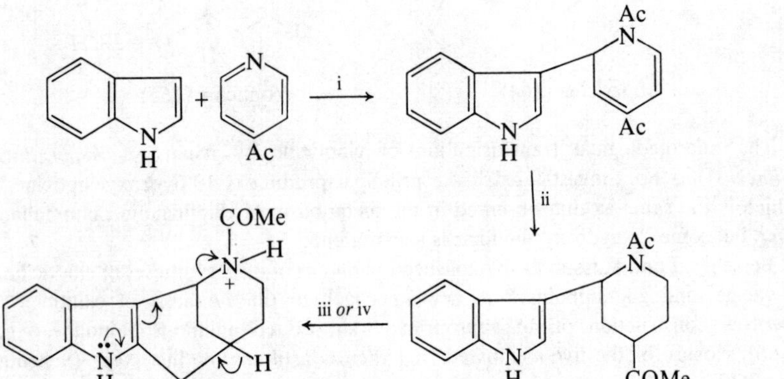

Olivacine (167)

Reagents: i, AcCl, CH$_2$Cl$_2$; ii, H$_2$, PtO$_2$; iii, TsOH, CHCl$_3$; iv, BF$_3$·Et$_2$O, CHCl$_3$; v, POCl$_3$; vi, Pd/C, heat

Scheme 25

(168) (not isolated), containing the uleine ring system, and therefore has potential for the synthesis of uleine itself. Acylative condensation of a 4-substituted pyridine with indole, followed by hydrogenation, gives a piperidylindole derivative (169), suitably functionalized to allow cyclization to the intermediate (168). Fragmentation of (168) then gives rise to a substituted carbazole (170), which can be converted into olivacine by conventional stages. This synthesis is essentially a variation on Husson's approach,[86] and its potential has recently been realized in a synthesis of (+)-20-*epi*-uleine (171) by Husson and his collaborators.[91] 2-Cyano-Δ^3-piperidines, such as (172), are stable equivalents of *N*-methyl-5,6-

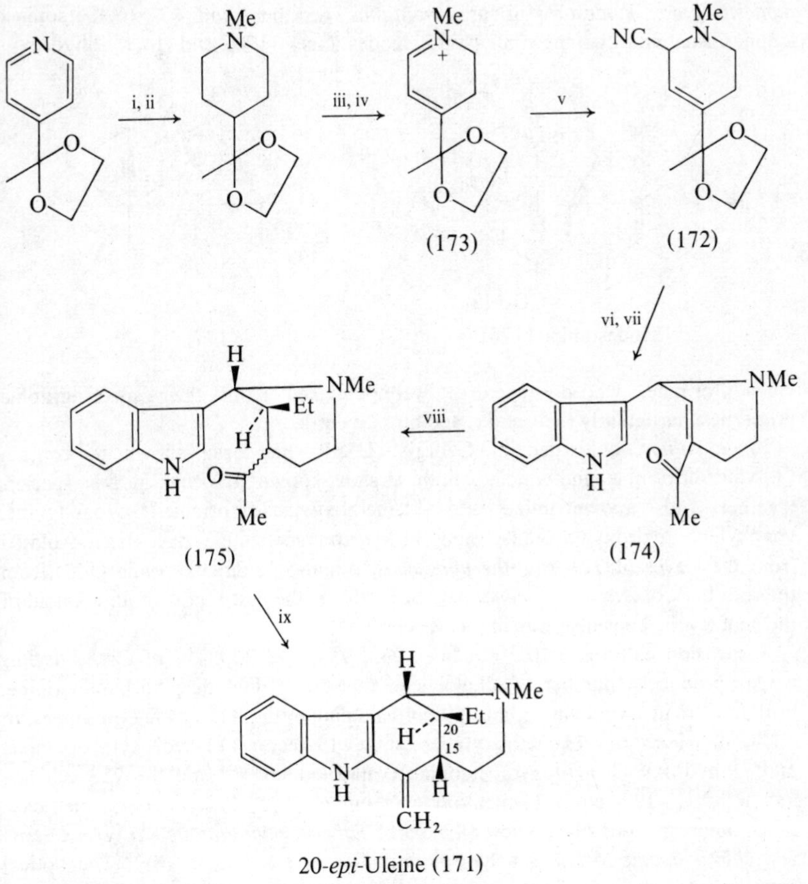

20-*epi*-Uleine (171)

Reagents: i, MeI, MeCN; ii, NaBH$_4$, MeOH; iii, *m*-ClC$_6$H$_4$CO$_3$H, CH$_2$Cl$_2$; iv, (CF$_3$CO)$_2$O, CH$_2$Cl$_2$; v, KCN, H$_2$O, at pH 4; vi, indole, AcOH, H$_2$O; vii, 10% HCl–H$_2$O; viii, EtMgBr, CuCl, THF, at 0 °C; ix, TsOH, CH$_2$Cl$_2$, N$_2$, heat

Scheme 26

[91] M. Harris, R. Besselièvre, D. S. Grierson, and H.-P. Husson, *Tetrahedron Lett.*, 1981, **22**, 331.

dihydropyridinium ions (173), which are required for condensation with indole. This reaction, followed by ketal hydrolysis, produced an indolyl-Δ^3-piperidine ketone (174), into which the required ethyl group could be introduced by 1,4-addition of an organocopper reagent. Finally, cyclization of the product (175) gave (±)-20-*epi*-uleine, fragmentation to a carbazole derivative being avoided under the mild conditions employed (Scheme 26).

Aspidospermine–Aspidofractine–Eburnamine Group. Tabersonine and venalstonidine are among the alkaloids recently extracted from the aerial parts of *Melodinus celastroides*.[40g] Tabersonine also occurs[62] in the leaves of *Hazunta modesta* var. *modesta* subvar. *montana*, together with 3-oxotabersonine, lochnericine, and two new alkaloids, modestanine (176) and 14,15-dihydroxy-

Modestanine (176)   (177)

vincadifformine, whose structures were deduced from their spectrographic properties, particularly their n.m.r. and mass spectra.

Vinca minor, cultivated in Georgia, USSR, has been shown to contain (±)-vincadifformine and vincine, both already known to occur in this species, together with apovincamine and 11-methoxyvincadifformine.[92] Voaphylline, voaphylline hydroxyindolenine, and 11-hydroxytabersonine have been isolated from the leaves of *Tabernanthe pubescens*,[43f] and 5,22-dioxokopsane (177) from the root bark of *Alstonia venenata* R. Br.;[93] this is the first report of an alkaloid of the heptacyclic kopsine group in this species.

Suspension cultures of the cell lines 943, 953, and 200 GW of *Catharanthus roseus* produce a number of alkaloids of this group, including hörhammerinine, hörhammericine, vindolinine, and 19-*epi*-vindolinine; the 943 cell line produces, in addition, 19-acetoxy-11-methoxytabersonine, 19-acetoxy-11-hydroxytabersonine, and 19-hydroxy-11-methoxytabersonine (vandrikidine),[42a,b] and the 953 cell line lochnericine, 19-acetoxy-11-methoxytabersonine, and 19-hydroxy-11-methoxytabersonine.[42b] Apodinine, a new alkaloid of *Tabernaemontana apoda* Wr. ex Sauv. (*Peschiera apoda* Mgf.), is a hydroxy-apodine of structure (178).[94] The optical rotation of apodinine was not recorded; hence it is not known to which stereochemical series [(+)- or (−)-vincadifformine] it belongs, and there is also no proposal as yet concerning the configuration at C-14.

19-*epi*-(+)-Echitoveniline (179), $[\alpha_D] = +462°$ (EtOH), is a new alkaloid from

[92] Z. V. Robakidze, M. M. Mujiri, V. Yu. Vachnadze, and K. S. Mujiri, *Khim. Prir. Soedin.*, 1980, 735.
[93] A. Chatterjee, D. J. Roy, and S. Mukhopadhyay, *Indian J. Chem., Sect. B*, 1979, **17**, 651.
[94] R. I. Lores, *Rev. CENIC, Cienc. Fis.*, 1979, **10**, 357 (*Chem. Abstr.*, 1981, **94**, 136 123).

Apodinine (178)

19-*epi*-(+)-Echitoveniline (179) R = —CO⟨benzene⟩OMe (OMe, OMe, OMe)

(+)-(19R)-Minovincinine (182) R = H
Echitovenine (183) R = Ac

(−)-Echitoveniline (180)

(19R)-19-Hydroxyquebrachamine (181)

the leaves of *Alstonia venenata* R. Br.[95] The u.v. spectrum of (179) indicates that it is composed of β-anilino-acrylate and 3,4,5-trimethoxybenzoate chromophores, and hence the alkaloid clearly bears a close structural resemblance to echitoveniline (180), which occurs in the fruits of the same plant. The high positive optical rotation of (179) indicates that it is related to (+)-vincadifformine, whereas echitoveniline belongs to the (−)-vincadifformine series. The two alkaloids, however, are not enantiomers, and must therefore have the same configuration at C-19. This was confirmed by degradation (by hydrolysis, decarboxylation, and reduction by NaBH₄ in MeOH) of (179) and (180), which gave the diastereoisomeric 19-hydroxyquebrachamine derivatives (181) and its C-20 epimer, respectively. Similarly, methanolysis of 19-*epi*-(+)-echitoveniline gave a base (182) which exhibited spectral properties closely similar to those of (−)-(19R)-minovincinine, prepared by methanolysis of (−)-echitoveniline; however, the two bases are not enantiomers, because, although they exhibit Cotton effects of opposite sign, the c.d. curves are not enantiomeric. Since the absolute stereochemistry of the parent vincadifformine skeleton dictates the sign of the Cotton effect, the two bases must be based on enantiomeric vincadifformine skeletons but possess the same (*i.e.* the R) configuration at C-19. Hence the new alkaloid is 19-*epi*-(+)-echitoveniline (179), and the identification of the methanolysis product (182) and its O-acetyl derivative (183) with the (+)-minovincinine and echitovenine also isolated from the fruits of *A. venenata* establishes the R configuration at C-19 in both these alkaloids.[95]

[95] P. L. Majumder, S. Joardar, B. N. Dinda, D. Bandyopadhyay, S. Joardar, and A. Basu, *Tetrahedron*, 1981, **37**, 1243.

The stem and root bark of *Hunteria elliottii* contain quebrachamine, vinca-difformine, aspidofractinine, kopsinine, eburnamine, isoeburnamine, eburnamenine, eburnamonine, and *O*-ethyleburnamine,[44e] and leaves and bark of *H. zeylanica* contain tuboxenine, eburnamine, isoeburnamine, eburnamenine, *O*-methyl-eburnamine, *O*-ethyleburnamine, and *O*-methylisoeburnamine.[43e] Δ^{14}-Vincamine has been found in the seeds of *Amsonia angustifolia*.[43g] The leaves and roots of *Vinca pusilla* have yielded lochnericine, vindorosine, and a new alkaloid, vincapusine (184), which is a desmethoxyvincarodine.[96] A closely related base, desmethoxy-carbonylvincapusine (185), is one of two minor alkaloids recently extracted from the leaves of *Voacanga africana* Stapf;[97] the second, also unnamed, is the epoxide (186) of a degraded aspidospermane base in which C-5 and C-6 of the original

Vincapusine (184) R = CO₂Me
(185) R = H

(186)

tryptamine ethanamine chain have been lost; this is the first recorded occurrence of a base having this ring system.

X-Ray data on vincamine 2-oxoglutarate have been reported[98a] and the structure of 10,12-dichloro-2,16-dihydro-16-hydroxy-2-methoxytabersonine (187) has been confirmed,[98b] also by the *X*-ray method.

Some 400 MHz n.m.r. data have been reported[99a] for 11-methoxytabersonine, vandrikidine, hazuntinine, and vandrikine. This study has established that C-19 in vandrikidine has the *R* configuration, and that Alkaloid M, isolated[99b] from the root bark of *Craspidospermum verticillatum*, is 19-hydroxyvandrikine (188).

(187)   (188)

[96] A. K. Mitra, A. Patra, and A. K. Mukhopadhyay, *Phytochemistry*, 1981, **20**, 865.
[97] N. Kunesch, J. Ardisson, J. Poisson, T. D. J. Halls, and E. Wenkert, *Tetrahedron Lett.*, 1981, **22**, 1981.
[98] (a) J. Fayos and M. Martinez-Ripoll, *Acta Crystallogr.*, Sect. B, 1981, **37**, 760; (b) J. Lamotte, L. Dupont, O. Dideberg, and G. Lewin, *ibid.*, 1980, **36**, 196.
[99] (a) M. Lounasmaa and S. K. Kan, *Acta Chem. Scand.*, Sect. B, 1980, **34**, 379; (b) C. Kan, A. Ahond, B. Mompon, and P. Potier, unpublished work, reported in ref. 99a.

The circular dichroism of alkaloids of the vincamine group has been discussed,[100] and in particular the influence of the chirality at C-21 and the nature of the substituent at C-16 (biogenetic numbering) on the c.d. spectra. Another Hungarian group has included vincamine and its stereoisomers in a study of the effect of stereochemistry on the mass-spectrometric behaviour of the indole alkaloids.[48b]

The vincadifformine–tabersonine group of alkaloids provide a rich field for the study of oxidative rearrangements. New work reported recently[101a] includes a study of the oxidation of 3-oxovincadifformine (189) with *m*-chloroperoxybenzoic acid (Scheme 27). The products were 3-oxovincamine (190), 3-oxo-16-*epi*-vincamine (191), and a dilactam (192), which is simply the result of oxidative fission of the 2,16 bond. Under carefully controlled conditions, the hydroxy-indolenine (193) was isolated, and it was subsequently shown to be an intermediate (as expected) in the formation of (190)—(192). Photochemical oxidative rearrangement gave similar results.

In an attempt to remove the carbonyl groups, the keto-dilactam (192) was reduced by potassium borohydride, which gave the alcohols (194); the hydroxy-group was then removed by standard methods [(194) → (195) → (196)]. Preferential reduction of the 3-oxo-group in the product (196) should then give desmethyl-vincatine or a stereoisomer. However, reduction of (196) by diborane gave a tricyclic oxindole (197), in which fission of the 7,21 bond had also occurred. This result prompted a reappraisal of the reduction of vincatine (198), which had earlier[101b] been reported to give a tetracyclic carbinolamine (199), identical[101c] with the product obtained by reduction of one stereoisomer of synthetic (200) (Scheme 28). For convenience, the reduction was attempted on the more accessible stereoisomer (201), obtained by total synthesis, and the product was shown conclusively to be the tricyclic oxindole-alcohol (202). It is therefore concluded that the common reduction product of vincatine and of the synthetic lactam (200) also has the gross structure (202).

The oxidative rearrangement of vincadifformine (203) to vincamine (204) is generally considered to proceed *via* the 16-hydroxyindolenine derivative (205). Additional evidence for the intermediacy of (205) has now been adduced, and the configuration at C-16 determined.[102] The N_b-oxides of (205) and (206), obtained respectively from vincadifformine and tabersonine, can be converted by obvious stages into the same indoline derivative (207); hence the stereochemistry of (205) is identical to that of (206). The N_a-methyl derivative (208), prepared by methylation and reduction of the N_b-oxide of (206), gave, with iodine–potassium iodate, a lactam (209), together with an aromatic iodo-derivative (210), when the reaction was prolonged. Evidently, ether formation between C-16 and C-15 can only occur if (208), and therefore (205) and (206), has the stereochemistry shown. A similar conclusion results from the oxidation of (208) with chromium trioxide in pyridine, which gives a small yield of the lactam (211), in which the six-membered ring D has been preserved (Scheme 29).[102]

[100] G. Tóth, O. Clauder, K. Gesztes, S. S. Yemul, and G. Snatzke, *J. Chem. Soc., Perkin Trans. 2*, 1980, 701.

[101] (*a*) G. Hugel, J. Y. Laronze, J. Laronze, and J. Lévy, *Heterocycles*, 1981, **16**, 581; (*b*) W. Döpke, H. Meisel, and H. W. Fehlhaber, *Tetrahedron Lett.*, 1969, 1701; (*c*) L. Castedo, J. Harley-Mason, and M. Caplan, *Chem. Commun.*, 1969, 1444.

[102] G. Hugel, G. Massiot, J. Lévy, and J. Le Men, *Tetrahedron*, 1981, **37**, 1369.

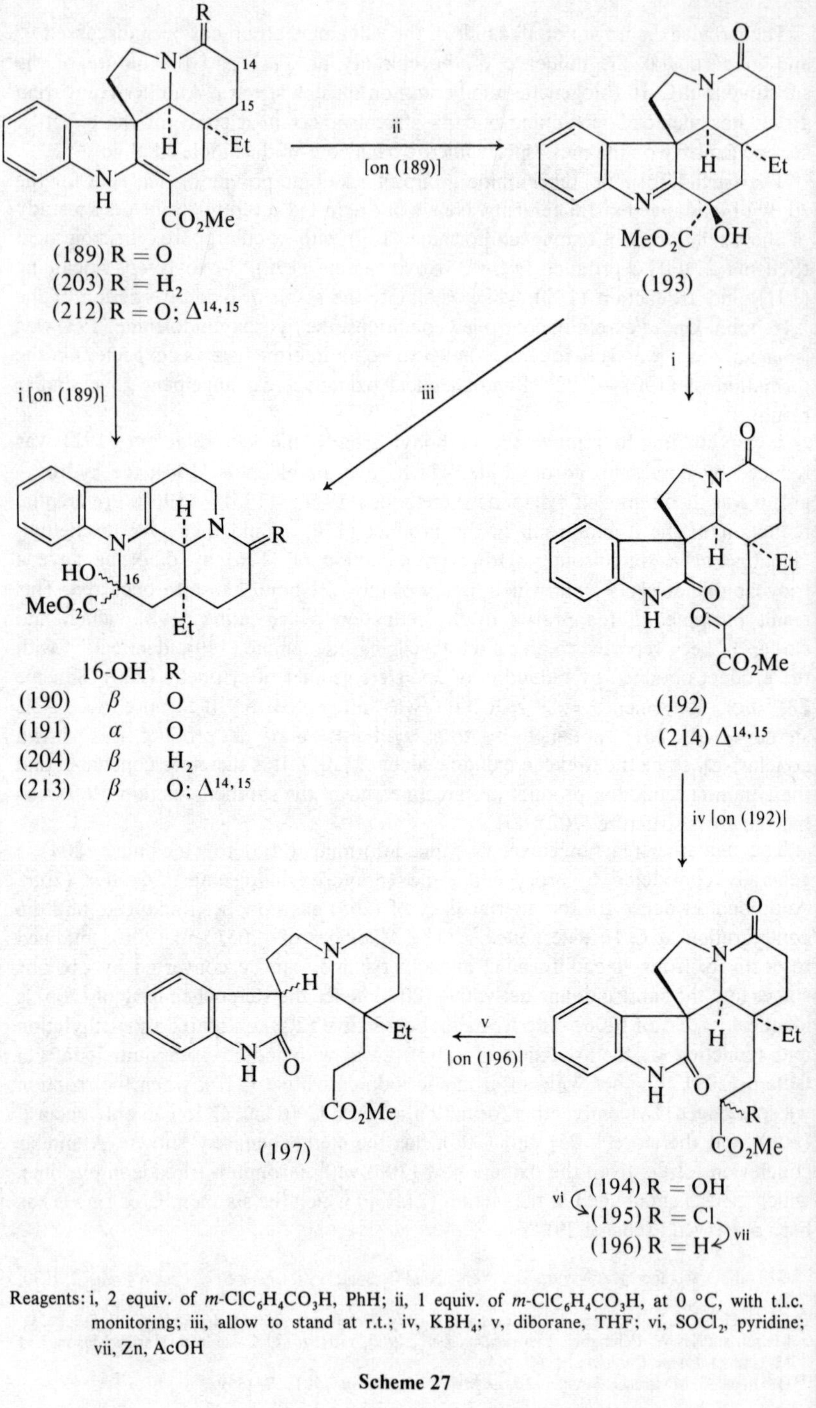

(189) R = O
(203) R = H$_2$
(212) R = O; $\Delta^{14,15}$

(193)

16-OH   R
(190) β O
(191) α O
(204) β H$_2$
(213) β O; $\Delta^{14,15}$

(192)
(214) $\Delta^{14,15}$

(197)

(194) R = OH
(195) R = Cl
(196) R = H

Reagents: i, 2 equiv. of *m*-ClC$_6$H$_4$CO$_3$H, PhH; ii, 1 equiv. of *m*-ClC$_6$H$_4$CO$_3$H, at 0 °C, with t.l.c. monitoring; iii, allow to stand at r.t.; iv, KBH$_4$; v, diborane, THF; vi, SOCl$_2$, pyridine; vii, Zn, AcOH

Scheme 27

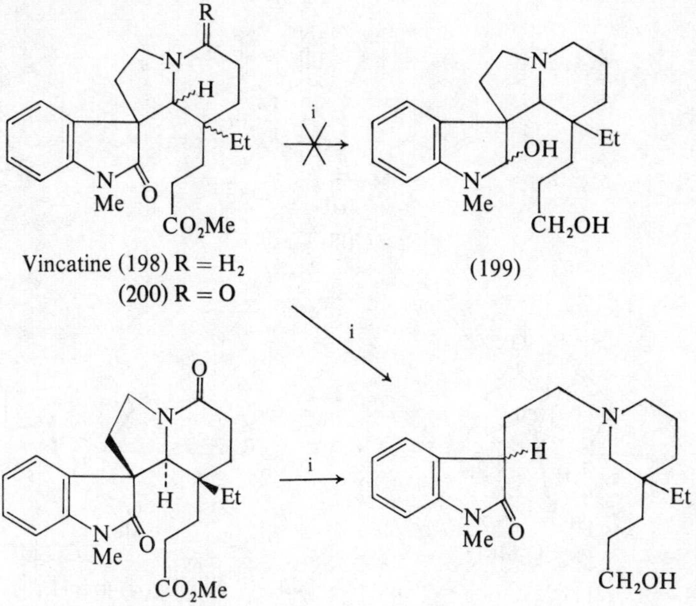

Vincatine (198) R = H₂
(200) R = O

(199)

(201)

(202)

Reagent: i, LiAlH₄

Scheme 28

The oxidative rearrangement of 3-oxotabersonine (212) to 3-oxo-Δ^{14}-vincamine (213) and the tetracyclic lactam (214) is analogous to the rearrangement of (189) to (190), (191), and (192).[103] Another product (215), resulting from the enlargement of ring B by an unspecified reagent, is also reported to be obtained; unfortunately, details of this work are at present inaccessible.

The influence of aromatic substituents (chlorine, bromine, or nitro-groups) on the ease of oxidative rearrangement of derivatives of vincadifformine into derivatives of

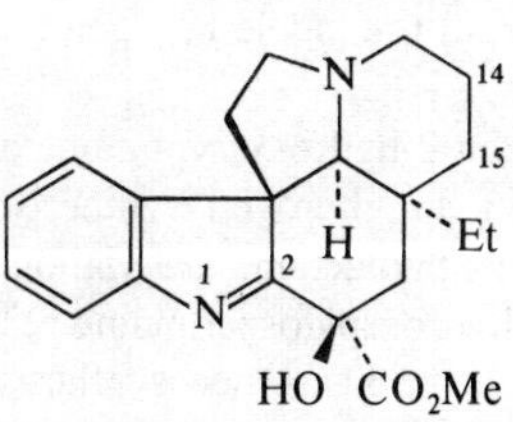

(205)
(206) $\Delta^{14,15}$
(207) 1,2-dihydro

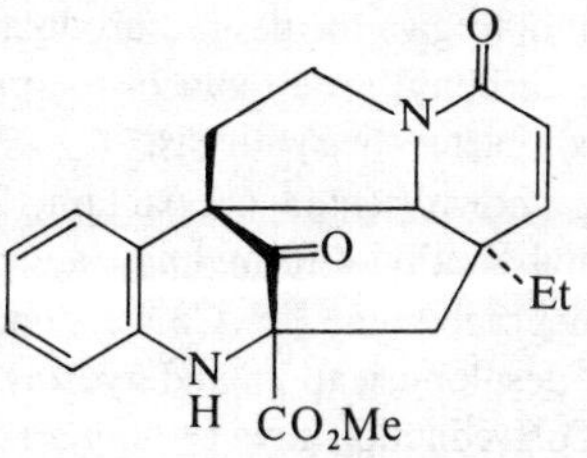

(215)

[103] N. Aimi, Y. Asada, S. Tanabe, S. Tsuge, Y. Watanabe, and S. Sakai, *Koen Yoshishu-Tennen Yuki Kagobutsu Toronkai, 22nd*, 1979, 540 (*Chem. Abstr.*, 1980, **93**, 26 614, 114 796).

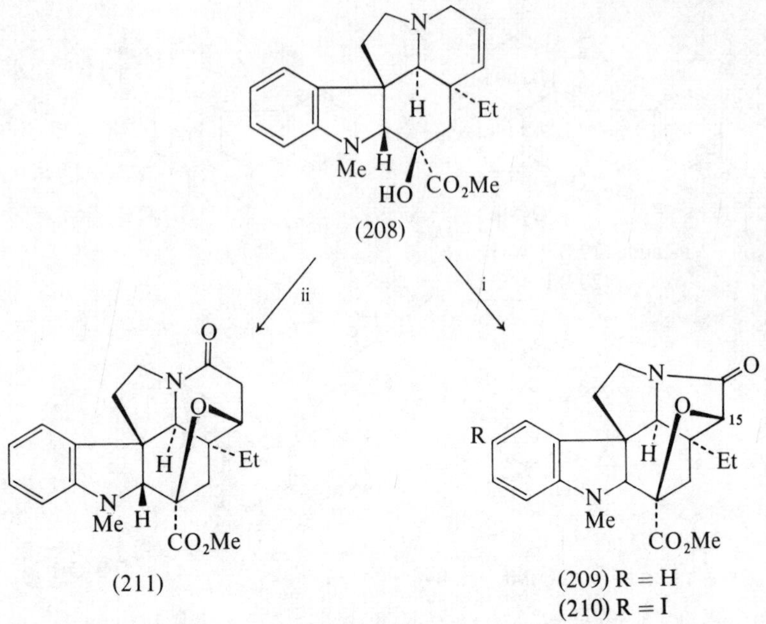

(208)

ii   i

(211)

(209) R = H
(210) R = I

Reagents: i, I$_2$, KIO$_3$, H$_2$O, AcOH, dioxan; ii, CrO$_3$, pyridine

Scheme 29

vincamine has been investigated.[104] In general, it appears that the rearrangement is discouraged by electron-attracting substituents.

The oxidation of vincamine by means of trifluoroperacetic acid simply gives the keto-lactam that results from oxidative fission of the indole 2,7 bond.[71]

The first synthesis of secodine (218)[105] involves (as its critical stage) an ingenious application of the Claisen orthoester rearrangement, in which the orthoester derived from the benzylic-type alcohol (216) and trimethyl β-methoxyorthopropionate (an acrylic ester equivalent) undergoes simultaneous rearrangement and elimination of methanol to give the desired product (217) in one step (Scheme 30). Removal of the amide carbonyl group and of the protecting group from the indole nitrogen atom then completes the synthesis.

The known pentacyclic lactam (219), prepared from 2-hydroxytryptamine and dimethyl 4-ethyl-4-formylpimelate, has been used in an improved synthesis of (±)-quebrachamine.[106a] Conversion of (219) into the thiolactam, acetylation to (220), desulphurization, and hydrolysis yielded 1,2-dehydroaspidospermidine (221), which on reduction gave (±)-quebrachamine (222) (Scheme 31). A new synthesis of the tetracyclic amino-alcohols (223) constitutes another formal synthesis of quebrachamine.[106b]

[104] G. Lewin, Y. Rolland, and J. Poisson, *Heterocycles*, 1980, **14**, 1915.

[105] S. Raucher, J. E. Macdonald, and R. F. Lawrence, *J. Am. Chem. Soc.*, 1981, **103**, 2419.

[106] (*a*) V. S. Giri, E. Ali, and S. C. Pakrashi, *J. Heterocycl. Chem.*, 1980, **17**, 1133; (*b*) S. Takano, C. Murakata, and K. Ogasawara, *Heterocycles*, 1981, **16**, 247.

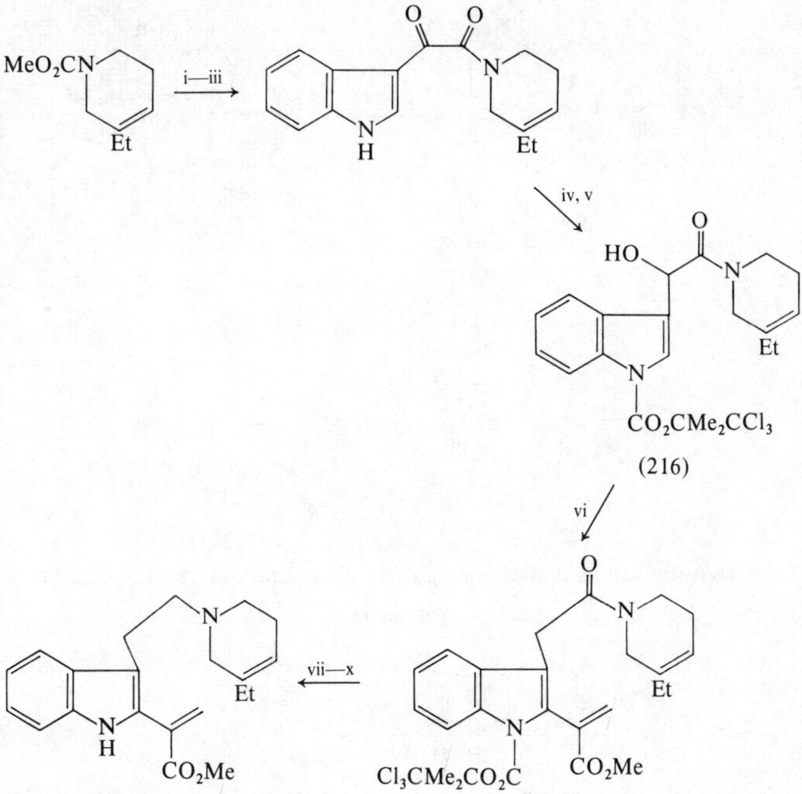

Reagents: i, MeLi; ii, HCl; iii, indole-3-glyoxyloyl chloride, NEt₃; iv, ClCO₂CMe₂CCl₃, NEt₃, CH₂Cl₂; v, NaBH₄; vi, MeOCH₂CH₂C(OMe)₃, mesitoic acid, argon, heat; vii, (MeOC₆H₄)₂-P₂S₄; viii, Et₃O⁺ BF₄⁻; ix, NaBH₄, MeOH, AcOH; Zn, MeOH, AcOH

Scheme 30

Kuehne's remarkable synthesis of the vincadifformine group of alkaloids has been extended,[107] and an alternative mechanism for the crucial complex stage in the synthesis has been proposed.

The condensation of the indolo-azepine (224) with aldehydes, *e.g.* (225), gives a mixture of epimeric bridged azepines (226) (not isolated), which can fragment [to (227)] and re-cyclize, provided that the initial aldehyde has hydrogen at the α-position. The product, obtained directly (in this example) in 85% yield from (224), is 3-oxovincadifformine (228). Alternatively, condensation of (224) with (225) under milder conditions gives the epimeric mixture of bridged indoloazepines (226), which can be isolated and benzylated. Fragmentation followed by

[107] M. E. Kuehne, T. H. Matsko, J. C. Bohnert, L. Motyka, and D. Oliver-Smith, *J. Org. Chem.*, 1981, **46**, 2002.

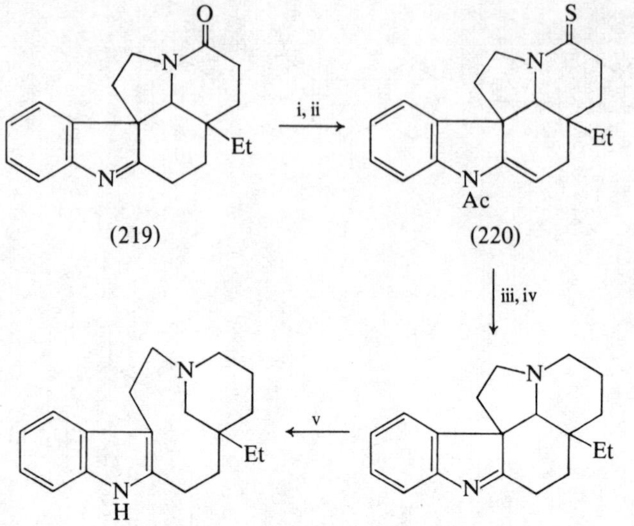

(219)   (220)

(222)   (221)

Reagents: i, P$_2$S$_5$; ii, Ac$_2$O, pyridine; iii, Ni, THF; iv, 6M-HCl, N$_2$, heat; v, KBH$_4$, H$_2$O, MeOH

Scheme 31

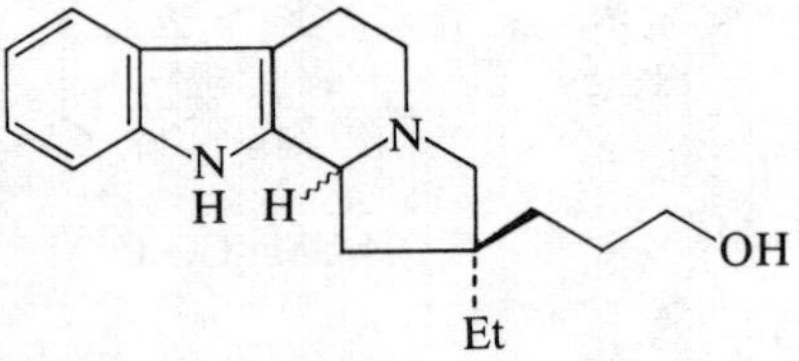

(223)

re-cyclization then gives a tetracyclic amino-ester (229), which, on debenzylation and cyclization, gives 3-oxovincadifformine (228) (Scheme 32).

The Kuehne synthesis has also been adapted to the preparation of 18-methylenevincadifformine (230).[108a] Unexpectedly, when (230) was hydrolysed by alkali, and the acid thus obtained was heated briefly in 3% aqueous hydrochloric acid, the only product that could be isolated was a diene-imine which has been assigned the structure (231). This is the first report of an intramolecular [4π + 2π] cycloaddition reaction involving an indolenine. When heated in benzene in the presence of toluene-*p*-sulphonic acid, 18-methylenevincadifformine (230) is smoothly hydrolysed and decarboxylated, with exclusive formation of the indolenine (232); this can be quantitatively transformed into (231) by heating with 3% aqueous hydrochloric acid (Scheme 33).

The indolenine (232) has been neatly used in a synthesis of (±)-strempeliopine (233), the laevorotatory enantiomer of which has recently been isolated from *Strempeliopsis strempelioides* K. Schum.[108b] Reductive rearrangement of (232) by

108 (*a*) J. Hájiček and J. Trojánek, *Tetrahedron Lett.*, 1981, **22**, 1823; (*b*) *ibid.*, p. 2927.

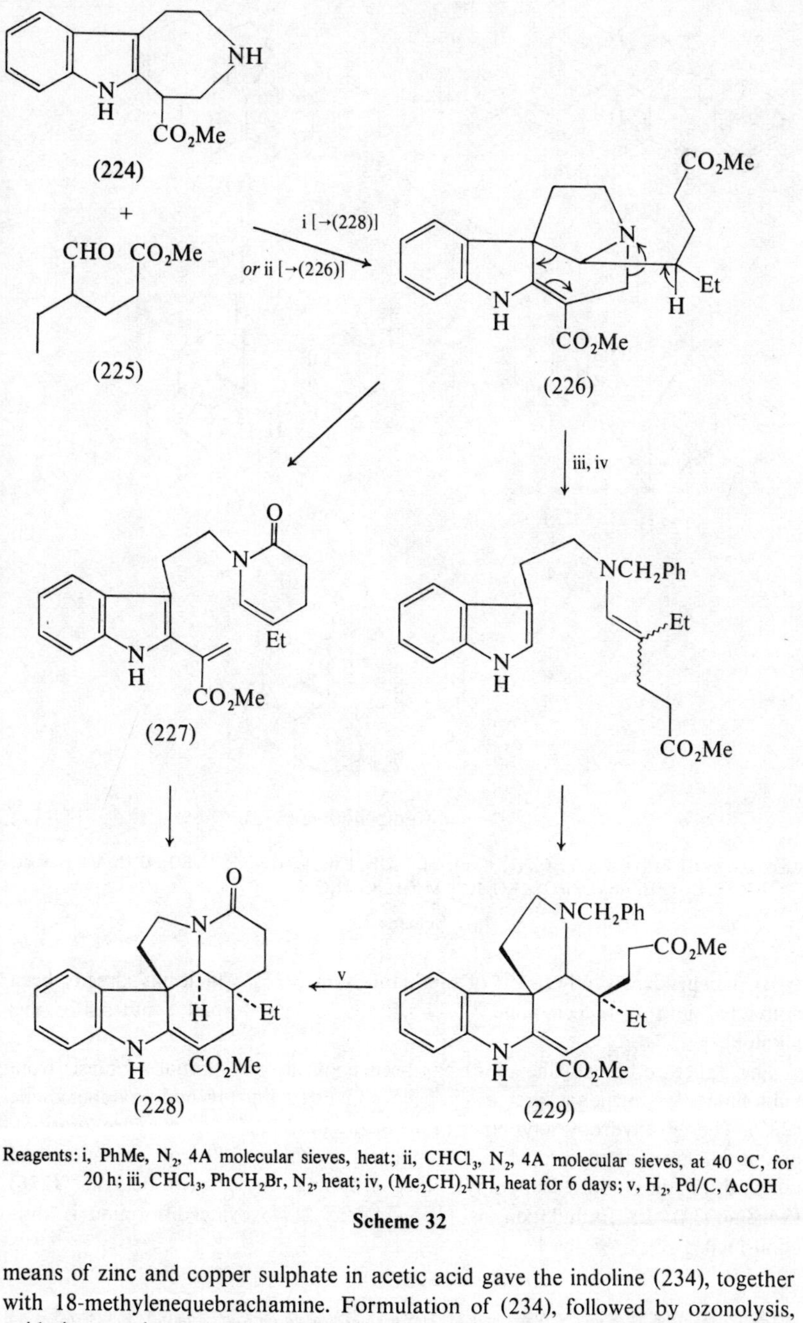

(224)

+

(225)

i [→(228)]

or ii [→(226)]

(226)

(227)

iii, iv

(228)

v

(229)

Reagents: i, PhMe, N_2, 4A molecular sieves, heat; ii, $CHCl_3$, N_2, 4A molecular sieves, at 40 °C, for 20 h; iii, $CHCl_3$, $PhCH_2Br$, N_2, heat; iv, $(Me_2CH)_2NH$, heat for 6 days; v, H_2, Pd/C, AcOH

Scheme 32

means of zinc and copper sulphate in acetic acid gave the indoline (234), together with 18-methylenequebrachamine. Formulation of (234), followed by ozonolysis, oxidative work-up, and hydrolysis of the *N*-formyl group, then gave (±)-strempeliopine (233) (Scheme 33).

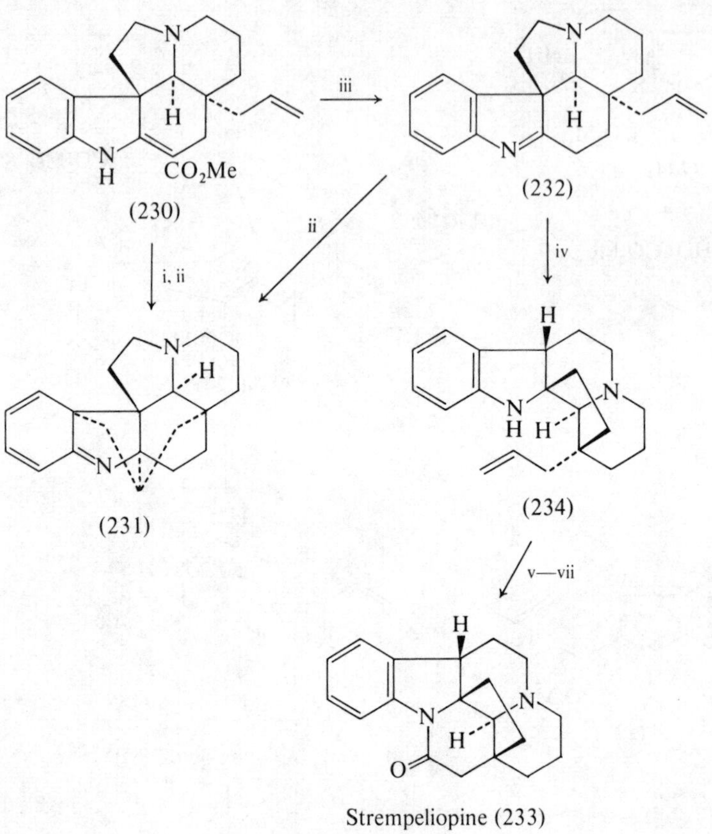

Strempeliopine (233)

Reagents: i, OH⁻, H₂O; ii, 3% HCl, H₂O, heat; iii, TsOH, PhH, heat; iv, Zn, CuSO₄, H₂O, AcOH, heat; v, HCO₂H, Ac₂O; vi, O₃, H₂O, HCl, MeOH; vii, H₂O₂

Scheme 33

An independent synthesis[109a] of the ketolactam (235), which has already been converted into (±)-tabersonine,[109b] constitutes a new formal synthesis of this alkaloid.

The structure of baloxine (236) has been confirmed by partial synthesis from vindolinine,[110] which has previously been converted into 19-hydroxytabersonine (237). The tetrahydropyranyl ether (238) of the (19S)-epimer, on hydroboration–oxidation, gave a mixture of C-14 epimeric alcohols (239); on oxidation and removal of the tetrahydropyranyl ether grouping, these gave baloxine (236) (Scheme 34). Its formulation as (19S)-hydroxy-14-oxovincadifformine is thus confirmed.

109 (a) T. Imanishi, H. Shin, N. Yagi, and M. Hanaoka, *Tetrahedron Lett.*, 1980, **21**, 3285; (b) F. E. Ziegler and G. B. Bennett, *J. Am. Chem. Soc.*, 1971, **93**, 5930; 1973, **95**, 7458.
110 C. Caron, L. Le Men-Olivier, M. Plat, and J. Lévy, *Heterocycles*, 1981, **16**, 645.

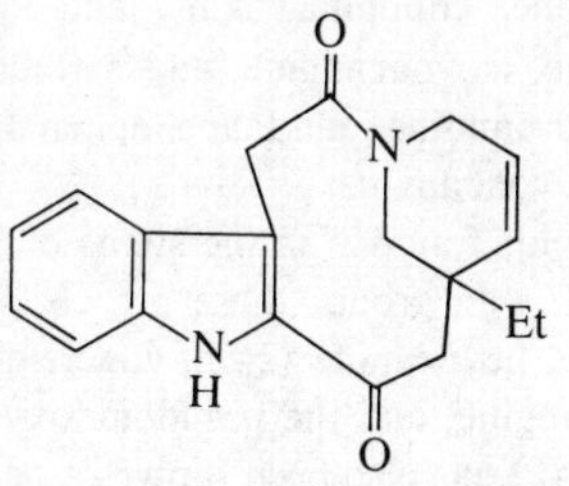

(235)

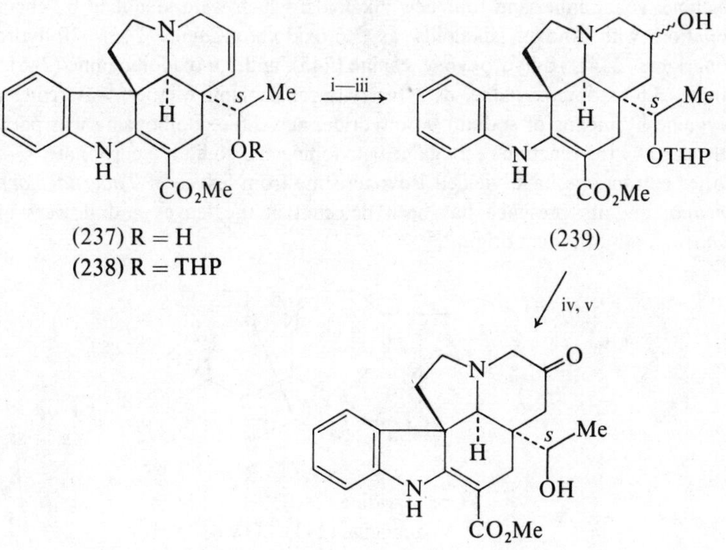

(237) R = H
(238) R = THP

(239)

iv, v

Baloxine (236)

Reagents: i, Et₃O⁺ BF₄⁻; ii, NaBH₄, THF; iii, H₂O₂, NaOH; iv, DMSO, Ac₂O; v, HCl, H₂O, EtOH

Scheme 34

New synthetic work in this sub-group includes syntheses of (±)-eburnamonine[111] and (±)-apovincamine.[112]

Catharanthine–Ibogamine–Cleavamine Group. Catharanthine is one of the alkaloids produced in suspension cultures of the 200 GW cell line of *Catharanthus roseus*,[42a] and ibogamine has been found in the stem and root bark of *Hazunta modesta* var. *modesta* subvar. *montana*.[62] Four *Ervatamia* species have been shown to contain several alkaloids of this group.[64] Thus, *E. orientalis* (R. Brown) Domin contains voacangine, conopharyngine, pandine, and pandoline, *E. daemeliana*

---

[111] K. Irie, M. Okita, T. Wakamatsu, and Y. Ban, *Nouv. J. Chim.*, 1980, **4**, 275; K. Irie and Y. Ban, *Heterocycles*, 1981, **15**, 201.
[112] B. Danieli, G. Lesma, and G. Palmisano, *Tetrahedron Lett.*, 1981, **22**, 1827.

Domin contains voacangine, conopharyngine, and iboxygaine, *E. obtusiuscula* Mgf. contains coronaridine, isovoacangine, and a trace of epipandoline, while *E. lifuana* Boiteau has yielded pandine, pandoline, epipandoline, conopharyngine, and traces of coronaridine, and voacangine.

Ten of the twelve alkaloids found[63c] in the stems of *Tabernaemontana olivacea* Müll. Arg. belong to the *iboga* group; these are coronaridine (240), voacangine (241), ibogamine, ibogaine, heyneanine (242), voacristine, the hydroxyindolenines of coronaridine and voacangine, and the pseudoindoxyl relatives of the same two alkaloids. Coronaridine (240) has also been shown to occur in *Catharanthus roseus* G. Don. (*Vinca rosea* L.)[113] and in *Tabernanthe pubescens*,[43f] in association with ibogamine, ibogaine, ibogaline, iboxygaine, voacangine, voacangine hydroxyindolenine, voacristine, and four new alkaloids, which were identified by chemical correlation with known alkaloids as 3,6-oxidoiboxygaine (243), 10-hydroxycoronaridine (244), 10-hydroxyheyneanine (245), and 3,6-oxidoibogaine (246). For example, 3,6-oxidoiboxygaine, as a benzylic-carbinolamine ether, was reduced to iboxygaine by means of sodium borohydride, and 3,6-oxidoibogaine was partially synthesized by the reaction of ibogaine with iodine and sodium bicarbonate.[43f]

Other extractions have yielded isovoacristine from leaves of *Tabernaemontana heyneana*, and its presence has been detected in the leaves and flowers of *T. coronaria*, both of Indian origin.[114]

	R¹	R²
Coronaridine (240)	H	H
Voacangine (241)	OMe	H
Heyneanine (242)	H	OH
10-Hydroxycoronaridine (244)	OH	H
10-Hydroxyheyneanine (245)	OH	OH

3,6-Oxidoiboxygaine (243) R = OH
3,6-Oxidoibogaine (246) R = H

[113] L. De Taeye, A. De Bruyn, C. De Pauw, and M. Verzele, *Bull. Soc. Chim. Belg.*, 1981, **90**, 83.
[114] P. G. Rao and B. P. Singri, *Indian J. Chem.*, Sect. B, 1979, **17**, 414.

A new alkaloid, 3-hydroxycoronaridine, has been found in *Tabernaemontana glandulosa* Stapf;[115] its structure was deduced from its spectra, particularly its mass spectrum, and confirmed by reduction with sodium borohydride, which gave coronaridine (240). The stems of a related species, *T. flavicans*, collected in Peru, contain ibophyllidine and ibophyllidine *N*-oxide.[116]

Three recent communications are concerned with the alkaloids of *T. albiflora* (Miq.) Pull., endemic to French Guyana.[117] Seven new alkaloids have been isolated; these include albifloranine (247),[117a] (+)-19-hydroxy-20-*epi*-pandoline (248), (+)-(20*R*)-18,19-dihydroxypseudovincadifformine (249),[117b] 19-hydroxy-ibophyllidine (250), (19*R*)-19-hydroxy-20-*epi*-ibophyllidine (251), (19*S*)-19-hydroxy-20-*epi*-ibophyllidine (252), and 18-hydroxy-20-*epi*-ibophyllidine (253).[117c]

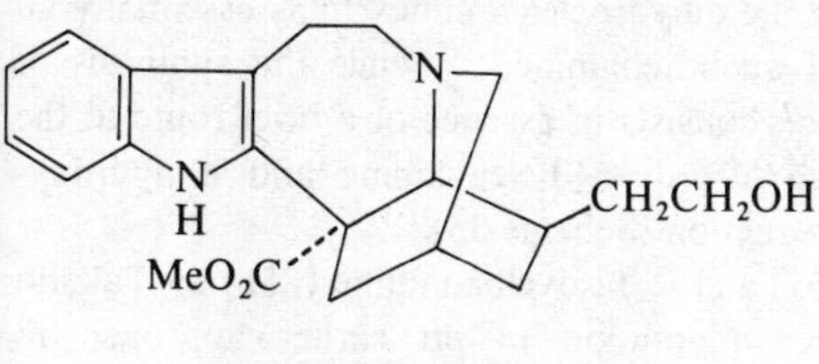

Albifloranine (247)

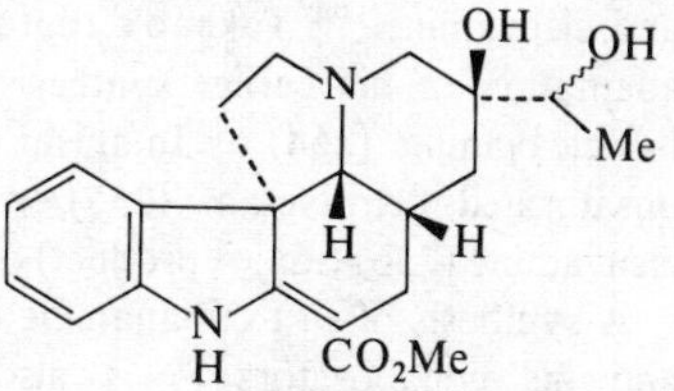

(248)

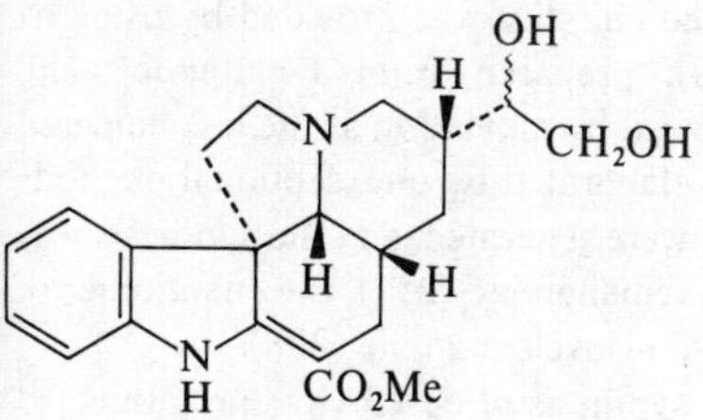

(249)

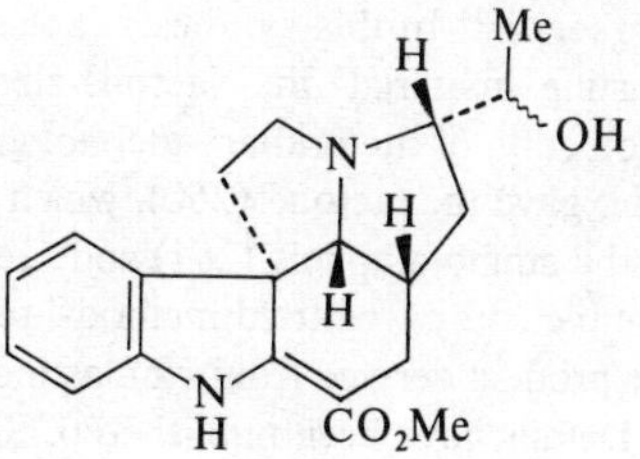

(250)

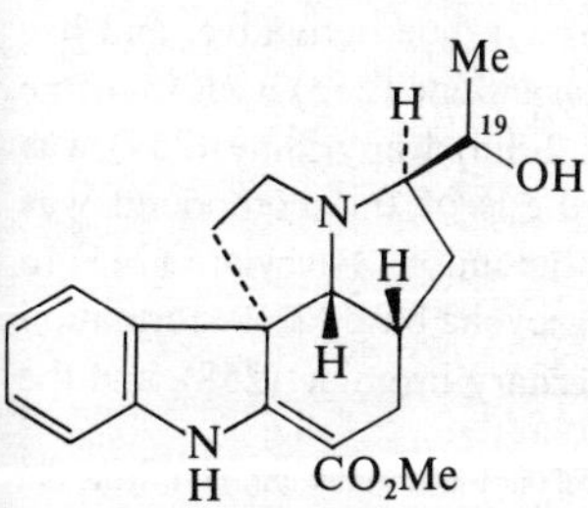

(251) 19*R*
(252) 19*S*

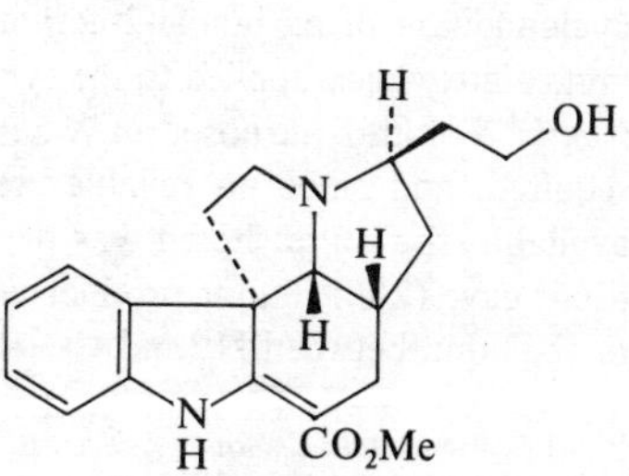

(253)

[115] H. Achenbach, B. Raffelsberger, and G. U. Brillinger, *Phytochemistry*, 1980, **19**, 2185.
[116] H. Achenbach and B. Raffelsberger, *Z. Naturforsch., Teil. B*, 1980, **35**, 1465.
[117] (a) C. Kan, H.-P. Husson, S. K. Kan, and M. Lounasmaa, *Planta Med.*, 1981, **41**, 72; (b) *ibid.*, p. 195; (c) *Tetrahedron Lett.*, 1980, **21**, 3363; (d) A. Henriques, C. Kan, H.-P. Husson, S. K. Kan, and M. Lounasmaa, *Acta Chem. Scand., Sect. B*, 1980, **34**, 509.

The structures of all seven alkaloids were deduced principally from a detailed examination of their 400 MHz ^{1}H n.m.r. spectra, and from comparison [in the case of (248) and (249)] with the spectra of pandoline and 20-*epi*-pandoline.

Five monomeric alkaloids isolated[117d] from the leaves of *Stenosolen hetero-phyllus* (Vahl) Mgf. have been identified as voacangine, voacangine hydroxy-indolenine, conoflorine, pandine, and pandoline.

An *X*-ray crystallographic study[118a] on capuronine acetate has confirmed the structure and stereochemistry previously deduced.[118b]

The oxidation of voacangine by trifluoroperacetic acid leads simply to voacangine hydroxyindolenine.[71]

New synthetic work in this area includes syntheses of desethyldihydro-cleavamine (desethylquebrachamine),[119a] 20αH- and 20βH-dihydrocleavamine,[119b] and cleavamine.[119c] Takano's route to the dihydrocleavamines[119b] is essentially an adaptation of his earlier synthesis of quebrachamine,[119d] while the synthesis of (+)-cleavamine (254) by Imanishi *et al.* consists in essence of a brief route to the unsaturated keto-lactam (255), which affords (±)-cleavamine and a hydroxy-cleavamine (256) (major product) on reduction (Scheme 35).[119c]

A synthesis of (+)-velbanamine (257) and (±)-isovelbanamine (258) by Takano and his collaborators[120a] is also an adaptation of an earlier approach to quebrachamine; however, it would appear to have been superseded already by an enantioselective synthesis of (−)-(257) and (+)-(258) by the same group of workers.[120b] In this synthesis (Scheme 36), the chirality was provided by using as starting material the lactone-alcohol (259), prepared from L-glutamic acid. Protection of the primary alcohol group, followed by alkylation at the less hindered side, gave the lactone (260), which was then elaborated by unexceptional methods to the amino-alcohols (261) and (262). These were separated, and the synthesis was completed by well-tried methods, to give (−)-velbanamine (257), the enantiomer of the product derived from vinblastine, and (+)-20-isovelbanamine (258).

Details have been published of Sundberg's synthesis of desethylcatharanthine,[121] and an independent new synthesis has also been reported.[122a] This new approach relies, for the formation of the complete skeleton, on a biogenetically patterned cycloaddition of an indole-2-acrylate to a 1,2-dihydropyridine derivative, and has immediately been applied to the synthesis of (±)-catharanthine (263) itself (Scheme 37).[122b] For this purpose, an *N*-substituted 3-ethyl-1,2-dihydropyridine (264) was required, and since no reliable method for the synthesis of this compound was available, the route shown was devised. Addition of the indole-2-acrylate (265) to (264) gave (266) (major product) and the desired tetracyclic base (267); formation of the bond between N_b and C-5 then gave the quaternary bromide (268), and the

[118] (a) C. Riche, *Acta Crystallogr.*, *Sect. B*, 1980, **36**, 1573; (b) I. Chardon-Loriaux and H.-P. Husson, *Tetrahedron Lett.*, 1975, 1845.
[119] (a) S. Takano, C. Murakata, and K. Ogasawara, *Heterocycles*, 1980, **14**, 1301; (b) S. Takano, M. Yonaga, S. Yamada, S. Hatakeyama, and K. Ogasawara, *ibid.*, 1981, **15**, 309; (c) T. Imanishi, A. Nakai, N. Yagi, and M. Hanaoka, *Chem. Pharm. Bull.*, 1981, **29**, 901; (d) S. Takano, S. Hatakeyama, and K. Ogasawara, *J. Am. Chem. Soc.*, 1976, **98**, 3022.
[120] (a) S. Takano, M. Hirama, and K. Ogasawara, *J. Org. Chem.*, 1980, **45**, 8729; (b) S. Takano, M. Yonaga, K. Chiba, and K. Ogasawara, *Tetrahedron Lett.*, 1980, **21**, 3697.
[121] R. J. Sundberg and J. D. Bloom, *J. Org. Chem.*, 1980, **45**, 3382.
[122] (a) C. Marazano, J. L. Fourrey, and B. C. Das, *J. Chem. Soc.*, *Chem. Commun.*, 1981, 37; (b) C. Marazano, M. T. Le Goff, J. L. Fourrey, and B. C. Das, *ibid.*, p. 389.

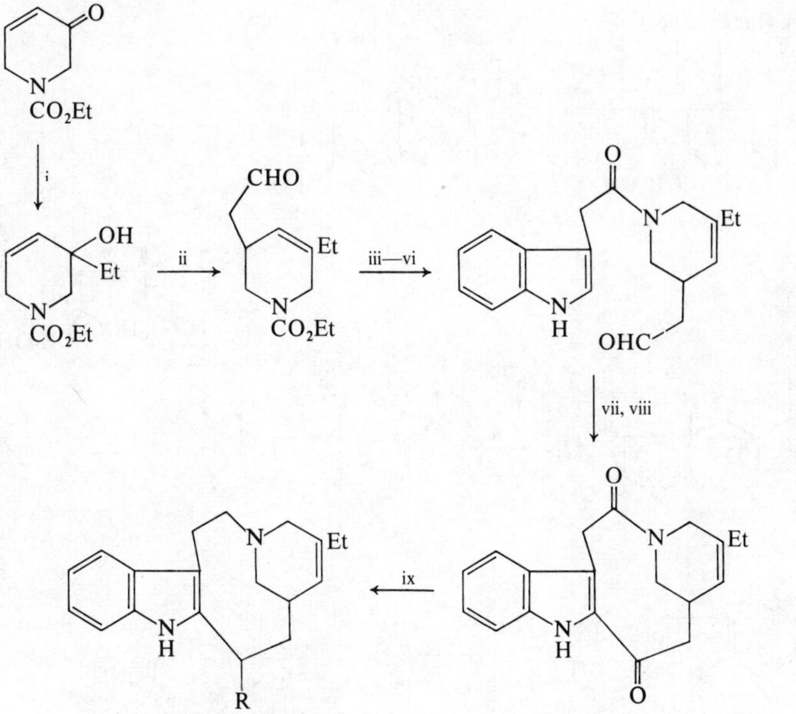

Cleavamine (254) R = H   (255)
(256) R = OH

Reagents: i, EtMgBr; ii, H₂C=CHOEt, Hg(OAc)₂; iii, (CH₂OH)₂, H⁺; iv, KOH, H₂O, EtOH; v, β-indolylacetyl chloride; vi, H⁺; vii, Ag₂O; viii, polyphosphate ester, CHCl₃, heat; ix, LiAlH₄, dioxan

Scheme 35

synthesis of (±)-catharanthine (263) was completed by dequaternization. Although the yields in some stages are low, this provides a very attractive route to the catharanthine ring-system.[122b]

The formal synthesis of (±)-catharanthine by Imanishi *et al.*[109a] consists in a new preparation of the pentacyclic ketoamide (269), which has previously been converted into catharanthine by Büchi *et al.*[123] The critical stage in this synthesis (Scheme 38) was the preparation of the quinuclidine ketone (270) by an intramolecular aldol reaction on the keto-aldehyde derived from the piperidine bis-acetal (271).

Re-synthesis[124a] of catharanthine (263) from the isoxazolidine (272), previously obtained by degradation of catharanthine,[124b] suggests that (272) could provide a

[123] G. Büchi, P. Kulsa, K. Ogasawara, and R. L. Rosati, *J. Am. Chem. Soc.*, 1970, **92**, 999.
[124] (a) R. Z. Andriamialisoa, N. Langlois, and Y. Langlois, *Heterocycles*, 1980, **14**, 1457; (b) Y. Langlois, F. Guéritte, R. Z. Andriamialisoa, N. Langlois, P. Potier, A. Chiaroni, and C. Riche, *Tetrahedron*, 1976, **32**, 945.

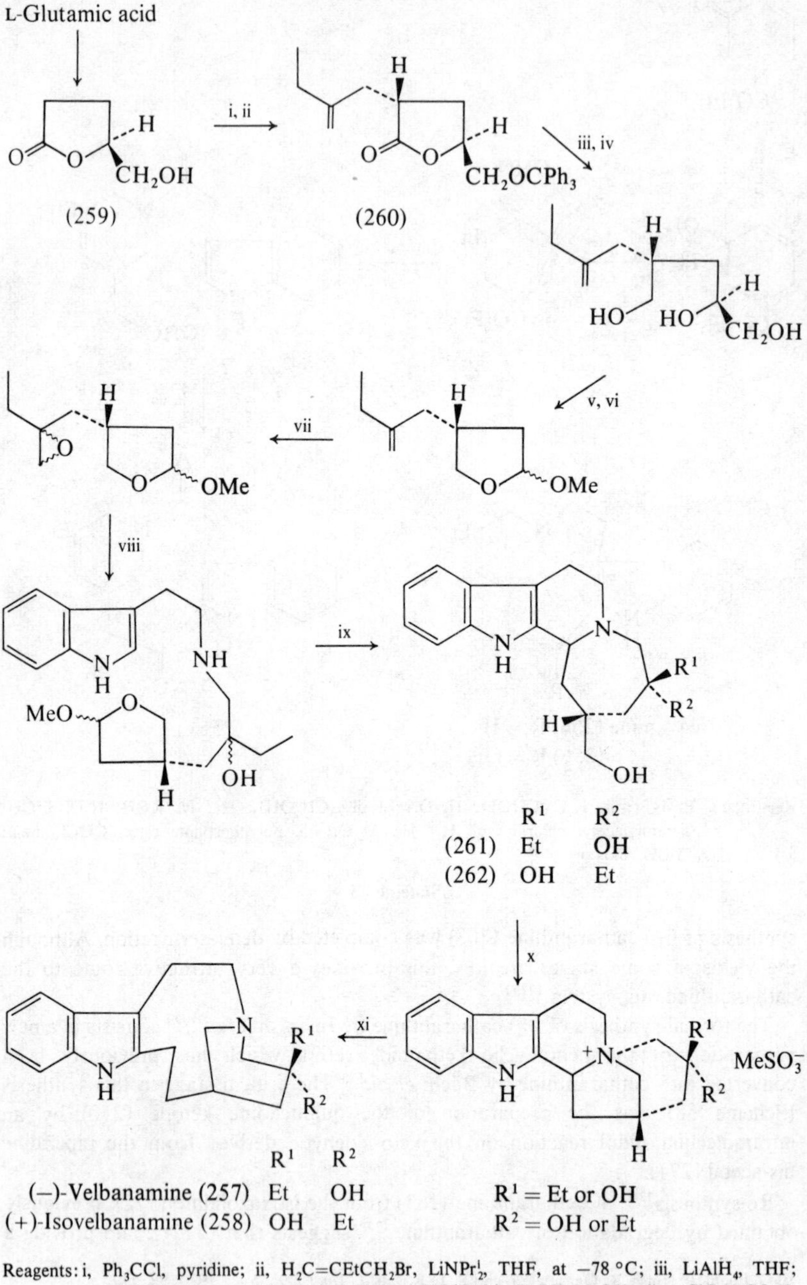
L-Glutamic acid
(259)
i, ii
(260)
iii, iv
v, vi
vii
viii
ix
R¹ R²
(261) Et OH
(262) OH Et
x
xi
MeSO₃⁻
R¹ R²
(−)-Velbanamine (257) Et OH
(+)-Isovelbanamine (258) OH Et
R¹ = Et or OH
R² = OH or Et

Reagents: i, Ph_3CCl, pyridine; ii, $H_2C{=}CEtCH_2Br$, $LiNPr^i_2$, THF, at −78 °C; iii, $LiAlH_4$, THF; iv, MeOH, HCl; v, $NaIO_4$, MeOH, H_2O; vi, $HC(OMe)_3$, TsOH, MeOH; vii, *m*-$ClC_6H_4CO_3H$, CH_2Cl_2; vii, tryptamine, MeOH, at 160 °C; ix, AcOH, H_2O; x, MesCl, pyridine; xi, Na, NH_3, EtOH

Scheme 36

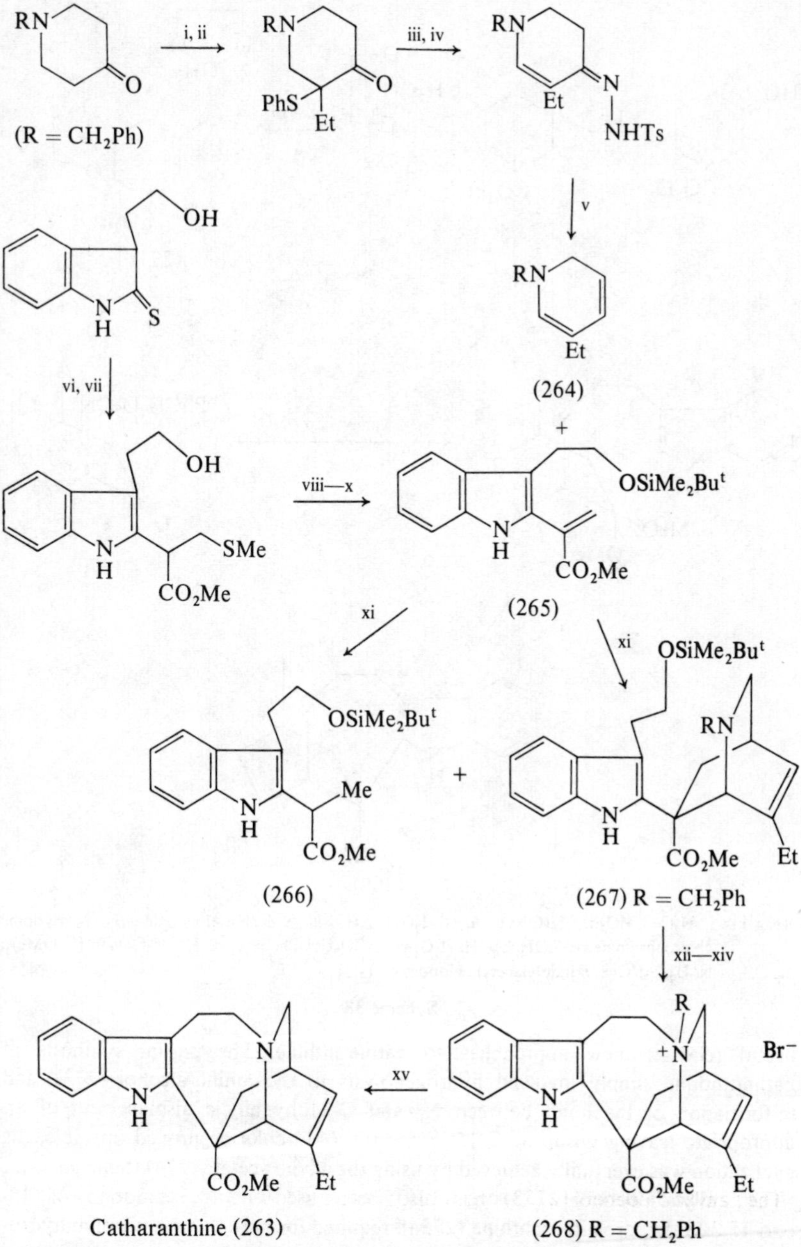

Reagents: i, Me₃SiNSiMe₃ Li⁺, Ph₂S₂, THF, HMPA; ii, NaH, EtI, THF, DMF; iii, TsNHNH₂, AcOH; iv, EtOAc, at 70 °C; v, BuLi, TMEDA, THF; vi, H₂C=CHCO₂Me, *hv*; vii, CH₂N₂; viii, BuᵗMe₂SiCl, NEt₃, CH₂Cl₂, DMAD; ix, *m*-ClC₆H₄CO₃H; x, PhMe, heat; xi, EtO₂, N₂, at 20 °C; xii, THF, AcOH, H₂O; xiii, MeSO₂Cl, NEt₃, CH₂Cl₂; xiv, LiBr, DMF, at 60 °C; xv, PrSH, LiH, HMPA, at 0 °C

Scheme 37

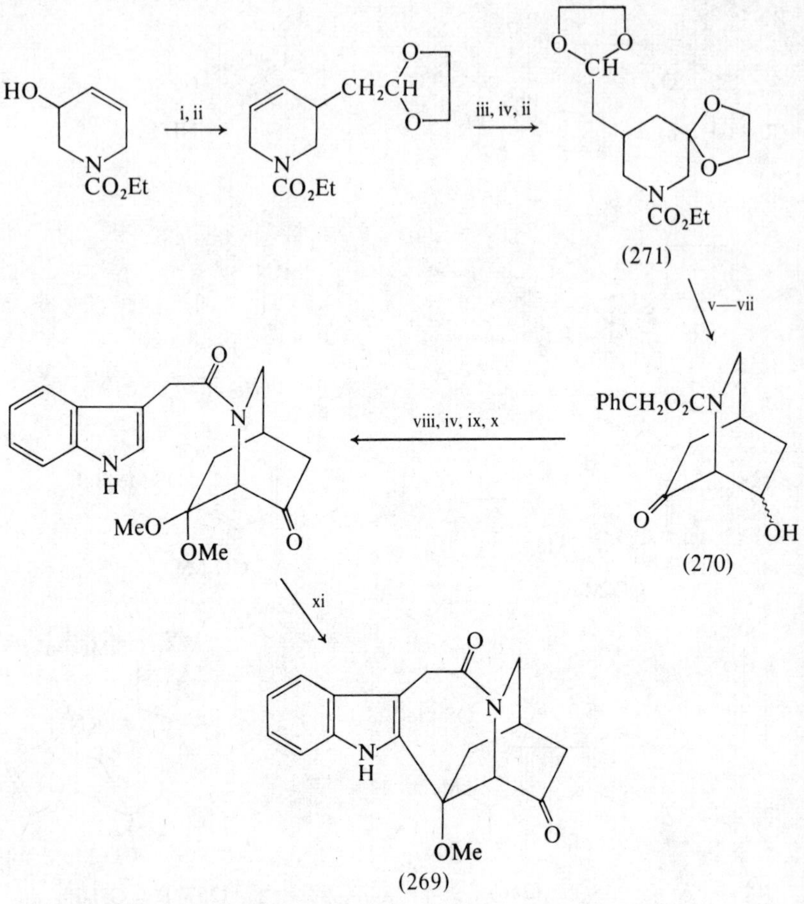

Scheme 38

Reagents: i, $H_2C=CHOEt$, $Hg(OAc)_2$; ii, $(CH_2OH)_2$, H^+; iii, hydroboration–oxidation; iv, pyridine chlorochromate; v, KOH, EtOH, H_2O; vi, $ClCO_2CH_2Ph$, base; vii, H^+, H_2O; viii, $HC(OMe)_3$; ix, H_2, Pd/C; x, β-indolylacetyl chloride; xi, TsOH

useful relay in new approaches to catharanthine. The partial synthesis of catharanthine simply involved hydrogenolysis to the amino-alcohol (273), and re-formation of the bond between N_b and C-21 by allylic displacement of an appropriate leaving group at C-15. Since the *C*-15-chloride proved unstable, the cyclization was eventually achieved by using the allylic acetate (274) (Scheme 39).

The allylic alcohol (273) has also been used in a preparation of 15-oxo-15,20*S*-dihydrocatharanthine (275a), required for partial synthesis of anhydro-vinblastine (*q.v.*).[125] Oxidation of (273) could be achieved by a variety of oxidizing agents, of which triphenylbismuth carbonate appeared to be the best (Scheme 39). Oxidation of the allylic alcohol group was followed by Michael addition of N_b to the

[125] N. Langlois, R. Z. Andriamialisoa, and Y. Langlois, *Tetrahedron*, 1981, **37**, 1951.

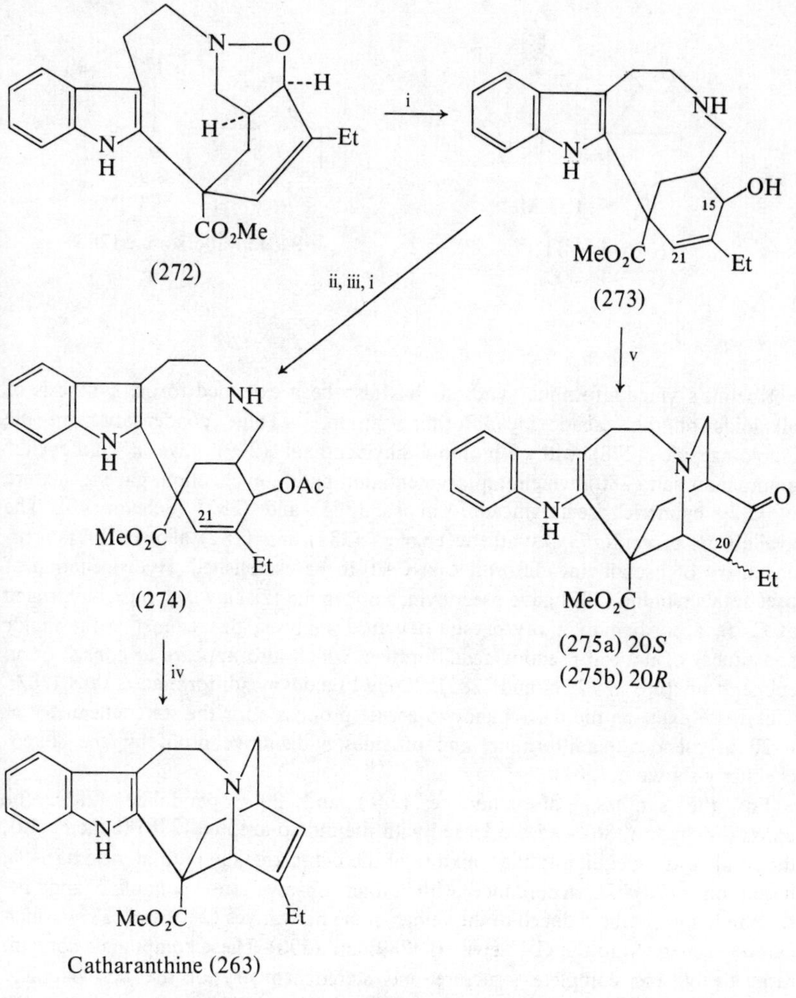

(272) → (273)

(274)

(275a) 20*S*
(275b) 20*R*

Catharanthine (263)

Reagents: i, H₂, Pd/C; ii, ClCO₂CH₂Ph; iii, acetylation; iv, [Pd(PPh₃)₄]; v, Ph₃BiCO₃, CH₂Cl₂

Scheme 39

enone system; although both C-20 epimers were obtained, the required 20*S*-epimer was fortunately the more stable, and could therefore be obtained on equilibration of the epimeric mixture.[125]

20-Desethyl-3-oxovincadifformine (276) has been synthesized[126] by appropriate modification of Lévy's route to 3-oxovincadifformine; alkylation of (276) by means of LDA and ethyl iodide then gave 3-oxopseudovincadifformine (277), of undetermined stereochemistry.

[126] J. Y. Laronze, D. Cartier, J. Laronze, and J. Lévy, *Tetrahedron Lett.*, 1980, **21**, 4441.

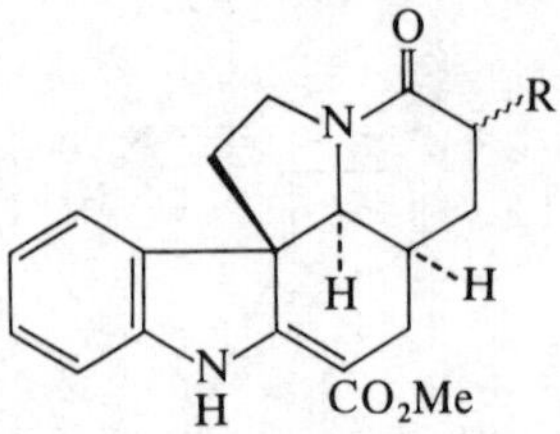

(276) R = H
(277) R = Et

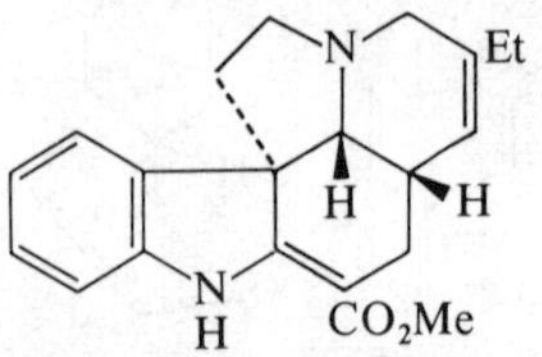

Pseudotabersonine (283)

Kuehne's vincadifformine synthesis has also been extended to the synthesis of alkaloids of the pseudovincadifformine group.[127] Thus, condensation of the indolo-azepine (278) with 5-bromo-4-ethylpentanal (279) gave a spirocyclic* ammonium salt (280), which, on fragmentation and re-cyclization, gave a mixture of C-20 epimeric pseudovincadifformines, (281) and (282) (Scheme 40). The availability of *crystalline* synthetic epimers (281) and (282) allowed the stereochemistry of pseudovincadifformine at C-20 to be established. Hydrogenation of pseudotabersonine (283) gave pseudovincadifformine (281), which has β-hydrogen at C-20, according to *X*-ray crystal-structure analysis; this epimer is the major constituent of natural pseudovincadifformine, which also appears to consist of an epimeric mixture of (281) and (282). 20-*epi*-Pseudovincadifformine is thus (282). This work explains the discrepancy in earlier proposals for the stereochemistry at C-20 in pseudovincadifformine, and provides a definitive proof for the stereochemistry shown in (281).

For the synthesis of pandoline (284) and 20-*epi*-pandoline (285), the epoxy-aldehyde (286) was condensed with the indolo-azepine (278) (Scheme 40); the result was an equimolecular mixture of the two alkaloids, in total yield of 64%, based on (278). In accordance with earlier observations, pandoline and 20-*epi*-pandoline can be reduced to the velbanamine derivatives (287) and (288), which can be isomerized to the C-16 epimers (289) and (290). These compounds contain, respectively, the complete structure and stereochemistry of the non-vindoline component of the oncolytic alkaloids leurosidine and vinblastine.

Finally, condensation of the indolo-azepine (278) with 4-chlorobutanal gave[107] an epimeric mixture of intermediate quaternary salts, formulated as (291), either of which, on reaction with triethylamine, suffered fragmentation and re-cyclization, with formation of desethylibophyllidine (292), which is the alkaloid recently isolated[14e] from *Tabernaemontana albiflora* (Scheme 41).

---

* For convenience, the intermediate is depicted here as a spirocyclic ammonium salt, but it should be noted that such an intermediate has only been rigorously established in the vincadifformine synthesis using α-substituted δ-halogeno-aldehydes; in other cases, an alternative mechanism may operate (see p. 223).

---

[127] M. E. Kuehne, C. L. Kirkemo, T. H. Matsko, and J. C. Bohnert, *J. Org. Chem.*, 1980, **45**, 3259.

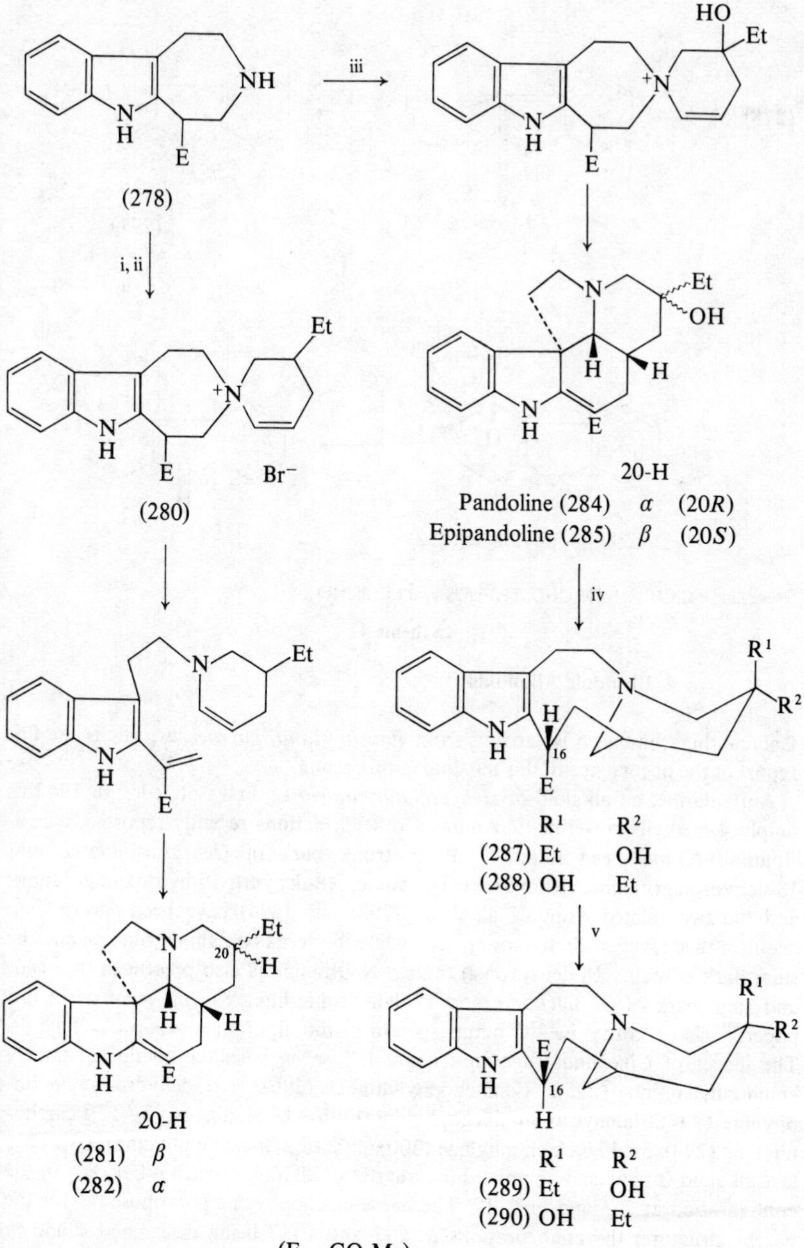

(278)

(280)

20-H
Pandoline (284) α (20R)
Epipandoline (285) β (20S)

20-H
(281) β
(282) α

	R^1	R^2
(287)	Et	OH
(288)	OH	Et

	R^1	R^2
(289)	Et	OH
(290)	OH	Et

$(E = CO_2Me)$

Reagents: i, BrCH$_2$CHEtCH$_2$CH$_2$CHO (279), MeOH; ii, NEt$_3$, at 40 °C; iii, H$_2$C—O—CEtCH$_2$CH$_2$CHO (286), N$_2$, MeOH, heat; iv, NaBH$_4$, AcOH; v, AcOH, heat

Scheme 40

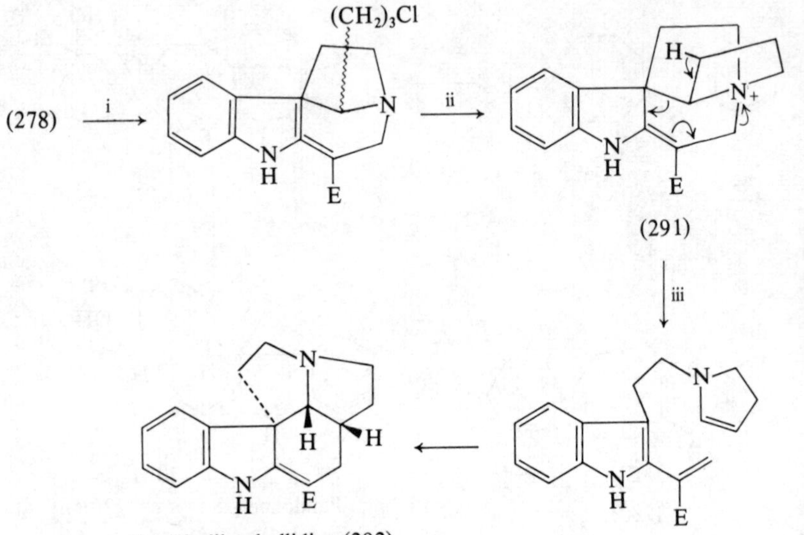

Desethylibophyllidine (292)

Reagents: i, ClCH$_2$CH$_2$CH$_2$CHO, THF; ii, 3 h, at r.t.; iii, NEt$_3$

Scheme 41

4 Bisindole Alkaloids

Calycanthine has been isolated[44d] from *Pausinystalia macroceras*; this is the first report of the occurrence of this alkaloid in this genus.

Auricularine, an alkaloid of *Hedyotis auricularia* L., first isolated[128] in 1942, is simply 4-methylborreverine.[129] Amongst other isolations recently reported, ochrolifuanine A has been found[43d] in the trunk bark of *Ochrosia moorei*, and 18-dehydronigritanine, 10-hydroxynigritanine, 18-dehydro-10-hydroxynigritanine, and the two related oxindole alkaloids (293) and (294) have been shown[40b] to occur in the leaves of *Strychnos barteri*, while the seeds contain nigritanine and the stem bark contains 18-dehydronigritanine. Nigritanine is also present in the seeds and stem bark of *S. nigritana*, and 18-dehydronigritanine in the stem bark; this paper[40b] also records the ^{13}C n.m.r. spectra of the nigritanine group of alkaloids. The leaves of *Cinchona ledgeriana* contain[44a] seven bisindole alkaloids, namely cinchophyllamine (295), isocinchophyllamine (296), tetradehydro-3α-cinchophylline (297), didehydro-3α-cinchophylline (298), 18,19-dihydro-3β-17β-cinchophylline (299), 3α,17β-cinchophylline (300), and 3β,17α-cinchophylline (301). This last alkaloid is new, and completes the quartet of alkaloids which differ only in the configuration at C-3 and/or C-17. The name cinchophylline is proposed[130] for the parent structure, the configurations at C-3 and C-17 being designated α and β. Carbon-13 n.m.r. data for all four bases (295), (296), (300), and (301) have been reported, and the four structural analogues of gross structure (302) have been

[128] A. N. Ratnagiriswaran and K. Venkatachalam, *J. Indian Chem. Soc.*, 1942, **19**, 389.
[129] K. K. Purushothaman and A. Sarada, *Phytochemistry*, 1981, **20**, 351.

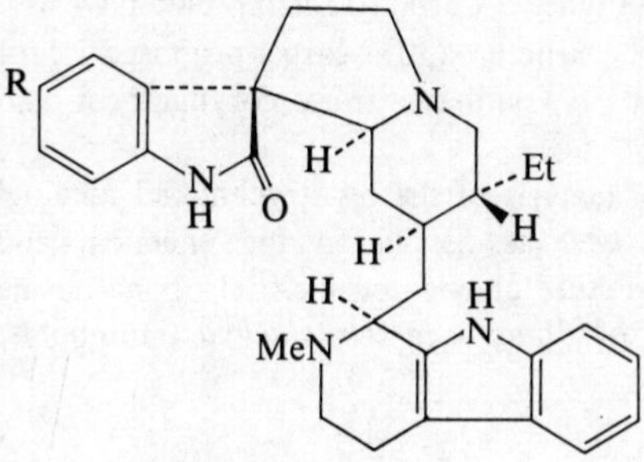

(293) R = H
(294) R = OH

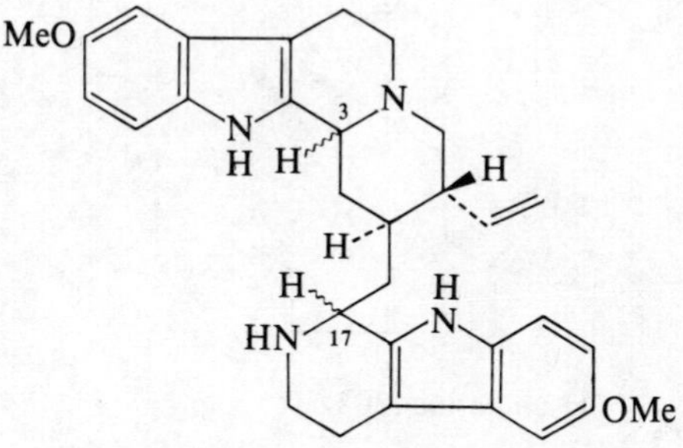

	3-H	17-H
(295)	β	β
(296)	α	α
(300)	α	β
(301)	β	α

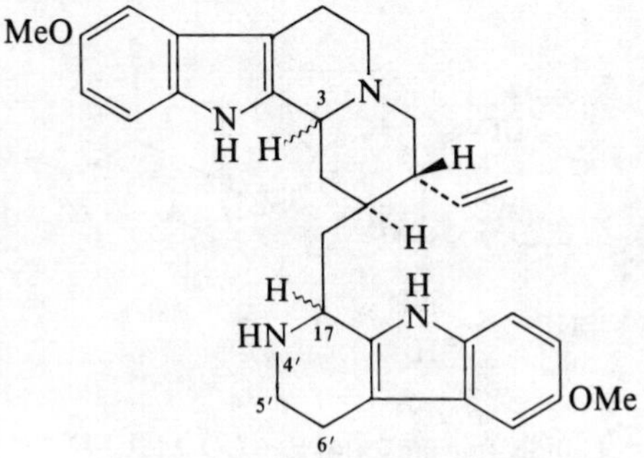

(297) 17,4′,5′,6′-tetrahydro; 3αH
(298) 17,4′-didehydro; 3αH
(299) 18,19-dihydro; 3βH, 17βH

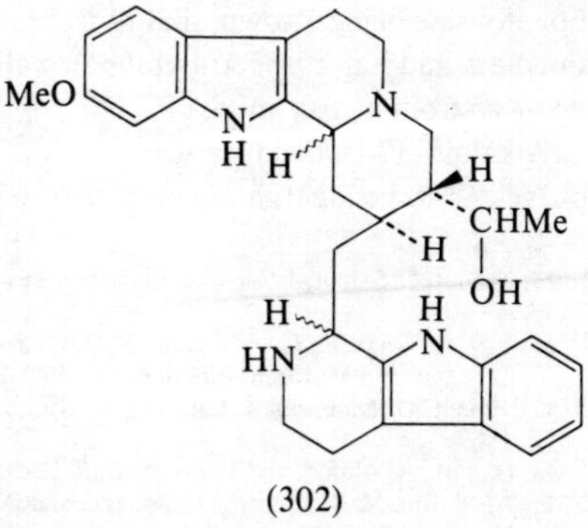

(302)

synthesized from tetraphyllicine and tryptamine.[130] Carbon-13 n.m.r. data have also been reported for usambarensine,[19c] and the structure (303), earlier proposed[131a] for usambarine, has now been proved by its synthesis from corynantheal and N_b-methyltryptamine.[131b]

Details of the X-ray crystal-structure analysis of the bis-strychninoid alkaloid sungucine have been published,[132] as well as details on the chemical and crystallographic determination of the structure of bonafousine.[133] Isobonafousine (304), another constituent of the leaves of *Bonafousia tetrastachya* (Humboldt,

Usambarine (303)

Isobonafousine (304) E = CO_2Me

Bonpland, et Kunth) Mgf., is composed of the same monomeric units, but the junction here is between C-12 of the hydroxycoronaridine component and a position adjacent to N_b in the hydrocanthine component. The structure of bonafousine being known, that of isobonafousine was determined by comparison of the mass and n.m.r. spectra of the two alkaloids, and the absolute configuration was deduced from its c.d. spectrum.[133]

Alkaloid T, isolated earlier[134a] from the stems and roots of *Hazunta costata*, proves[134b] to be tabernaelegantine A (305), and Alkaloid X, a new alkaloid from the

[130] M. Zeches, F. Sigaut, L. Le Men-Olivier, J. Lévy, and J. Le Men, *Bull. Soc. Chim. Fr.*, Part 2, 1981, 75.
[131] (a) L. J. G. Angenot, C. A. Coune, M. J. G. Tits, and K. Yamada, *Phytochemistry*, 1978, 17, 1687; (b) E. Seguin and M. Koch, *Planta Med.*, 1979, 37, 175 (*Chem. Abstr.*, 1980, 92, 111 208).
[132] L. Dupont, O. Dideberg, J. Lamotte, K. Kambu, and L. Angenot, *Acta Crystallogr.*, Sect. B, 1980, 36, 1669.
[133] M. Damak, A. Ahond, and P. Potier, *Bull. Soc. Chim. Fr.*, Part 2, 1980, 490.
[134] (a) A. M. Bui, M. M. Debray, P. Boiteau, and P. Potier, *Phytochemistry*, 1977, 16, 703; (b) M. Urrea, A. Ahond, A. M. Bui, and P. Potier, *Bull. Soc. Chim. Fr.*, Part 2, 1981, 147.

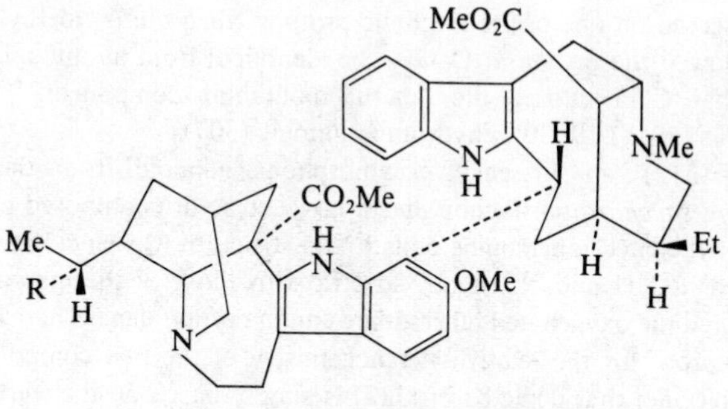

Tabernaelegantine A (305) R = H
(19R)-Hydroxytabernaelegantine A (306) R = OH

same source, is readily diagnosed, from its mass and n.m.r. spectra, to be a
19-hydroxy-tabernaelegantine A (306), in which C-19 has the *R* configuration.[134b]
The second new alkaloid (307) is composed of ibogamine and tabernaemontanine
units (mass and n.m.r. spectra), a deduction that was confirmed by its synthesis
from ibogamine (308) and tabernaemontaninol (309) in acid solution. A decision
between positions 10 and 11 as the point of attachment of the tabernaemontanine
component to the ibogamine component could not be made, either on the basis of
the n.m.r. spectra or on the mode of synthesis. Hence the *N*-methyl derivative (310)
was synthesized from tabernaemontaninol and N_a-methylibogamine (311). A n.O.e.

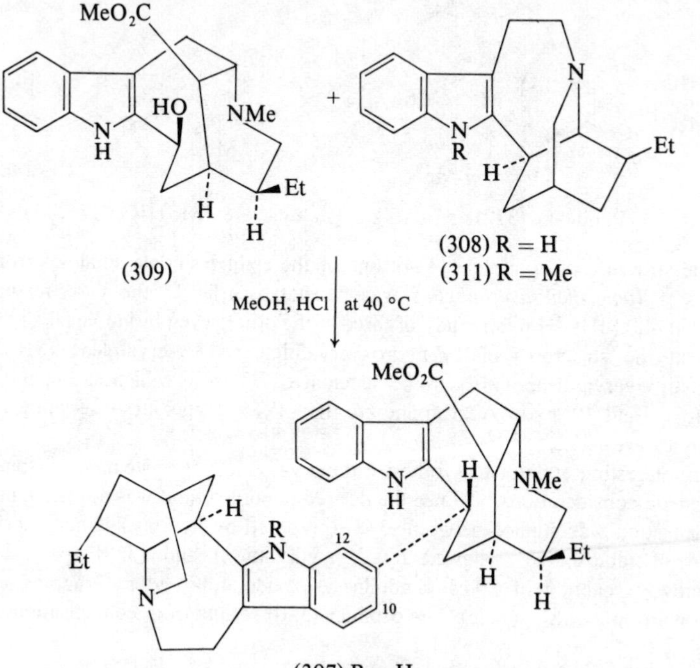

(309)

(308) R = H
(311) R = Me

MeOH, HCl at 40 °C

(307) R = H
(310) R = Me

that was observed on one of the aromatic protons when the N_a-methyl group was irradiated allowed the proton at C-12 to be identified; from its multiplicity, it was concluded that C-11 carried the tabernaemontanine component.[134b] This new alkaloid is thus (20'S)-19',20'-dihydrotabernamine (307).

Pandicine (312), an air-sensitive, amorphous alkaloid from the leaves of *Pandacastrum saccharatum* Pichon, has a novel structure composed of macroline and highly oxygenated tabersonine units.[135] Its structure was elucidated mainly by comparison of its ^{1}H and [13]C n.m.r. spectra with those of the macroline unit in villalstonine and the oxygenated tabersonine unit in cryophylline. There is at present no definitive proof for the relative stereochemistry of the two components of the pandicine molecule; that depicted in (312) is simply based on the stereochemistry deduced for related monomers, namely talcarpine and hazuntinine. Presumably the ease of oxidation of pandicine is due to the facile formation of the quinonoid species (313).

Pandicine (312)   (313) [R as in (312)]

The structure of ervafoline (314), one of the eight bisindole alkaloids from the leaves of *Stenosolen heterophyllus*, was elucidated earlier by the X-ray method. A detailed 400 MHz [1]H n.m.r study of three of the other seven bisindole alkaloids has allowed the structures of 19'-hydroxyervafoline (315), ervafolene (316), and 19'-hydroxyervafolene (317) to be elucidated,[117d] from which it appears that ervafoline and 19'-hydroxyervafoline are the 14',15'-β-epoxides of ervafolene and 19'-hydroxyervafolene.[117d]

An interesting attempt to synthesize the ervafoline ring system was inspired by biogenetic considerations.[136] Since the desired pandoline unit was not available, the condensation was studied between (318) (prepared by hydrolysis and decarboxylation of tabersonine, followed by N_a-methylation) and (319) (prepared by Polonovski reaction of 20-*epi*-pandoline N_b-oxide, followed by trapping of the immonium ion with cyanide). The product (320) results from condensation of the

[135] C. Kan Fan, G. Massiot, B. C. Das, and P. Potier, *J. Org. Chem.*, 1981, **46**, 1481.
[136] A. Henriques and H.-P. Husson, *Tetrahedron Lett.*, 1981, **22**, 567.

Ervafoline (314) R = H
19'-Hydroxyervafoline (315) R = OH
Ervafolene (316) R = H; $\Delta^{14',15'}$
19'-Hydroxyervafolene (317) R = OH; $\Delta^{14',15'}$

enamine (318) with the immonium ion derived from (319), followed by formation of a carbinolamine ether. The stereochemistry proposed for (320) is that which results if the condensation takes the most favoured steric course.

Kutney *et al.*[137] have investigated the synthesis of alkaloids of the voacamine type by applying the Büchi condensation to the reaction between perivinol or dregaminol and voacangine or catharanthine. For example, the condensation between perivinol and voacangine gave desmethylvoacamine (321), which has been found in *Tabernaemontana accedens*, and its 9'-isomer, which is *N*-desmethyl-voacamidine. This communication also includes a considerable amount of n.m.r. data on bisindole alkaloids of the voacamine, tabernaelegantine, and vinblastine type, which is most useful as an aid to structure elucidation.

Vinblastine has been found in the roots of *Catharanthus trichophyllus* (Bak.) Pichon, apparently for the first time.[63d] Catharanthamine (322), a new alkaloid[138] related to vinblastine, and with anti-tumour activity, is the first alkaloid of the group to contain oxygen at C-17'; its structure was deduced from a comparison of its ^{13}C n.m.r. spectrum with that of vinblastine. Proton n.m.r. data for vinblastine and leurosine are the subject of two further communications,[139a,b] and ^{13}C n.m.r. data form the subject of the third.[139c]

15-Oxo-15,20*S*-dihydrocatharanthine (275a), prepared as described above, has been employed in a new synthesis of anhydrovinblastine (323). Stereospecific reduction of (275a), followed by acetylation and formation of the N_b-oxide, gave an

[137] J. P. Kutney, A. Horinaka, R. S. Ward, and B. R. Worth, *Can. J. Chem.*, 1980, **58**, 1829.
[138] A. El-Sayed and G. A. Cordell, *J. Nat. Prod.*, 1981, **44**, 289.
[139] (*a*) A. De Bruyn, L. De Taeye, and M. J. O. Anteunis, *Bull. Soc. Chim. Belg.*, 1980, **89**, 629; (*b*) A. De Bruyn, L. De Taeye, R. Simonds, and C. De Pauw, *ibid.*, 1981, **90**, 185; (*c*) E. Wenkert, E. W. Hagaman, N. Wang, G. E. Gutowski, and J. C. Miller, *Heterocycles*, 1981, **15**, 255.

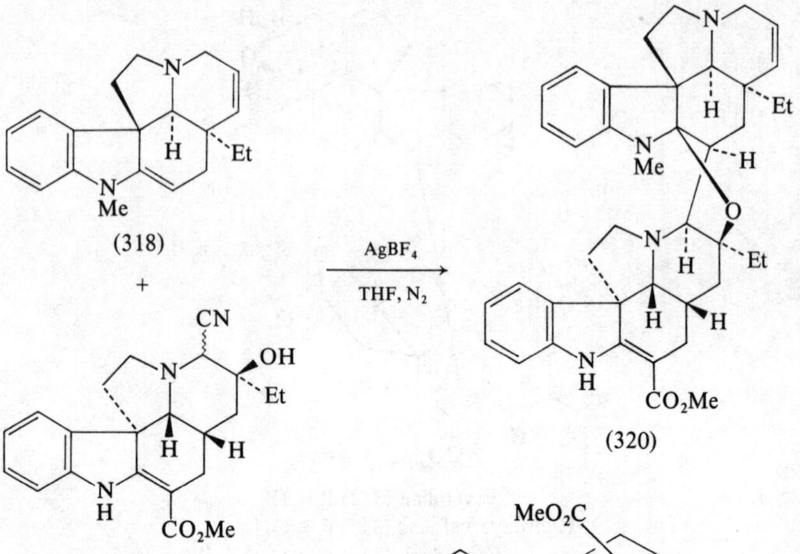

(318)

+

(319)

$\xrightarrow{\text{AgBF}_4}$ THF, N_2

(320)

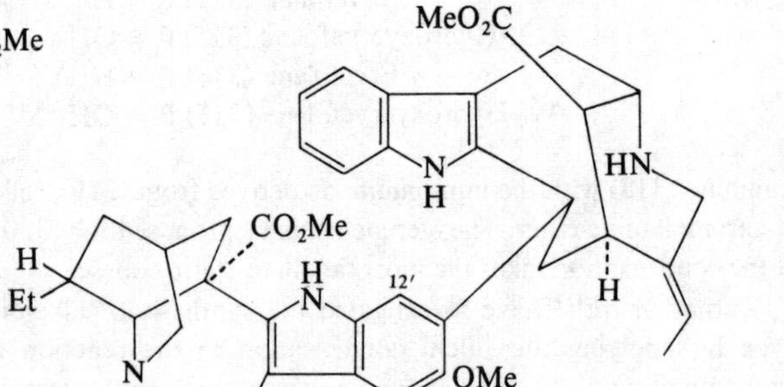

N-Desmethylvoacamine (321)

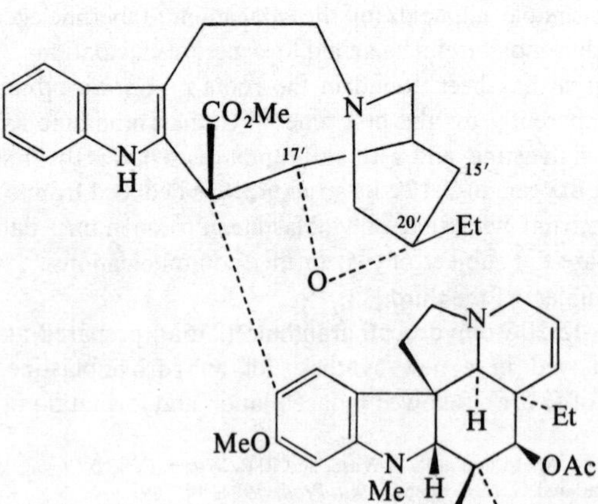

Catharanthamine (322)

intermediate (324), which was condensed with vindoline under the usual Polonovski–Potier conditions. The resulting enamine (325) lost the allylic acetate group at C-15 before it suffered reduction; the product was thus 15,20-anhydrovinblastine (323) (Scheme 42).[125]

In trifluoroacetic acid, 15-oxo-15,20-dihydrocatharanthine (275a + b) fragments, presumably *via* the enol of the C-7-protonated species, to give the epimeric tetracyclic keto-esters (326). Hydrogenation of the (16S)-epimer of (326), followed by reaction with N-chlorobenzotriazole and condensation with vindoline in acid solution, gives a bisindole species (327); unfortunately, this has the undesired 16'R configuration (Scheme 43).[140]

Further investigations have been carried out into the direct preparation of 5'-nor-alkaloids from parent alkaloids of the vinblastine group.[141] The 7'-chloro-indolenines (328) and (329) derived from anhydrovinblastine and leurosine fragment in the presence of silver fluoroborate to give, after re-cyclization, the 5'-nor-bases (330) and (331). In the formation of (328) and (329), some chlorination also occurs at position 12, and this can, under appropriate conditions, form the major reaction. Products similar to (330) and (331) were also obtained from the analogous reactions with anhydrovincristine and leuroformine.

The preparation of (331) described here is superior to the route which involves an application of the Polonovski reaction;[14f] in fact, the major product of this latter reaction proves to be 21'-hydroxyleurosine (332), identical with the alkaloid of this structure recently isolated from *Catharanthus ovalis*.[14g] Disproportionation of (332) readily occurs; hence the products of the reaction include 21'-oxoleurosine (333) and leurosine.[141]

5 Biogenetically Related Quinoline Alkaloids

Cinchona Group.—The relationship between conformation and antimalarial activity in the quinine series has been discussed.[142a] Apparently there is no intramolecular hydrogen-bond in the active alkaloids; conversely, the inactive ones exhibit intramolecular hydrogen-bonding.

Further reports on asymmetric synthesis in the presence of *Cinchona* alkaloids have been made.[142b,c] For example, hydrogenation of methyl pyruvate with a platinum–alumina catalyst containing quinine gives (+)-(R)-methyl lactate in 87% optical yield.[142b] Asymmetric induction with optical yields up to 36 and 26% has been observed in the Michael addition of thiols and nitro-alkanes to $\alpha\beta$-unsaturated ketones in the presence of quaternary salts derived from the *Cinchona* alkaloids.[142c]

Camptothecin.—Hutchinson has contributed a detailed review of the chemistry, biosynthesis, and pharmacology of camptothecin and its derivatives.[143]

[140] R. Z. Andriamialisoa, N. Langlois, and Y. Langlois, *Heterocycles*, 1981, **15**, 245.
[141] R. Z. Andriamialisoa, N. Langlois, Y. Langlois, and P. Potier, *Tetrahedron*, 1980, **36**, 3053.
[142] (a) B. J. Oleksyn and L. J. Lebioda, *Pol. J. Chem.*, 1980, **54**, 755; (b) Y. Orito, S. Imai, and S. Niwa, *Nippon Kagaku Kaishi*, 1980, 670 (*Chem. Abstr.*, 1980, **93**, 113 912); (c) S. Colonna, A. Re, and H. Wynberg, *J. Chem. Soc., Perkin Trans. 1*, 1981, 547.
[143] C. R. Hutchinson, *Tetrahedron*, 1981, **37**, 1047.

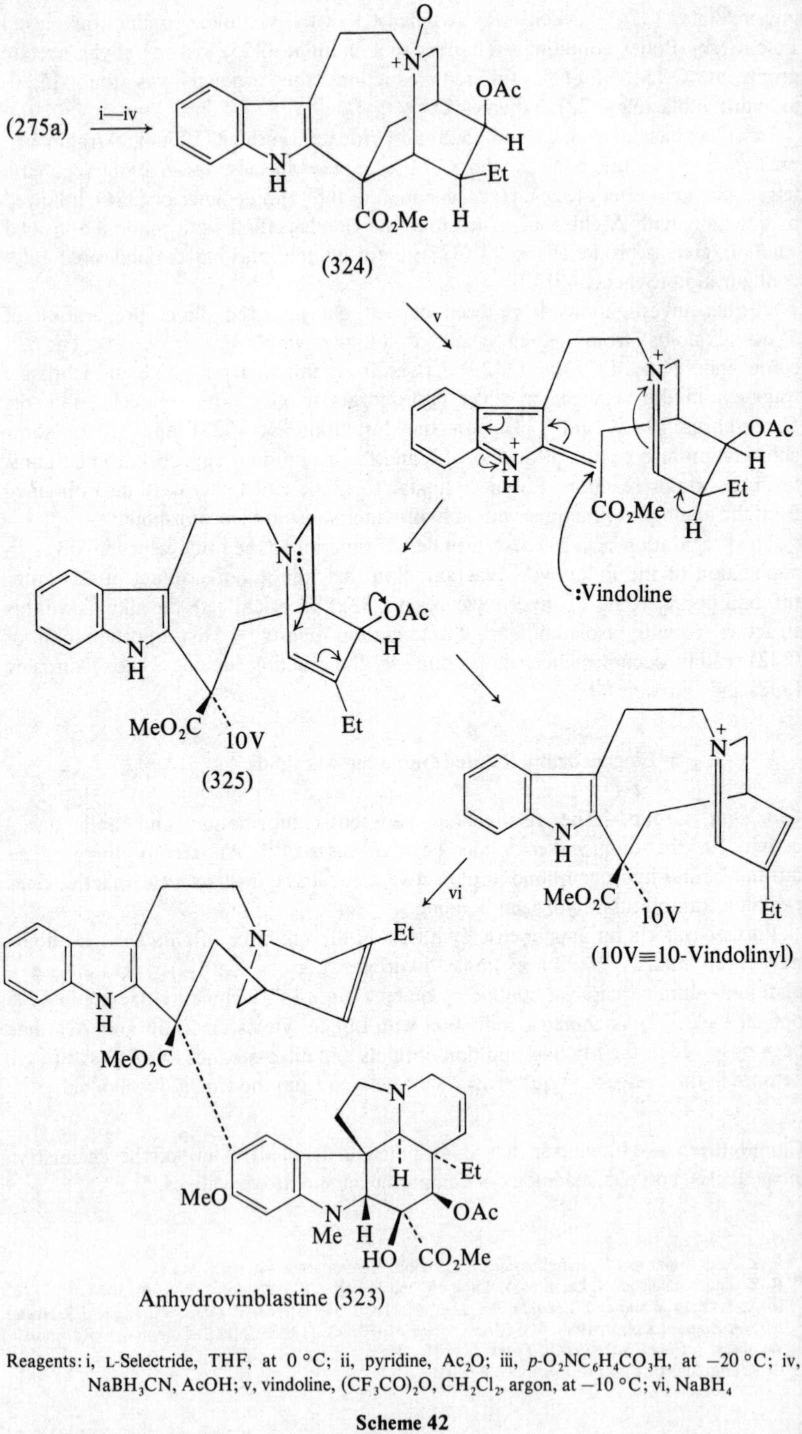

(275a) $\xrightarrow{\text{i—iv}}$ (324)

(325)

(10V≡10-Vindolinyl)

Anhydrovinblastine (323)

Reagents: i, L-Selectride, THF, at 0 °C; ii, pyridine, Ac$_2$O; iii, p-O$_2$NC$_6$H$_4$CO$_3$H, at −20 °C; iv, NaBH$_3$CN, AcOH; v, vindoline, (CF$_3$CO)$_2$O, CH$_2$Cl$_2$, argon, at −10 °C; vi, NaBH$_4$

Scheme 42

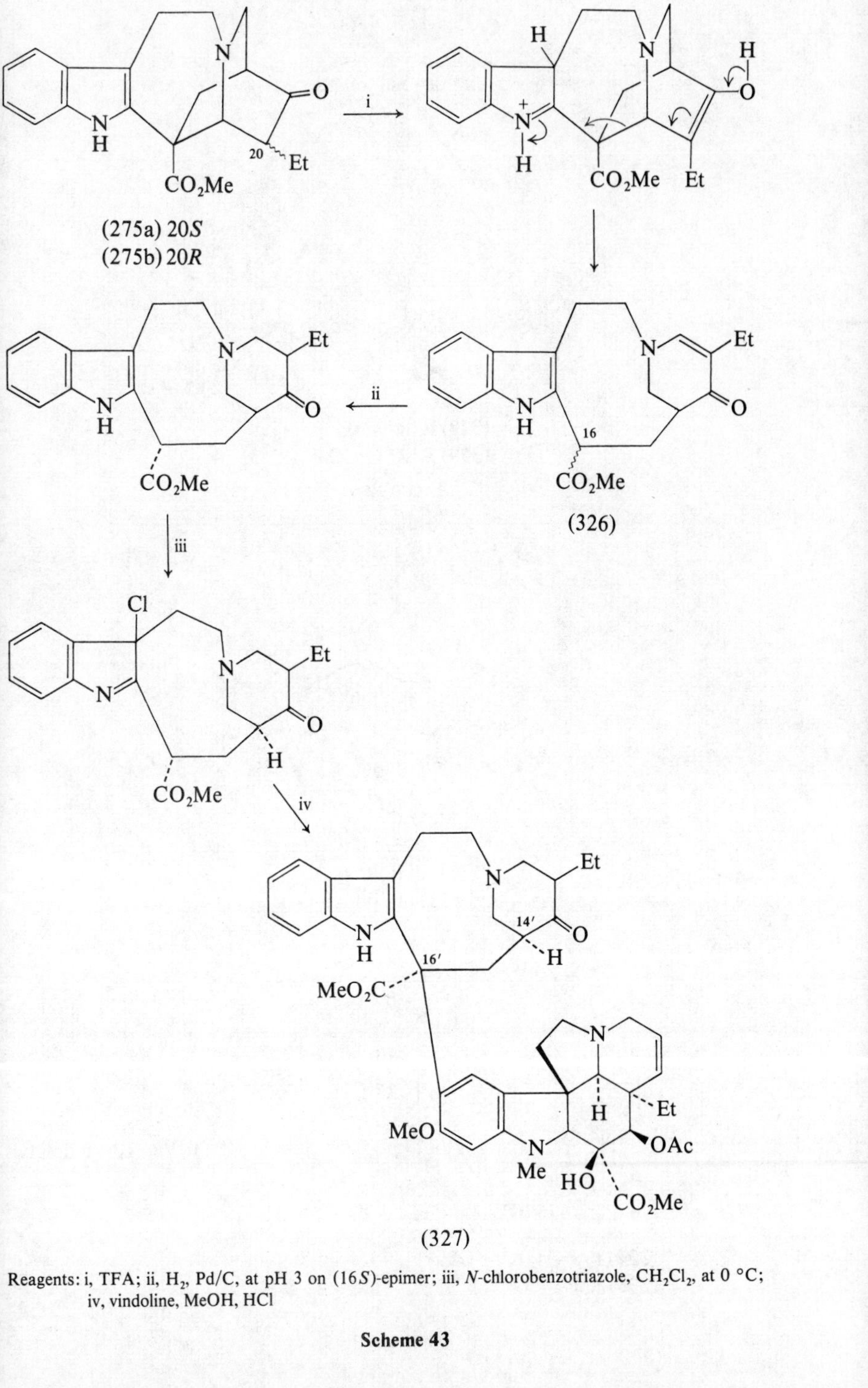

(275a) 20*S*
(275b) 20*R*

(326)

(327)

Reagents: i, TFA; ii, H₂, Pd/C, at pH 3 on (16*S*)-epimer; iii, *N*-chlorobenzotriazole, CH₂Cl₂, at 0 °C; iv, vindoline, MeOH, HCl

Scheme 43

(328) R^{1}R^2 = Δ
(329) R^{1}R^2 = —O—

(330) R^{1}R^2 = Δ
(331) R^{1}R^2 = —O—

(332) R^1 = H, R^2 = OH
(333) R^{1}R^2 = O

(10V ≡ 10-Vindolinyl)

A new synthesis (Scheme 44)[144] of the tetracyclic pyridone (334) constitutes a formal synthesis of (±)-camptothecin (335), since (334) has already been converted into camptothecin by Winterfeldt *et al.*[145] 10-Methoxycamptothecin (336) was similarly synthesized.

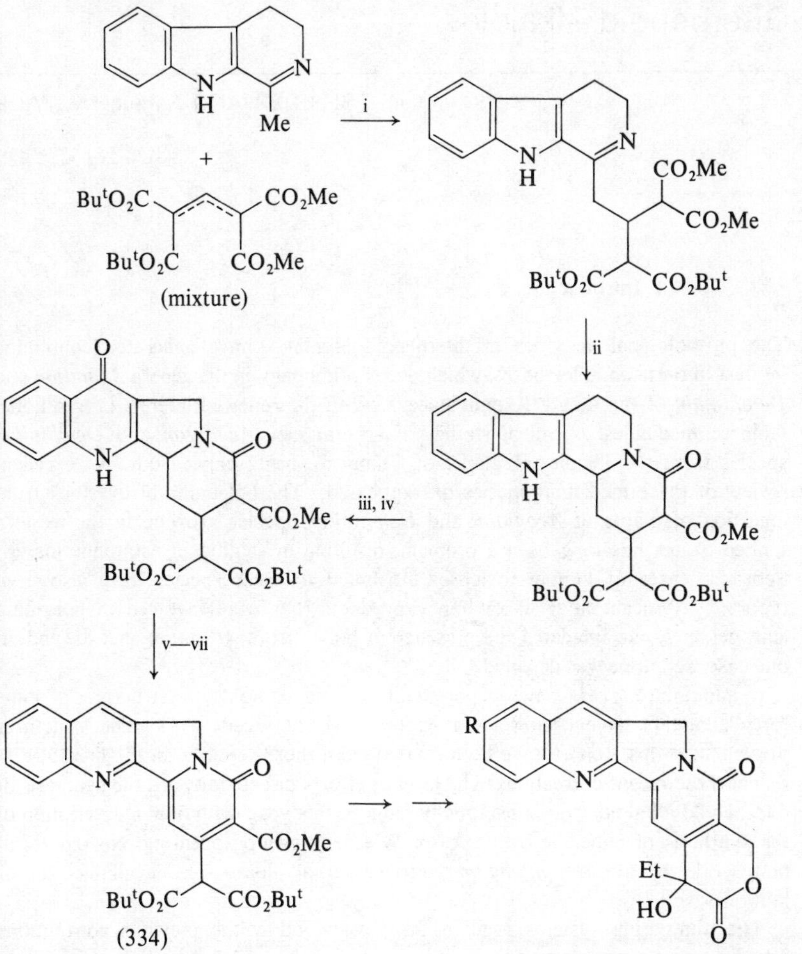

Camptothecin (335) R = H
10-Methoxycamptothecin (336) R = OMe

Reagents: i, THF, at r.t.; ii, NaBH₄; iii, *hv*/O₂, at r.t., Rose Bengal, MeOH; iv, NaHCO₃, H₂O; v, SOCl₂, DMF; vi, Pd/BaSO₄, H₂, MeOH; vii, DDQ, dioxan, heat

Scheme 44

[144] T. Kametani, T. Ohsawa, and M. Ihara, *Heterocycles*, 1980, **14**, 951; *J. Chem. Soc., Perkin Trans. 1*, 1981, 1563.
[145] E. Winterfeldt, T. Korth, D. Pike, and M. Boch, *Angew. Chem., Int. Ed. Engl.*, 1972, **11**, 289.

13

Diterpenoid Alkaloids

BY S. William PELLETIER AND Samuel W. PAGE

1 Introduction

The physiological activities of diterpenoid alkaloids have generated continuing interest in these complex bases, which occur principally in the genera *Aconitum* and *Delphinium* of the family Ranunculaceae. With the renewed interest in traditional Chinese medicines, chemical studies of several new *Aconitum* and *Delphinium* species from the Peoples Republic of China are being carried out. An excellent review of these medicinal species has appeared.[1] The poisoning of livestock from ingestion of plants of *Aconitum* and *Delphinium* species growing in the western United States has long been a problem, resulting in significant economic losses.[2] Sporadic cases of human toxicoses attributed to these species have also been reported. A recent analysis of honey products that were involved in poisoning episodes in Japan indicated the presence of pollen from *Aconitum* species, and, in one case, aconitine was detected.[3]

In addition to several newly reported diterpenoid alkaloids, the structures of some forty alkaloids have been revised on the basis of recent work. The structures presented in this Report have been revised from those reported in the literature to reflect these recent corrections. The level of efforts directed toward the synthesis of diterpenoid alkaloids was substantially reduced this year, with only a description of the synthesis of napelline by Professor Wiesner's group appearing. No reports of new work on the *Daphniphyllum* diterpenoid alkaloids were available to our laboratories.

The numbering systems used in this review follow the previous conventions for the aconitine, lycoctonine, atisine, and veatchine skeletons, as indicated in structures A, B, C, and D, respectively.

[1] N. G. Bisset, *J. Ethnopharmacology*, 1981, **4**, 247.
[2] E. H. Cronin, D. B. Nielsen, and N. J. Nadsen, *J. Range Management*, 1976, **29**, 364.
[3] Y. Saito, A. Mitsura, K. Sasaki, M. Satake, and M. Uchiyama, *Eisei Shikensho Hokoku*, 1980, No. 98, p. 532.

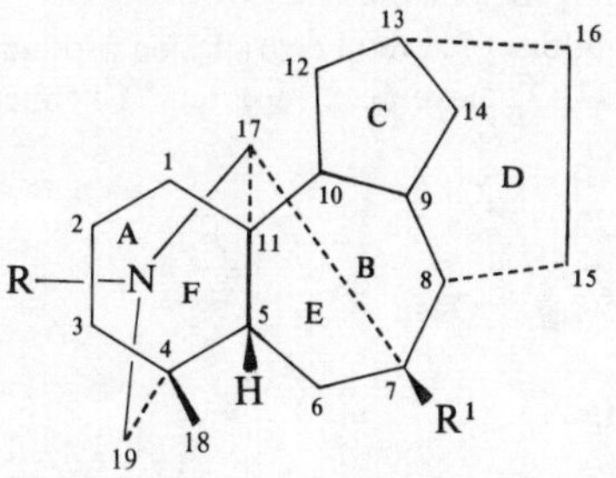

(A) Aconitine skeleton, $R^1 = H$
(B) Lycoctonine skeleton, $R^1 = OR^2$

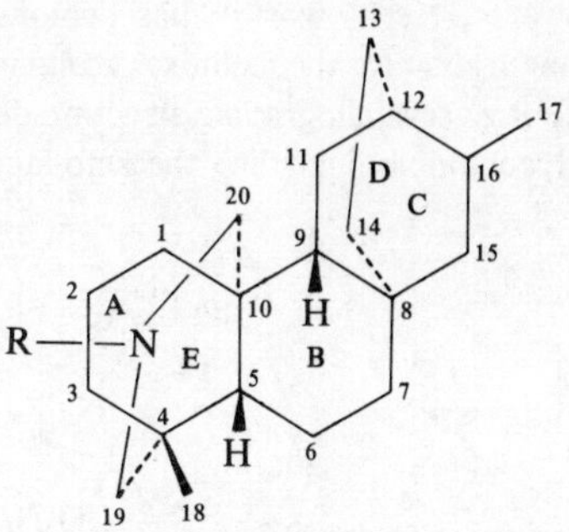

(C) Atisine skeleton

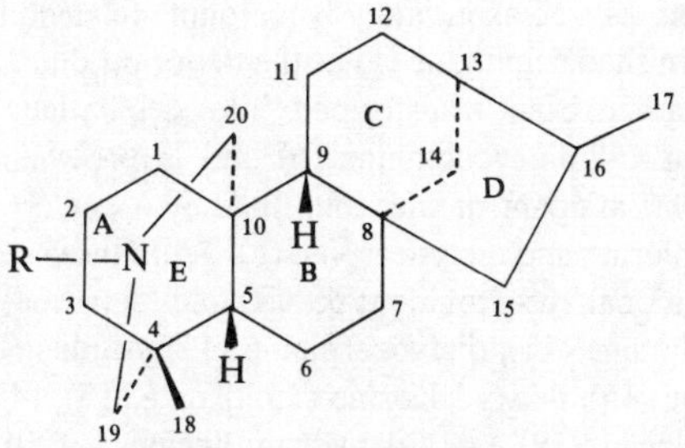

(D) Veatchine skeleton

2 Structure Revisions and General Studies

Revision of the Structures of Thirty-seven Lycoctonine-related Diterpenoid Alkaloids.—Lycoctonine was assigned structure (1) on the basis of an X-ray crystallographic analysis of (2) in 1956.[4] Since that time, the structures of most of the lycoctonine-type alkaloids have been based on correlations with lycoctonine.

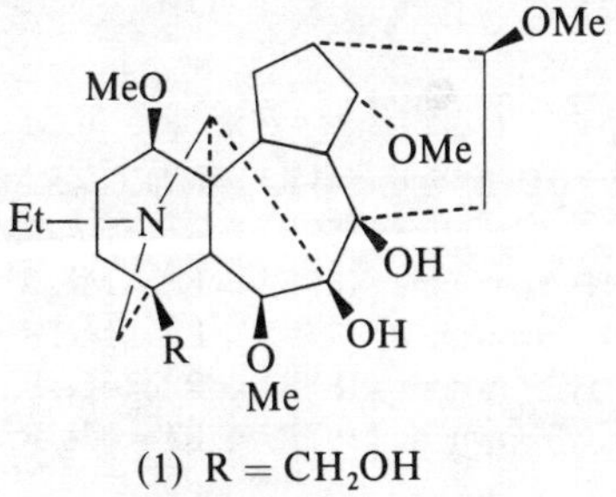

(1) R = CH₂OH
(2) R = H

[4] M. Przybylska and L. Marion, *Can. J. Chem.*, 1956, **34**, 185.

However, recent work[5,6] has demonstrated that the original assignment of a β-configuration for the methoxyl group at C-1 in lycoctonine was in error.

X-Ray crystallographic structure determinations of several degradation products of lycoctonine, including the keto-lactam acid (3), have been reported.[5] Chemical

(3)

and X-ray studies have established the α-configuration of the methoxyl group at C-1 in lycoctonine itself, as well as in many lycoctonine-related alkaloids.[6] Methylation of delsoline (4) furnished delphatine (5) as the major product. Since lycoctonine and browniine had earlier been transformed into delphatine by methylation,[7] the methoxyl group at C-1 in lycoctonine (6) and in browniine (7) must be in the α-configuration. This assignment was confirmed by X-ray crystallographic analysis of browniine perchlorate and dictyocarpine (8). From these results, the structures of the following additional lycoctonine-type alkaloids have been revised: elatine (9) deltaline (10), deltamine (11), dictyocarpinine (12), delpheline (13), delcorine (14), 6-dehydrodelcorine (15), deoxydelcorine (16), ilidine (17), 14-acetylbrowniine (18), 14-benzoylbrowniine (19), 14-dehydrobrowniine (20), 7,18-di-O-methyl-lycoctonine (21), delcaroline (22), demethylene-eldelidine (23), ambiguine (24),

Delsoline (4) $R^1 = H, R^2 = R^3 = Me$
Delphatine (5) $R^1 = R^2 = R^3 = Me$
Lycoctonine (6) $R^1 = R^2 = Me, R^3 = H$
Browniine (7) $R^1 = R^3 = Me, R^2 = H$
14-Acetylbrowniine (18) $R^1 = R^3 = Me, R^2 = Ac$
14-Benzoylbrowniine (19) $R^1 = R^3 = Me, R^2 = Bz$

[5] M. Cygler, M. Przybylska, and O. E. Edwards, *Acta Crystallogr., Sect. A, Suppl.*, 1981, **37**, Abstr. 09.2-32, on p. C-211.

[6] S. W. Pelletier, N. V. Mody, K. I. Varughese, J. A. Maddry, and H. K. Desai, *J. Am. Chem. Soc.*, 1981, **103**, 6536.

[7] S. W. Pelletier, R. S. Sawhney, H. K. Desai, and N. V. Mody, *J. Nat. Prod. (Lloydia)*, 1980, **43**, 395.

Dictyocarpine (8) $R^1 = Ac, R^2 = OH, R^3 = H$
Deltaline (10) $R^1 = Ac, R^2 = OH, R^3 = Me$
Deltamine (11) $R^1 = H, R^2 = OH, R^3 = Me$
Dictyocarpinine (12) $R^1 = R^3 = H, R^2 = OH$
Delpheline (13) $R^1 = R^2 = H, R^3 = Me$

Elatine   (9) $R^1 = H, \beta\text{-}OMe; R^2 = Me; R^3 = -C$
Delcorine (14) $R^1 = H, \beta\text{-}OH; R^2 = R^3 = Me$
6-Dehydrodelcorine (15) $R^1 = O; R^2 = R^3 = Me$
Deoxydelcorine (16) $R^1 = H_2; R^2 = R^3 = Me$
Ilidine (17) $R^1 = O; R^2 = H; R^3 = Me$

14-Dehydrobrowniine (20) $R^1 = H; R^2 = O$
7,18-Di-*O*-methyl-lycoctonine (21) $R^1 = Me; R^2 = H, \alpha\text{-}OMe$

Delcaroline (22) R^1 = Me, R^2 = H, R^3 = CH$_2$OMe
Demethylene-eldelidine (23) R^1 = H, R^2 = R^3 = Me

Ambiguine (24) R^1 = R^3 = Me, R^2 = Ac
Delectinine (25) R^1 = R^2 = R^3 = H
Delectine (26) R^1 = R^2 = H, R^3 = COC$_6$H$_4$-2-NH$_2$
14-Acetyldelectine (27) R^1 = H, R^2 = Ac, R^3 = COC$_6$H$_4$-2-NH$_2$
N-Acetyldelectine (28) R^1 = R^2 = H, R^3 = COC$_6$H$_4$-2-NHAc
Ajadine (29) R^1 = H, R^2 = Ac, R^3 = COC$_6$H$_4$-2-NHAc

Ajacusine (30) R^1 = H, R^2 = Bz, R^3 =

delectinine (25), delectine (26), 14-acetyldelectine (27), *N*-acetyldelectine (28), ajadine (29), ajacusine (30), tricornine (31), anthranoyl-lycoctonine (32), ajacine (33), avadharidine (34), *N*-(3-hydroxycarbonylpropionyl)anthranoyl-lycoctonine (35), delsemine (36), methyl-lycaconitine (37), lycaconitine (38), septentriodine (39), and septentrionine (40).[6]

In view of the fact that no C$_{19}$ diterpenoid alkaloids with a β-methoxyl group are now known to occur in Nature, the β-configuration for the hydroxyl groups at C-1 in such bases as talatizidine (41) and delphirine (42) is interesting.

With the more advanced systems now available for the collection, refinement,

Tricornine (31) R = Me
Anthranoyl-lycoctonine (32) R = C$_6$H$_4$-2-NH$_2$
Ajacine (33) R = C$_6$H$_4$-2-NHAc
Avadharidine (34) R = C$_6$H$_4$-2-NHCOCH$_2$CH$_2$CONH$_2$
N-(3-Hydroxycarbonylpropionyl)-
anthranoyl-lycoctonine (35) R = C$_6$H$_4$-2-NHCOCH$_2$CH$_2$CO$_2$H
Delsemine (36) R= C$_6$H$_4$-2-NHCOCH(Me)CH$_2$CONH$_2$ and
C$_6$H$_4$-2-NHCOCH$_2$CH(Me)CONH$_2$

Methyl-lycaconitine (37) R = 2-C$_6$H$_4$—N

Lycaconitine (38) R = 2-C$_6$H$_4$—N

Septentriodine (39) R = C$_6$H$_4$-2-NHCH$_2$CH$_2$CO$_2$Me

O=C—C$_6$H$_4$-2-NHCH$_2$CH$_2$CO$_2$Me
Septentrionine (40)

Talatizidine (41) R = H
Delphirine (42) R = OMe

and manipulation of *X*-ray crystallographic data, such mistakes in structure assignment are highly unlikely. Indeed, they are quite rare even in the earlier literature. Were it not for the aura of 'infallibility' of *X*-ray crystallographic structure analyses, the error in the structure of lycoctonine would probably have been recognized a number of years ago.

Revisions Based on ¹³C N.M.R. Analysis.—The revision of the structures of five C_{19} diterpenoid alkaloids on the basis of the ¹³C n.m.r. spectral data has been reported.[8,9] Yunusov and co-workers,[10] relying predominantly on mass-spectral data, had proposed structure (43) for acomonine ($C_{25}H_{41}NO_7$), isolated from the roots of *Aconitum monticola*. Based partially on chemical correlations with acomonine, they later assigned structures to iliensine (44) ($C_{24}H_{39}NO_7$)[11] and 14-dehydroiliensine (45),[12] isolated from *Delphinium biternatum*. These were the first reported C_{19} diterpenoid alkaloids lacking an oxygen function at C-1.

In examining authentic samples of acomonine and iliensine that were provided by Dr. M. S. Yunusov, the ¹³C n.m.r. spectra for these compounds were found to be

(43) R = H, α-OMe
(44) R = H, α-OH
(45) R = O
(46) R = H, α-OBz

[8] S. W. Pelletier and N. V. Mody, *Tetrahedron Lett.*, 1981, **22**, 207.
[9] N. V. Mody, S. W. Pelletier, and N. M. Mollov, *Heterocycles*, 1980, **14**, 1751.
[10] V. Nezhevenko, M. S. Yunusov, and S. Yu. Yunusov, *Khim. Prir. Soedin.*, 1975, 389.
[11] M. S. Yunusov, V. E. Nezhevenko, and S. Yu. Yunusov, *Khim. Prir. Soedin.*, 1975, 770.
[12] B. T. Salimov, M. S. Yunusov, and S. Yu. Yunusov, *Khim. Prir. Soedin.*, 1978, 106.

identical with those of delsoline (4) and delcosine (47), respectively.[8] All the physical constants and the mass and [1]H n.m.r. spectral data confirmed the identity of acomonine and iliensine with delsoline and delcosine, respectively. Consequently, 14-dehydroiliensine (45) corresponds to 14-dehydrodelcosine (48) and 14-benzoyliliensine (46) corresponds to 14-benzoyldelcosine (49). Comparisons of the reported data[12] support this conclusion. Therefore, the names 'acomonine' and 'iliensine' should no longer be used.

The structure of cammaconine, isolated from plants of *Aconitum variegatum*, has been revised to (50) on the basis of the [13]C n.m.r. data.[9] Mollov and co-workers[13] had assigned structure (51) to cammaconine from the results of a chemical correlation with isotalatizidine (52) and [1]H n.m.r. and mass-spectral analyses. Cammaconine is therefore identical with the hydrolysis product of aconorine (53), isolated from *Aconitum orientale* Mill.[14]

Delsoline   (4) R = H, α-OMe
Delcosine (47) R = H, α-OH
14-Dehydrodelcosine (48) R = O
14-Benzoyldelcosine (49) R = H, α-OBz

Cammaconine (50) R^1 = R^2 = Me, R^3 = H
(51) R^1 = R^3 = Me, R^2 = H
Isotalatizidine (52) R^1 = R^3 = H, R^2 = Me
Aconorine (53) R^1 = R^2 = Me, R^3 = COC$_6$H$_4$-2-NHAc

[13] M. A. Khaimova, M. D. Palamareva, N. M. Mollov, and V. P. Krestev, *Tetrahedron*, 1971, **27**, 819.
[14] V. A. Tel'nov, M. S. Yunusov, S. Yu. Yunusov, and B. Sh. Ibragimov, *Khim. Prir. Soedin.*, 1975, 814.

Computer-assisted Analysis of ^{13}C N.M.R. Spectra and Prediction of Structures for the C_{19} Diterpenoid Alkaloids.—A logical extension of work on the expanding ^{13}C n.m.r. spectral data-base for the C_{19} diterpenoid alkaloids was the application of computer methods for handling these data. This task has been accomplished in collaborative efforts by researchers at the University of Georgia (USA) and at Stanford University.[15] These studies utilized recently developed computer programs[16-18] and a data-base of the ^{13}C n.m.r. resonances with the substructures which characterize the constitutional and stereochemical environments of the resonating carbons of ninety-three C_{19} diterpenoid alkaloids and their derivatives. Unknown ^{13}C n.m.r. spectra are analysed by comparing the observed resonances with those of the data-base and retrieving a set of substructures. These substructures are refined, using an iterative interpretation procedure, to deduce detailed portions of the unknown structure. Proposed structures are then generated and evaluated by predicting their spectra and comparing each predicted spectrum of the unknown with the observed spectrum. The candidate spectra are then ordered in rank. These methods were evaluated for the correction of the structures of 'iliensine' and 'acomonine' described in the previous section and for the structure determinations of glaucephine, glaucerine, and glaucenine, described in Section 4. These programs can quickly limit the possible structures of complex *Aconitum*-type bases to a choice of only two or three. In addition, these computer methods facilitate the storage and retrieval of the ^{13}C n.m.r. data.

3 Alkaloids from *Aconitum* Species

Screening of European *Aconitum* Species by T.L.C.—Staehelin and Katz[19,20] have reported phytochemical studies of several species and subspecies of the genus *Aconitum* in Europe. The following plants were screened by t.l.c.: *A. variegatum* (3 subspecies), *A. paniculatum* (2 subspecies), *A. toxicum*, *A. napellus* (9 subspecies), *A. pentheri* (2 subspecies), and *A. angustifolium*. Extracts of *A. variegatum*, *A. paniculatum*, and *A. toxicum* showed mainly two different t.l.c. patterns, which correlated with geographic regions rather than with botanical classification.[20] The plants collected east of the massif of the St. Gothard and the Pyrenees showed a pattern with the major alkaloid at $R_F = 0.76$. From plants collected in the Alps west of St. Gothard, the major alkaloid in the t.l.c. analyses occurred at $R_F = 0.28$. This alkaloid, $C_{31}H_{35}NO_7$ (m.pt 265—268 °C), was isolated and partially characterized. This base corresponds in melting point to paniculatin, reported in 1921 by Brunner from *A. paniculatum*.[21] From the ^{1}H and ^{13}C n.m.r., i.r., and mass-spectral data, the partial structure (54) was assigned for paniculatin.

[15] J. Finer-Moore, N. V. Mody, S. W. Pelletier, N. A. B. Gray, C. W. Crandell, and D. H. Smith, *J. Org. Chem.*, 1981, **46**, 3399.

[16] N. A. B. Gray, C. W. Crandell, J. G. Nourse, D. H. Smith, M. L. Dageforde, and C. Djerassi, *J. Org. Chem.*, 1981, **46**, 703.

[17] N. A. B. Gray, J. G. Nourse, C. W. Crandell, D. H. Smith, and C. Djerassi, *Org. Magn. Reson.*, 1981, **15**, 375.

[18] R. E. Carhart, D. H. Smith, N. A. B. Gray, J. G. Nourse, and C. Djerassi, *J. Org. Chem.*, 1981, **46**, 1708.

[19] A. Katz and E. Staehelin, *Pharm. Acta Helv.*, 1979, **54**, 253.

[20] E. Staehelin and A. Katz, *Pharm. Acta Helv.*, 1980, **55**, 221.

[21] G. E. Brunner, *Diss. ETH Zurich*, 1921, 60.

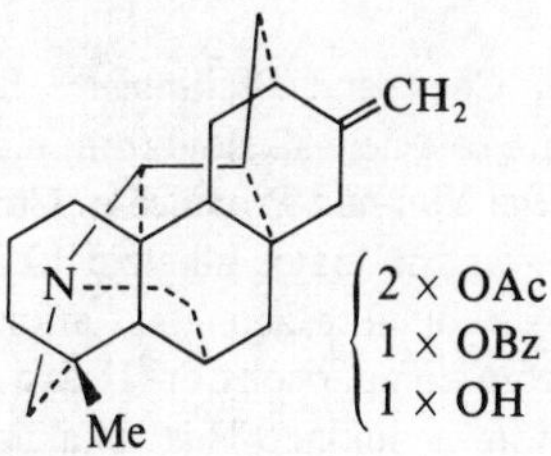

Paniculatin (54)

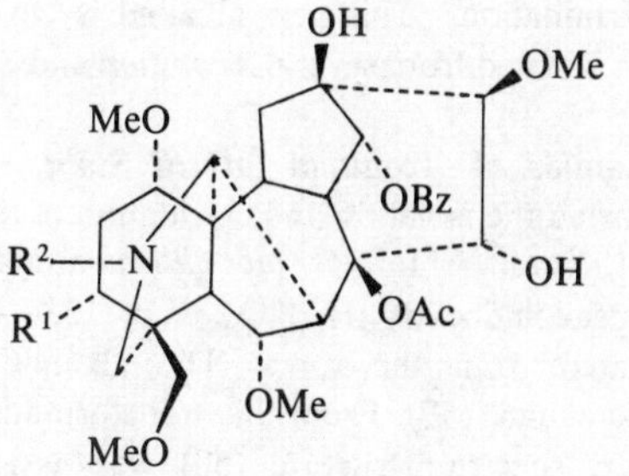

$2 \times$ OAc
$1 \times$ OBz
$1 \times$ OH

Aconitine (55) $R^1 = OH$, $R^2 = Et$
Mesaconitine (56) $R^1 = OH$, $R^2 = Me$
Hypaconitine (57) $R^1 = H$, $R^2 = Me$
Deoxyaconitine (79) $R^1 = H$, $R^2 = Et$

Extracts of *A. napellus*, *A. pentheri*, and *A. angustifolium* exhibited three main types of t.l.c. pattern.[19] Aconitine (55) and mesaconitine (56) were identified as the two major t.l.c. spots in *A. napellus* ssp. *fissurae* and *superbum*. *Aconitum angustifolium* contained mainly mesaconitine (56) and hypaconitine (57), with only small amounts of aconitine. The t.l.c. patterns from extracts of all of the plants indicated the presence of several other alkaloids. From these patterns, the following subspecies of *A. napellus* could be further distinguished: *napellus*, *corsicum*, *neomontanum*, *vulgare*, *hians*, and *tauricum*. Commercial samples of 'aconitine' were analysed and shown to be a mixture of mesaconitine and hypaconitine, containing only traces of aconitine. This result suggests that this 'aconitine' was extracted from *A. angustifolium*. Mesaconitine (56) and hypaconitine (57) were isolated from these samples, and their structures were confirmed by [13]C n.m.r. spectral studies.

Alkaloids of *Aconitum delphinifolium* DC.—Benn and co-workers[22] have isolated a new alkaloid, delphinifoline ($C_{23}H_{37}NO_7$; m.pt 218—220 °C), from this species. Structure (58) was assigned on the basis of the ^{1}H and [13]C n.m.r., i.r., and mass-spectral data and confirmed by an *X*-ray crystallographic structure

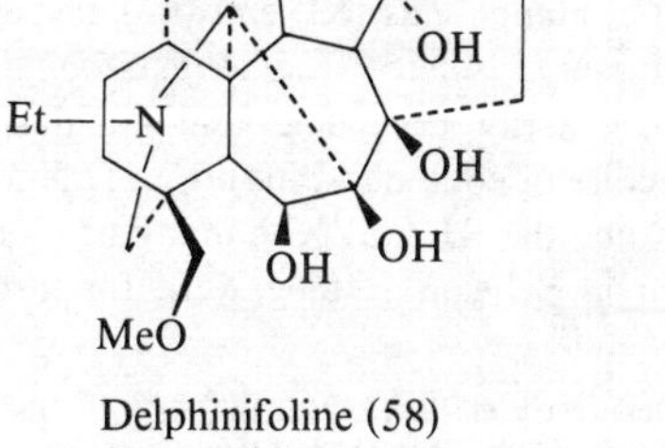

Delphinifoline (58)

[22] V. N. Aiyar, P. W. Codding, K. A. Kerr, M. H. Benn, and A. J. Jones, *Tetrahedron Lett.*, 1981, **22**, 483.

determination. This new alkaloid is closely related to delcosine (47), which has been isolated from several *Aconitum* and *Delphinium* species.

Alkaloids of *Aconitum forestii* Stapf.—Wei-Shin Chen and Breitmaier[23] have reported the isolation and elucidation of the structure of a new alkaloid from plants of *A. forestii* Stapf var. *albo-villosum* collected in the Yun-nan Province of China. Foresaconitine ($C_{35}H_{49}NO_9$; m.pt 153—154 °C) was the major alkaloid (0.2%) isolated from the roots. The alkaline hydrolysis of foresaconitine afforded chasmanine (59). From this transformation and from comparisons of ${}^1H$ and ${}^{13}C$ n.m.r. spectra, structure (60) was assigned to foresaconitine. This alkaloid is identical with villmorrianine C, which is described on page 264.

Chasmanine (59) $R^1 = R^2 = H$
Foresaconitine (60) $R^1 = Ac$, $R^2 = COC_6H_4$-4-OMe

Alkaloids of *Aconitum heterophyllum* Wall.—The nature of acid-catalysed rearrangement of hetisine (61), an alkaloid isolated from *A. heterophyllum* and *Delphinium cardinale*, eluded researchers for a number of years. A compound (m.pt 278—279.5 °C) with the same molecular formula as hetisine, *i.e.* $C_{20}H_{27}NO_3$, was isolated from *A. heterophyllum*. This same compound was formed by treatment of hetisine with 5% sulphuric acid.[24] Wiesner and co-workers[25] observed an acid-catalysed rearrangement in the formation of a methiodide of hetisine. They proposed structure (62) for this rearrangement product.

A recent investigation[26] of this problem has shown that the ${}^{13}C$ and ${}^1H$ n.m.r. spectral data are not consistent with structure (62), and has revealed that a novel rearrangement occurs, resulting in an adamantane-type system. Treatment of hetisine (61) with 10% aqueous HCl or 5% aqueous sulphuric acid under reflux gave a 95:5 mixture of compounds (63) and (64), respectively. Compound (63) is identical with the compound isolated from *A. heterophyllum*. The structure of (63) was confirmed by an *X*-ray crystallographic analysis ($R = 0.046$ on 1379 reflections). The structure of compound (64) ($C_{20}H_{27}NO_3$; m.pt 288—290 °C) was assigned from the i.r. and the ${}^1H$ and ${}^{13}C$ n.m.r. data. The proposed rearrangement pathways are shown in Scheme 1; they were supported by deuterium-labelling

[23] W.-S. Chen and E. Breitmaier, *Chem. Ber.*, 1981, **114**, 394.
[24] A. J. Solo and S. W. Pelletier, *J. Am. Chem. Soc.*, 1959, **81**, 4439.
[25] K. Wiesner, Z. Valenta, and L. G. Humber, *Tetrahedron Lett.*, 1962, 621.
[26] S. W. Pelletier, N. V. Mody, J. Finer-Moore, A.-M. M. Ateya, and L. C. Schramm, *J. Chem. Soc., Chem. Commun.*, 1981, 327.

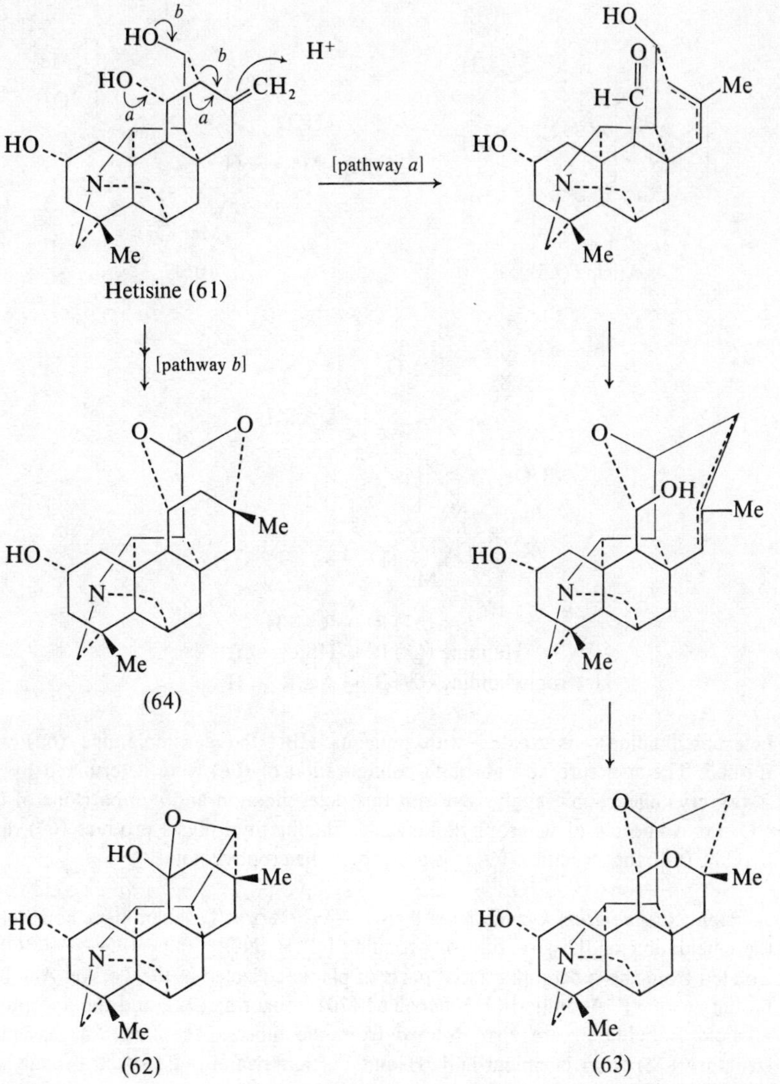

Scheme 1

studies. Since isolation of the alkaloids from the plant material involved an acid extraction, the alkaloid (63) from *A. heterophyllum* is probably an artifact.

Alkaloids of *Aconitum heterophylloides* Stapf.—Atisine (65) and a new amorphous alkaloid, heterophylloidine, $C_{23}H_{29}NO_4$, have been isolated from the roots of this rare species from the Himalayan region of India.[27] When a methanolic solution of

[27] S. W. Pelletier, N. V. Mody, J. Finer-Moore, and H. K. Desai, *Tetrahedron Lett.*, 1981, **22**, 313.

Atisine (65)   (66)

(67) $R^1 = R^2 = H$
Hetidine (68) $R^1 = H, R^2 = OH$
Heterophylloidine (69) $R^1 = Ac, R^2 = H$

heterophylloidine was treated with aqueous HBr, the carbinolamine (66) was formed. The structure and absolute configuration of (66) were determined by an X-ray crystallographic analysis. From this determination and comparisons of the ^{13}C n.m.r. spectra of heterophylloidine, its alkaline hydrolysis product (67), and hetidine (68), the structure (69) was assigned to heterophylloidine.

Alkaloids of *Aconitum karakolicum* Rapcs.—Workers at Tashkent have completed the elucidation of the structure of aconifine ($C_{34}H_{47}NO_{12}$; m.pt 195—197 °C),[28] isolated from the tubers and aerial parts of plants collected in the Terskei Ala-Tau mountain range. Aconitine (55), napelline (70), songorine (71), and the aporphine alkaloid isoboldine were also isolated from the tubers. Aconifine was assigned structure (72) from chemical and ^1H and ^{13}C n.m.r. and mass-spectral analyses. Alkaline hydrolysis of (72) afforded benzoic acid and the amino-alcohol aconifidine (73). Acetylation of (72) with acetic anhydride in pyridine gave the monoacetate (74), while acetylation with acetyl chloride gave a tetra-acetate (75). Vacuum pyrolysis of aconifine gave two main products. One of these could be saponified to yield pyroaconifidine (76). Oxidation of aconifine with $KMnO_4$ produced oxoaconifine (77) and N-de-ethylaconifine (78).

[28] M. Sultankhodzhaev, L. V. Beshitaishvili, M. S. Yunusov, M. R. Yagudaev, and S. Yu. Yunusov, *Khim. Prir. Soedin.*, 1980, 665.

Napelline (70) R = H, α-OH
Songorine (71) R = O

Aconifine (72) $R^1 = R^2 = R^3 = R^4 = H$
(74) $R^1 = Ac, R^2 = R^3 = R^4 = H$
(75) $R^1 = R^2 = R^3 = R^4 = Ac$

(73) $R^1 = R^2 = H, R^3 = Et$
(78) $R^1 = Ac, R^2 = Bz, R^3 = H$

(76) $R^1 = R^2 = H; R^3 = O; R^4 = H_2$
(77) $R^1 = Ac; R^2 = Bz; R^3 = H, α-OH; R^4 = O$

[(79) is with structure (55)]

Alkaloids of *Aconitum kusnezoffii* Reichb.—Researchers in the Peoples Republic of China have reported studies on the alkaloids of *Aconitum kusnezoffii* Reichb. that had been collected in Inner Mongolia.[29] The pharmacological effects of extracts of these plants were studied in several animal systems.[30] The known alkaloids aconitine (55), mesaconitine (56), hypaconitine (57), and deoxyaconitine (79) were identified on the basis of comparison of their physical and chemical properties with those of authentic samples. In addition, a new base, beiwutine ($C_{33}H_{45}NO_{12}$; m.pt 196—198 °C), was reported. This base formed a perchlorate salt (m.pt 255—257 °C) and a tetra-acetate derivative (m.pt 242—244 °C). From the ^{1}H and ^{13}C n.m.r., i.r., and mass-spectral data, structure (80) was proposed for beiwutine.

[29] Y.-G. Wang, Y.-L. Zhu, and R.-H. Zhu, *Yao Hsueh Hsueh Pao*, 1980, **15**, 526.
[30] S.-F. Liu and Y.-Z. Yang, *Yao Hsueh Hsueh Pao*, 1980, **15**, 520.

Beiwutine (80)

Dynamics of the Accumulation of the Alkaloids of *Aconitum leucostomum* Worosch.—Studies of the alkaloid content of *Aconitum leucostomum* Worosch. during different phases of growth and of plants from different locations in the Soviet Union have resulted in several interesting observations.[31] Earlier reports have noted the co-occurrence of diterpenoid and isoquinoline alkaloids in *Aconitum* and *Delphinium* species.[32-34] In the current study, the total alkaloid content of the epigeal plant material was observed to decrease from 0.87% of the dry weight in the rosette-leaf phase to 0.09% during the fruit-bearing phase. Most notably, the ratio of diterpenoid to isoquinoline alkaloids changed from 82:18 in the rosette-leaf phase to 35:57 in the fruit-bearing phase. These determinations were based on the g.l.c. data. The diterpenoid alkaloids lappaconitine (81) and lappaconidine (82) were isolated from these plants and identified by spectral comparisons. The following isoquinoline alkaloids were identified: corydine, *O*-methylarmepavine, *N*-demethylcolletine, and glaunidine. From the plant materials collected at different growth sites, all of it in the rosette-leaf phase, the total alkaloids varied from 0.25 to 0.51%, and the ratio of diterpenoid alkaloids to isoquinoline alkaloids varied from 98:2 to 63:34.

Lappaconitine (81) R^1 = Me, R^2 = COC$_6$H$_4$-2-NHAc
Lappaconidine (82) R^1 = R^2 = H

[31] M. G. Zhamierashvili, V. A. Tel'nov, M. S. Yunusov, S. Yu. Yunusov, A. Nigmatullaev, and K. Taizhanov, *Khim. Prir. Soedin.*, 1980, 805.
[32] H. Guinaudeau, M. Leboeuf, and A. Cavé, *Lloydia*, 1975, **38**, 275.
[33] T. Kosuge and M. Yokota, *Chem. Pharm. Bull.*, 1976, **24**, 176.
[34] B. T. Salimov, N. D. Abdullaev, M. S. Yunusov, and S. Yu. Yunusov, *Khim. Prir. Soedin.*, 1978, 235.

Alkaloids of *Aconitum nagarum* Stapf.—Researchers at the Shanghai Institute of Materia Medica[35] have studied the alkaloids of the roots of *A. nagarum* Stapf var. *lasiandrum* W. T. Wang, which is the classical Chinese folk-medicine 'Xuan-Wu'. Four crystalline alkaloids were isolated and designated as Xuan-Wu 1 ($C_{22}H_{33}NO_2$; m.pt 253 °C), Xuan-Wu 2 ($C_{22}H_{31}NO_3$; m.pt 202 °C), Xuan-Wu 3 ($C_{24}H_{39}NO_6$; m.pt 160 °C), and Xuan-Wu 4 ($C_{26}H_{41}NO_7$; m.pt 198—202 °C). Comparisons of their physical, i.r. and ^{1}H n.m.r. spectral, t.l.c., and melting-point data demonstrated that these bases were identical with the previously reported[36,37] bullatines A, G, B, and C, respectively. From these data, Xuan-Wu 2 (Bullatine G) and Xuan-Wu 3 (Bullatine B) were identical with songorine (71) and neoline (83), respectively. Xuan-Wu 4 (Bullatine C) was assigned structure (84), corresponding to 14-acetylneoline. Data for the nitrate and perchlorate salts and the diacetate derivative (85) of Xuan-Wu 4 were reported.

Neoline (83) $R^1 = R^2 = R^3 = H$
Bullatine C (14-Acetylneoline) (84) $R^1 = R^2 = H$, $R^3 = Ac$
(85) $R^1 = R^2 = R^3 = Ac$

Alkaloids of *Aconitum soongoricum* Stapf.—Further studies of the alkaloids of the roots of *A. soongoricum* Stapf have been reported.[38] In addition to songorine (71), neoline (83), and songoramine (86), a new base, $C_{24}H_{31}NO_4$, was isolated. Alkaline

Songoramine (86) $R^1 = H$, $R^2 = Et$
15-Acetylsongoramine (87) $R^1 = Ac$, $R^2 = Et$
Norsongoramine (117) $R^1 = R^2 = H$

[35] H.-C. Wang, D.-Z. Zhu, Z. Y. Zhao, and R.-H. Zhu, *Hua Hsueh Hsueh Pao*, 1980, **38**, 475.
[36] J. H. Chu, S. T. Fang, and W.-K. Huang, *Hua Hsueh Hsueh Pao*, 1964, **30**, 139.
[37] J. H. Chu and S. T. Fang, *Hua Hsueh Hsueh Pao*, 1965, **31**, 222.
[38] M. G. Zhamierashvili, V. A. Tel'nov, M. S. Yunusov, and S. Yu. Yunusov, *Khim. Prir. Soedin.*, 1980, 733.

hydrolysis of this alkaloid afforded songoramine (86). From the ^{1}H n.m.r., i.r., and mass-spectral data, this new compound was identified as 15-acetylsongoramine (87).

Alkaloids of *Aconitum vilmorrianum* Kom.—Studies of the Chinese herbal drug 'Huang-tsao-wu' (*Aconitum vilmorrianum* Kom.) have been reported by workers at the Kunming Institute of Botany.[39] Three alkaloids were isolated from this species and designated as vilmorrianine A ($C_{35}H_{49}NO_{10}$; m.pt 182—184 °C), vilmorrianine C ($C_{35}H_{49}NO_9$; m.pt 156—157 °C), and yunaconitine ($C_{35}H_{49}NO_{11}$; m.pt 141—143 °C). From the i.r., u.v., ^{1}H n.m.r., and mass-spectral data, these compounds were assigned structures (88), (60), and (89), respectively. Vilmorrianine B, also a constituent of this plant, had earlier been shown to be identical with karakoline (90), previously isolated from *A. hemsleyanum*.[40] Recently, vilmorrianine C and vilmorrianine D have been reported to be identical with foresaconitine (60)[23] and sachaconitine (91),[41] respectively. The structure of vilmorrianine A has subsequently been confirmed by the synthesis described in the next paragraph.

Vilmorrianine A (88) $R^1 = OH$, $R^2 = H$
Yunaconitine (89) $R^1 = R^2 = OH$
Vilmorrianine C (Foresaconitine) (60) $R^1 = R^2 = H$

Vilmorrianine B (Karakoline) (90) R = H
Sachaconitine (91) R = Me

[39] C.-R. Yang, X.-J. Hao, D.-Z. Wang, and J. Zhou, *Hau Hsueh Hsueh Pao*, 1981, **39**, 147.
[40] S.-Y. Chen, *Hua Hsueh Hsueh Pao*, 1979, **37**, 15.
[41] T.-R. Yang, X.-J. Hao, and J. Chow, *Yun-nan Chih Wu Yen Chiu*, 1979, **1**, 41 (*Chem. Abstr.*, 1980, **93**, 46 909).

Alkaloids of *Aconitum yesoense* Nakai.—Sakai and co-workers[42] have reported the isolation and elucidation of the structure of three new C_{19} diterpenoid alkaloids from *Aconitum yesoense* Nakai. These minor bases, ezochasmanine, ezochasmaconitine, and anisoezochasmaconitine, were isolated together with pseudo-kobusine (92), chasmanine (59), and jesaconitine (93). Ezochasmanine (94) ($C_{25}H_{41}NO_7$; m.pt 115—118 °C) formed a mixture of the monoacetate (95) and the diacetate (96) when treated with acetic anhydride in pyridine. The reaction of (94) with acetic anhydride and *p*-TsOH at 80—90 °C yielded the triacetate (97). Comparisons of the ^1H and ^{13}C n.m.r. data of (94), (95), (96), and (97) with that of chasmanine (59) permitted the assignment of structure (94) for ezochasmanine. Furthermore, ezochasmanine was converted into chasmanine as outlined in Scheme 2. This route was also used to prepare 8-acetyl-14-anisoylezochasmanine (88), which corresponds to vilmorrianine A. Ezochasmaconitine ($C_{34}H_{47}NO_8$; m.pt 163—165 °C) was shown to be 8-benzoyl-14-acetylchasmanine (98) from spectral analyses and from the conversion of chasmanine into (98). Anisoezochasm-aconitine ($C_{35}H_{49}NO_9$; m.pt 136—138.5 °C) was shown to be the 8-anisoyl-

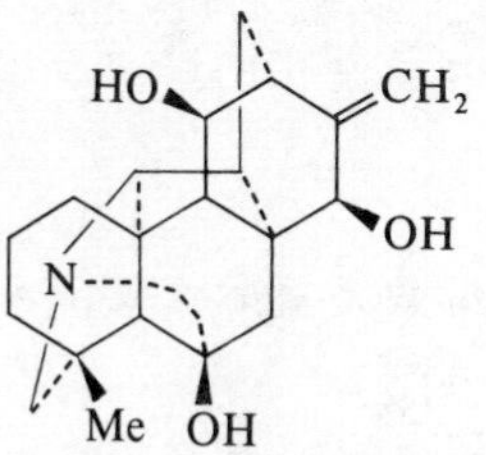

Pseudokobusine (92)

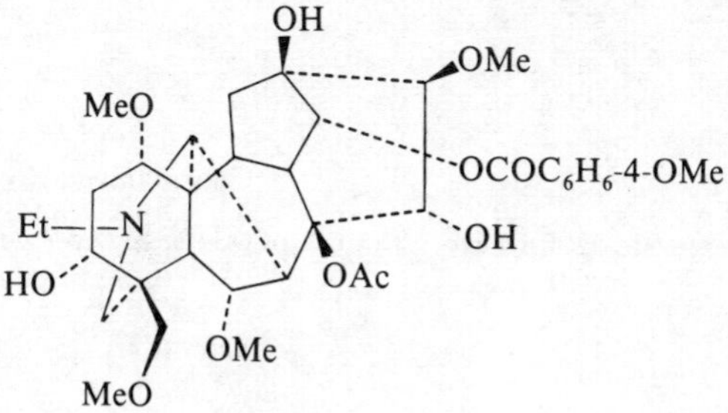

Jesaconitine (93)

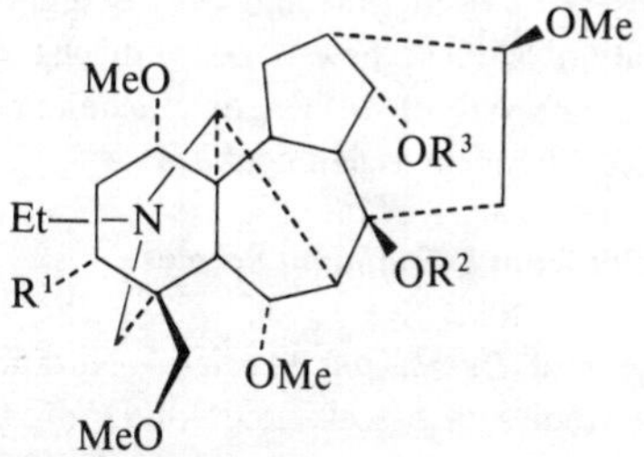

Ezochasmanine (94) $R^1 = OH, R^2 = R^3 = H$
(95) $R^1 = AcO, R^2 = R^3 = H$
(96) $R^1 = AcO, R^2 = H, R^3 = Ac$
(97) $R^1 = AcO, R^2 = R^3 = Ac$
Ezochasmaconitine (98) $R^1 = H, R^2 = Bz, R^3 = Ac$
Anisoezochasmaconitine (99) $R^1 = H, R^2 = COC_6H_4-4-OMe, R^3 = Ac$

[42] H. Takayama, M. Ito, M. Koga, S. Sakai, and T. Okamoto, *Heterocycles*, 1981, **15**, 403.

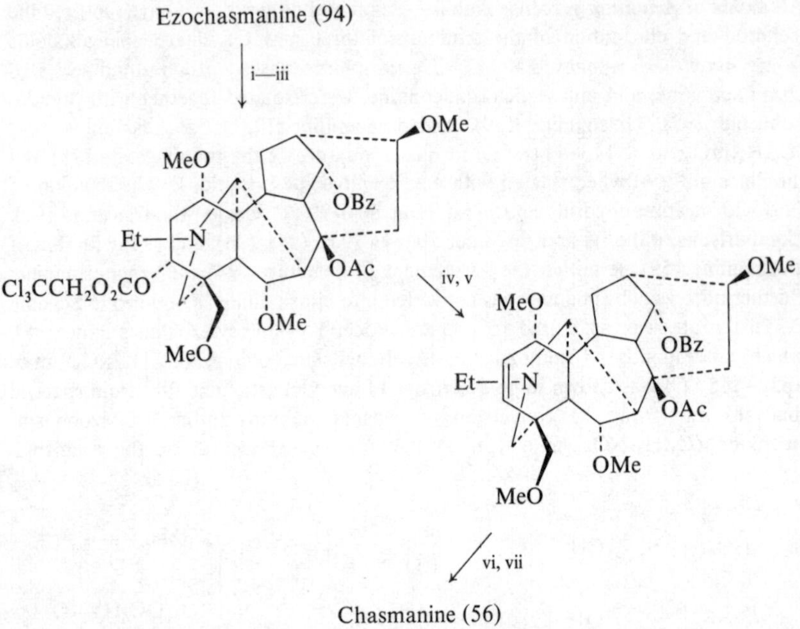

Ezochasmanine (94)

Chasmanine (56)

Reagents: i, BzCl, pyridine, ii, Cl₃CCH₂OCOCl, iii, Ac₂O, p-TsOH; iv, Zn, AcOH; v, SO₂Cl₂; vi, H₂, Pt; vii, OH⁻

Scheme 2

14-acetyl derivative (99) of chasmanine from the ^{1}H and ^{13}C n.m.r. spectral data and by the conversion of chasmanine into (99). The presence of the aroyl groups at C-8 and acetyl groups at C-14 in (98) and (99) is unique in the C_{19} diterpenoid alkaloids. The correlation of these new bases with chasmanine and the previous X-ray crystallographic analysis of 14-benzoylchasmanine hydrochloride[43] establishes the indicated stereochemical assignments.

4 Alkaloids from *Delphinium* Species

Phytochemical Analyses of *Delphinum* Species.—Narzullaev and Sabirov[44] have described preliminary studies of the alkaloids of *D. oreophilum*, *D. leptocarpum*, and *D. batalinii*. From the aerial parts of plants of *D. oreophilum* that were collected during the flowering stage, an alkaloid fraction comprising 1.51% of the dry weight of the plant material was isolated. From this mixture, lycoctonine (6) and methyl-lycaconitine (37) were isolated as their perchlorate salts. The roots contained 1.1% by weight of total alkaloids, from which (6) and (37) were also isolated. The plants of *D. leptocarpum* and *D. batalinii* afforded 0.25 and 0.5% by weight of total alkaloids, respectively.

43 S. W. Pelletier, W. H. DeCamp, and Z. Djarmati, *J. Chem. Soc., Chem. Commun.*, 1976, 253.
44 A. S. Narzullaev and S. S. Sabirov, *Khim. Prir. Soedin.*, 1981, 250.

Alkaloids of *Delphinium bicolor* Nutt.—Codding and Kerr[45] have published a full report of the structure and conformation of Alkaloid A hydroiodide, as determined by an *X*-ray crystallographic analysis. Alkaloid A was isolated from plants of *Delphinium bicolor* Nutt. that were collected in the Rocky Mountains. The preliminary account of this study[46] was reviewed in Vol. 11, p. 206. The structure of this hydroiodide was shown to be (100), with a final $R = 0.039$ for 2579 reflections. Ring A was shown to be in a boat form, as has previously been observed in C_{19} diterpenoid alkaloids with oxygenated substituents at C-1 and a protonated nitrogen atom. In crystal structures of related bases in which the nitrogen atom is not protonated, ring A exists in the chair form. Rings B and E were shown to be in chair conformations, while ring D is in a boat form and ring C has an envelope conformation, with C-14 at the flap. Comparisons of the torsional angles from the *X*-ray crystallographic data of Alkaloid A and those previously reported for condelphine (101),[47] chasmanine (59),[48] and delphisine (102)[49] show that these compounds have rigid skeletons and that the observed conformations are not sensitive to packing considerations or to the counter-ion.

Alkaloid A (100) R^1 = β-OAc, R^2 = OH, R^3 = Me
Condelphine (101) R^1 = H, R^2 = OH, R^3 = CH$_2$OMe
Delphisine (102) R^1 = α-OMe, R^2 = OAc, R^3 = CH$_2$OMe

Alkaloids of *Delphinium carolinianum* Walt.—Delcaroline ($C_{25}N_{41}NO_8$; amorphous; m.pt of perchlorate is 160—162 °C) was the major alkaloid isolated from whole plants of *D. carolinianum* Walt., which is native to the south-eastern United States.[50] The known bases ajaconine (103) and browniine (7) were also isolated from these plants. Acetylation of delcaroline with acetic anhydride in pyridine gave the monoacetate (104). The structure (22) for delcaroline was assigned primarily on the basis of comparison of the ^{13}C n.m.r. spectra of (22), its monoacetate (104), browniine (7), and dictyocarpine (8). Thus delcaroline, with a hydroxyl group at C-10 and a methoxymethylene group at C-4, represents a novel substitution pattern among the known C_{19} diterpenoid alkaloids.

[45] P. W. Codding and K. A. Kerr, *Acta Crystallogr., Sect. B.*, 1981, **37**, 379.
[46] P. W. Codding, K. A. Kerr, M. N. Benn, A. J. Jones, S. W. Pelletier, and N. V. Mody, *Tetrahedron Lett.*, 1980, **21**, 127.
[47] S. W. Pelletier, W. H. DeCamp, D. L. Herald, Jr., S. W. Page, and M. G. Newton, *Acta Crystallogr., Sect. B*, 1977, **33**, 716.
[48] W. H. DeCamp and S. W. Pelletier, *Acta Crystallogr., Sect B*, 1977, **33**, 722.
[49] S. W. Pelletier, Z. Djarmati, S. Lajsic, and W. H. DeCamp, *J. Am. Chem. Soc.*, 1976, **98**, 2617.
[50] S. W. Pelletier, N. V. Mody, and R. C. Desai, *Heterocycles*, 1981, **16**, 747.

Ajaconine (103)

Delcaroline (22) R = H
            (104) R = Ac

Alkaloids of *Delphinium glaucescens* Rybd.—A comprehensive investigation[51] of the alkaloids of *D. glaucescens* Rybd. has resulted in the isolation of nine known and five new C_{19} diterpenoid alkaloids. Extraction of the aerial parts of this plant, which has been implicated in numerous cases of poisoning of cattle in the western United States, gave a mixture of crude alkaloids in 0.75% yield. The known bases reported (in decreasing order of abundance) were lycoctonine (6), dictyocarpine (8), browniine (7), 14-dehydrobrowniine (20), methyl-lycaconitine (37), delcosine (47), dictyocarpinine (12), deltaline (10), and anthranoyl-lycoctonine (32). Glaudelsine, $C_{36}H_{48}N_2O_{10}$, an amorphous solid, was assigned structure (105) primarily on the basis of the ^1H and ^{13}C n.m.r. spectral data for glaudelsine and the ^1H n.m.r. spectrum of its hydrolysis product (106). The structures of glaucenine (107) ($C_{31}H_{47}NO_9$; amorphous; m.pt of perchlorate is 227.5—232.5 °C), glaucerine

Glaudelsine (105) $R^1 = H$, $R^2 = Me$, $R^3 = CO-$

(106) $R^1 = R^3 = H$, $R^2 = Me$

Glaucedine (110) $R^1 = R^3 = Me$, $R^2 = CO-\overset{Me}{\underset{H}{C}}-Et$

[51] S. W. Pelletier, O. D. Dailey, Jr., N. V. Mody, and J. D. Olsen, *J. Org. Chem.*, 1981, **46**, 3284.

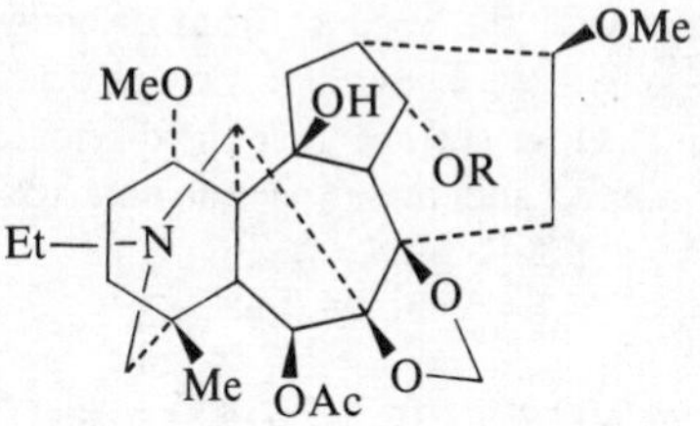

Glaucenine (107) R = CO—C—Et
                          |
                          Me (above), H (below)

Glaucerine (108) R = COCHMe₂
Glaucephine (109) R = Bz

(108) ($C_{30}H_{45}NO_9$; amorphous), glaucephine (109) ($C_{33}H_{43}NO_9$; amorphous), and glaucedine (110) ($C_{30}H_{49}NO_8$; m.pt 117—120 °C) were determined by n.m.r., i.r., and mass-spectral analyses and confirmed by the synthesis of these esters from dictyocarpine and browniine. Thus, glaucerine, glaucenine, and glaucephine are the 14-isobutyryl, 14-(2-methylbutyryl), and 14-benzoyl esters of dictyocarpine, respectively. Glaucedine is the 14-(2-methylbutyryl) ester of browniine. These isobutyryl and 2-methylbutyryl esters are the first to be reported amongst the diterpenoid alkaloids. Dictyocarpinine, which is the hydrolysis product of dictyocarpine, has not previously been isolated from natural sources.

Alkaloids of *Delphinium iliense* Huth.—Yunusov and co-workers[52] have isolated a new alkaloid, delcoridine, $C_{25}H_{35}NO_7$, (111), from the aerial parts of *D. iliense* Huth that was collected in the Trans-Ili Ala-Tau. Delcorine (14), 6-dehydro-delcorine (15), lycoctonine (6), deltaline (eldeline) (10), ilidine (17), browniine (7), and dictyocarpinine (12) were also isolated from these plants. Methylation of delcoridine with methyl iodide and sodium hydride in dioxan gave 6-*O*-methyl-

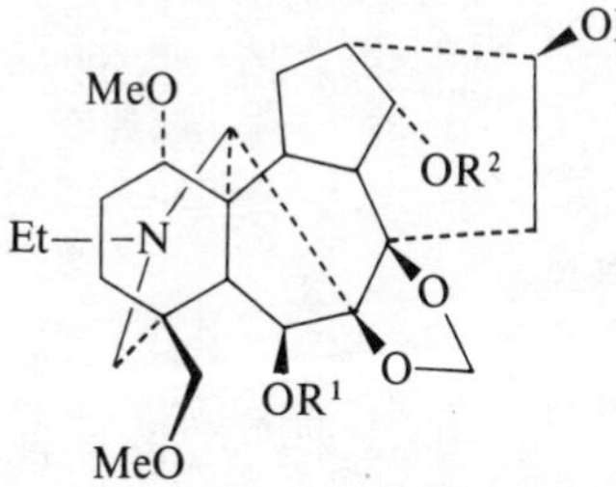

Delcoridine (111) R¹ = R² = H
            (112) R¹ = R² = Me
            (113) R¹ = R² = Ac

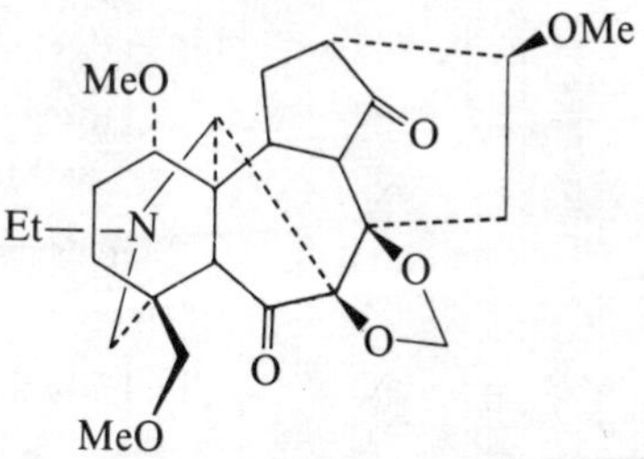

(114)

[52] M. G. Zhamierashvili, V. A. Tel'nov, M. S. Yunusov, and S. Yu. Yunusov, *Khim. Prir. Soedin.*, 1980, 663.

delcorine (112), identical with the product from delcorine (14). Acetylation of delcoridine with acetic anhydride in pyridine yielded delcoridine diacetate (113). Oxidation of (111) with CrO_3 in acetone gave the diketone (114). On the basis of these and the i.r., ^{1}H n.m.r., and mass-spectral data, delcoridine was assigned structure (111).

Alkaloids of *Delphinium staphisagria* L.—An *X*-ray crystallographic structure determination of staphisine, using direct methods, has been reported.[53] The structure (115) for staphisine had been assigned from the *X*-ray diffraction study of its monomethiodide.[54] However, the sample of staphisine used in this study was subsequently shown to be a mixture of staphisine and staphidine (116).[55] These two alkaloids were separated, and the pure staphisine was used for a more accurate *X*-ray crystallographic analysis. These data were refined to $R = 0.049$ for 2529 observed reflections. From the crystal structure, the difference in the ^{1}H n.m.r. chemical shift that is observed for the *N*-methyl group in staphisine relative to its resonance in the demethoxylated companion alkaloid staphidine appears to result from the steric interaction of the methoxyl group with the *N*-methyl group.

Staphisine (115) R = OMe
Staphidine (116) R = H

[(117) is with structure (86)]

[53] S. W. Pelletier, W. H. DeCamp, J. Finer-Moore, and I. V. Mićović, *Acta Crystallogr.*, *Sect. B*, 1980, **36**, 3040.
[54] S. W. Pelletier, A. H. Kapadi, A. H. Wright, S. W. Page, and M. G. Newton, *J. Am. Chem. Soc.*, 1972, **94**, 1754.
[55] S. W. Pelletier, N. V. Mody, Z. Djarmati, I. V. Mićović, and J. K. Thakkar, *Tetrahedron Lett.*, 1976, 1055.

Alkaloids of *Delphinium tamarae* Kem. Nath.—Yunusov and co-workers[56] investigated the alkaloids of the roots of *D. tamarae* that had been collected during the dormant season. The total alkaloids comprised 2.02% of the dry weight of the plant material. In addition to methyl-lycaconitine (37), lycoctonine (6), and anthranoyl-lycoctonine (32), a new base, $C_{20}H_{25}NO_3$ (m.pt 286 °C) was isolated. From the ^{1}H n.m.r., i.r., and mass-spectral data, this new alkaloid was determined to be norsongoramine (117). This assignment was confirmed by the ethylation of (117) with ethyl iodide and potassium carbonate to afford songoramine (86).

5 Alkaloids from *Garrya* Species

Alkaloids of *Garrya ovata*.—The full report of the alkaloids of *Garrya ovata* var. *lindheimeri* Torr. has appeared.[57] The isolation and elucidation of the structure of two new alkaloids, ovatine (118) and lindheimerine (119), are described in detail. Garryfoline (120) and cuauchichicine (121), which had earlier been isolated from *G. laurifolia*, were also isolated from *G. ovata*. From the ^{13}C n.m.r. spectral studies and an *X*-ray crystallographic analysis, the structure of cuauchichicine was determined to be the C-16 and C-20 epimer of the originally assigned structure (122).[58] The previously reported[58] correlation of cuauchichicine with (−)-'β'-dihydrokaurene was re-investigated. From these studies it was concluded that the epimerization at C-16 in cuauchichicine, which resulted in the previous incorrect assignment of its structure, most probably occurs during the Wolff–Kishner

Ovatine (118) R = Ac
Garryfoline (120) R = H

Lindheimerine (119)

Cuauchichicine (121)

(122)

[56] L. V. Beshitaishvili, M. N. Sultankhodzhaev, K. S. Mudzhiri, and M. S. Yunusov, *Khim. Prir. Soedin.*, 1981, 199.
[57] S. W. Pelletier, N. V. Mody, and H. K. Desai, *J. Org. Chem.*, 1981, **46**, 1840.
[58] H. Vörbrueggen and C. Djerassi, *J. Am. Chem. Soc.*, 1962, **84**, 2990.

reduction step in the degradation to dihydrokaurene. Cuauchichicine is the only known 'normal-type' oxazolidine-ring-containing C_{20} diterpenoid alkaloid which does not exist as a mixture of epimers at C-20.

6 Synthetic Studies

A New Synthesis of Napelline.—Workers at New Brunswick have reported an excellent stereospecific synthesis of napelline (70), based on their 'fourth generation methods' that were developed in the synthesis of several delphinine-type alkaloids.[59] These methods had previously been explored with model compounds, and employ the rearrangement of a 'denudatine' system to a 'napelline' system.[60] The starting material (123) for this synthesis was prepared by the route analogous to that used for the preparation of 13-deoxydelphonine.[60,61]

Compound (123) was reduced with $LiBH_4$ to yield the alcohol (124). Heating (124) with 6M-HCl removed the protecting groups to afford (125). Oxidation of (125) with $Tl(NO_3)_3$ gave the o-quinone acetal (126). The reaction of (126) with

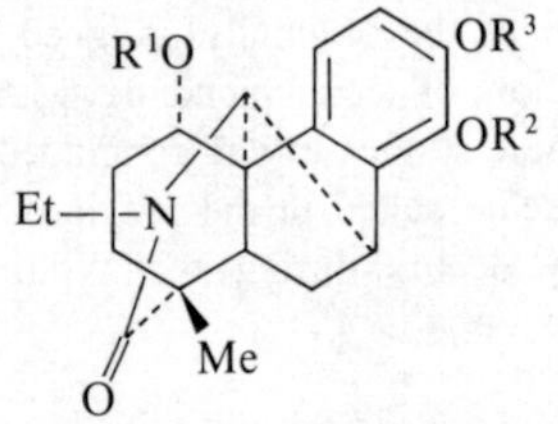

(123) $R^1 = THP$, $R^2 = CH_2CO_2Me$, $R^3 = CH_2OMe$
(124) $R^1 = THP$, $R^2 = CH_2CH_2OH$, $R^3 = CH_2OMe$
(125) $R^1 = R^3 = H$, $R^2 = CH_2CH_2OH$

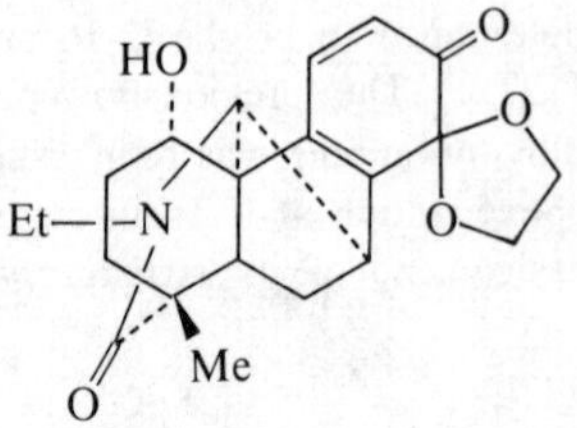

(126)

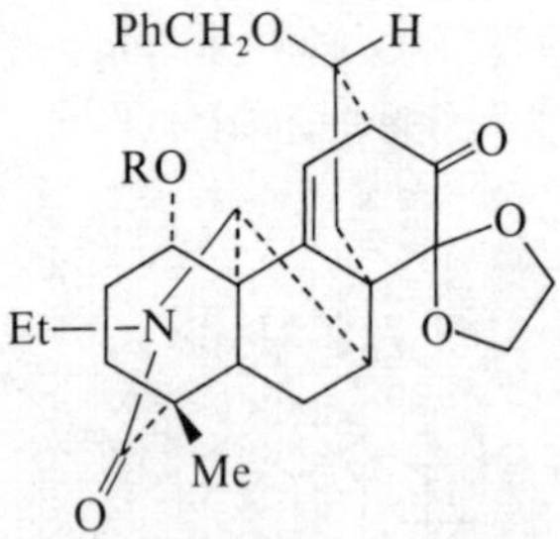

(127) $R = H$
(128) $R = THP$

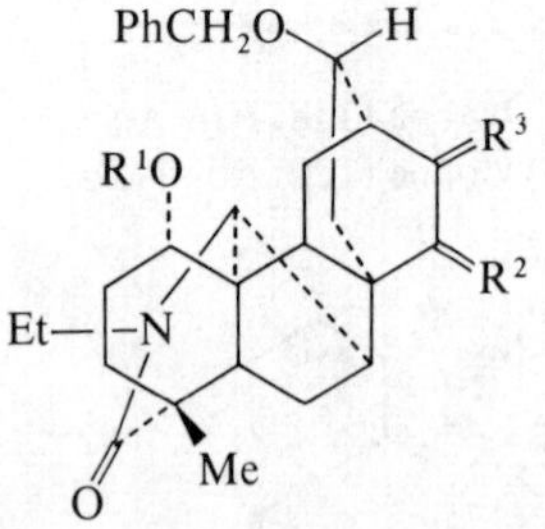

(129) $R^1 = THP$, $R^2 = \overset{O}{\underset{O}{\diagup}}$, $R^3 = \overset{OH}{\underset{CH_2SiMe_3}{}}$

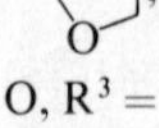

(130) $R^1 = H$, $R^2 = O$, $R^3 = CH_2$

[59] S. P. Sethi, K. S. Atwal, R. M. Marini-Bettolo, T. Y. R. Tsai, and K. Wiesner, *Can. J. Chem.*, 1980, **58**, 1889.
[60] K. S. Atwal, R. Marini-Bettolo, I. H. Sanchez, T. Y. R. Tsai, and K. Wiesner, *Can. J. Chem.*, 1978, **56**, 1102.
[61] K. Wiesner, S. Ito, and Z. Valenta, *Experientia*, 1958, **14**, 167.

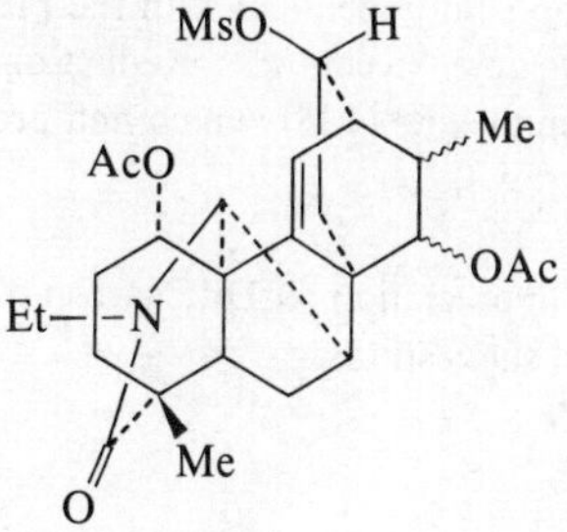

(131)

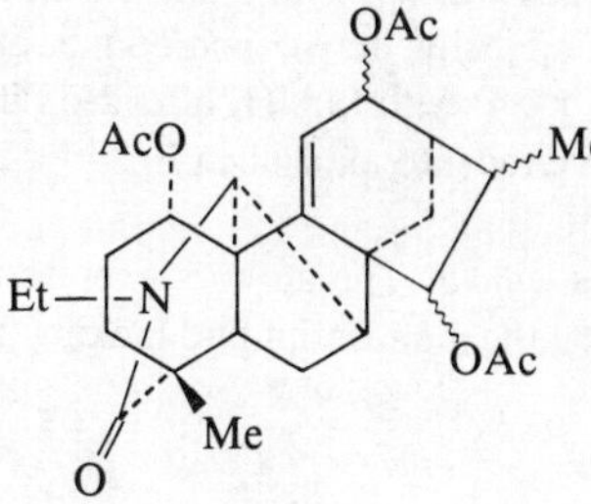

(132)

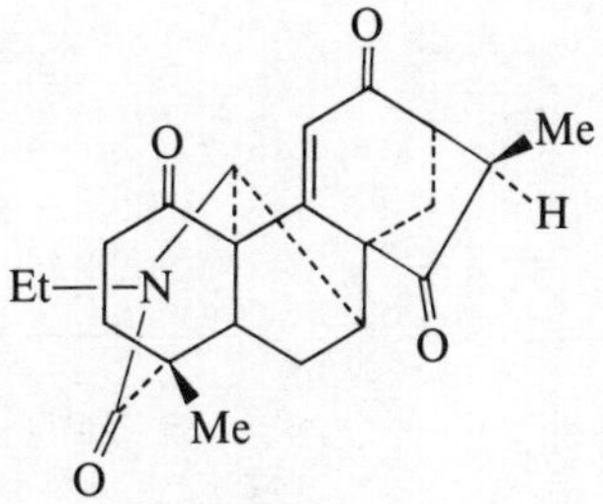

(133)

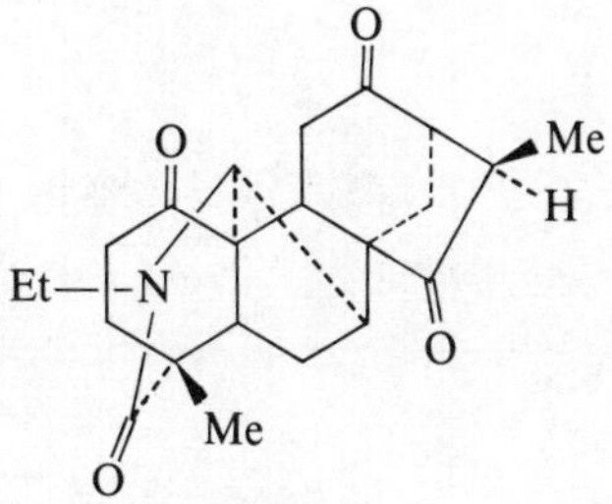

(134)

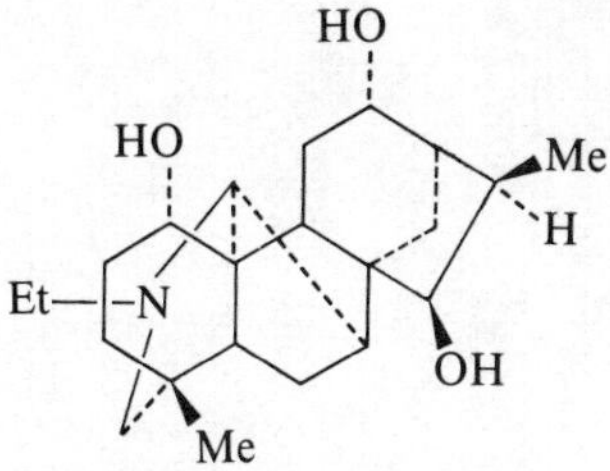

Dihydronapelline (135)

benzyl vinyl ether gave the adduct (127), which was quantitatively converted into the tetrahydropyranyl derivative (128). The latter was treated with trimethyl-silylmethylmagnesium chloride to afford the epimeric alcohols (129). Heating (129) with 70% HClO$_4$ in THF afforded the $\alpha\beta$-unsaturated ketone (130) in 85% yield. Reduction of (130) with LiBH$_4$, followed by acetylation, hydrogenolysis of the benzyl group, and mesylation, gave (131). This 'denudatine' system was heated at reflux in glacial acetic acid to afford the 'napelline' system (132) in 95% yield. Saponification of this triacetate, followed by oxidation with CrO$_3$ and pyridine in CH$_2$Cl$_2$, yielded the unsaturated keto-lactam (133). Hydrogenation of (133) gave

(134), which was spectrally and chromatographically identical with the previously reported optically active compound of the same structure.[61] Reduction of the racemate (134) with $LiAlH_4$ afforded dihydronapelline (135), which had previously been converted into napelline (70).[62,63]

Acknowledgment. The authors express their appreciation to Dr. Naresh V. Mody for reading the manuscript and making helpful suggestions.

[62] K. Wiesner, P. T. Ho, and C. S. J. Tsai, *Can. J. Chem.*, 1974, **52**, 2353.
[63] K. Wiesner, P. T. Ho, C. S. J. Tsai, and Y. K. Lam, *Can. J. Chem.*, 1974, **52**, 2355.

14

Steroidal Alkaloids

BY D. M. HARRISON

1 *Buxus* Alkaloids

Dev and co-workers have described the partial synthesis of cyclobuxophyllinine-M (1a), and of the related alkaloids cyclobuxophylline-K (1b) and buxanine-M (1c), from the acetoxy-ketone (2).[1] In a formal sense, this work completes the total synthesis of each of these alkaloids, since the acetoxy-ketone (2) was prepared by the degradation of cycloartenol,[2] while the latter has been the subject of a total synthesis.[3]

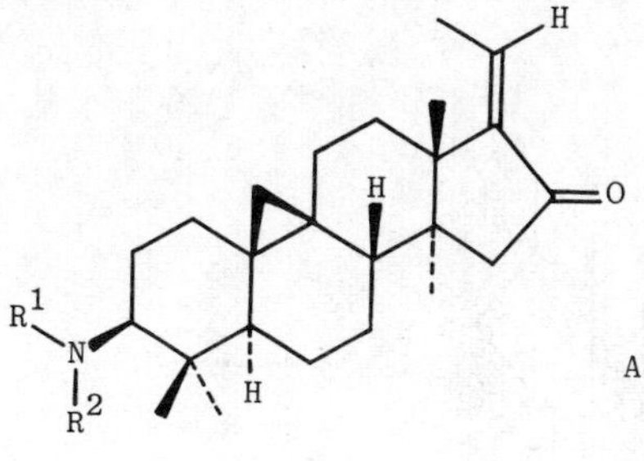

(1) a; $R^1 = H$, $R^2 = Me$

    b; $R^1 = R^2 = Me$

    c; $R^1 = Me$, $R^2 = COPh$

    d; $R^1 = R^2 = H$

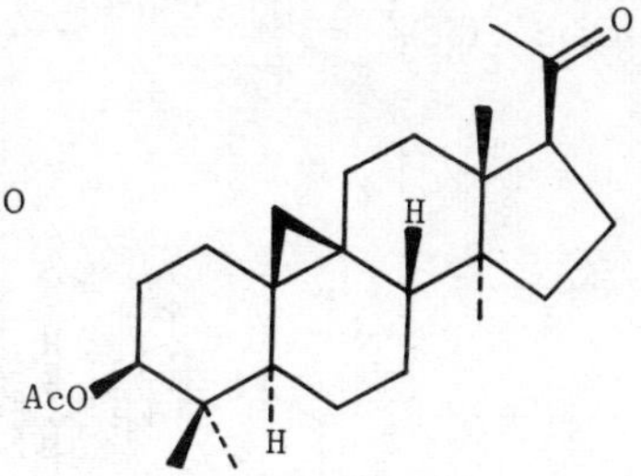

(2)

Two new alkaloids, cyclobuxophylline-O (1d) and buxithienine-M (3), have been isolated from leaves of *Buxus sempervirens* var. *rotundifolia*.[4] Also isolated were the known alkaloids buxamine-E, buxaminol-B, buxpiine-K, buxtauine-M, cyclo-bullatine-A, cyclobuxine-D, cycloprotobuxine-C, and cyclovirobuxine-D (4),

[1] M. C. Desai, J. Singh, H. P. S. Chawla, and S. Dev, *Tetrahedron*, 1981, **37**, 2935.
[2] A. S. Narula and S. Dev, *Tetrahedron*, 1971, **27**, 1119.
[3] D. H. R. Barton, D. Kumari, P. Welzel, L. J. Danks, and J. F. McGhie, *J. Chem. Soc. C*, 1969, 332.
[4] Le thi Thien Huong, Z. Votický, and V. Paulík, *Collect. Czech. Chem. Commun.*, 1981, **46**, 1425.

(3')   (4)

together with eleven unidentified bases. The structures indicated for the new alkaloids were determined by spectroscopic studies. Confirmation of these assignments was obtained by the N-methylation of cyclobuxophylline-O (1d), which gave the known alkaloid cyclobuxophylline-K (1b), and by the correlation of buxithienine-M (3) with cyclovirobuxine-D (4).[4]

In the first investigation of the species, leaves of *B. harlandi* have yielded the new alkaloid buxamine-B (5a) together with the known bases buxamine-E (principal alkaloid), buxpiine-K, buxtauine-M, cyclobuxine-D, and cycloprotobuxines-C and D. The structure proposed for buxamine-B (5a) was suggested by the spectroscopic data and confirmed by the preparation of the N-acetyl derivative (5b), and also by the preparation of the N-methyl derivative, which was identical to buxamine-A (5c).[5]

(5) a; R = H
    b; R = Ac
    c; R = Me

2 Alkaloids of the Apocynaceae

Extraction of the leaves of *Holarrhena curtisii* that had grown in Thailand yielded the known constituent holacurtine (6a) together with a new alkaloid.[6] The latter was assigned the structure N-desmethylholacurtine (6b) on the basis of spectroscopic comparison with holacurtine. The ¹H n.m.r. spectrum of N-acetylholacurtine (6c) showed two resonances each for the O-methyl, N-methyl, and secondary C-methyl groups, which was interpreted to mean that the amide group

[5] A. Vassová, Z. Votický, J. Černik, and J. Tomko, *Chem. Zvesti*, 1980, **34**, 706 (*Chem. Abstr.*, 1981, **94**, 44 051).
[6] J. R. Cannon, E. L. Ghisalberti, and V. Lojanapiwatna, *J. Sci. Soc. Thailand*, 1980, **6**, 81 (*Chem. Abstr.*, 1980, **93**, 217 922).

(6) a; $R^1 = H$, $R^2 = Me$

b; $R^1 = R^2 = H$

c; $R^1 = Me$, $R^2 = Ac$

d; $R^1 = H$, $R^2 = Ac$

was subject to restricted rotation. No anomalies were observed in the ^1H n.m.r. spectrum of *N*-acetyl-*N*-desmethylholacurtine (6d).[6]

5,6-Dihydro-5,6-dihydroxyconessine has been prepared by the action of aqueous nitrous acid on conessine (7a).[7] Treatment of conessine with BrCN yields the *N*-cyano-*N*-desmethyl derivative (7b).[8] The latter has been converted (by standard methods) into derivatives (7c), (7d), and (7e).[9]

(7) a; $R = Me$

b; $R = CN$

c; $R = CONH_2$

d; $R = CH_2NH_2$

e; $R = C(NH_2)=NH$

In a study of the Hofmann–Löffler reaction, the *N*-chloro-amine (8a) was irradiated in trifluoroacetic acid solution and underwent photolysis to yield the pyrrolidine (9a) together with the 20*S*-isomer (9b) in a ratio 8:1. The starting amine (8b) was prepared by hydrogenation of (8c), followed by von Braun demethylation and hydrolysis of the *N*-cyano-amine (8d) which was so formed. The ^1H n.m.r. spectrum of (8b) and of related amines in the presence of lanthanide shift reagents was discussed.[10]

Details are now available of the reactions of peracids and of hydrogen peroxide with the imines (10) and (11). The reactions of the product oxaziridines and nitrones with the same two oxidants were also discussed.[11] The reactions of sodium borohydride, benzoyl chloride, and toluene-*p*-sulphonic acid with the hydroxy-nitrones (12) and (13) have been reported.[12] The c.d. spectra of 2-substituted pyrrolidines have been compared with that of conanine (14).[13]

[7] M. I. Quresi and M. A. Khan, *Egypt. J. Chem.*, 1977 (publ. 1978), **20**, 103 (*Chem. Abstr.*, 1980, **93**, 239 773).

[8] S. Siddiqui and R. H. Siddiqui, *J. Indian Chem. Soc.*, 1934, **11**, 787 (*Chem. Abstr.*, 1935, **29**, 2960).

[9] S. Siddiqui and B. S. Siddiqui, *Z. Naturforsch., Teil. B*, 1980, **35**, 1049.

[10] G. Van de Woude, M. Biesemans, and L. van Hove, *Bull. Soc. Chim. Belg.*, 1980, **89**, 993 (*Chem. Abstr.*, 1981, **94**, 121 781).

[11] H. Dadoun, J. P. Alazard, and X. Lusinchi, *Tetrahedron*, 1981, **37**, 1525.

[12] J. P. Alazard, H. Dadoun, and X. Lusinchi, *Tetrahedron*, 1981, **37**, Suppl. 1, p. 41.

[13] B. Ringdahl, W. E. Pereira, and J. C. Craig, *Tetrahedron*, 1981, **37**, 1659.

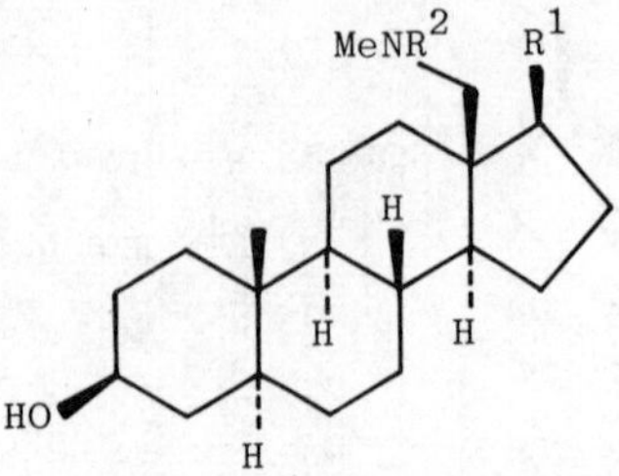

(8) a; R^1 = Et, R^2 = Cl

    b; R^1 = Et, R^2 = H

    c; R^1 = CH=CH$_2$, R^2 = Me

    d; R^1 = Et, R^2 = CN

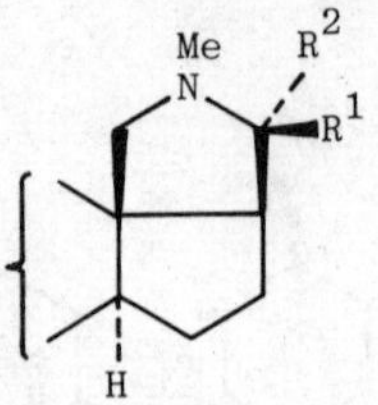

(9) a; R^1 = H, R^2 = Me

    b; R^1 = Me, R^2 = H

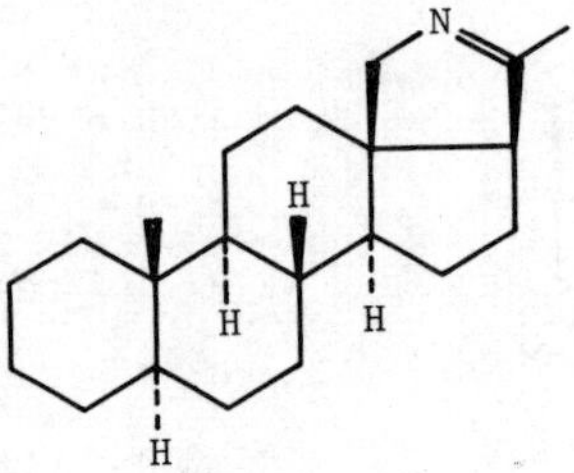

(10)

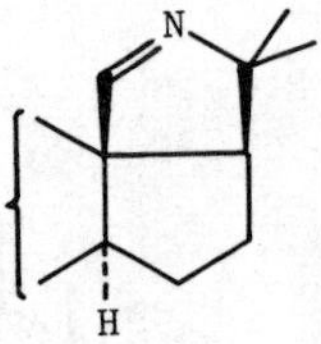

(11)

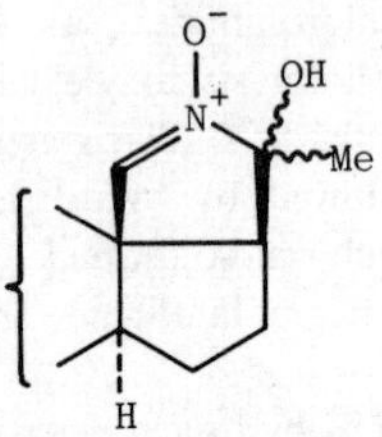

(12)

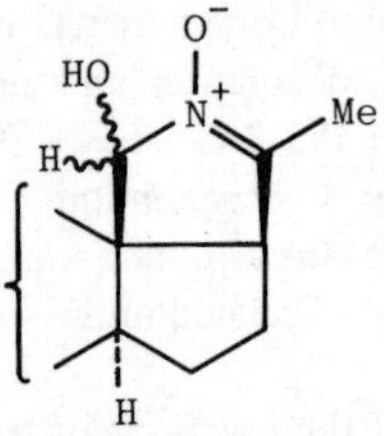

(13)

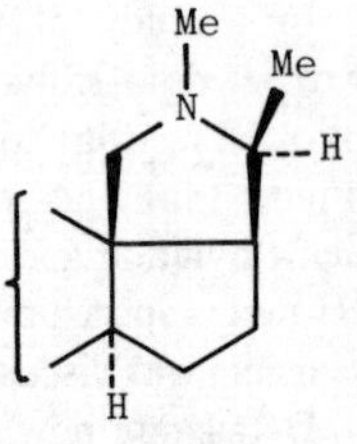

(14)

Acid-catalysed hydrolysis of the carbamate (15a)[14] gave the amino-pregnane (15b).[15] The latter amine underwent a condensation reaction with 3,5-di-t-butyl-*o*-quinone to yield an anil which gave progesterone (15c) on hydrolysis with an acetate buffer.[16] Other amino-pregnane derivatives have been prepared by degradation of solasodine[17] (*vide infra*).

[14] L. O. Krbechek, U.S. P. 4 251 450 (*Chem. Abstr.*, 1981, **94**, 192 571).

[15] L. O. Krbechek, U.S. P. 4 252 730 (*Chem. Abstr.*, 1981, **94**, 192 570).

[16] L. O. Krbechek, E. B. Spitzner, and J. P. Clark, U.S. P. 4 252 732 (*Chem. Abstr.*, 1981, **94**, 192 569).

[17] B. U. Khodzhaev, R. Shakirov, and S. Yu. Yunusov, *Khim. Prir. Soedin.*, 1980, 370 (*Chem. Abstr.*, 1980, **93**, 186 666).

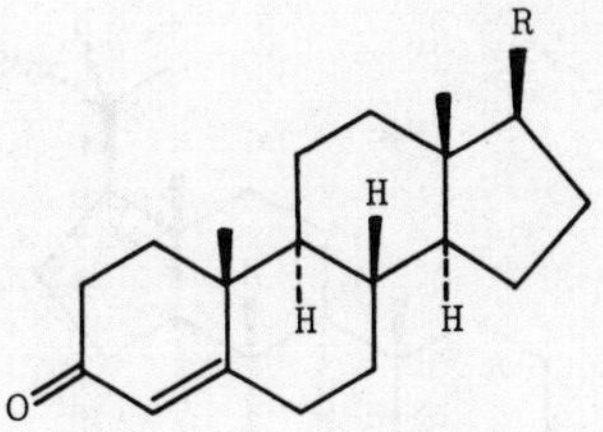

(15) a; R = CHMeNHCO$_2$Me

      b; R = CHMeNH$_2$

      c; R = C=O
            Me

3 *Solanum* Alkaloids

Solasodine (16a) is an important starting material for the preparation of 3β-acetoxypregna-5,16-dien-20-one (17).[18] Hydrogenation of the latter over Raney nickel gave the epimeric pregnenols (18a), which were converted (by standard reactions) into the amines (18b) and into the dimethylamino-pregnanes (19a) and (19b).[17] The 3β-amino-spirosolanes (20a), (20b), and (20c) have been prepared

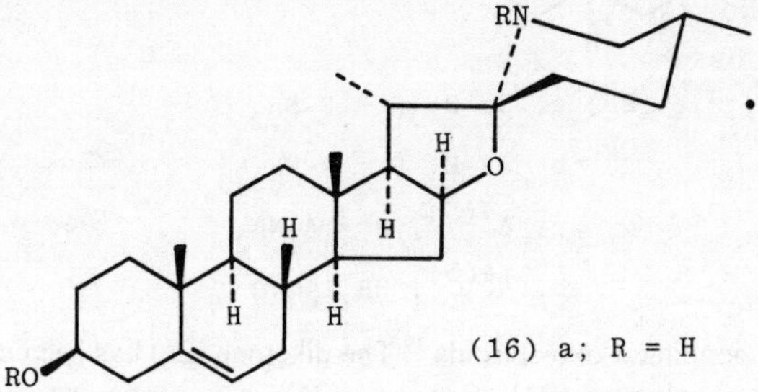

(16) a; R = H

      b; R = Ac

      c; R = Ts

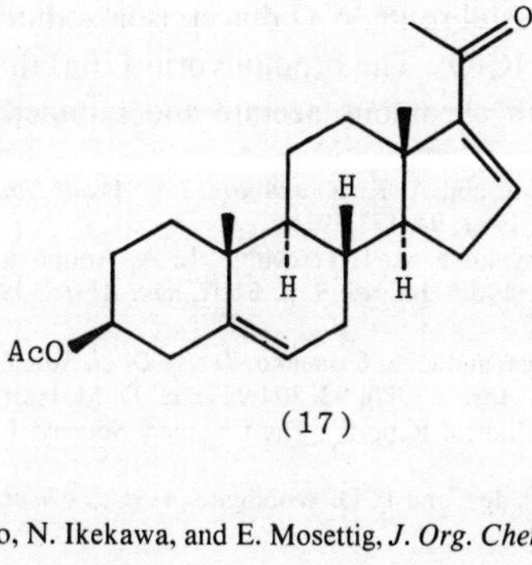

(17)

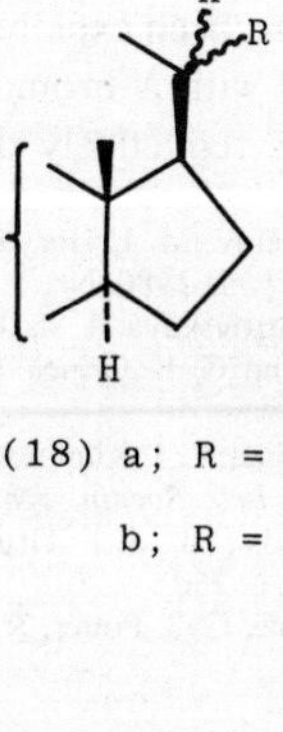

(18) a; R = OH

      b; R = NH$_2$

[18] Y. Sato, N. Ikekawa, and E. Mosettig, *J. Org. Chem.*, 1960, **25**, 783.

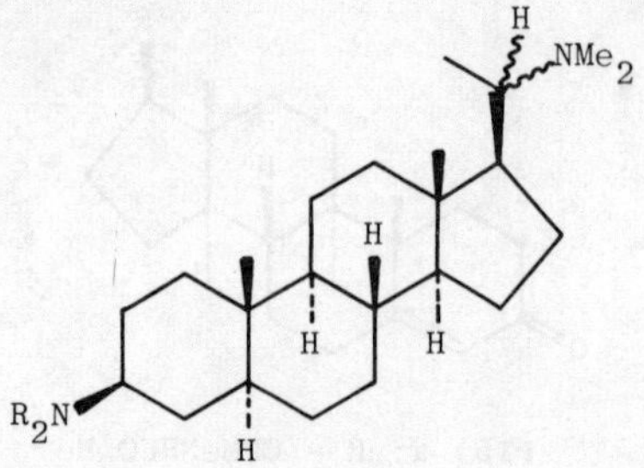

(19) a; R = H

b; R = Me

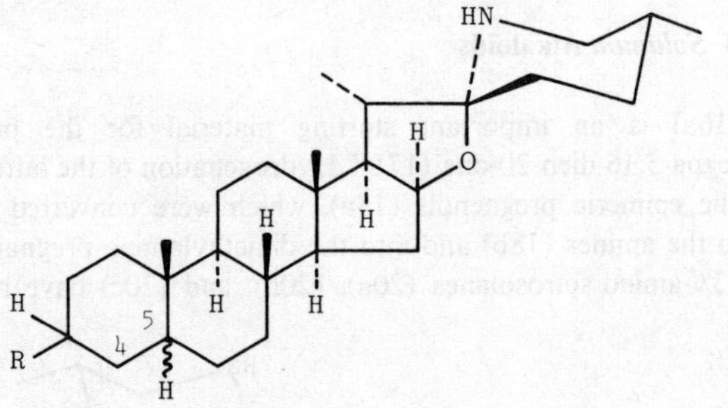

(20) a; 5α-H, R = β-NH$_2$

b; 5β-H, R = β-NH$_2$

c; Δ$^{4(5)}$, R = β-NH$_2$

d; Δ$^{4(5)}$, HR is =O

from the corresponding 3-oxo-steroids.[19] The diketone (21) has been converted into the heterocyclic analogues (22), (23), and (24), in a continuing search for new physiologically active steroids derived from solasodine.[20] The iodination and autoxidation reactions of diacetylpseudosolasodine (25) have been described.[21]

The partial synthesis from solasodine of nitrogen-containing analogues of ecdysone has been explored.[22] In the successful route, N,O-diacetylsolasodine (16b) was treated with N-bromoacetamide and HClO$_4$. The bromohydrin (26a) that was formed was reductively debrominated with chromous acetate and butanethiol to

[19] M. P. Irismetov, M. I. Goryaev, V. S. Bazalitskaya, and A. K. Kairgalieva, Izv. Akad. Nauk Kaz. SSR, Ser. Khim., 1980, No. 5, p. 81 (Chem. Abstr., 1981, **94**, 121 791).

[20] L. M. Morozovskaya, L. I. Klimova, V. A. Kobyakova, A. I. Terekhina, L. A. Antipova, O. N. Kruglova, and G. S. Grinenko, Khim. Farm. Zh., 1980, **14**, No. 9, p. 61 (Chem. Abstr., 1981, **94**, 47 609).

[21] G. G. Malanina, L. I. Klimova, L. M. Morozovskaya, and G. S. Grinenko, Tezisy Dokl. Sov. Indiiskii Simp. Khim. Prir. Soedin., 5th, 1978, p. 51, (Chem. Abstr., 1980, **93**, 204 927); cf. D. M. Harrison, in 'The Alkaloids', ed. M. F. Grundon, (Specialist Periodical Reports), The Chemical Society, London, 1978, Vol. 8, p. 250.

[22] R. C. Cambie, G. J. Potter, R. W. Read, P. S. Rutledge, and P. D. Woodgate, Aust. J. Chem., 1981, **34**, 599.

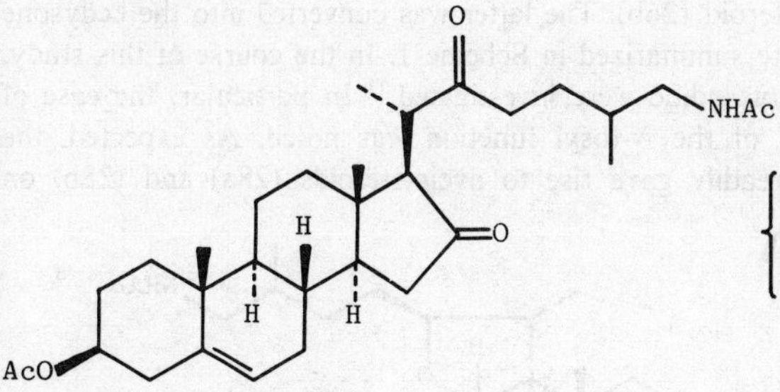

(21)

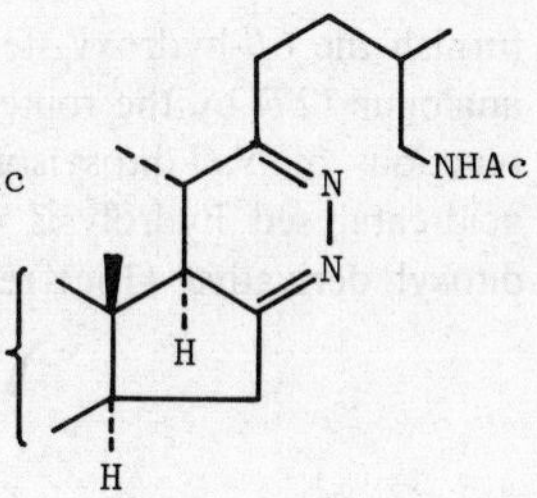

(22)

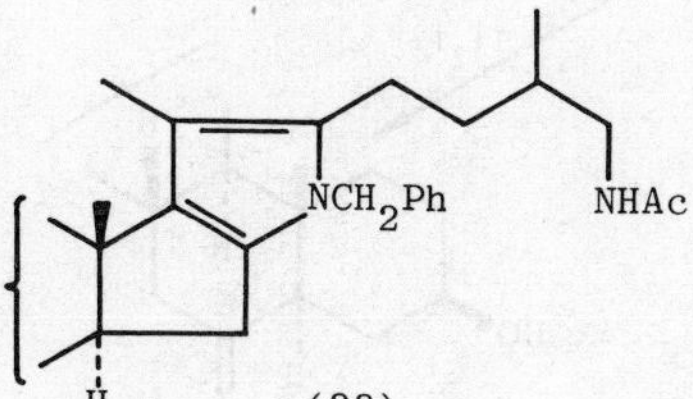

(23)

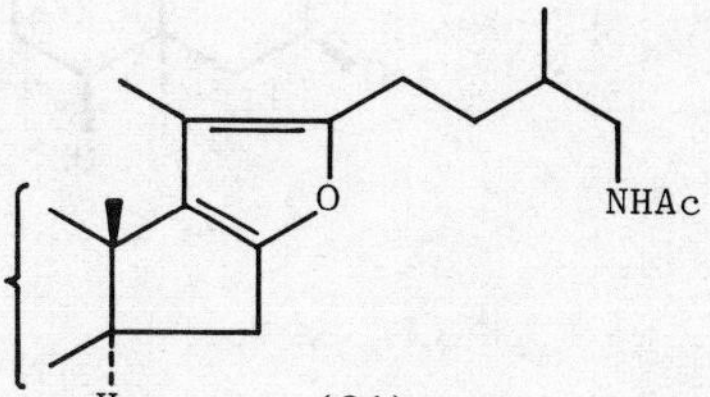

(24)

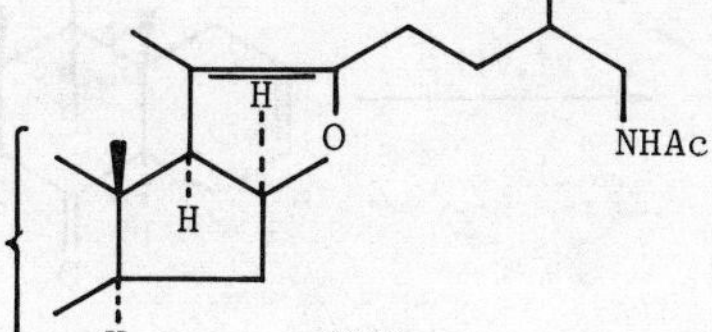

(25)

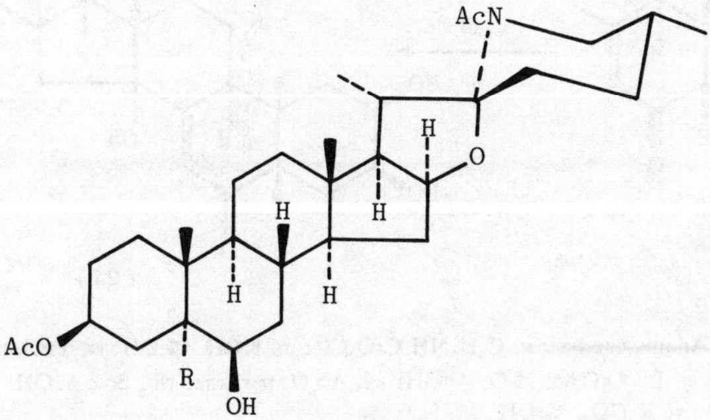

(26) a; R = Br

b; R = H

furnish the 6β-hydroxy-steroid (26b). The latter was converted into the ecdysone analogue (27) by the route summarized in Scheme 1. In the course of this study, reactions of N,O-ditosylsolasodine were investigated.[23] In particular, the ease of acid-catalysed hydrolysis of the N-tosyl function was noted. As expected, the ditosyl derivative (16c) readily gave rise to cyclo-steroids (28a) and (28b) on

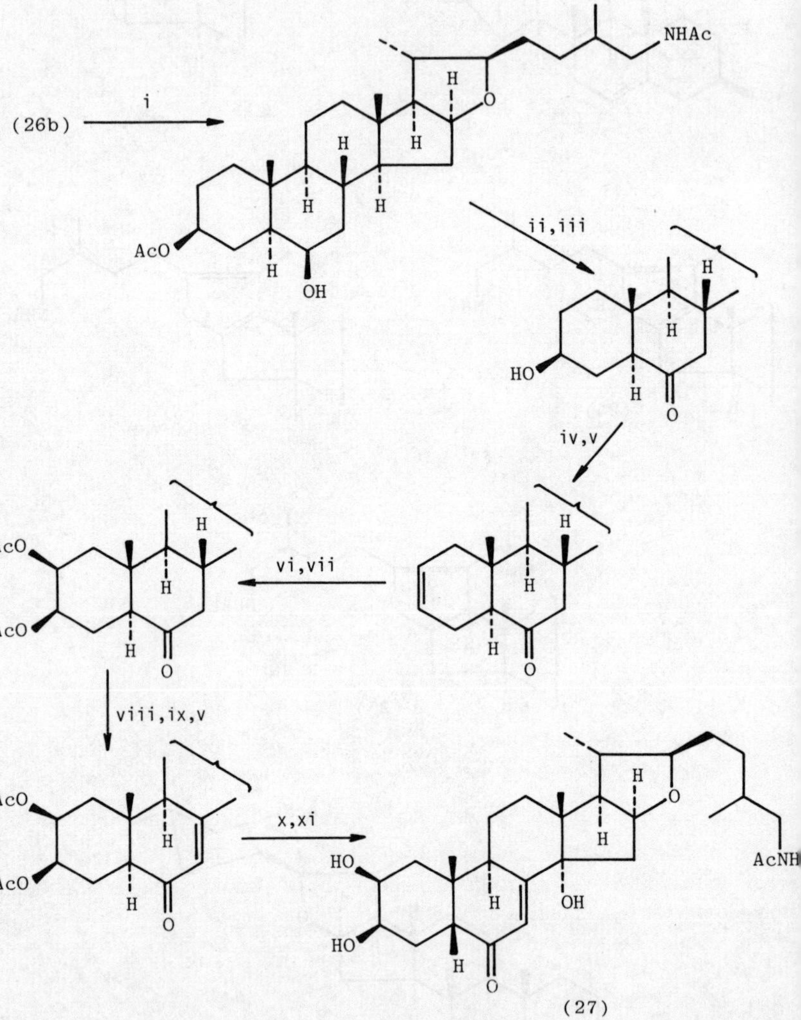

Reagents: i, H_2, Adams catalyst; ii, $C_5H_5\overset{+}{N}H$ CrO_3Cl^-; iii, KOH, MeOH; iv, TsCl, pyridine; v, LiBr, DMF; vi, I_2, AgOAc, H_2O, AcOH; vii, Ac_2O, pyridine; viii, Br_2, AcOH; ix, HBr, AcOH; x, SeO_2; xi, K_2CO_3, MeOH

Scheme 1

[23] R. C. Cambie, G. J. Potter, P. S. Rutledge, and P. D. Woodgate, *Aust. J. Chem.*, 1981, **34**, 829.

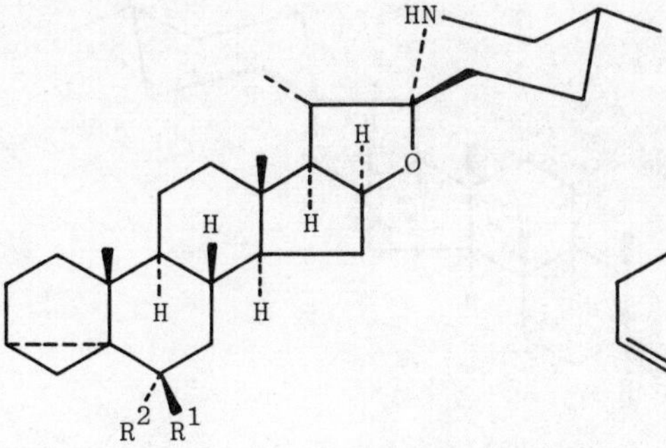

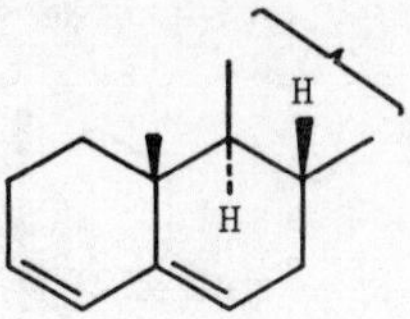

(28) a; R^1= OH, R^2= H   (29)

  b; R^1R^2= O

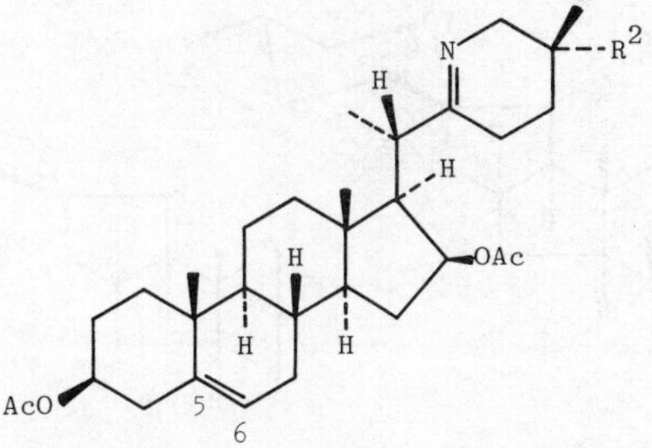

(30) a; R^1= H, R^2= Me

  b; R^1= Me, R^2= H; 5,6-dihydro

solvolysis with acetate in aqueous acetone or in dimethyl sulphoxide, respectively. In both cases, the conjugated diene (29) was a minor product, while in the latter case only the enone (20d) was also formed. In contrast, solvolysis of (16c) in acetic acid gave exclusively the ring-E-cleaved diacetoxy-derivative (30a).[23]

The previously described preparations of *N*-methyltomatidine and of the solasodine analogue have been re-investigated.[24] Thus treatment of tomatidine (31a) with zinc chloride and acetic anhydride gave the ring-E-cleaved diacetoxy-compound (30b), which was converted into the *N*-methiodide salt. The latter was treated with methanolic potassium hydroxide to yield a single product which was identified by [13]C n.m.r. as the 22S-derivative *N*-methyltomatidine (31b). When the same sequence of reactions was applied to solasodine, two *N*-methyl compounds, (32) and (33), were formed in a ratio of 3:2 and were identified unambiguously by [13]C n.m.r. The difference in behaviour between tomatidine and solasodine was rationalized by conformational arguments.

The isolation of solasodine from *Solanum khasianum* has been reviewed.[25] The

[24] H. E. Gottlieb, I. Belic, R. Komel, and M. Mervic, *J. Chem. Soc., Perkin Trans. 1*, 1981, 1889.
[25] N. S. Sharma and S. Varghese, *Indian Drugs*, 1980, **18**, 1 (*Chem. Abstr.*, 1981, **94**, 1967).

(31) a; R = H

b; R = Me

(32)   (33)

solasodine contents of immature fruits of eleven species grown in Panama were determined as follows: *S. asperum* and *S. umbellatum* (0.44%), *S. subinerine* and *S. rugosum* (0.33%), *S. hayesii* (0.31%), *S. acerosum* (0.23%), *S. intermedium* (0.16%), *S. accrescens* (0.06%), *S. ochraeo-ferrugineum* (0.02%), and *S. jamaicense* and *S. siparunoides* (0.01%).[26] Solasodine has been isolated from green berries of *S. americanum*,[27] *S. marginatum* (2.5%),[28] and *S. nigrum* (5%).[29] The variation with time of solasodine production by the latter species has been studied.[30] Solasodine has been isolated from *S. mammosum*[31] and *S. viarum*,[31, 32] and from tissue cultures of *S. jasminoides*.[33] The metabolism of solanidine in potato tuber slices and in cell suspension cultures has been studied.[34]

[26] M. P. Gupta, M. Correa, A. A. Soto, M. B. Gonzalez, and T. D. Arias, *Rev. Latinoam. Quim.*, 1980, 11, 133 (*Chem. Abstr.*, 1981, 94, 44 084).

[27] I. Mathe, H. V. Mai, and I. Mathe, *Acta Agron. Acad. Sci. Hung.*, 1980, 29, 227 (*Chem. Abstr.*, 1980, 93, 91 916).

[28] A. Sanabria Galindo, *Rev. Colomb. Cienc. Quim.-Farm.*, 1980, 3, No. 4, p. 29 (*Chem. Abstr.*, 1981, 94, 205 545).

[29] B. Bose and C. Ghosh, *J. Inst. Chem.* (*India*), 1980, 52, 83 (*Chem. Abstr.*, 1980, 93, 66 325).

[30] I. Mathe, I. Mathe, and H. Van Mai, *Herba Hung.*, 1979, 18, 143 (*Chem. Abstr.*, 1980, 93, 146 452).

[31] A. K. Chakravarty, C. R. Saha, T. K. Dhar, and S. C. Pakrashi, *Indian J. Chem., Sect. B*, 1980, 19, 468 (*Chem. Abstr.*, 1980, 93, 182 811).

[32] A. R. Pingle and V. R. Dnyansagar, *Indian Drugs*, 1980, 17, 366 (*Chem. Abstr.*, 1981, 94, 12 812).

[33] S. C. Jain, P. Khanna, and S. Sahoo, *J. Nat. Prod.*, 1981, 44, 125 (*Chem. Abstr.*, 1981, 94, 117 859).

[34] S. F. Osman, R. M. Zacharius, and D. Naglak, *Phytochemistry*, 1980, 19, 2599.

The following topics have been reviewed: the glyco-alkaloids of the Solanaceae[35] and of the genus *Solanum*;[36] the constituents of the genus *Solanum*;[37] and the biological effects of the solanidine glycosides.[38] A sterol-binding assay for potato glyco-alkaloids has been described.[39]

The glyco-alkaloid solamargine has been isolated from fresh fruits of *S. albicaule* (0.5%)[40] and, together with solasonine, from *S. marginatum*.[41] Solanine, solamargine, β-solamargine, and solasodine have been isolated from *S. dubium*.[42] Thirteen steroidal glycosides were isolated from root bark of *S. macrocarpum*; on acid hydrolysis of the glycoside mixture, ten aglycons were recovered, including solasodine and tomatidine.[43] Glyco-alkaloids have been isolated from *S. melongena*.[44] Potato blossoms[45] and tomato herbage[46] have been described as sources of the appropriate glyco-alkaloids. The distribution of α-tomatine in the tomato plant has been studied at different growth phases.[47] A study has been reported on the effect of the production of glyco-alkaloids by *S. chacoense* on its resistance to the Colorado potato beetle. Clones of *S. chacoense* in which commersonine or dehydrocommersonine was the major foliar glyco-alkaloid were more resistant than clones that were rich in solanine and chaconine.[48]

The mass-spectral fragmentation of solasodine has been discussed.[49] A g.l.c. method has been described for the analysis of potato glyco-alkaloids.[50] A study has been reported on the separation of *Solanum* and *Veratrum* alkaloids by h.p.l.c.[51]

4 *Fritillaria* and *Veratrum* Alkaloids

A variety of new alkaloids have been isolated from plants of the genus *Fritillaria*. Extraction of the aerial parts of *Fritillaria camtschatcensis* furnished solanidine, solasodine, tomatidenol, and two new aglycons, *i.e.* hapepunine and anrakorinine.[52]

[35] S. F. Osman, *Recent Adv. Phytochem.*, 1980, **14**, 75.
[36] U. Mahmood and R. S. Thakur, *Curr. Res. Med. Aromat. Plants*, 1980, **2**, 142 (*Chem. Abstr.*, 1981, **94**, 188 615).
[37] J. K. Bhatnagar, R. K. Jaggi, and A. Bhandari, *Pharmacos*, 1980, **24**, 44 (*Chem. Abstr.*, 1980, **93**, 182 749).
[38] V. A. Drinyaev, E. V. Troshko, V. A. Artyushkova, and L. I. Kulida, *Biol. Nauki* (*Moscow*), 1980, No. 10, p. 5 (*Chem. Abstr.*, 1981, **94**, 754).
[39] J. G. Roddick, *Phytochemistry*, 1980, **19**, 2455.
[40] V. Ahmad, S. F. Ali, and R. Ahmad, *Planta Med.*, 1980, **39**, 186.
[41] A. Sanabria Galindo, *Rev. Colomb. Cienc. Quim.-Farm.*, 1979, **3**, No. 3, p. 89 (*Chem. Abstr.*, 1980, **93**, 146 283).
[42] Y. M. El Kheir and M. H. Salih, *Fitoterapia*, 1979, **50**, 255 (*Chem. Abstr.*, 1980, **93**, 66 105).
[43] M. M. Shabana, Y. W. Mirhom, and S. H. Hilal, *Egypt. J. Pharm. Sci.*, 1978 (publ. 1980), **19**, 77 (*Chem. Abstr.*, 1981, **94**, 205 410).
[44] K. L. Bajaj, G. Kaur, and M. L. Chadha, *J. Plant Foods*, 1979, **3**, 163 (*Chem. Abstr.*, 1980, **93**, 44 301).
[45] R. J. Bushway, E. S. Barden, A. W. Bushway, and A. A. Bushway, *Am. Potato J.*, 1980, **57**, 175 (*Chem. Abstr.*, 1980, **93**, 41 000).
[46] Z. Kowalewski, K. Drost-Karbowska, and M. Szaufer, *Herba Pol.*, 1980, **26**, 157 (*Chem. Abstr.*, 1981, **94**, 188 642).
[47] L. K. Klyshev and A. Zh. Dosymbaeva, *Izv. Akad. Nauk Kaz. SSR, Ser. Biol.*, 1980, No. 4, p. 76 (*Chem. Abstr.*, 1980, **83**, 201 110).
[48] S. L. Sinden, L. L. Sanford, and S. F. Osman, *Am. Potato J.*, 1980, **57**, 331 (*Chem. Abstr.*, 1980, **93**, 128 913).
[49] M. Hasan, M. U. Zubair, and K. Ali, *Islamabad J. Sci.*, 1978, **5**, 11 (*Chem. Abstr.*, 1981, **94**, 121 786).
[50] R. R. King, *J. Assoc. Off. Anal. Chem.*, 1980, **63**, 1226 (*Chem. Abstr.*, 1981, **94**, 1688).
[51] I. R. Hunter, M. K. Walden, and E. Heftmann, *J. Chromatogr.*, 1980, **198**, 363.
[52] K. Kaneko, U. Nakaoka, M. Tanaka, N. Yoshida, and H. Mitsuhashi, *Phytochemistry*, 1981, **20**, 157.

The *N*-methyl-22,26-epiminocholestane structure (34a) was earlier assigned to hapepunine in a preliminary communication.[53] The mass spectrum of the related base anrakorinine showed ions with m/z 445 (M^+), 414 ($M^+ - CH_2OH$), and 112 [base peak; assigned to the ion (35)].[52] The ^{1}H n.m.r. spectrum of anrakorinine closely resembled that of hapepunine, but lacked a methyl singlet at $\delta 0.96$ and displayed instead an AB quartet ($J = 12$ Hz) with chemical shifts centred at $\delta 3.62$ and $\delta 3.88$. Thus anrakorinine was assigned the 18-hydroxy-structure (34b); this assignment was confirmed by its conversion into the primary tosyl derivative (34c), which yielded hapepunine when submitted to reduction with lithium aluminium hydride.[52]

(34) a; R = H   (35)

      b; R = OH

      c; R = OTs

Bulbs of *F. camtschatcensis* yielded solanidine (36a) together with traces of hapepunine (34a), veralkamine, an unidentified dihydroxy-spirosolane, and a new aglycon, *i.e.* camtschatcanidine (36b).[54] The structure assigned to camtschatcanidine was deduced from the spectroscopic data, and in particular by comparison of its ^{13}C n.m.r. spectrum with that of solanidine. The structure and stereochemistry proposed were proven by the reduction (by LiAlH$_4$) of the *O*-acetyl-*O*-tosyl derivative (36c) of camtschatcanidine to yield solanidine.

Fresh bulbs of *F. thunbergii* yielded a new alkaloid (37) together with the known constituents verticine (38a) and verticinone (38b).[55] The new alkaloid was assigned the structure (37) by spectroscopic comparison of the natural product and its derivatives with jervine. The oriental drug 'Bai-mo', prepared from the same bulbs by bleaching in the sun, was also investigated in the same study. As well as verticine (38a) and verticinone (38b), the drug afforded the *N*-oxides (39a) and (39b), whose structures were established by their reduction (by Ph$_3$P) to verticine and verticinone, respectively, and by their formation by oxidation (by H$_2$O$_2$) of the latter bases. Two other artefacts, isolated as their diacetyl derivatives, were assigned the epoxide structures (40) and (41) on the basis of spectroscopic studies.[55]

[53] K. Kaneko, U. Nakaoka, M. W. Tanaka, N. Yoshida, and H. Mitsuhashi, *Tetrahedron Lett.*, 1978, 2099.

[54] K. Kaneko, M. Tanaka, U. Nakaoka, Y. Tanaka, N. Yoshida, and H. Mitsuhashi, *Phytochemistry*, 1981, **20**, 327.

[55] J. Kitajima, N. Noda, Y. Ida, K. Miyahara, and T. Kawasaki, *Heterocycles*, 1981, **15**, 791.

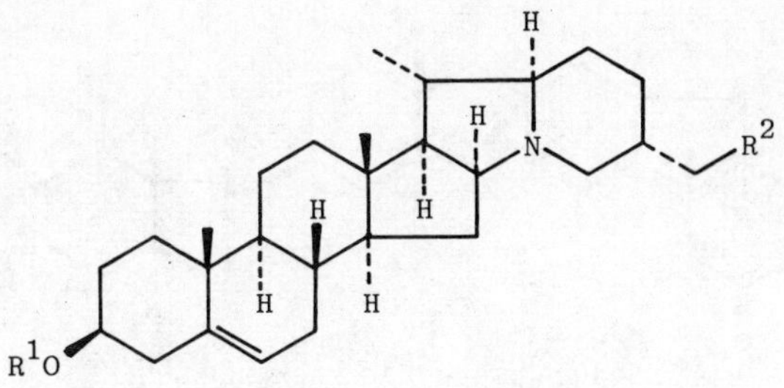

(36) a; $R^1 = R^2 = H$

b; $R^1 = H$, $R^2 = OH$

c; $R^1 = Ac$, $R^2 = OTs$

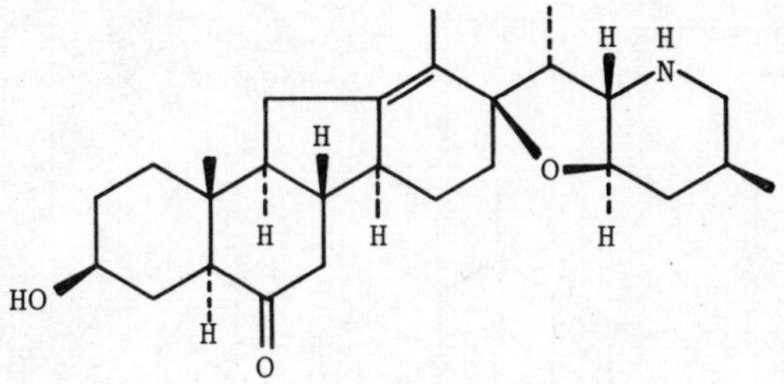

(37)

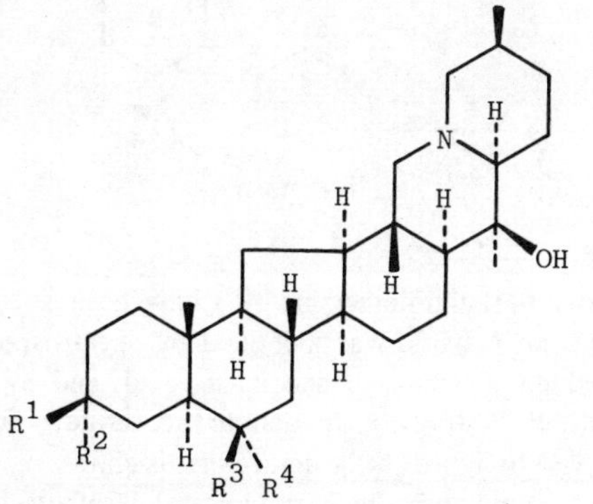

(39)

(38) a; $R^1 = R^4 = OH$, $R^2 = R^3 = H$

b; $R^1 = OH$, $R^2 = H$, $R^3R^4 = O$

c; $R^1 = R^3 = H$, $R^2 = R^4 = OH$

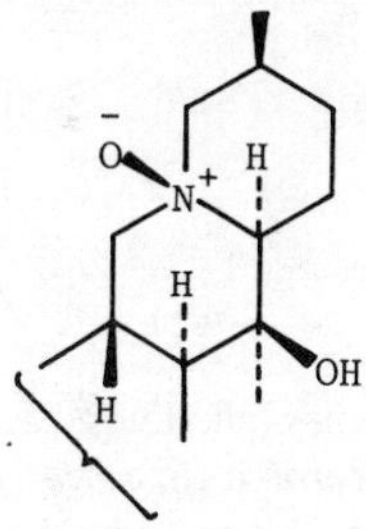

Kaneko has reported the isolation of isobaimonidine (38c), which is a new epimer of verticine (38a), from aerial parts of *F. thunbergii*.[56] The structure and stereochemistry of isobaimonidine were established by its synthesis from verticine.

[56] K. Kaneko, N. Naruse, K. Haruki, and H. Mitsuhashi, *Chem. Pharm. Bull.*, 1980, **28**, 1345.

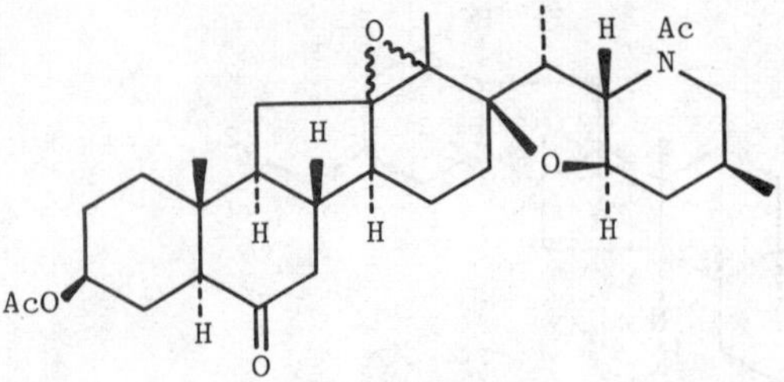

(40) 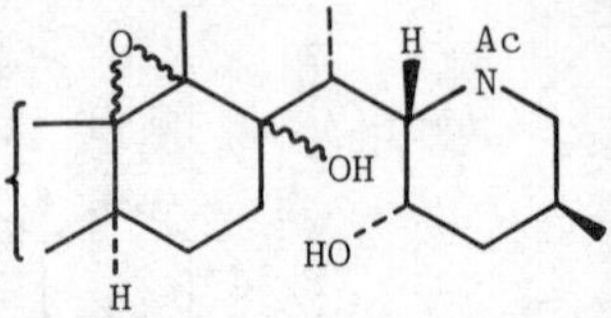   (41)

The same workers have isolated another new cevanine alkaloid, fritillarizine, from aerial parts of *F. verticillata*.[57] Fritillarizine (42a) formed a monoacetyl derivative (42b) and gave the enone (43) on Oppenauer oxidation. The structure shown for fritillarizine (42a) was assigned on the basis of spectroscopic studies and confirmed by its synthesis from verticinone (38b).[57]

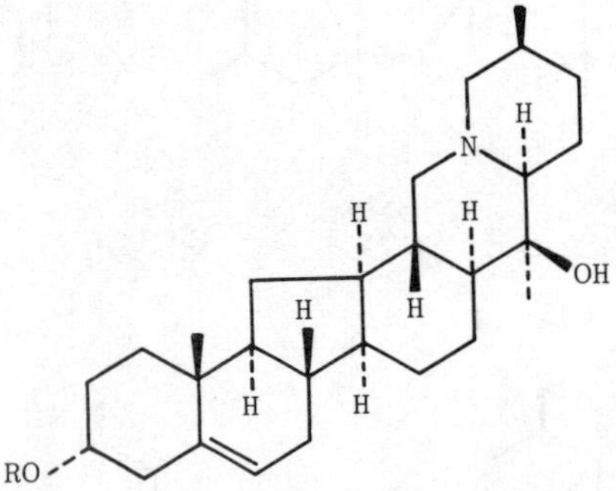

(42) a; R = H   (43)

b; R = Ac

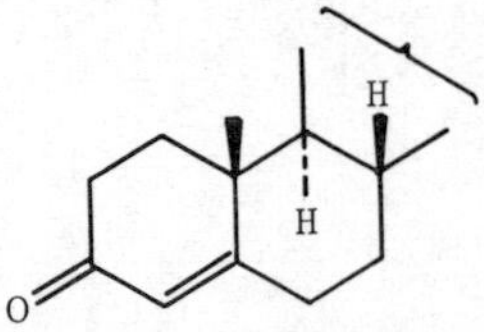

Two new alkaloids, severine *N*-oxide (44) and korsemine (45), have been isolated from *Korolkowia sewerzowii*.[58] Severine *N*-oxide was identified by spectroscopic methods, by the products of its reduction with zinc and acetic acid, and by its preparation by treatment of severine[59,60] (previously denoted ceverine[60]) with hydrogen peroxide. Korsemine (45) was assigned the structure that is shown on the basis of a spectroscopic study of the alkaloid and of its tetra-*O*-acetyl derivative and from the formation of the aglycon korsevine on acid-catalysed hydrolysis.[58]

[57] K. Kaneko, N. Naruse, M. Tanaka, N. Yoshida, and H. Mitsuhashi, *Chem. Pharm. Bull.*, 1980, **28**, 3711.
[58] K. Samikov, D. U. Abdullaeva, R. Shakirov, and S. Yu. Yunusov, *Khim. Prir. Soedin.*, 1979, 823 (*Chem. Abstr.*, 1980, **93**, 72 049).
[59] D. U. Abdullaeva, K. Samikov, R. Shakirov, and S. Yu. Yunusov, *Khim. Prir. Soedin.*, 1977, 671 (*Chem. Abstr.*, 1978, **88**, 170 356).
[60] D. M. Harrison, in 'The Alkaloids', ed. M. F. Grundon, (Specialist Periodical Reports), The Chemical Society, London, 1979, Vol. 9, p. 249.

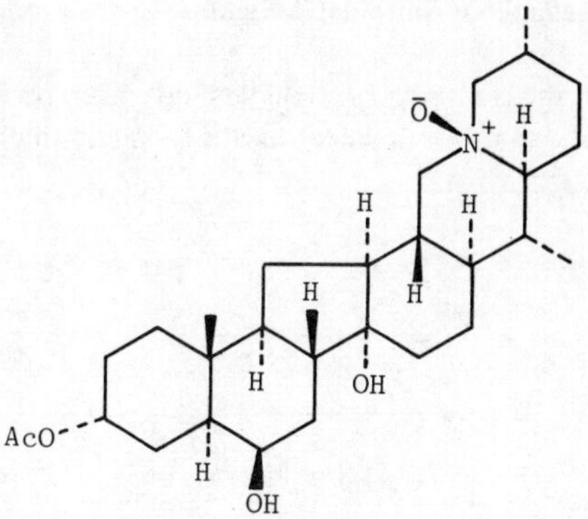

(44)

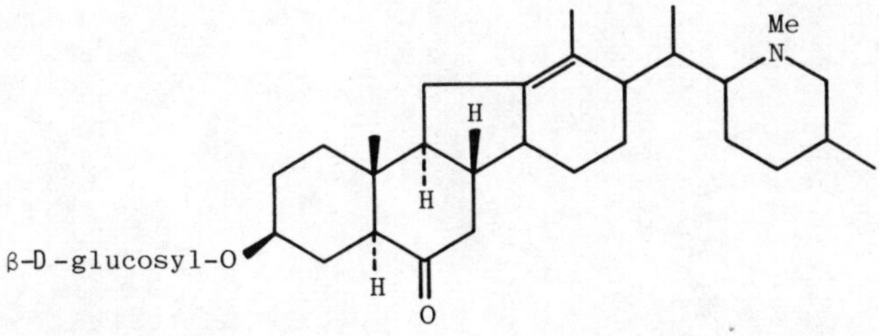

(45)

Veratrum album subsp. *lobelianum* has yielded a range of alkaloids.[61,62] Amongst these, 3-*O*-acetyl-15-*O*-veratroylgermine, 15-*O*-veratroylgermine, veralobine, veramarine, verarine, verazine, veralkamine, veralinine, and veramine were isolated for the first time from underground parts of the plant.[61] Aerial parts furnished verazine, veramiline, and four new alkaloids, *i.e.* veratine, veratimine, veradanine, and veravine.[62]

The mechanism of formation of the c-nor-D-homo-steroids from solanidanine alkaloids has been discussed[63] and the toxic principles of *V. album* have been reviewed.[64] The transformation of jervine into c/D-*trans* c-nor-D-homo-steroids has been described.[65] The ^{13}C n.m.r. spectra of four ceveratrum alkaloids have been recorded and interpreted.[66]

[61] J. Tomko and A. Vassova, *Farm. Obz.*, 1981, **50**, 115 (*Chem. Abstr.*, 1981, **94**, 205 391).
[62] A. Vassova and J. Tomko, *Acta Fac. Pharm. Univ. Comenianae*, 1979, **34**, 151 (*Chem. Abstr.*, 1981, **94**, 171 035).
[63] K. Kaneko, N. Kawamura, M. Tanaka, and H. Mitsuhashi, *Tennen Yuki Kagobutsu Toronkai Koen Yoshishu, 22nd*, 1979, 55 (*Chem. Abstr.*, 1980, **93**, 65 937).
[64] K. Von der Dunk, *PTA Prakt. Pharm.*, 1980, **9**, 72 (*Chem. Abstr.*, 1980, **93**, 65 934).
[65] H. Suginome, H. Ono, and T. Masamune, *Bull. Chem. Soc. Jpn.*, 1981, **54**, 852.
[66] F. A. Carey, W. C. Hutton, and J. C. Schmidt, *Org. Magn. Reson.*, 1980, **14**, 141.

5 Miscellaneous Steroidal Alkaloids

Synthetic studies on the *Salamander* alkaloids have been reviewed.[67] The synthesis of 18-amino-steroids that are related to the alkaloid batrachotoxin has been discussed.[68]

[67] K. Oka, *Yakugaku Zasshi*, 1980, **100**, 227 (*Chem. Abstr.*, 1980, **93**, 46 951).

[68] N. K. Levchenko, G. M. Segal, and I. V. Torgov, *Tezisy Dokl. Sov. Indiiskii Simp. Khim. Prir. Soedin.*, *5th*, 1978, p. 48 (*Chem. Abstr.*, 1980, **93**, 186 661); *cf.* N. K. Levchenko, A. P. Sviridova, G. M. Segal, and I. V. Torgov, *Bioorg. Khim.*, 1978, **4**, 1651 (*Chem. Abstr.*, 1979, **90**, 187 205); D. M. Harrison, in 'The Alkaloids', ed. M. F. Grundon, (Specialist Periodical Reports), The Royal Society of Chemistry, London, 1981, Vol. 10, p. 240.

15
Miscellaneous Alkaloids

BY J. R. LEWIS

1 Muscarine and Oxazolidone Alkaloids

α-D-Glucose is the starting material for a synthesis of the iodides of epiallo-muscarine (1; R = H), isoepiallomuscarine (2), and 3-hydroxyepiallomuscarine (1; R = OH). Deoxygenation of the sugar was achieved by tosylation and reduction by LiAlH$_4$, the diacetal subsequently being oxidized to a carboxylic acid, which was converted into an amide. Reduction of the amide yielded the dimethylamine.[1] In a chelation-controlled synthesis of (±)-muscarine (5), cyclopentadiene was converted (by photo-oxygenation) into 3,5-dihydroxypent-2-ene; upon protection of hydroxyl groups and ozonization, this gave the cyclic hydrate (3), which, after acetylation was converted into the lactone (4) by stereocontrolled Grignard addition of MeMgBr and then oxidation. This lactone reacted with dimethylamine to form the amide, which was transformed into (±)-muscarine by conventional methods.[2] In studying the cause of the magenta colour produced by the pentacyanoammoni-ferrate reagent on extracts of jack beans (*Canavalia ensiformis*), it has been found that both L-canavanine and 3-isoxazolidone (6) give positive results. This is the first time that (6) has been reported as a natural product.[3]

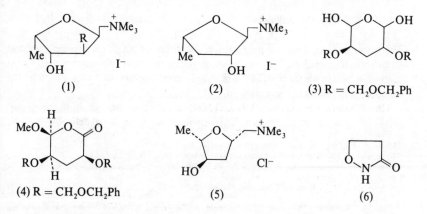

(1)   (2)   (3) R = CH$_2$OCH$_2$Ph

(4) R = CH$_2$OCH$_2$Ph   (5)   (6)

[1] P.-C. Wang and M. M. Joullié, *J. Org. Chem.*, 1980, **45**, 5359.
[2] W. C. Still and J. A. Schneider, *J. Org. Chem.*, 1980, **45**, 3375.

2 Imidazole Alkaloids

Natural products that contain the imidazole nucleus have been found in such diverse sources as mushrooms and broiled fish. In the former, the fresh fruiting bodies of *Clitocybe acromelalga* contain the new amino-acid betaine clithioneine (7);[4] in the latter case, the presence of 2-amino-3-methylimidazo[4,5-*f*]quinoline (8) has been confirmed by its synthesis from quinoline-5,6-diamine.[5] The seeds of *Butea monosperma* contain three alkaloids, one of which has now been identified[6] as 1-*N*-acetyl-2-oxo-4-methoxy-3*H*,5*H*-imidazole (9). It is possible to detect pilocarpine (10; R^1 = H, R^2 = Et) and isopilocarpine (10; R^1 = Et, R^2 = H) by h.p.l.c. at sensitivities down to 300 ng; also detectable were their degradation products, pilocarpic and isopilocarpic acids.[7]

A synthesis of the imidazole alkaloid dolichotheline (11) (and of other amide-containing alkaloids) has been reported; 3-acyl-thiazolidine-3-thiones are used in an aminolysis procedure.[8]

(7)

(8)

(9)   (10)   (11)

3 Peptide Alkaloids

The occurrence of severine palmitate in the fruits of *Atalantia monophylla* and *Hesperethusa crenulata*,[9] as well as its previous report in other citrus plants, has enabled a re-examination of its n.m.r. and mass spectra to be made, with the consequent revision of its structure to (12).

The isobutylamide (13; R = $CH_2CH_2CHMe_2$) has been isolated from the epigeal parts of *Ottonia ovata*,[10] and it is thought to arise biogenetically by the

[3] M. Sugii, H. Miura, and K. Nagata, *Phytochemistry*, 1981, **20**, 451.

[4] K. Konno, H. Shirahama, and T. Matsumoto, *Tetrahedron Lett.*, 1981, **22**, 1617.

[5] H. Kasai, S. Nishimura, K. Wakabayashi, M. Nagao, and T. Sugimura, *Proc. Jpn. Acad., Ser. B*, 1980, **56**, 382 (*Chem. Abstr.*, 1981, **94**, 30 643).

[6] B. Mehta and M. M. Bokadia, *Chem. Ind. (London)*, 1981, 98.

[7] J. J. O'Donnell, R. Sandman, and M. V. Drake, *J. Pharm. Sci.*, 1980, **69**, 1096.

[8] Y. Nagao, K. Seno, T. Miyasaka, K. Kawabata, S. Takao, and E. Fujita, *Tennen Yuki Kagobutsu Toronkai Koen Yoshishu, 22nd.*, 1979, 554 (*Chem. Abstr.*, 1980, **93**, 204 909).

[9] D. L. Dreyer, J. F. Rigod, S. C. Basa, P. Mahanty, and D. P. Das, *Tetrahedron*, 1980, **36**, 827.

[10] R. Haensel, A. Leuschke, and F. Bohlmann, *Planta Med.*, 1980, **40**, 161.

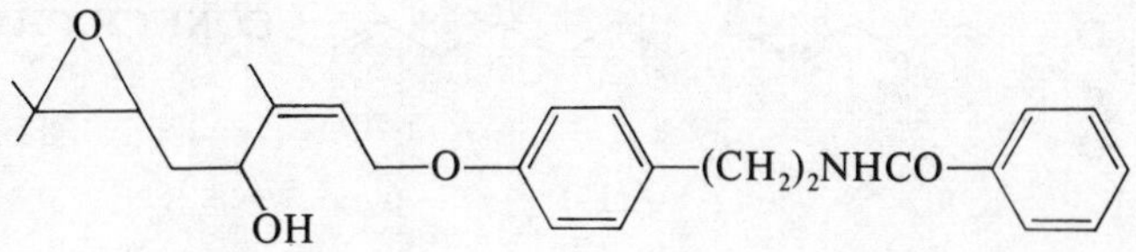

(12)

transformation of piperovatine (13; R = CH_2CHMe). A synthesis of the cytotoxic spermidine metabolites of the soft coral *Sinularia brongersmai* has been achieved, using a hexahydropyrimidine ring as both a protecting group and a latent *N*-methyl function. Thus (14; R = O) was converted into (14; R = H_2) and thence into the amide (15), which, upon reductive cleavage with formic acid, gave the natural product (16); reduction of its double-bond gave the other metabolite.[11] This

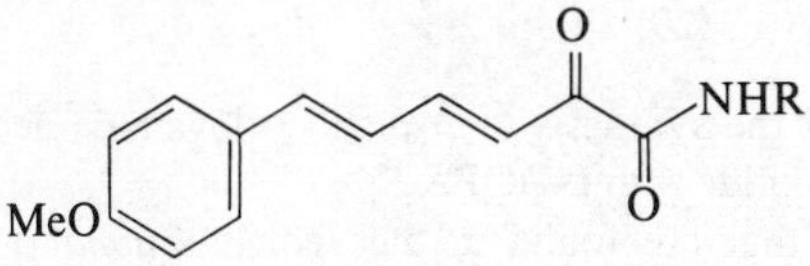

(13)

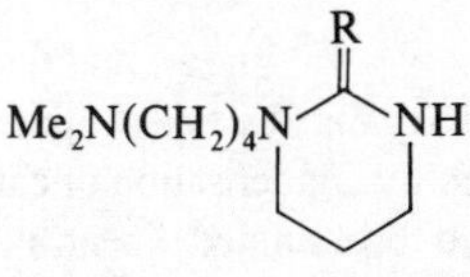

(14)

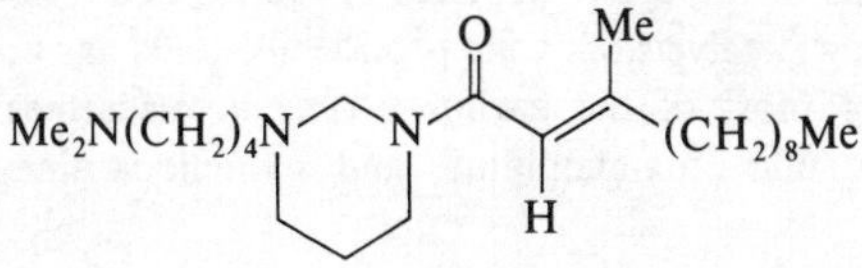

(15)

$$Me(CH_2)_8C(Me) = CHCONH(CH_2)_3\overset{\overset{\displaystyle Me}{|}}{N}(CH_2)_4NMe_2$$

(16)

technique has also been used to synthesize celacinnine, celabenzine, and maytenine.[12]

A synthesis of piperstachine (17) has been achieved by Wittig condensation of the hexanal (19) with ylide (18) followed by removal of the protecting group and oxidation to the aldehyde (20); a subsequent Wittig condensation with ethyl diethylphosphonoacetate gave the conjugate ester, which, on hydrolysis and condensation with $Me_2CHCH_2NH_2$, gave (17).[13]

Sharma's group has also synthesized guineensine, which is the alkaloid produced by both *Piper guineense* and *P. officinarum*.[14] The isolation of *trans*- and *cis*-(−)-clovamides and their deoxy-analogues from the bark of *Dalbergia*

[11] K. Chantrapromma, J. S. McManis, and B. Ganem, *Tetrahedron Lett.*, 1980, **21**, 2605.
[12] J. S. McManis and B. Ganem, *J. Org. Chem.*, 1980, **45**, 2041.
[13] O. P. Vig, S. D. Sharma, and S. S. Bari, *Indian J. Chem., Sect. B*, 1980, **19**, 276.
[14] O. P. Vig, S. D. Sharma, O. P. Sood, and S. S. Bari, *Indian J. Chem., Sect. B*, 1980, **19**, 350.

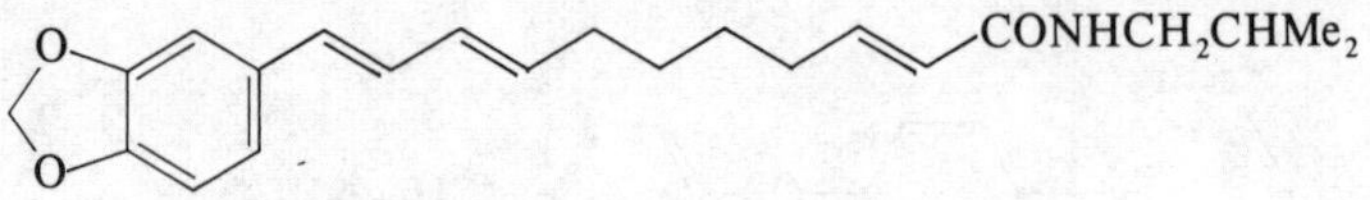

(17)

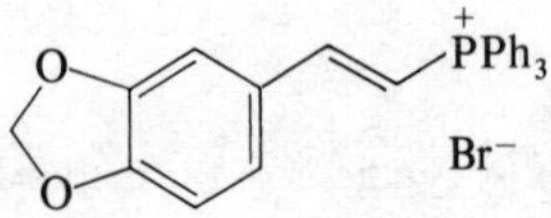

(18)

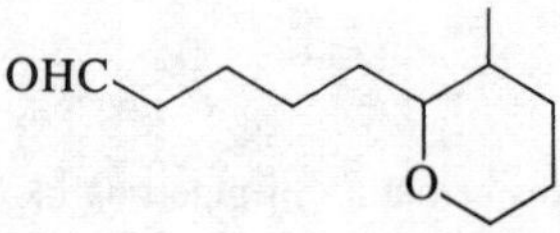

(19)

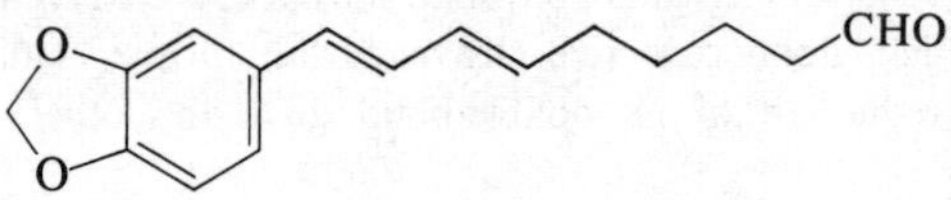

(20)

melanoxylon has been reported and the synthesis of *trans*-(−)-clovamide accomplished by condensation of caffeyl chloride with L-DOPA.[15]

Two brominated linear peptides that are found in the sponge *Cliona celata*, celenamide A (21; R = CH_2CHMe_2) and celenamide B (21; R = $CHMe_2$), bear close structural relationships to the cyclic peptides integerrin and lasiodine A that occur in plants.[16] A ^{13}C n.m.r. analysis of cyclic peptide alkaloids has been reported and the chemical shifts of most of the carbon atoms in discarines A and B, lasiodines A and B, pandamine, pandaminine, and hymenocardine have been assigned.[17]

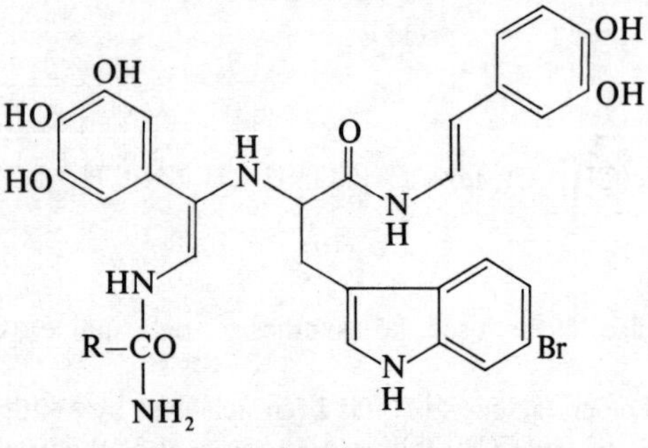

(21)

Since all naturally occurring phencyclopeptines have the same stereochemistry at C-9, a synthetic route involving peptide cyclization of the *p*-nitrophenyl ester (22) to give a single diastereoisomeric cyclic peptine (23) suggests a feasible route to the total synthesis of these natural products.[18] Pubescine A (24), obtained from the

[15] F. R. van Heerden, E. V. Brandt, and D. G. Roux, *Phytochemistry*, 1980, **19**, 2125.
[16] R. J. Stonard and R. J. Andersen, *J. Org. Chem.*, 1980, **45**, 3687.
[17] M. Païs, F.-X. Jarreau, M. G. Sierra, O. A. Mascaretti, E. A. Ruveda, C.-J. Chang, E. W. Hagaman, and E. Wenkert, *Phytochemistry*, 1979, **18**, 1869.
[18] D. Goff, J. C. Lagarias, W. C. Shih, M. P. Klein, and H. Rapoport, *J. Org. Chem.*, 1980, **45**, 4813.

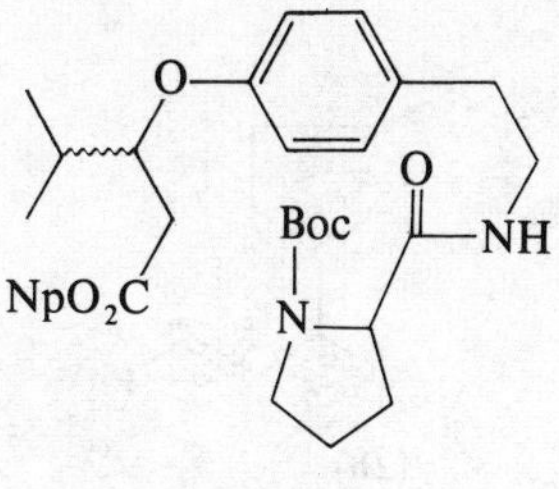

(22) 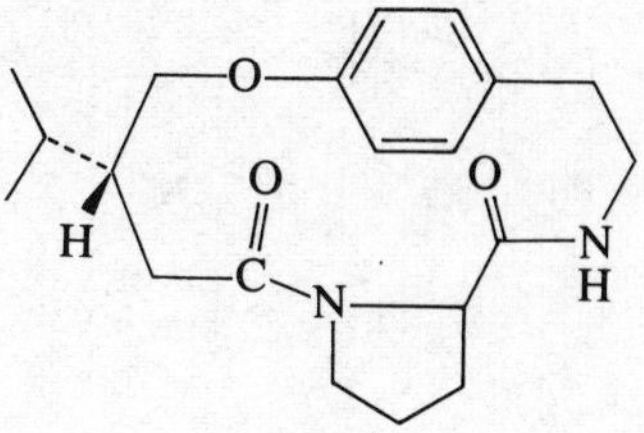 (23)

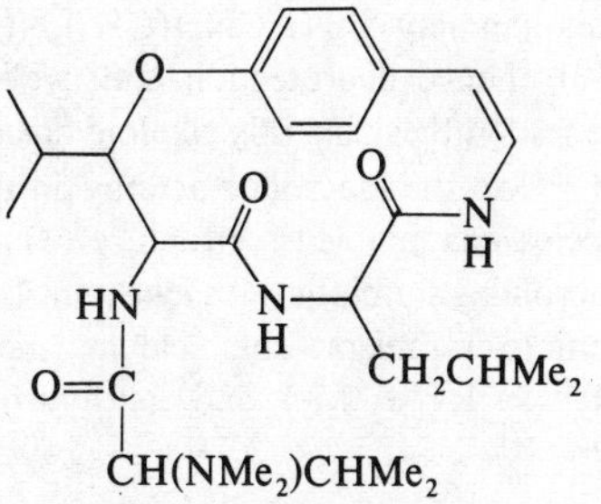

(24)

Rhamnaceous plant *Discaria pubescens*, differs from melonovine A only in its melting point and its optical rotation, and is thus considered to be diastereo-isomeric with respect to the amino-acid chain.[19]

The absolute configurations of two spermine alkaloids, *O*-methylorantine and aphelandrine, obtained from *Chaenorhinum minus* and *C. villosum*, differ only in configuration at C-31,[20] as shown in the partial structures (25) and (26).

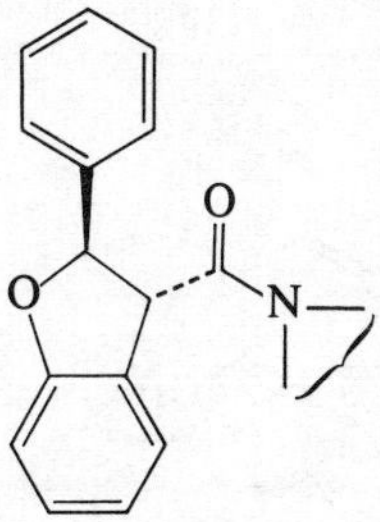

(25) 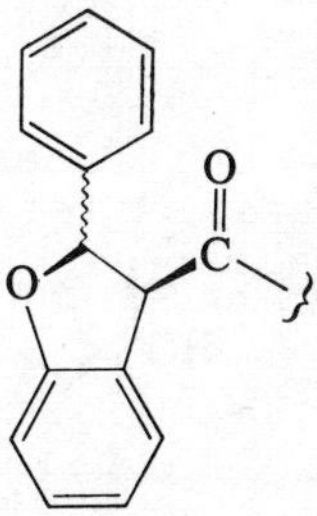 (26)

A total synthesis of (±)-dihydroperiphylline (27) has been achieved, starting from piperidazine and utilising a ring-expansion reaction by a transamidation with the imino-ether (28) and a β-lactam.[21] This procedure has also been used to synthesize celacinnine.[22]

[19] R. Tschesche, D. Hillebrand, and I. R. C. Bick, *Phytochemistry*, 1980, **19**, 1000.
[20] P. Daetwyler, H. Bosshardt, S. Johne, and M. Hesse, *Helv. Chim. Acta*, 1979, **62**, 2712.
[21] H. H. Wasserman and H. Matsuyama, *J. Am. Chem. Soc.*, 1981, **103**, 461.
[22] H. H. Wasserman, R. P. Robinson, and H. Matsuyama, *Tetrahedron Lett.*, 1980, **21**, 3493.

(27)   (28)

The *Lunaria* alkaloid codonocarpine (29) has been regiospecifically synthesized by coupling of the spermidine moiety $Bu^tO_2CNH(CH_2)_4N(CO_2Bu^t)(CH_2)_3NH_2$ with the carboxylic acid (30), thence deprotection and cyclization;[23] the use of the transamidation procedure for synthesizing this alkaloid has also been reported.[24] The incorporation of ^2H and ^{13}C in the macrolide actamycin *via* the Lepetit strain E/784 of a *Streptomyces* species has enabled a structure (31) to be proposed[25,26] for this antibiotic. Two macrolide antibiotics, machbecin I and II, have been obtained from a *Nocardia* micro-organism; chemical and spectral evidence has suggested structures (32; $R^1 = R^2 = Me$) and its hydroquinone analogue[27]

(29)   (30)

(31)

[23] M. J. Humora, D. E. Seitz, and J. Quick, *Tetrahedron Lett.*, 1980, **21**, 3971.

[24] Y. Nagao, K. Seno, and E. Fugila, *Tetrahedron Lett.*, 1980, **21**, 4931.

[25] M. S. Allen, I. A. McDonald, and R. W. Rickards, *Tetrahedron Lett.*, 1981, **22**, 1145.

[26] I. A. McDonald and R. W. Rickards, *Tetrahedron Lett.*, 1981, **22**, 1149.

[27] M. Muroi, K. Haibara, M. Asai, and T. Kishi, *Tetrahedron Lett.*, 1980, **21**, 309.

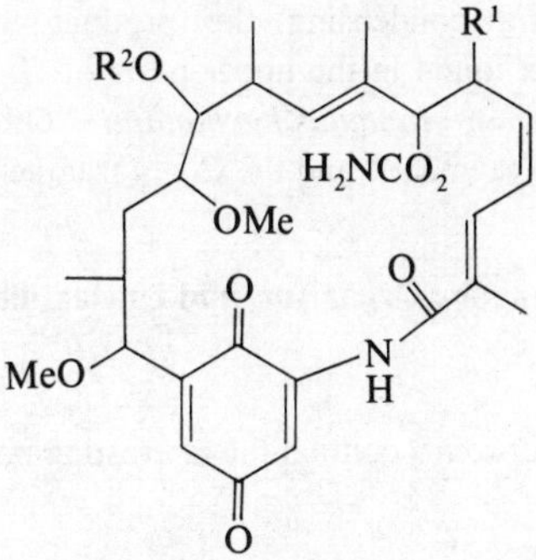

(32)

respectively. The structurally related macrolide antibiotic herbimycin (32; R^1 = OMe, R^2 = Me) has potent herbicidal activity and is isolated from *Streptomyces hygroscopicus*;[28] its desmethyl derivative[29] herbimycin B (32; R^1 = OMe, R^2 = H) also possesses a similar (but lower) herbicidal activity, together with an activity against tobacco mosaic virus.

Two cyclic peptides that were isolated from the marine tunicate *Lissoclinum patella*,[30] named ulicyclamide and ulithiacyclamide, have structures (33) and (34).

Maytansinoids continue to be isolated or synthesized in attempts to improve anti-tumour activity, and a review of the results of their preclinical and clinical testing has appeared.[31] A variety of maytansinol esters has been prepared by

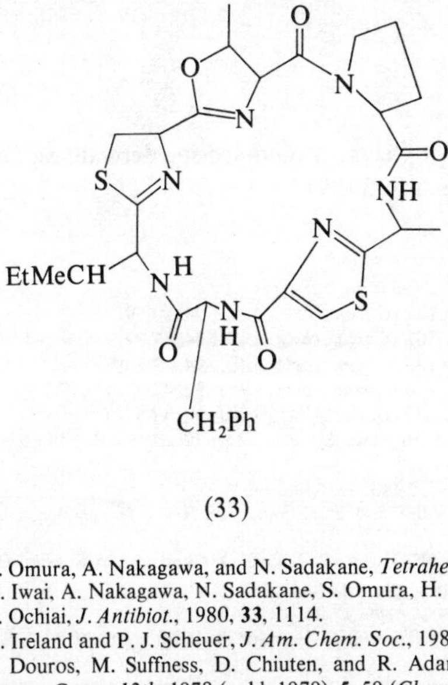

(33)

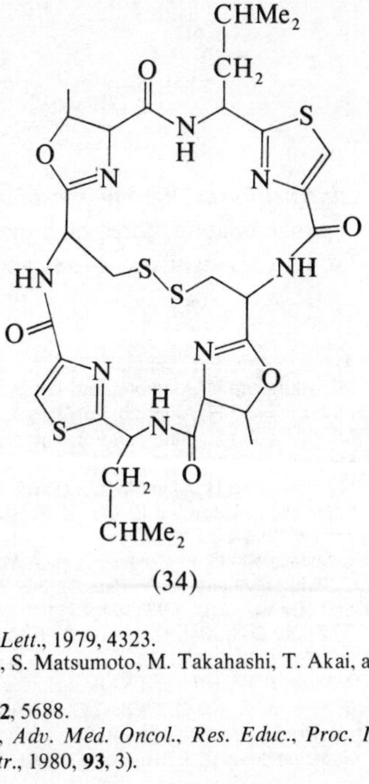

(34)

[28] S. Omura, A. Nakagawa, and N. Sadakane, *Tetrahedron Lett.*, 1979, 4323.
[29] Y. Iwai, A. Nakagawa, N. Sadakane, S. Omura, H. Oiwa, S. Matsumoto, M. Takahashi, T. Akai, and Y. Ochiai, *J. Antibiot.*, 1980, **33**, 1114.
[30] C. Ireland and P. J. Scheuer, *J. Am. Chem. Soc.*, 1980, **102**, 5688.
[31] J. Douros, M. Suffness, D. Chiuten, and R. Adamson, *Adv. Med. Oncol., Res. Educ., Proc. Int. Cancer Congr. 12th*, 1978 (publ. 1979), **5**, 59 (*Chem. Abstr.*, 1980, **93**, 3).

reducing ansamitocin and condensing the product with various acids.[32-37] Maytansine itself has been found in the upper parts of *Gymnosporia diversifolia*[38] and its *N*-methyl derivative in *Maytenus buchananii*.[39] Other synthetic approaches have produced the C_1–C_8 fragment[40] and the C_9–N fragment.[41]

4 Alkaloid-containing Organisms and Unclassified Alkaloids

Agaricus bisporus
The gill tissues of this mushroom[42] contain the interesting imine quinone (35).

Agrobacterium tumefaciens
The crown-gall tumour that is induced by certain strains of this micro-organism contains a number of interesting amino-acids, but agropine (36), a novel bicyclic derivative of glutamic acid, has recently been isolated and is present in quantities of up to 7% of the dry weight of the tumour.[43]

(35)   (36)

Aplysina fistularis (*Verongia fistularis*)
This sponge contains three high-molecular-weight bromotyrosine derivatives,[44] *i.e.* fistularin-1 (37), fistularin-2 (38), and fistularin-3 (39).

[32] H. Akimoto, Eur. Pat. Appl. 21 178 (*Chem. Abstr.*, 1981, **94**, 208 922).
[33] O. Miyashita and H. Akimoto, Eur. Pat. Appl. 21 177 (*Chem. Abstr.*, 1981, **94**, 208 923).
[34] O. Miyashita and H. Akimoto, Eur. Pat. Appl. 11 302 (*Chem. Abstr.*, 1981, **94**, 47 372).
[35] O. Miyashita and H. Akimoto, Eur. Pat. Appl. 10 735 (*Chem. Abstr.*, 1981, **94**, 47 369).
[36] O. Miyashita and H. Akimoto, Eur. Pat. Appl. 11 276 (*Chem. Abstr.*, 1981, **94**, 47 370).
[37] O. Miyashita and H. Akimoto, Eur. Pat. Appl. 11 277 (*Chem. Abstr.*, 1981, **94**, 47 371).
[38] Z.-S. He, Y.-L. Chou, G.-E. Ma, R.-S. Xu, and Q.-M. He, *Tzu Jan Tsa Chih*, 1980, **3**, 639 (*Chem. Abstr.*, 1981, **94**, 20 299).
[39] A. T. Sneden and G. L. Beemsterboer, *J. Nat. Prod.*, 1980, **43**, 637.
[40] B.-C. Pan, H. Zhang, C.-Y. Pan, Y. Shu, Z.-F. Wang, and Y.-S. Gao, *Hua Hsueh Hsueh Pao*, 1980, **38**, 502 (*Chem. Abstr.*, 1981, **94**, 174 940).
[41] Q.-T. Zhou, D.-L. Bai, H.-L. Sun, Y.-C. Yang, Y.-H. Du, Y.-B. Xu, X.-Q. Chen, and Y. S. Gao, *Hua Hsueh Hsueh Pao*, 1980, **38**, 507 (*Chem. Abstr.*, 1981, **94**, 174 941).
[42] P. D. Mize, P. W. Jeffs, and K. Boekelheide, *J. Org. Chem.*, 1980, **45**, 3540.
[43] D. T. Coxon, A. M. C. Davies, G. R. Fenwick, R. Self, J. L. Fermin, D. Lippin, and N. F. James, *Tetrahedron Lett.*, 1980, **21**, 495.
[44] Y. Gopichand and F. J. Schmitz, *Tetrahedron Lett.*, 1979, 3921.

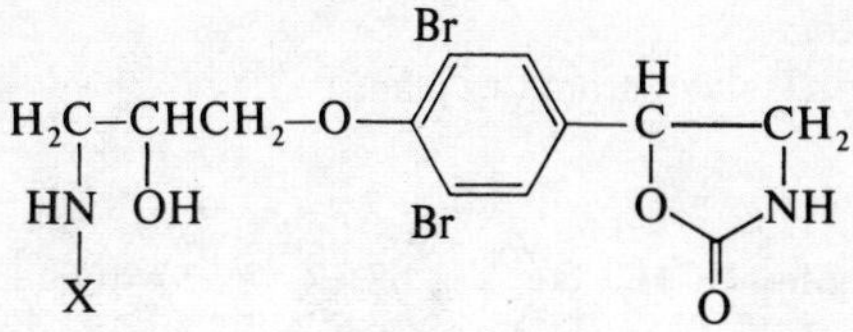

(37)

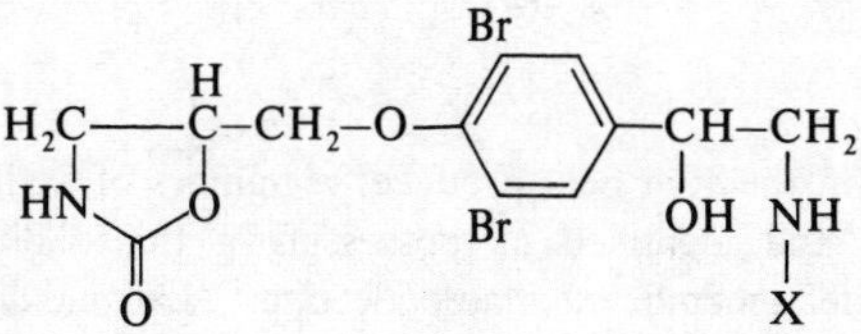

(38)

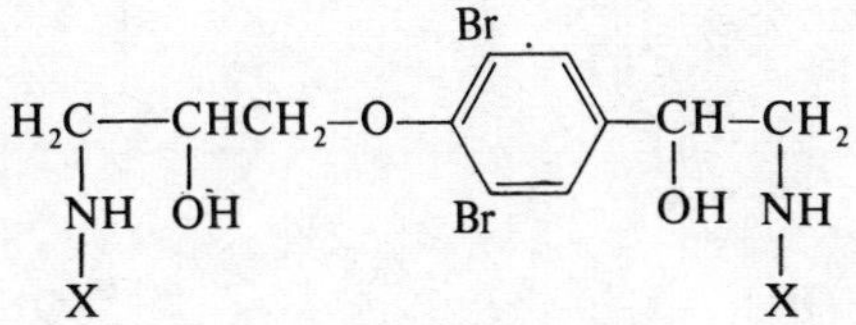

(39)

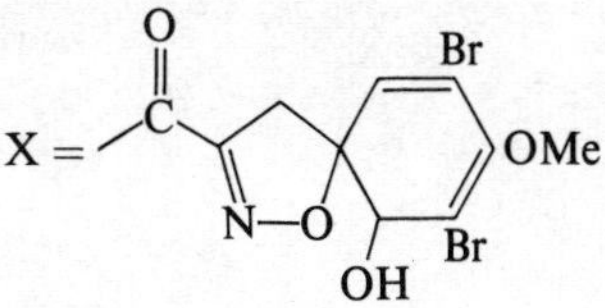

Halocynthia roretzi

A series of mycosporine-like amino-acids have been isolated from this ascidian;[45] the newest (and seventh) is (40).

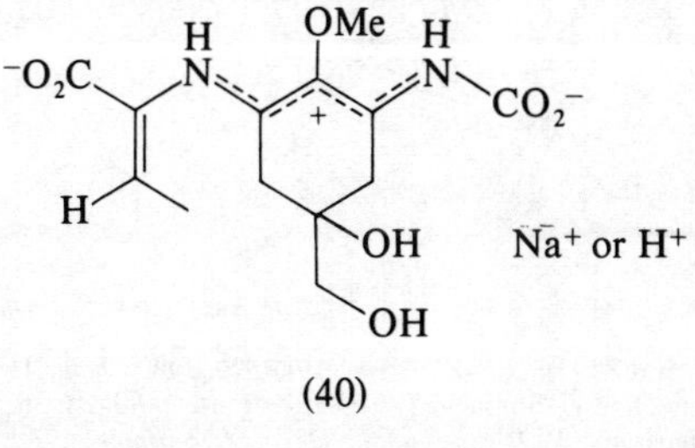

(40)

[45] J. Kobayashi, H. Nakamura, and Y. Hirata, *Tetrahedron Lett.*, 1981, **22**, 3001.

Hymenocallis arenicola
Amisine (41) has been isolated from this plant.[46]

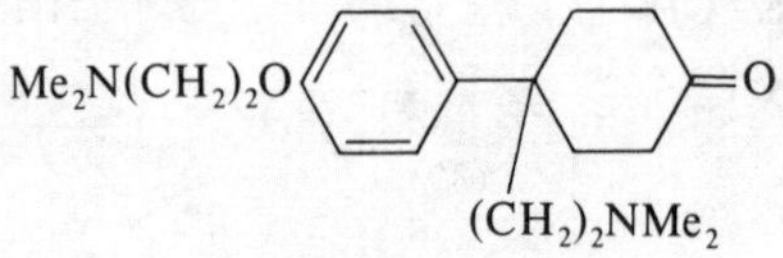

$$Me_2N(CH_2)_2O \quad \text{—} \quad \bigcirc \quad \text{—} \quad =O$$

$$(CH_2)_2NMe_2$$

(41)

Latrunculia magnifica
This sponge, found in the Red Sea, produces a number of toxins, three of which have been isolated and identified as possessing a thiazolidinone ring that is connected to a fourteen-membered macrolide ring.[47] Latrunculins A and B have been assigned structures (42) and (43), and latrunculin C is a stereoisomer of (42).

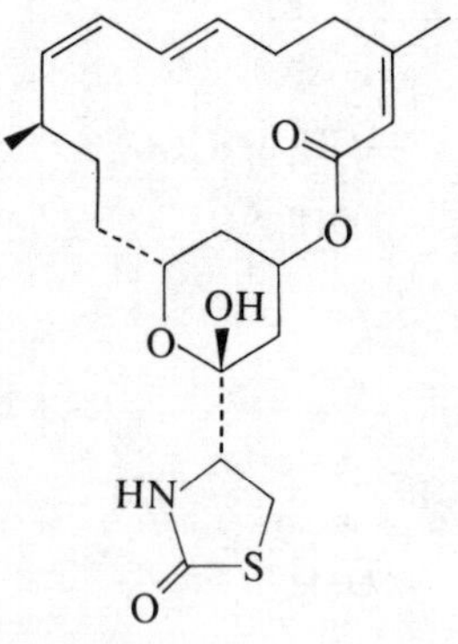

(42) 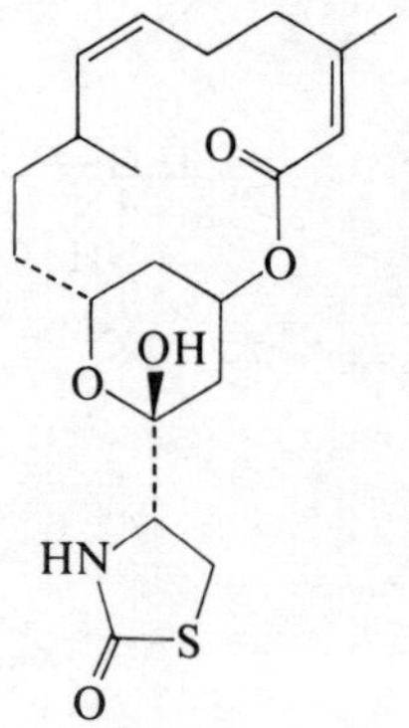   (43)

Myxococcus fulvus
An antibiotic, myxothiazole (44), has been obtained from this fungus.[48]

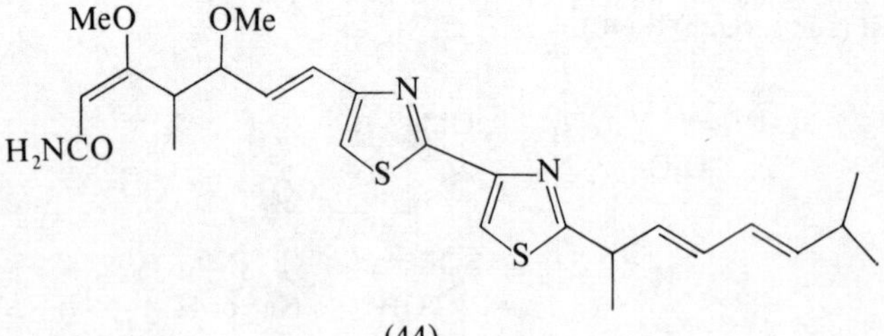

(44)

[46] W. Doeppe, E. Sewerin, and Z. Trimino, *Z. Chem.*, 1980, **20**, 298 (*Chem. Abstr.*, 1981, **94**, 30 972).
[47] Y. Kashman, A. Groweiss, and U. Shmueli, *Tetrahedron Lett.*, 1980, **21**, 3629.
[48] W. Trowitzsch, G. Reifenstahl, W. Wray, and K. Gerth, *J. Antibiot.*, 1980, **33**, 1480 (*Chem. Abstr.*, 1981, **94**, 119 398).

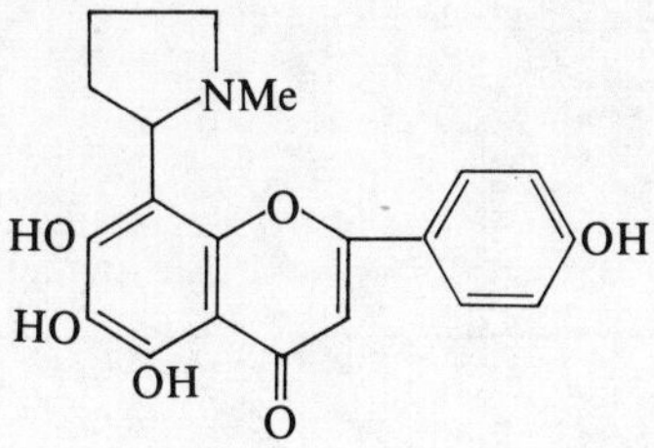

(45)

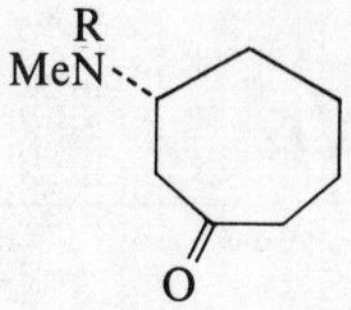

(46)

Phyllospadix iwatensis
This sea grass[49] contains the flavones hispidulin and luteolin as well as the flavonoidal alkaloid phyllospadine (45).

Physalis peruviana
(+)-Physoperuvine (46; R = H) and its *N*-methyl analogue (46; R = Me), as well as racemic (46; R = H), have been obtained from roots of this plant.[50]

Tedania digitata
This marine sponge produces a pharmacologically active isoguanosine derivative (47).[51]

Zanthoxylum culantrillo
A new alkaloid, culantraramine (48), has been obtained from the stems of this Rutaceous plant.[52]

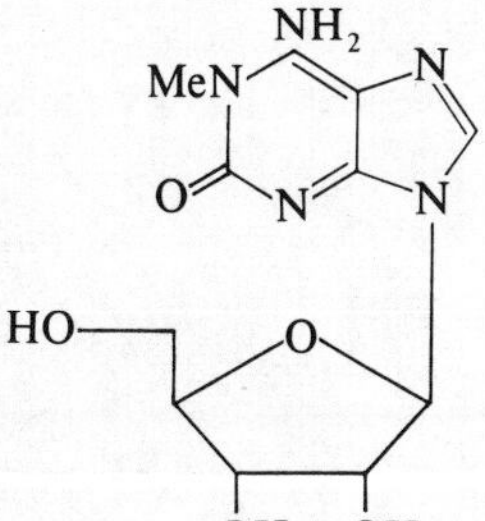

(47)

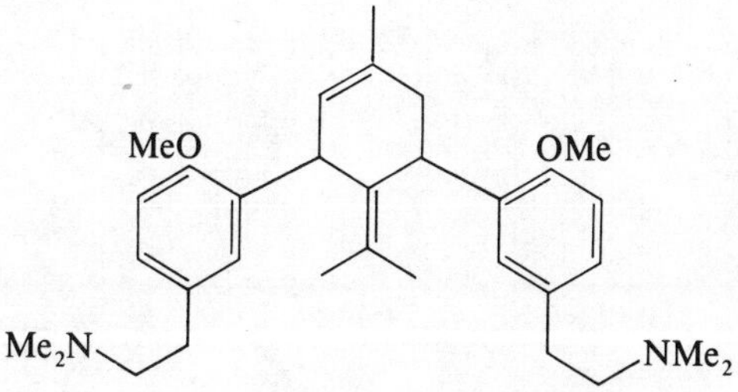

(48)

[49] M. Takagi, S. Funahashi, K. Ohta, and T. Nakabayashi, *Agric. Biol. Chem.*, 1980, **44**, 3019 (*Chem. Abstr.*, 1981, **94**, 117 768).
[50] M. Sahai and A. B. Ray, *J. Org. Chem.*, 1980, **45**, 3265.
[51] A. F. Cook, R. T. Bartlett, R. P. Gregson, and R. J. Quinn, *J. Org. Chem.*, 1980, **48**, 4020.
[52] J. Swinehart and F. R. Sturmitz, *Phytochemistry*, 1980, **19**, 1219.

Errata

Author Index